THE MORPHODYNAMICS OF THE WADDEN SEA

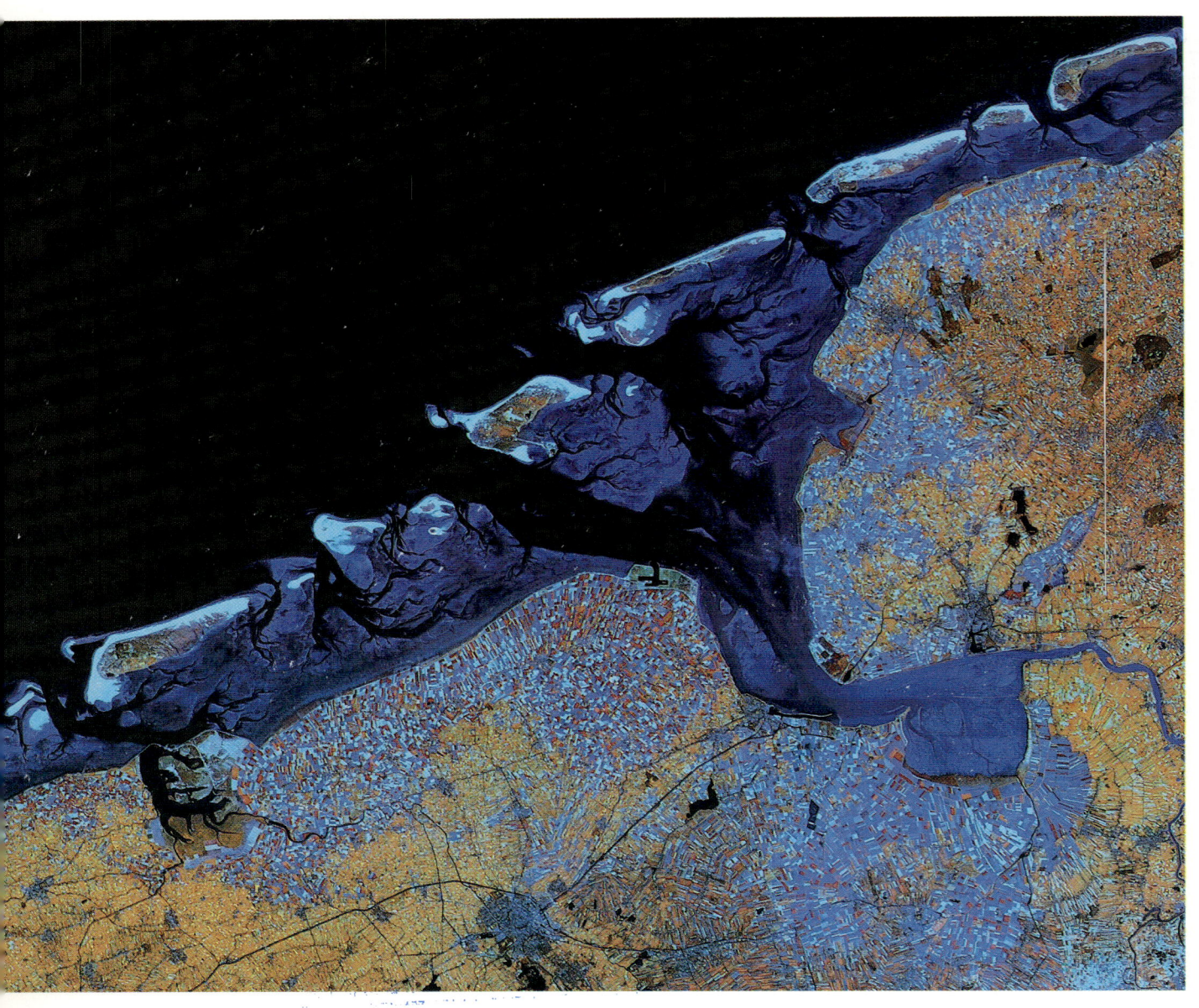

Satellite image of the coast between Schiermonnikoog and Langeoog; Landsat 5 TM image of 22.8.1984, Track 197, Frame 023 (DFVLR).

THE MORPHODYNAMICS OF THE WADDEN SEA

By
JÜRGEN EHLERS
Geological Survey, Hamburg, Germany

A.A.BALKEMA / ROTTERDAM / BROOKFIELD / 1988

Published by
A.A.Balkema, P.O. Box 1675, 3000 BR Rotterdam, Netherlands
A.A.Balkema Publishers, Old Post Road, Brookfield, VT 05036, USA

ISBN 90 6191 679 8
© 1988 A.A.Balkema, Rotterdam
Printed in the Netherlands

CONTENTS

PREFACE	VII
1. INTRODUCTION	1
2. NATURAL PRECONDITIONS	5
2.1 Pre-existing relief	5
2.2 Transgression	9
2.3 Tidal range	13
2.4 Storm surges	17
3. BARRIER ISLAND DEVELOPMENT	23
4. RECENT GEOMORPHOLOGICAL PROCESSES	33
4.1 Currents	33
4.2 Wind	40
4.3 Waves	46
4.4 Ice	54
4.5 Biogenic processes	59
5. MORPHODYNAMIC UNITS	65
5.1 Tidal flats	65
5.1.1 Subdivision of the tidal flats	65
5.1.2 Landform changes	73
5.2 Tidal inlets	78
5.2.1 Subdivision of the tidal inlets	78
5.2.2 Migration of shoals	90
5.2.3 Migration of the deep bars	108
5.2.4 Megaripple migration	111
5.2.5 Expansion of the tidal deltas	130
5.3 Barrier islands	134
5.3.1 Subdivision of the islands	134
5.3.2 Beach processes	135
5.3.3 Dune formation	150
5.3.4 Washover areas	174
5.3.5 Natural accretion	179
5.3.6 Coastal erosion	188
5.3.7 Coastal protection	195
6. HISTORICAL DEVELOPMENT	211
7. REGIONAL DESCRIPTIONS	221
7.1 Denmark	221
7.1.1 Skallingen	221

7.1.2 Fanø	224
7.1.3 Mandø	229
7.1.4 Rømø	233
7.1.5 Jordsand	236
7.2 Schleswig-Holstein	238
7.2.1 Sylt	238
7.2.2 Amrum	248
7.2.3 Föhr	252
7.2.4 Halligen	256
7.2.5 Pellworm	275
7.2.6 Nordstrand	275
7.2.7 Eiderstedt	276
7.2.8 The mobile islands off Dithmarschen	280
7.3 Niedersachsen	285
7.3.1 Scharhörn	285
7.3.2 Neuwerk	288
7.3.3 Knechtsand	288
7.3.4 Mellum	288
7.3.5 Minsener Oog	290
7.3.6 Wangerooge	291
7.3.7 Spiekeroog	298
7.3.8 Langeoog	302
7.3.9 Baltrum	305
7.3.10 Norderney	311
7.3.11 Juist	316
7.3.12 Memmert	318
7.3.13 Lütje Hörn	320
7.3.14 Borkum	320
7.4 The Netherlands	326
7.4.1 Rottum	326
7.4.2 Schiermonnikoog	329
7.4.3 Engelsmanplaat	333
7.4.4 Ameland	334
7.4.5 Terschelling	340
7.4.6 Griend	344
7.4.7 Vlieland	347
7.4.8 Texel	351
8. SUMMARY	361
REFERENCES	363
INDEX	381

PREFACE

This book was stimulated by discussions with my academic teacher Prof. Dr. H. Mensching (Geographisches Institut der Universität Hamburg). The investigations are based on experience gained during the geomorphological mapping for the Wangerooge sheet.

The fieldwork was undertaken between 1980 and 1985, mainly concentrating on the Norderney–Baltrum area. Additional investigations were carried out on Texel, Vlieland, Terschelling, Ameland, Schiermonnikoog, Borkum, Langeoog, Spiekeroog, Wangerooge, in St.Peter-Ording, on the Halligen, on Amrum, Sylt, Rømø and Fanø supplemented by short visits to Juist, Föhr and Skallingen and day trips to the tidal flats around Mellum and Arensch. Besides this fieldwork, large scale maps, aerial photographs and satellite imagery were also evaluated.

The core of the work is the investigation of shoal and megaripple migration on the East Frisian tidal deltas. F. Böker of the Niedersächsisches Landesamt für Bodenforschung kindly helped me to effect aerial photographic surveys of the Wichter Ee ebb delta on four successive low tides. Dipl.-Geol. A. Iwanoff assisted us during the surveys. Gerd Göhlsdorf helped in monitoring megaripple migration in the field on the Wichter Ee flood delta. I am very grateful for all their assistance.

I had stimulating discussions with my colleagues from the Niedersächsisches Landesamt für Bodenforschung on numerous occasions, in particular with Dr. H. Streif, Dr. J. Barckhausen and Dr. J. Hanisch, to whom I want to express my special thanks. Valuable advice was provided by Prof. Dr. H.-E. Reineck and Dr. F. Wunderlich (Senckenberg-Institut in Wilhelmshaven) and Prof. R. Köster (Geologisches Institut der Universität Kiel), who critically reviewed an earlier draft of the manuscript. Dr.-Ing. G. Luck (Forschungsstelle für Insel- und Küstenschutz, Norderney) discussed the sand migration problems with me. For valuable information and comments concerning the formation of the North Frisian Wadden Sea I wish to thank Dr. D. Hoffmann (Institut für Ur- und Frühgeschichte der Universität Kiel). Dr. G. Linke (Geologisches Landesamt Hamburg) kept me informed of his research in the Neuwerk–Scharhörn area.

Dr. Th. Poetsch and Mrs. Wulff (Geographisches Institut der Universität Hamburg) greatly assisted with grain-size analyses and prepared the thin sections. Dr. K. Figge (Deutsches Hydrographisches Institut, Hamburg) provided a number of additional grain-size analyses.

Many colleagues helped me to obtain important literature or to organize scientific contacts, whilst others provided valuable information about specific aspects of the book. Discussions with H.-J. Jürgens, Wangerooge, were especially informative concerning the historical development of the island and its coastal protection problems.

The work was also influenced profoundly by Prof. Dr. G. Hillmer, Prof. Dr. I. Valeton, Prof. Dr. E. Grimmel, Prof. Dr. D. Pohl and Prof. Dr. W. Zahel (Universität Hamburg).

Mr. W. ter Karapetian helped me to solve all problems concerning my word-processing system. Maps and diagrams were drawn by the author, who also took all the ground photographs. Aerial photographs were kindly supplied by the following institutions: the Fototheek Topografische Dienst, Emmen (Netherlands); the Rijkswaterstaat, Meetkundige Dienst (Netherlands); the Niedersächsisches Landesverwaltungsamt, Landesvermessung, Hannover; the Vermessungsamt Hamburg; the Landesvermessungsamt Schleswig-Holstein, Kiel; the Geodætisk Institut, Copenhagen. Special thanks are also due to Vermessungsbüro N. Rüpke, Hamburg, for taking reasonably-priced aerial photographs.

Satellite imagery was provided by the European Space Agency (ESA), Rome, and by the Deutsche Forschungs- und Versuchsanstalt für Luft- und Raumfahrt (DFVLR), Oberpfaffenhofen. I wish to express my sincere thanks to Mr. Lichtenberger (ESA) and Mr. Dinger (DFVLR) who helped me to obtain the best possible material. Dr. W. Rosenthal and Dr. H. Günther (Max-Planck-Institut für Meteorologie, Hamburg) helped me to obtain radar imagery of the German North Sea coast, taken during the MARSEN experiment in 1979.

The manuscript of this book was critically reviewed by ir. J.H. de Reus and ing. R. Reenders from Rijkswaterstaat, Adviesdienst Hoorn and Directie Groningen, Meet- en Adviesdienst, as well as by Dr. Margot Jespersen of the Geographical Institut of the University of Copenhagen. I am particularly grateful to Dr. J. Scourse (University College of North Wales, Department of Oceanography, Gwynedd) who reviewed the English of the manuscript, and to Dr. E. Otvos (Gulf Coast Research Laboratory, Ocean Springs) whose critical comments helped to weed out some of the major mistakes. Belinda Gooden and Gay Wilson helped to correct the manuscript, and a decisive final check was made by Dr. Philip Gibbard (University of Cambridge, Sub-Department of Quaternary Research), to whom I wish to express my sincere thanks.

The book would never have been written without the constant help and encouragement of Uta, my wife, who has spent the the last few years' holidays, some in winter and in the most inclement weather, on the North Sea coast to assist me in the fieldwork.

Hamburg, January 1986.

CHAPTER 1

INTRODUCTION

In recent years the Wadden Sea has been the subject of extensive geoscientific research. A synopsis has been provided by Wolff (1983), and the geological evolution of the entire North Sea area was covered by Oele, Schüttenhelm & Wiggers (1979). Extensive examination of archive material and the comparison of old maps have revealed the historical changes of the major landforms i.e. islands and tidal inlets, from the 17th century to the present day. Investigations by the 'Forschungsstelle für Insel- und Küstenschutz' (Research station for island- and coastal protection) on Norderney were especially important in this respect (Homeier, 1962; Luck, 1975). These studies have largely concentrated on the problems of coastal protection.

Research into the landforms of the Wadden Sea area using a sedimentological approach began with the work of van Veen (1950) and van Straaten (1950). Numerous other studies have followed including, for example, Reineck (1963) and Jakobsen (1964). A new approach to the interpretation of the morphodynamics of tidal inlets was recently introduced by researchers in the USA (e.g. Hubbard, 1975; Hayes, 1975). Nummedal & Penland (1981) and FitzGerald, Penland & Nummedal (1984) have applied these methods to the North Sea coast.

The landforms in the Wadden Sea, especially those near the tidal inlets are subject to continuous change. Investigations by the 'Forschungsstelle für Insel- und Küstenschutz' (Norderney) provided first estimates of the magnitudes of the processes involved (Homeier & Kramer, 1957; Homeier & Luck, 1971; Luck & Witte, 1974). The problem of sand movement along the German North Sea coast was investigated by a large-scale research programme of the 'Deutsche Forschungsgemeinschaft' (DFG, 1979), but this has only partially answered the questions raised.

During the course of geomorphological field mapping of the Wangerooge sheet (Ehlers & Mensching, 1982) I came to realize the importance of both the major and minor landforms of the Wadden Sea for the interpretation of their dominant formative processes. At the same time, my observations led me to suspect that the remodelling of the landscape occurred at a much faster rate than previously assumed. I was stimulated by Hanisch's (1981) detailed research on sediment transport in the tidal inlet between Spiekeroog and Wangerooge. As a result I decided to continue this line of research.

The first objective was to obtain at least two detailed landform surveys within a short time span. For this purpose aerial photographs of Wichter Ee Inlet, between Norderney and Baltrum, were taken over a period of 39 days. Interpretation of these photographs indicated large-scale landform changes. Despite the short time interval it was not possible, in all cases, to follow every development stage of each shoal (Ehlers, 1984).

The conclusions drawn from interpretation of the aerial photographs form the basis of this study. The processes were investigated in detail in one case (Wichter Ee) by measurements of megaripple migration and aerial photographs taken during four successive low tides. These results led to the conclusions on sediment transport in other parts of the Wadden Sea. Tietze's (1983) seismic investigations of the Hever region have proved especially helpful in the context of these wider interpretations.

The study begins with an introduction outlining the essential background to recent geological processes. This is followed by a discussion of the events leading to the formation of the barrier islands. Because of the lack of reliable data this part is largely hypothetical in character. In the next chapter the recent geomorphological processes are briefly discussed.

The main part of the book is dedicated to three major morphodynamic units 'tidal flats', 'tidal inlets' and 'barrier islands'. The whole Wadden Sea region from Den Helder to Blåvandshuk (fig.

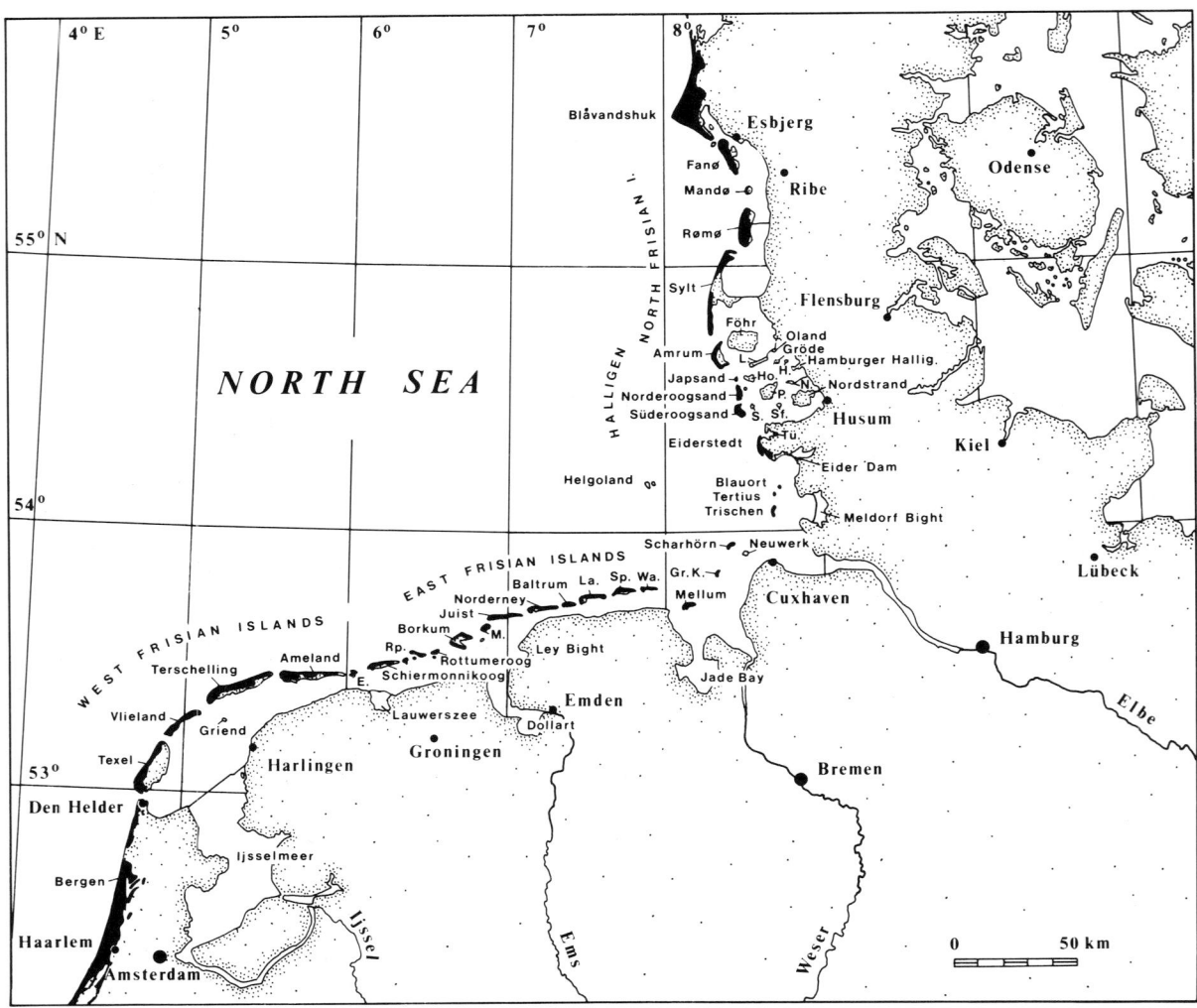

Figure 1: Location map. The coastal dunes are shown in black. L = Langeneß, H = Habel, Ho = Hooge, N = Nordstrandischmoor, P = Pellworm, S = Süderoog, Sf = Südfall, Tü = Tümmlauer Bucht, Gr.-K. = Großer Knechtsand, Wa = Wangerooge, Sp = Spiekeroog, La = Langeoog, M = Memmert, Rp = Rottumerplaat, E = Engelsmanplaat.

1) is considered through map and aerial photograph interpretation in a 'regional description'.

In contrast to Gripp's (1956: 152) definition, the 'Wadden Sea' is here regarded as not only the intermittently submerged part of the coastal area, but the whole area including tidal inlets and islands (cf. Wolff, 1983). This is in accordance with the more recent usage of the term.

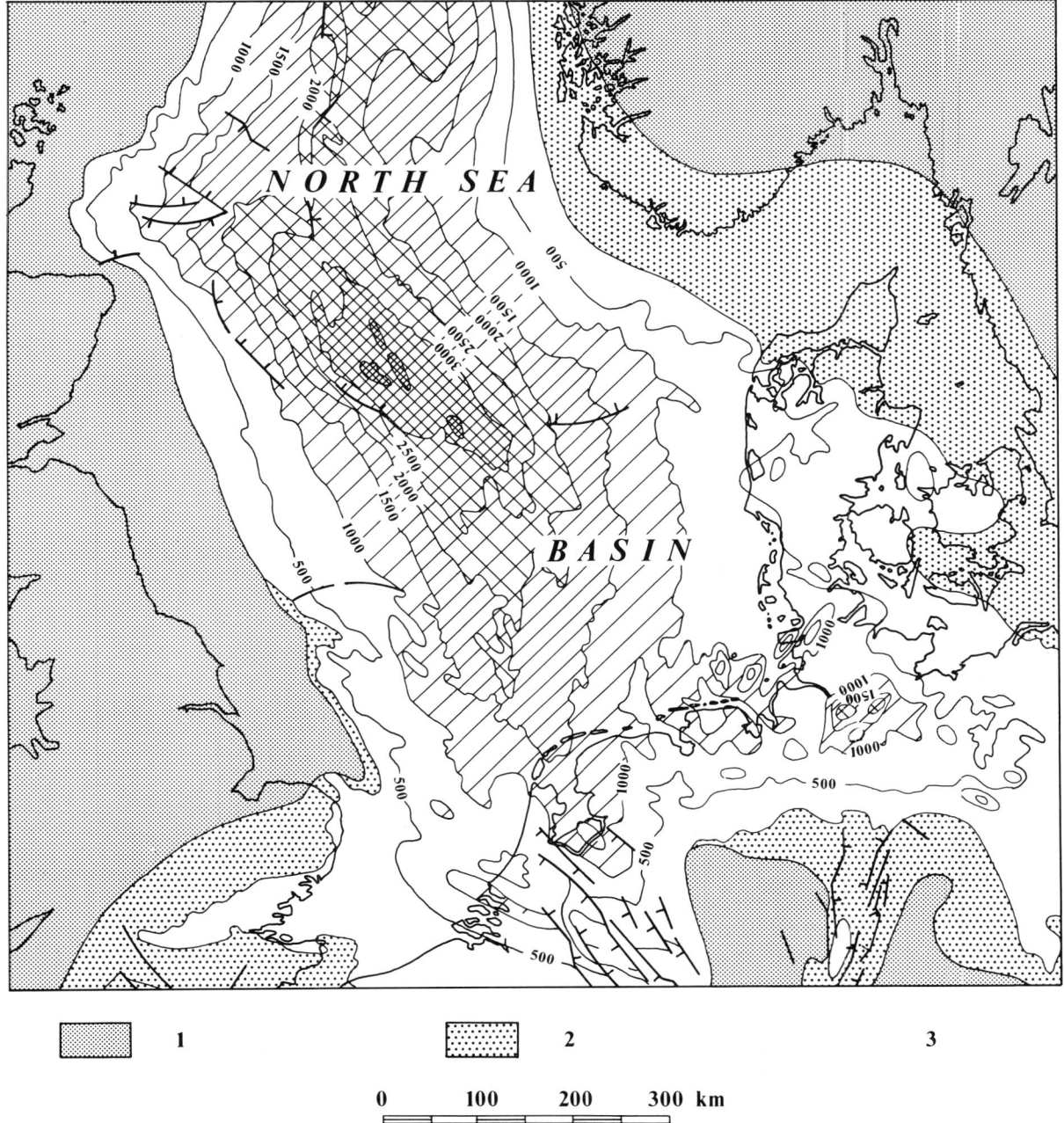

Figure 2: Thickness of the Tertiary and Quaternary sediments in the North Sea sedimentary basin (after Ziegler, 1975); thickness in metres. 1 = continental areas, 2 = areas of non-deposition, 3 = fault.

CHAPTER 2

NATURAL PRECONDITIONS

2.1 PRE-EXISTING RELIEF

The present North Sea is part of a subsiding basin which includes major parts of Britain, the Netherlands, North Germany, Denmark and parts of Poland. Since the Tertiary subsidence has been concentrated within the present North Sea area. The total thickness of the Tertiary deposits exceeds 3000 m in the central parts of the basin (fig. 2) (Ziegler, 1975, 1982; Ziegler & Louwerens, 1979).

Subsidence was initiated not later than in the Late Palaeozoic. This was not a continuous process but was interrupted by phases of uplift. Not all the stages of the younger history of the earth are therefore represented in the North Sea marine record, and some stages are represented by terrestrial sediments. The red Bunter Sandstone rocks of the Isle of Helgoland, for instance, were formed during a period of terrestrial sedimentation. Subsidence has, however, been dominant. The thickness of the sedimentary rocks in Ostfriesland (East Frisia) is about 4500 m and in Dithmarschen about 6000 m.

Subsidence or sea level rise?

According to Müller (1917 a: 78), speculation about widespread coastal subsidence can be traced back to Forchhammer (1837: 56; *op. cit.* in Müller, 1917). Meyn (1876: 751) assumed a total subsidence of about 3–9 m for the 'Younger Alluvium'. This hypothesis was, however, strongly questioned by others. Wahnschaffe (1901: 249) rejected the evidence for recent subsidence.

Discussion about the causes of the sea level changes observed along the German North Sea coast were dominated for over 50 years by the works and ideas of Schütte, and his concept of coastal subsidence (cf. Hartung, 1964). Beginning with the observation that plough marks were found below sea level around the Oberahnesche Felder in the Jade Bay he concluded that coastal subsidence must have had occurred quite recently.

His first attempt to calculate the rate of subsidence in 1908 produced too high a figure, which he later had to revise (Schütte, 1927: 352).

A few years earlier Schucht (1903: 76; summarized: 1910) had concluded that the North Sea coast had subsided by at least 20 m since the end of the last glaciation. His evidence constituted forest beds and peat layers covered with marine clay. However, Schucht did not believe this subsidence had continued up to the present. His opinion was shared by Behrmann (1921: 91).

Schütte's ideas, however, prevailed against early criticism. The situation in North Frisia seemed to support his concept of continuous subsidence (Schütte, 1929). He was able to construct a subsidence curve for the Jade-Weser area, in which he estimated a subsidence rate of 5.1 m since 0 A.D. (Schütte, 1933: figs. 2 and 3). Schütte later summarized the results of his investigations in his book 'Sinkendes Land an der Nordsee?' (Subsiding Land at the North Sea?) in 1939.

The eustatic consequences of melting glaciers and ice sheets on sea level was recognized relatively late (Ramsay, 1930; Penck, 1933). Even after the recognition of this relationship it was regarded as insignificant as a causal mechanism for recent sea level changes (Dienemann & Scharf, 1931: 322 and 385).

Like Schütte, Wolff (e.g. 1923, 1939) assumed that coastal subsidence was the reason for the apparent rise in sea level. Similarly, Heck (1936: 80) attributed the sea level changes in North Frisia to tectonic movements. Dewers (1940: 39; 1941: 354) was the first to seriously question the assumption that eustatic sea level changes were negligible today.

The coastal subsidence theory was soon refuted after the Second World War. Dechend (1954, 1956: 312) continued to calculate subsidence rates of 2–2.5 cm per century, but Dittmer, who had also supported the subsidence hypothesis in 1938 (p. 141), then changed his mind (1948). Further investigations in the Schleswig-Holstein coastal area

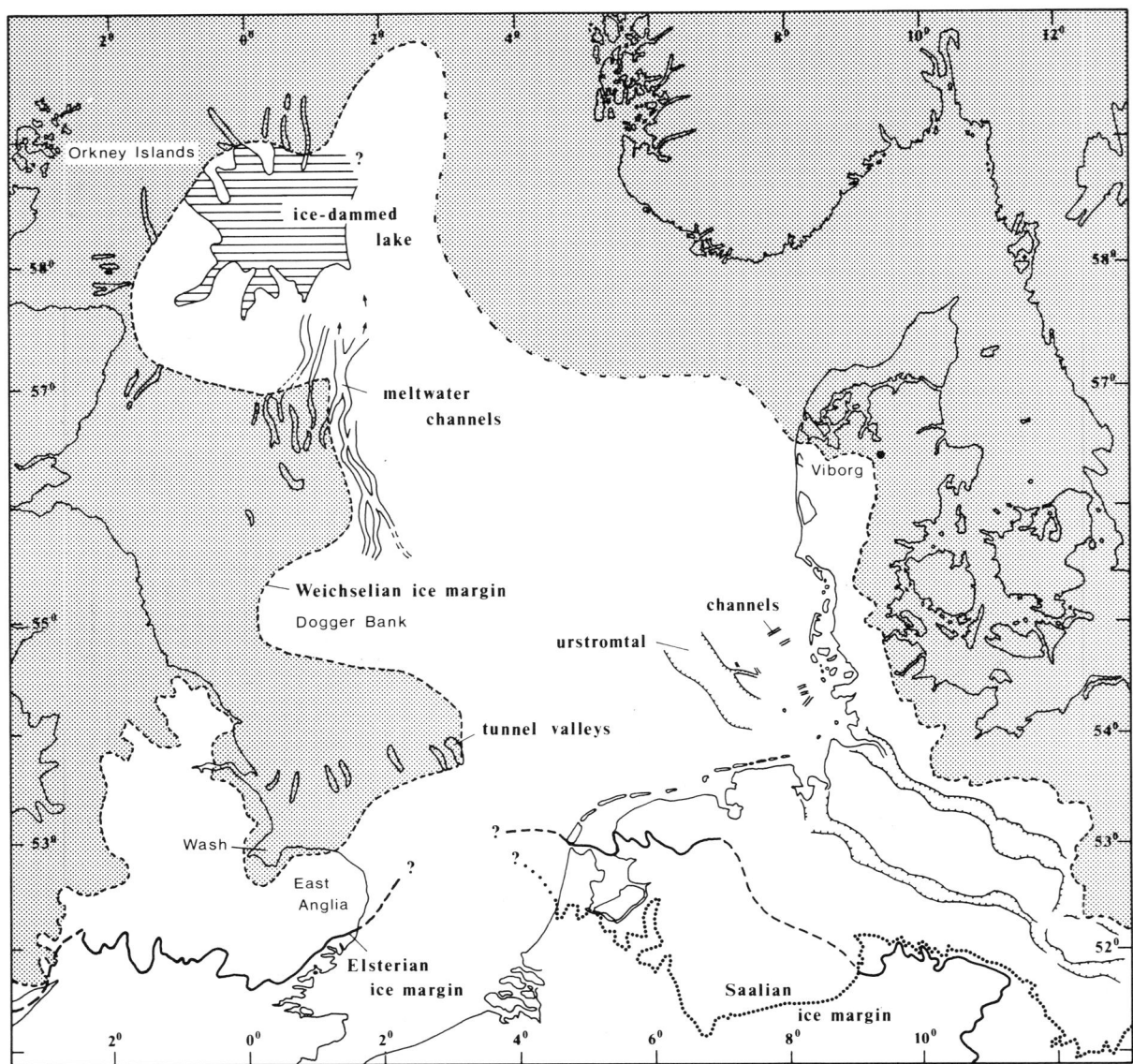

Figure 3: Maximum extent of the Quaternary glaciations in the North Sea area (after Borth-Hoffmann, 1980; Bowen, 1984; Ehlers, Meyer & Stephan, 1984; Figge, 1983; Jansen et al., 1979a, b; Prasad, 1983; West, 1967; Woldstedt, 1958; Zagwijn, 1979; Zagwijn & Staalduinen, 1975).

demonstrated that it was unneccessary to invoke tectonic movements to explain the transgression history (Dittmer, 1952; 1960). The repetition of the North Sea Coast Triangulation between 1948-59 failed to indicate any measurable subsidence (Gronwald, 1960), and the critics of the subsidence theory finally prevailed (Dittmer, 1960; Bantelmann, 1960; Gripp, 1964: 80; Sindowski & Streif, 1974).

Whether or not subsidence is active at present is uncertain (cf. Sindowski & Streif, 1974). Lassen, Linke & Braasch (1984: 107) assume a tectonic subsidence rate of about 0.1 mm per year, an amount which is clearly below the limits of empirical measurement.

The large-scale epirogenetic processes in the North Sea Basin have been locally modified by halokinetic movements. In the Wadden Sea area the influence of salt domes can be traced from Borkum (Borkum salt dome) to Eiderstedt (Ol-

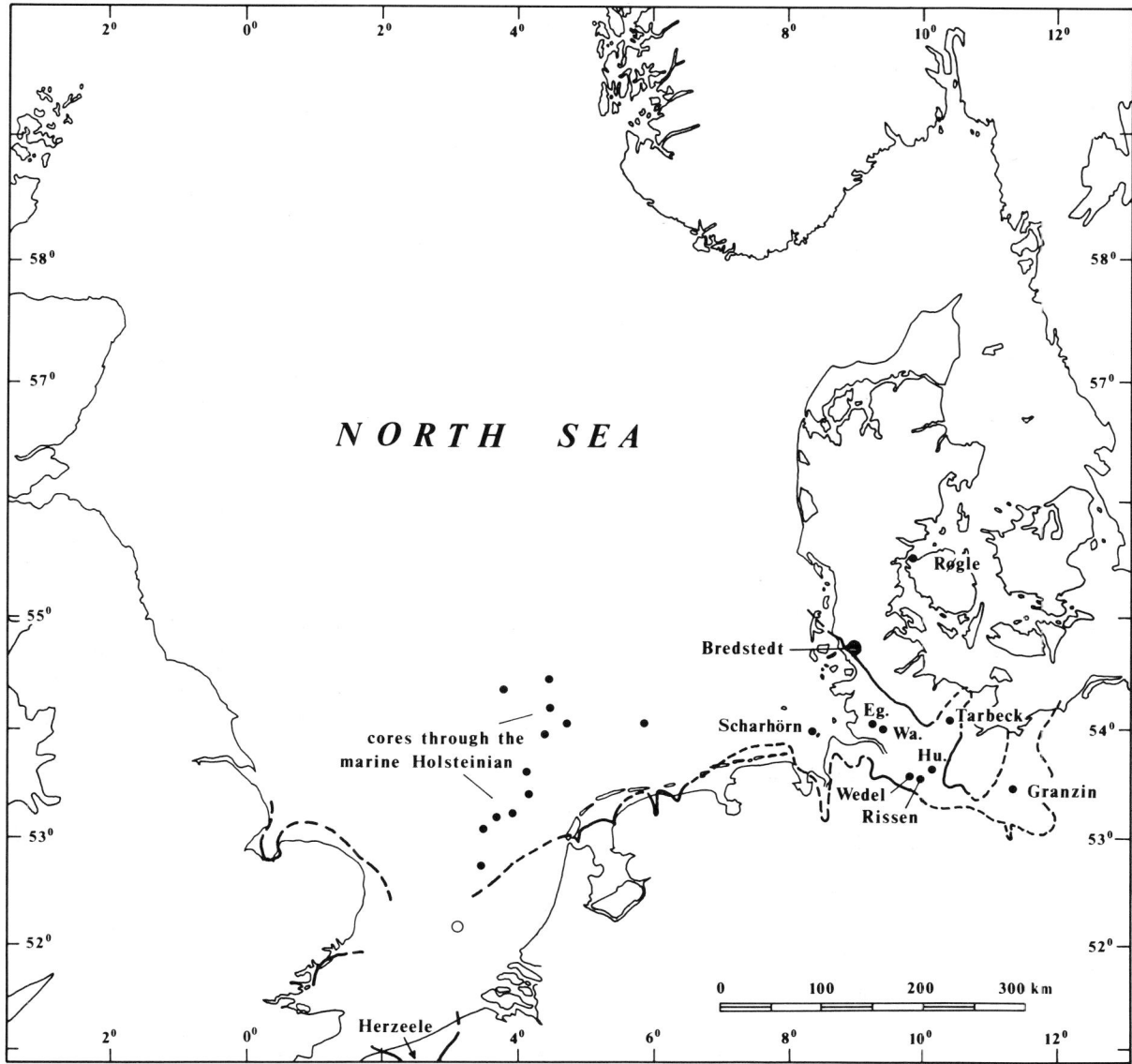

Figure 4: Maximum extent of the Holsteinian transgression in the North Sea area (after Heck, 1947; Linke, 1970; Sjørring, 1983; Woldstedt & Duphorn, 1974; Zagwijn, 1979). Black dots: drillholes with marine Holsteinian. Circles: drillholes without marine Holsteinian. Shaded area near Bredstedt: Brackish Holsteinian. Eg. = Eggstedt; Wa. = Wacken; Hu. = Hummelsbüttel.

denswort salt dome). The penetration of the covering sediments by the salt domes began in the Keuper (Borkum, Juist-West, Juist-Ost, Norderney, Wichter Ee, Langeoog and Spiekeroog salt domes) and continued through the Liassic (Scharhörn–Eversand–Mellum), Dogger (Wangerooge), Cretaceous (Oldenswort-Süd) until the Tertiary and Quaternary (Borkum-Nord, Harle-Riff, Roter Sand–Feuerschiff Elbe 1, Oldenswort-Mitte and Oldenswort-Nord) (Jaritz, 1973). The most visible signs of halokinetic movements are the uplifted Bunter rocks of Helgoland (Schmidt-Thomé, 1982). Apart from this single exception no direct halokinetic influence on the present configuration of the shoreline can be demonstrated.

In the Wadden Sea area, the salt domes and the halokinetically-uplifted Mesozoic strata are normally found at great depth. For example, the top of the Zechstein salt beneath Wangerooge lies at a depth of about -400 m (Sindowski & Streif, 1974).

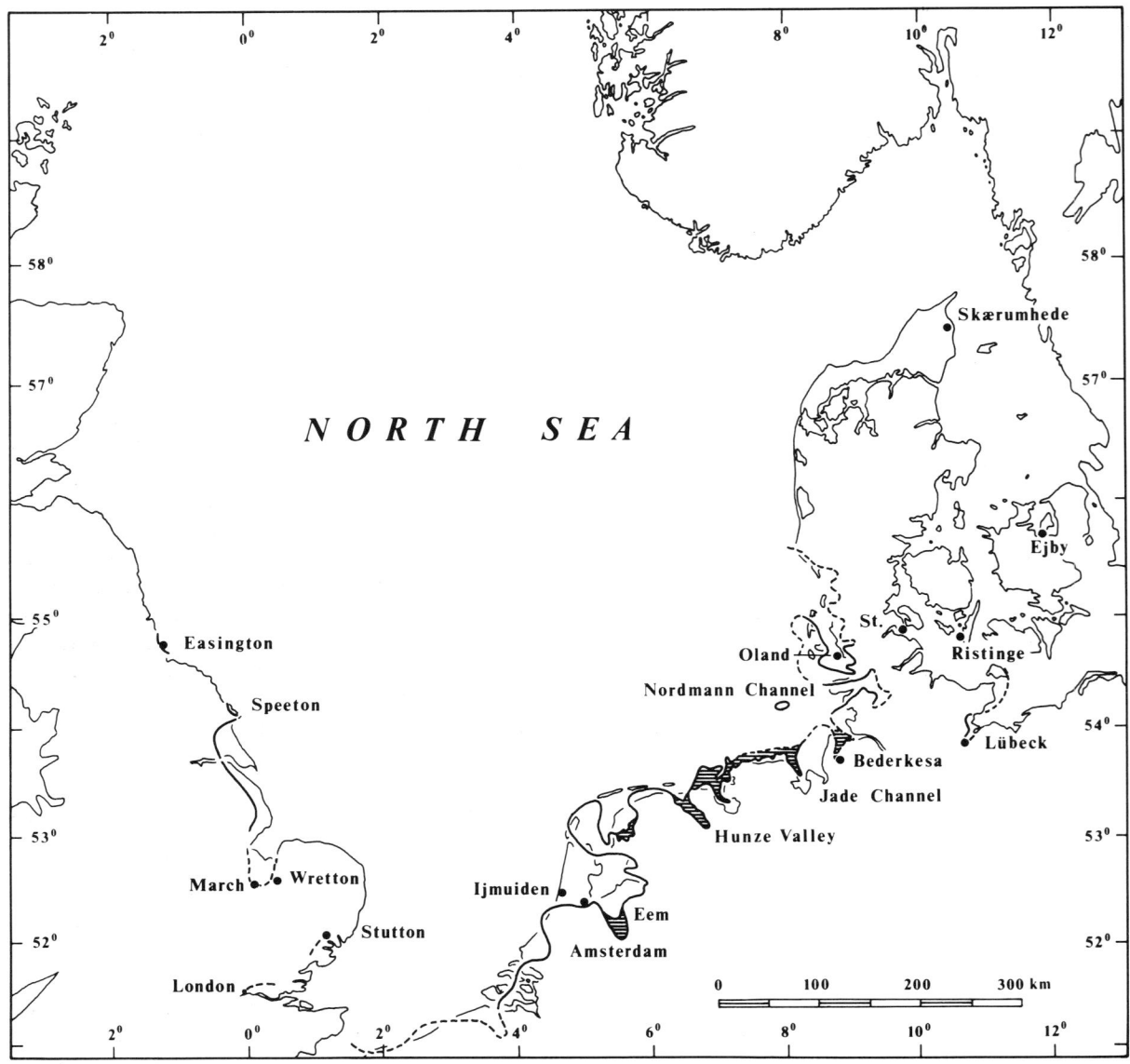

Figure 5: Maximum extent of the Eemian transgression in the North Sea area (after Jardine, 1979; Jelgersma, Oele & Wiggers, 1979; Krog, 1979; Merkt, 1984; Woldstedt & Duphorn, 1974). Dots: marine Eemian. Shaded areas: brackish Eemian.

The relief in the present Wadden Sea area prior to the Holocene transgression was influenced mainly by the Pleistocene glaciations. During the Quaternary the North Sea Basin was totally or partially covered by the ice sheets of at least three glaciations (Elsterian, Saalian and Weichselian). In the Netherlands and in England there seems to be evidence for at least one other glaciation preceeding the three mentioned above. Traces of this glaciation have not so far been identified in North Germany (Ehlers, Meyer & Stephan, 1984).

The extent of these glaciations varied considerably (fig. 3): During the Elsterian glaciation (Anglian) the glaciers reached their maximum extension in England. This ice advance reached the area north of London and forced the River Thames to change its course (Gibbard, 1977). In the Netherlands only two questionable finds of Elsterian till have so far come to light (Zandstra, 1977, 1983). The existence of buried channels over 300 m deep in the northern Netherlands (ter Wee, 1983) seems to indicate that at least the northernmost parts of the country were affected by the Elsterian glaciation. In the Emsland region of Niedersachsen (Lower Saxony) Elsterian till has been found, but only in a few localities (Meyer, 1970).

During the Saalian the ice sheet reached its maximum extension in the Netherlands (Zagwijn & van Staalduinen, 1975; de Jong & Maarleveld, 1983). In England the limit of the Saalian (Wolstonian) glaciation is found much further to the north, approximately in the Wash area. East Anglia was not overridden by the Saalian glaciation.

The Weichselian glaciation in North Germany did not cross the Elbe River. In Northern Denmark, at Viborg, the Weichselian ice sheet reached farther to the west and affected the area of the present North Sea (Sjørring, 1983: 174). There was certainly no contact with the Weichselian glaciers of Britain in the Dogger Bank area, as suggested by Valentin (1957). Recent investigations in the northern North Sea demonstrated the temporary existence of a major ice-dammed lake in front of the ice-sheet (Jansen, 1976: 34). The glaciers of Norway and Scotland were in contact with each other east of the Orkney Islands. The margins of the Weichselian glaciation can be mapped on the sea floor by tracing the ends of the tunnel valleys (Flinn, 1967).

It can be assumed that the Pleistocene landforms found on land were formerly also present on the floor of the North Sea. Seismic investigations in the offshore area have demonstrated not only the existence of the tunnel valleys mentioned above, but also the thrust moraines of the older glaciations (Borth-Hoffmann, 1980; Figge, 1983; Prasad, 1983). Assuming that during the major glaciations global sea level was lowered by something over 100 m during these periods, the floor of the North Sea would have formed an extension of the land. The extension of the Elbe ice-marginal valley can be traced offshore to the Dogger Bank area (Figge, 1980).

The resultant glacial landforms controlled the extent of the interglacial transgressions. The Holsteinian Sea intruded far inland in North Germany following the course of the Elsterian tunnel valleys. Estimates on the extent of the transgression in North Germany as indicated in the map (fig. 3), according to Woldstedt & Duphorn (1974), are now rather out of date. In the Hamburg region (Grube, 1982) and further to the west a much more complex picture has evolved. Subsequent erosion has played little part in the present confinement of the marine Holsteinian sediments to Elsterian channels. The Holsteinian transgression that reached Hamburg and Lauenburg seems to have been restricted to the channel areas.

The distribution of the marine Eemian sediments is better documented than those of the Holsteinian (fig. 4). The Eemian transgression was in most places less extensive than the maximum extent during the Holocene transgression. The major embayments of the present North Sea (Zuider Zee, Lauwerszee, Dollart, Jadebusen) all had Eemian predecessors.

The map shows only the maximum extent of the Eemian transgression; it provides no information on the actual configuration of the coastline at any given time, because very little is known about coastal development during this period (fig. 5). In contrast to the vast embayments of the Holsteinian Sea, the Eemian bays were smaller, and mostly occupied former valleys (Sindowski, 1965). Since the Eemian was probably about as long as the Holocene, a similar coastal configuration may have evolved. The Eemian Sea was in all likelihood fringed by a barrier coast at a sea level lower than the present and further to seaward. Heck (1947) discovered lagoonal Eemian deposits at Bredstedt (Schleswig-Holstein). Von der Brelie (1959) investigated the diatom fauna of the Eemian described by Dechend (1951) from the Norderney area. He found that the sediments represented a tidal flat facies comparable to that of the present Wadden Sea. The barrier itself was not discovered, being situated further seaward and removed by marine erosion during the Holocene.

In contrast to the 'tectonically stable' English east coast where Eemian marine sediments suggest a significantly higher Eemian sea level of around + 7.5 m O.D. (Jardine, 1979: 165), in the area of this investigation both the Holsteinian and Eemian marine sediments lie considerably below present sea level. This is attributed to the general trend of subsidence in this region (Linke, personal communication) which is considered significant during the time spans of over 100,000 or 200,000 years. However, during the short period of the Holocene transgression (less than 10,000 years) the influence of subsidence is negligible.

2.2 TRANSGRESSION

As a result of the vast amounts of water bound up in the large continental ice sheets, during the Pleistocene glaciations, global sea level was peri-

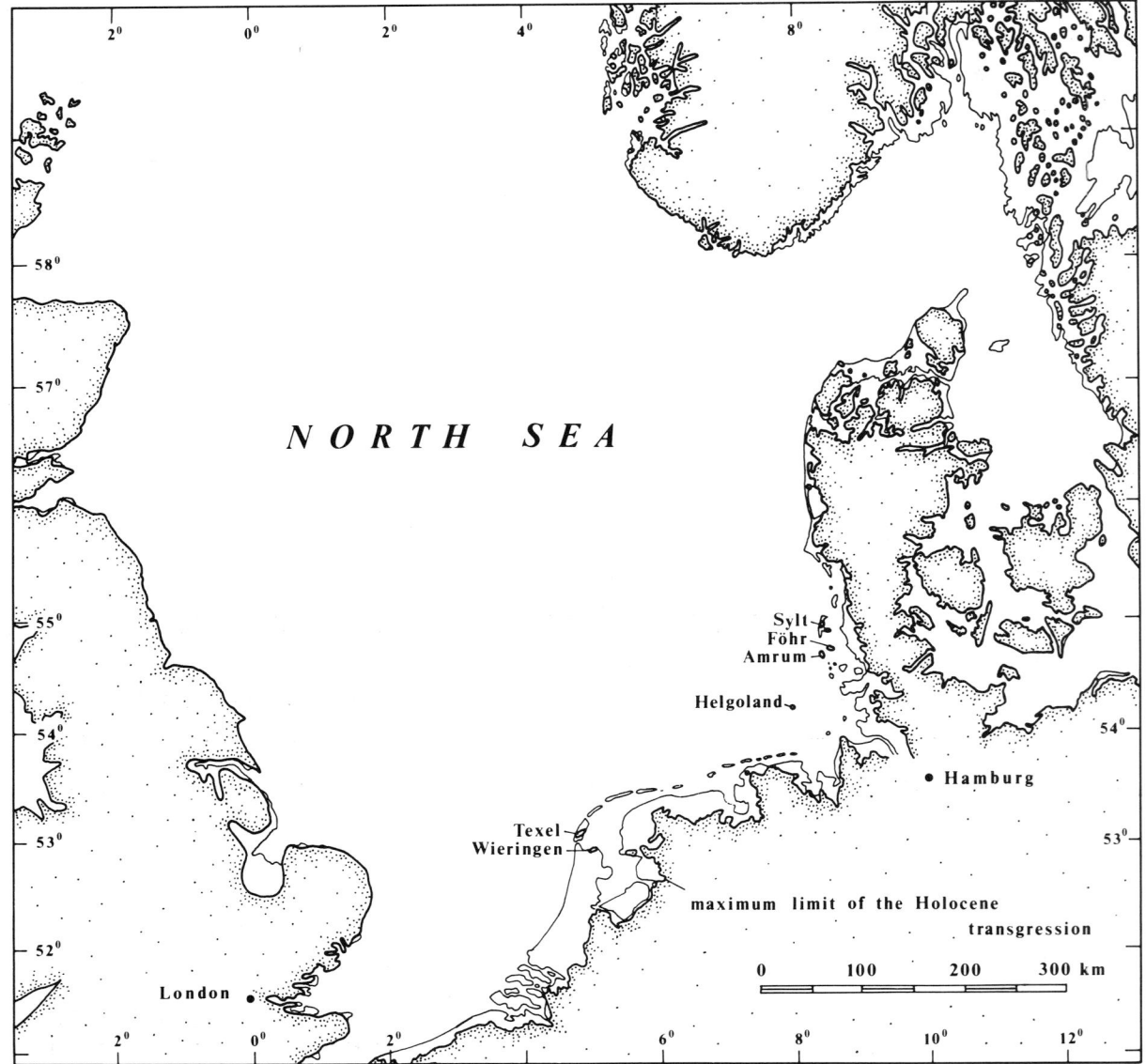

Figure 6: Maximum extent of the Holocene transgression in the North Sea area (after Oele, Schüttenhelm & Wiggers, 1979).

odically lowered by more than 100 metres. With the exception of the Norwegian trough, the whole North Sea area formed an extension of the mainland during stadials. At the end of these stadials sea level started to rise gradually once more.

Before the transgression could reach the level of the present coastline, the changing drainage conditions led to an increase in coastal wetland areas resulting in widespread peat growth. The so-called 'basal peat' was formed. This peat, the lowermost member of the Holocene stratigraphic sequence, has been recovered in boreholes throughout the whole coastal area.

If one assumes that the melting of the Pleistocene ice sheets has been the cause of this rise of the sea level, one might conclude that a continuous but gradually decreasing rise would result during the Postglacial. This is not the case. From Denmark to the Netherlands young salt marshes fringe the North Sea shore, consisting of young, Holocene marine sediments - evidence of a marked regression (fig. 6).

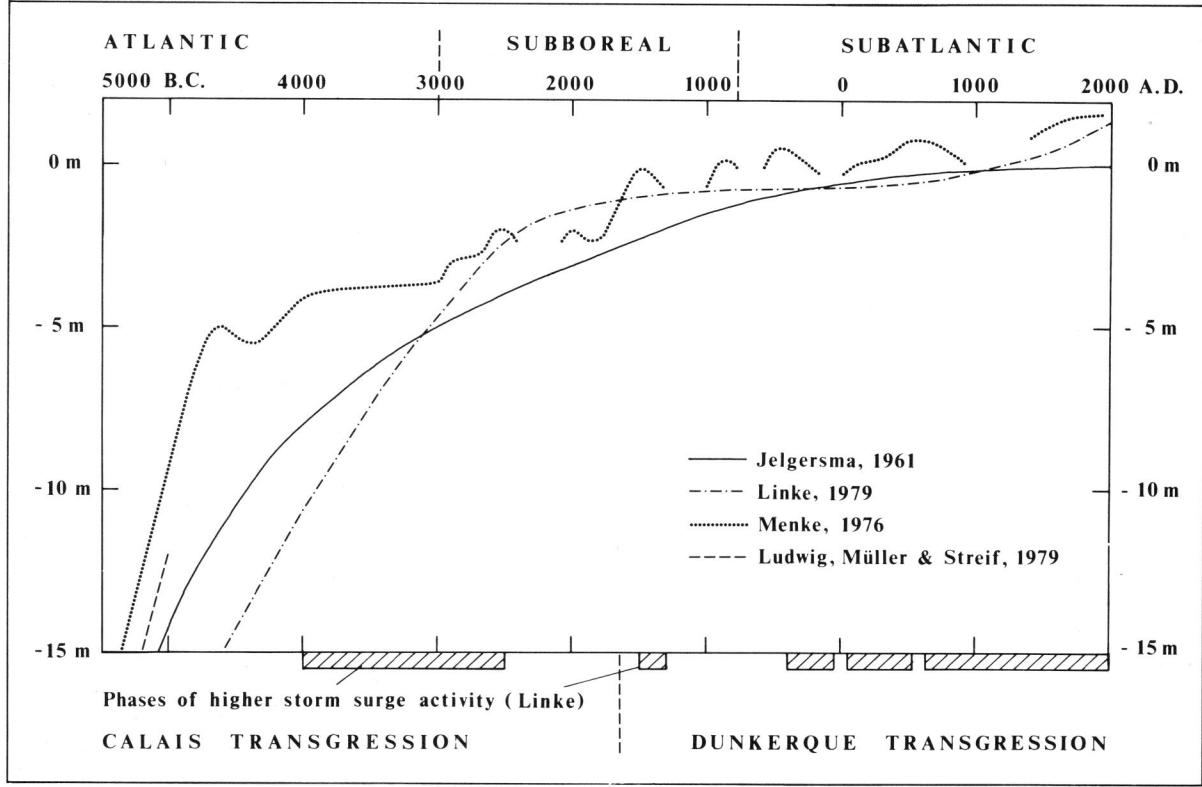

Figure 7: The Holocene transgression in the North Sea area (after Jelgersma, 1961; Linke, 1979; Menke, 1976; Ludwig, Müller & Streif, 1979).

The Postglacial sea level rise has been subdivided into two major phases of transgression:
1. the Calais Transgression and
2. the Dunkerque Transgression

'Transgression' and 'regression' must not be regarded as 'real' oscillations of sea level, but only as an extension of the sea at the expense of the land and *vice versa* (Jelgersma & Ente, 1977: 27).

In a relatively unstable coastal area it is difficult to identify the real causes of a relative rise in sea level. While some workers prefer a sequence of numerous transgressions and regressions (e.g. Müller, 1962; Menke, 1976; Tooley, 1978), there is some evidence from the Dutch and German North Sea coastal areas that the rise in sea level was continuous (Jelgersma, 1961; Linke, 1979; Ludwig, Müller & Streif, 1979, 1981) (fig. 7). This model is in accordance with the evidence for Postglacial climatic evolution.

The Weichselian deglaciation has proved difficult to date precisely. Lundqvist (1980: 236) concludes that the Scandinavian ice sheet finally melted not later than 8,000–9,000 B.P. (cf. Lundqvist, 1981; 1984); for the complete melting of the Laurentide ice sheet a date of about 5,000 B.P. is assumed (Andrews, 1982). Under the influence of the mild climate of the Atlantic period parts of the circumpolar ice caps also partially melted. This process declined towards the end of the Atlantic and therefore the course of Postglacial sea level rise was considerably decreased not later than 3,000 B.C. Subsequently increasing freshwater influence led to renewed peat development along the coast.

The Dunkerque Transgression (Dunkerque 0) started along the Dutch coast around 1700 B.C., and also later affected the German and Danish coasts. This renewed transgression can no longer be explained by eustatic processes only; at its onset the sea level was already established near its current position (fig. 7). Thus, Bakker (1954) suggested that this second transgression was possibly due to increased storm surge activity. This interpretation was later adopted by Jelgersma & Ente (1977) and Linke (1979, 1982).

The rates of transgression differed between different coastal sectors. In the Netherlands the two major transgressive periods are subdivided into four (Calais Transgression) and three (Dunkerque Transgression) transgressive phases, separated by regressions. These phases cannot be directly correlated with Menke's (1976) scheme which was developed for the Schleswig-Holstein coast. One of the reasons for this is the different configuration of the coastal regions. In more protected areas even minor regressions will lead to the growth of peat layers which may be missing in more exposed localities.

Dating of the transgressions relies on radiocarbon dates from peat and other organic matter. Future revisions are anticipated in the dates of the transgressive phases.

As the young Holocene deposits, especially peat, gyttja and marine clay, are susceptible to sediment compaction, local transgressions can be attributed to this process. Areas with thick Holocene deposits therefore tend to exhibit more complicated sedimentary sequences than areas with thin Holocene sediment cover. Lassen, Linke & Braasch (1984) have been able to demonstrate that a significant proportion of the relative sea level rise registered at the Cuxhaven gauge (monitored since 1841; Rohde, 1977) has resulted from compaction of the underlying sediments.

Where historical evidence is lacking it is difficult to date the processes documented in the sedimentary record. Only relatively few samples have been radiocarbon-dated so far and as a result stratigraphical correlations have relied heavily on the interpretation of the lithological units. According to Streif (1971) the marine Holocene sediments of the Niedersachsen North Sea coastal area can be subdivided into three lithogenetic units:
1. tidal flat sequence
2. lagoonal sequence
3. channel fill sequence

The channel fill sequence is assigned to the first transgression and the lagoonal sequence is correlated with the subsequent regression. Sediment movement along the south-eastern North Sea coast is assumed to have begun not later than this time. The sand available on the foreshore was concentrated into shoals by coast-normal wave attack, creating a barrier of shoals from which finally the recent islands have evolved (Jelgersma & Ente, 1977: 30).

The transgression has continued up to the present day. Gauge measurements along the German North Sea coast over the last 100 years have generally shown a relative sea level rise of 20–30 cm per century (Kramer, 1983). However, new investigations have revealed that this estimate may have been too low. Measurements of the mean high tide of 12 gauges along the German North Sea coast over the last 25 years (1959-83) have demonstrated an average rise of 64 ±15 cm per century (Jensen, 1984: 509). Führböter & Jensen (1985) caution against extrapolations of these values into the future because it is uncertain whether the recent trends will continue. Something that can be stated with certainty, however, is that during the last 25 years much of the safety reserve planned in dyke construction assuming a much slower sea level rise has already been expended.

The world-wide mass balance of glaciers and ice sheets is now negative. Ice melt contributes to the rise in sea level. At present the strongest effects are caused by the retreat of small glaciers and ice caps. The Greenland ice sheet may also possibly contribute to this rise in sea level, whilst the mass budget of the Antarctic ice sheet is regarded as balanced or even slightly positive.

Estimated mass balance of glaciers and ice sheets at the present time (from U.S. Department of Energy, 1985: 2):

Ice mass	Average mass balance (water equiv.) (m/year)	Effect on sea level (mm/year)
Glaciers and small ice caps	− 1.2 ± 0.7	+ 0.5 ± 0.3
Greenland ice sheet	+ 0.02 ± 0.08	− 0.1 ± 0.4
Antarctic ice sheet	+ 0.02 ± 0.02	− 0.6 ± 0.6

The sea level rise caused by mass balance changes is supplemented by volumetric changes of sea water as a consequence of rising temperature, the so-called steric effect. Measurements undertaken off the Bermudas over the last 20 years indicate that these volumetric changes are of an order of magnitude sufficient to explain the present rise in sea level without the need to invoke any mass balance changes (U.S. Department of Energy, 1985: 19).

In general it is assumed that sea level rise will accelerate in the future. The continuing increase of the CO_2 content of the atmosphere threatens to result in a global warming, only the initial effects of which have so far been felt. The precise conse-

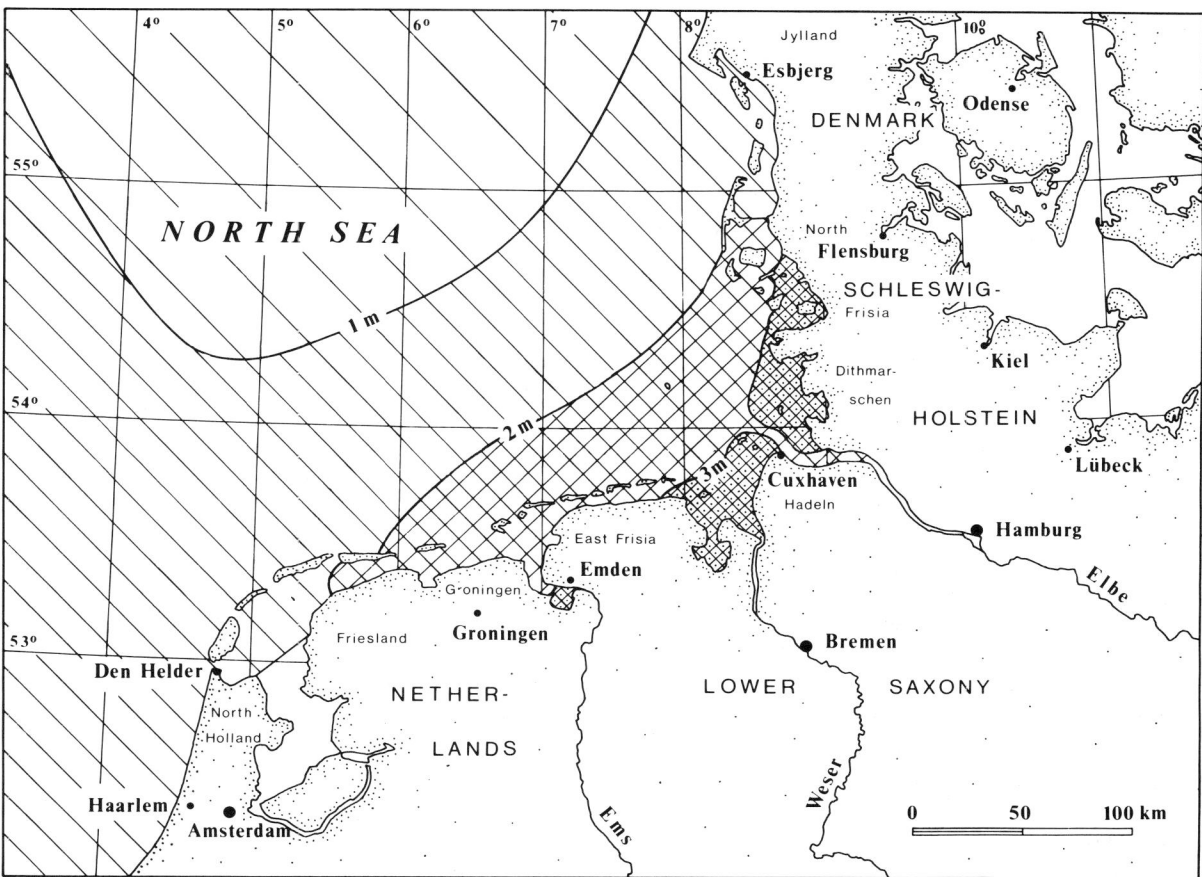

Figure 8: Lines of equal tidal range at spring tide (after Reineck, 1978 a; H. Postma, 1982).

quences of this enhanced 'greenhouse effect' are hard to predict. In a recent report of the National Research Council, Carbon Dioxide Assessment Committee (1983) the consequences of glacier melt on sea level rise until the year 2100 have been calculated. The most likely increase was estimated to be about 70 cm. The Environmental Protection Agency (1983) even predicted increases of 144–217 cm. In both these calculations the consequences of the steric effect were ignored (*op. cit.* in: U.S. Department of Energy, 1985: 7).

2.3 TIDAL RANGE

The general morphological outline of the southeastern North Sea coast, excluding wave action, is a result of the variation in tidal range (Streif, 1978). The influence of the tidal range on the coastal morphology was pointed out by Hayes (1975, 1979) and Dijkema (1980). Where the tidal range is small, a continuous, straightened shoreline is to be found, characterized by a broad belt of coastal dunes. This system is found along most of the coasts of Belgium and Holland as well as along the Jutlandic North Sea coast.

Where the mean tidal range exceeds 1.35 m (Den Helder: 1.37 m, Esbjerg: 1.44 m; Dijkema, 1980: 76) the continuous dune belt is replaced by chains of barrier islands, i.e. dune islands, separated from each other by tidal inlets. This system occurs from Texel to the mouth of the Jade Bay in the south and from Fanø to Süderoogsand in the north. The sice of the barrier islands decreases towards the inner parts of the German Bight in relation to the increase in tidal range (Dijkema, 1980: 77).

Where the mean tidal range exceeds 2.90 m barrier islands are absent. The inner parts of the Meldorf Bight record a maximum tidal range of 3.44 m and the inner parts of the Jade Bight 3.73 m

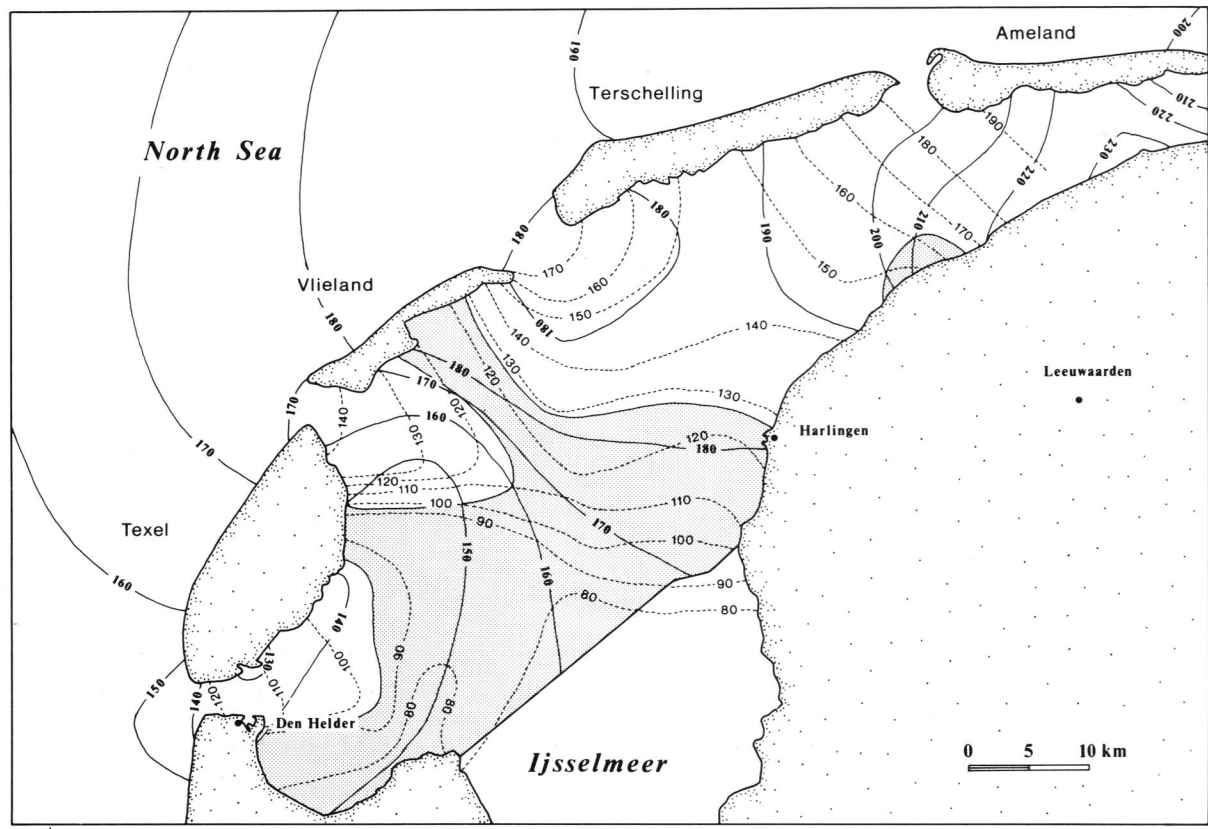

Figure 9: The mean tidal range in the south-western part of the Wadden Sea before (broken lines) and after (solid lines) the damming of the Zuider Zee. The tidal range is given in centimetres. Shaded area: Tidal range difference of over 50 cm (Source: Klok & Schalkers, 1980).

(fig. 8). The Wadden Sea thus belongs exclusively to the microtidal (tidal range: 0–2 m) and mesotidal (tidal range: 2 - 4 m) categories devised by Davies (1964).

Whilst the tidal range outside the barrier islands increases regularly towards the inner German Bight and the Elbe and Weser estuaries, in the Wadden Sea the picture is slightly modified (fig. 9). The greatest tidal range is met in the close vicinity of the inlets, while in the lee of the islands minimum ranges are found. The simultaneously low current velocities lead to increased sedimentation in these areas. This local effect, however, is only felt where the barrier islands are of considerable length, like the West Frisian Islands. Behind the East Frisian Islands and the small islands east of Ameland it can be negligible (cf. Siefert et al., 1985).

The coastal system, however, cannot be regarded as static. In the inner parts of the German Bight no barrier islands have so far been formed, but on the broad tidal flats a number of high lying sandy shoals occur (e.g. Blauort, Knechtsand etc.). Some of these shoals appear to be developing into barrier islands (Mellum, Scharhörn, Trischen).

Changes of the drainage areas

The local tidal range changes if the sice of the tidal basin is changed. There are a number of examples of such changes which, during the last centuries, have occurred mainly as results of land reclamation works and dam construction. A well-documented example showing the consequences of dam building was the damming of the Zuider Zee. Klok & Schalkers (1980) have evaluated the long-term measurements of the Rijkswaterstaat. According to their results the consequences can be subdivided into four groups:

1. Changes of the tidal range

Before the damming of the Zuider Zee the tidal wave reflected from the southern bank of the Zuider Zee interfered with the next tidal wave, resulting in a landward decrease of the tidal range down to a minimum of 20 cm off Urk. Construction of the 'Afsluitdijk' considerably reduced the sice of the drainage basin, reducing also the interference effect. As a consequence the tidal range increases in a landward direction today (fig. 9). It took about 5 years before the tidal range was adjusted to the changed conditions (Klok & Schalkers, 1980: 5).

2. Changes of the tidal currents

The damming of the Zuider Zee altered the progress direction of the tidal wave within the Wadden Sea. Whilst the tidal wave entering the lagoon through the Marsdiep and Eijerlandse Gat inlets was formerly directed roughly north-southwards, this has now changed to a west-eastward progression. The increase in tidal range resulted at the same time in an increase in current velocities. As no zero measurements were made before the damming, the order of magnitude of the changes can only be estimated. Klok & Schalkers (1980: 6) assume that the changes anticipated by the 'Staatscommissie Zuider Zee' were roughly correct:

Marsdiep (at Den Helder)	26 % increase
Eijerlandse Gat	10 % increase
Vliestroom	19 % increase

3. Rise of the storm surge heights

The narrowing of the tidal basin resulted in a marked increase of the storm flood levels. Because of lacking comparability of real storm surge events, again only the results of the model calculations can be quoted here. An assumed repetition of the 1894 storm surge would 1954 have resulted in the following rise of the water levels:

Den Helder	ca. 50 cm increase
Terschelling	ca. 40 cm increase
Harlingen	ca. 100 cm increase

4. Morphological changes

It is quite clear that the above mentioned changes should also result in morphological changes. These are already caused by the changed catchment areas of the inlets:

Inlet	Before	After the damming
Marsdiep	600	700 million m³
Eijerlandse Gat	200	170 million m³
Vlie	700	630 million m³
Amelander Gat	250	250 million m³

The morphological changes, however, take place only over long periods of time and are very hard to anticipate. Repeated measurements of channel cross sections in the area influenced by the damming indicate that changes are still occurring. The continuous decrease in size of the Amsteldiep and Wierbalg channels is certainly caused by the damming. For other morphological changes reasons other than the damming may also be considered (Klok & Schalkers, 1980: 9).

On the islands the increase of the tidal range and rise of the storm flood level has led to a lack of accretion. This is felt most severely on the small Isle of Griend, where the increase in tidal range of about 40 cm has drastically increased erosion in the west and simultaneously stopped accretion in the east (Brouwer et al., 1950). The continued existence of the bird island can now only be secured by strong coastal protection measures.

Land reclamation and dyking influences the tidal range in the whole Wadden Sea area. Fig. 9 demonstrates that the consequences of the IJsselmeer dam are felt as far to the east as the Isle of Ameland, where they meet the effects of the Lauwersmeer dam.

Falling of the mean low water level

Independently from these influences other factors exist, which contribute to changing tidal ranges. The observed strong increase in tidal range in the German Bight during recent years cannot be attributed to any significant changes of drainage areas.

Jensen's (1984) investigations have shown that the increase is not only the result of rising Mean High Water levels, but that simultaneously the Mean Low Water levels have fallen considerably.

Period	Mean rise in m/100 years		
	MLW	MHW	Tidal range
1884–1983 (100 years)	0.03	0.25	0.22
1934–1983 (50 years)	−0.04	0.33	0.35
1959–1983 (25 years)	−0.13	0.64	0.76

Source: Jensen (1984: 513).

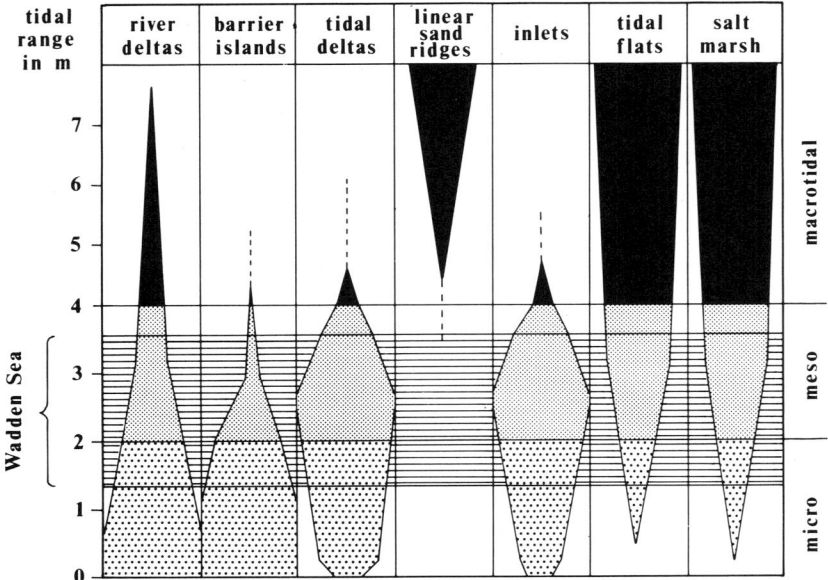

Figure 10: Morphological variations of coastal plain shorelines depending on different tidal range (after Hayes, 1975).

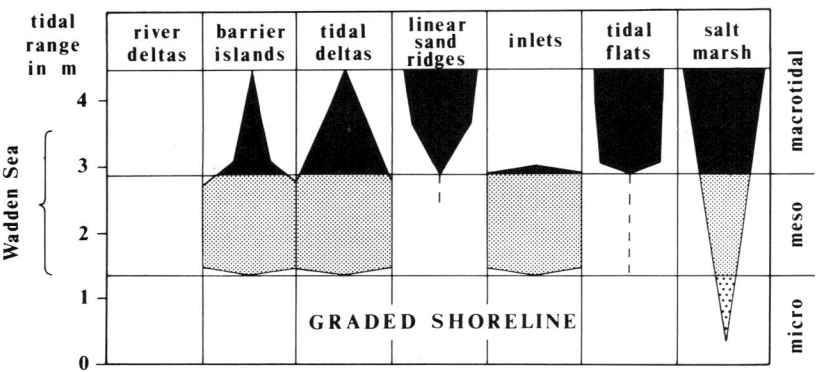

Figure 11: Morphological variations of the south-eastern North Sea shorelines depending on different tidal range with respect to local peculiarities.

The cause of this development remains unclear, the consequences for the morphodynamics of the Wadden Sea being clearly visible. The increased tidal range leads to higher current velocities in the tidal creeks and inlets, resulting in increased erosion (Führböter & Jensen, 1985). One possible morphological result would be an extension of the tidal deltas, as it can be observed already in a number of inlets (cf. section 5.2.5).

The importance of the tidal range on the shoreline configuration is obvious. Hayes (1975, 1979) has presented a model whereby the morphology of coastal plain shorelines of tide-dominated coasts can be explained as a result of differences in tidal ranges (fig. 10). The model has to be modified slightly where wave action becomes important (cf. FitzGerald, Penland & Nummedal, 1984: 356). The validity of this model for the North Sea coast is discussed below.

Some differences are clearly visible. Recent salt marsh formation is restricted to a few very small areas in the Wadden Sea, and river deltas are totally missing, so that these two points have to be changed from the original model.

The boundaries drawn between the different tidal range classes seem to be drawn rather arbitrarily by Davies, and they are not of much use when applied to the North Sea coast. As mentioned above, the tidal range in the Wadden Sea nowhere exceeds 3.72 m. Thus the macrotidal area of Davies (1964) is not represented here. However, the linear sand ridges, which in Hayes' model occur only in the macrotidal area, dominate the region of tidal flats between the Eiderstedt Peninsula and Jade Bay. The macrotidal landforms are found here at a tidal range of over 2.90 m. If Hayes' scheme shall be applied to the North Sea area, the boundaries between macro-, meso- and microtidal areas have to be defined differently in order to correspond to the morphological situation. This is shown in fig. 11.

Local peculiarities

In addition the coastline of the study area exhibits a number of peculiarities which cannot be reconciled with Hayes' simplified model, only the main ones are listed below:

1. Barrier island length does not decrease regularly with increasing tidal range. The system is modified in all localities where instead of easily erodible Holocene sands, older, more compacted sediments outcrop. The Isle of Sylt with its core of Pliocene and Pleistocene sediments is able to resist erosional attack quite successfully and as a result extends further seaward than the rest of the North Frisian coast. The exceptional length of the island (38 km) results from the littoral drift divide in front of the island core. As a result, two spits, each over 10 km long, have formed at both ends of the island core, one in a northerly, the other in a southerly direction.

2. Between Amrum and Eiderstedt the barrier islands are especially poorly developed, and deep tidal inlets reach far inland. The seaward islands (Japsand, Norderoogsand, Süderoogsand) are pure sandbanks, bearing no dunes. The same type of barrier is found between Rømø and Fanø as well as between Schiermonnikoog and Rottum and between Ameland and Schiermonnikoog. Reworking of the sediments has been too strong to allow natural stabilization of the islands.

3. On the Eiderstedt Peninsula the dune barrier is directly linked to the mainland marsh. Marsh growth was promoted here by the old beach ridges of Garding and Tholendorf, which are interpreted as former spits belonging to a drowned Pleistocene core in the area of the present Norderhever inlet (Menke, 1976).

However, it must be remembered that the present morphology is only partly a result of freely acting natural forces, and one should be cautious not to draw far-reaching conclusions. It has to be taken into consideration that the chain of barrier islands formerly did not end at Texel. In the Middle Ages the islands of Huisduinen and Callantsoog existed further to the south. They were to be connected later with the Dutch mainland through human interference (Westenberg, 1956). The tidal range at Den Helder was also lower before the damming of the Zuider Zee (see above). The barrier islands in those times extended into areas of smaller tidal range.

It can be concluded that a model similar to that of Hayes can help to elucidate the dominant controlling factors in coastal morphodynamics. In each particular case, however, a broad range of local peculiarities have to be taken into consideration.

2.4 STORM SURGES

Under the influence of strong winds the tides of the North Sea can rise to extreme heights. The resulting storm surges cannot be generally defined by the height of their water levels since these vary from place to place as a function of wind direction and local tidal range. For the German North Sea coast, therefore, a 'light storm surge' is defined as the maximum water level reached between 10 times a year and once every two years. A 'severe storm surge' occurs once every 2 to 20 years, and the peak water levels of a 'very severe storm surge' are reached once in every 20 years or even more infrequently. At the Cuxhaven gauge, for example, the different classes of storm surge reached the following water levels (Petersen & Rohde, 1977):

Storm surge heights at the Cuxhaven gauge:

Light storm surge	2.62 – 3.58 m above NN*
Severe storm surge	3.59 – 4.32 m above NN
Very severe storm surge	over 4.33 m above NN

*NN = German ordnance datum (mean sea level).

The height of the water level at the different gauges depends on their position within the Wadden Sea area. Towards the outer margins of the Wadden Sea the tidal range is much smaller than in the inner areas (Postma, 1982: 13). This difference is augmented during storm surges, when wind pressure promotes higher water tables along the coast. Petersen & Rohde (1977: 19) have measured the height differences for four profiles through the Wadden Sea off Schleswig-Holstein:

Height differences of the water tables between Hörnum (Sylt) and Südwesthörn (24 km):

Normal MTHw (mean tidal high water)	34 cm difference
Storm surge of 17.2.1962	64 cm difference
Storm surge of 3.1.1976	110 cm difference

Height difference of the water tables between Amrum and Dagebüll (23 km):

Normal MTHw	10 cm difference
Storm surge of 17.2.1962	45 cm difference
Storm surge of 3.1.1976	57 cm difference

Height differences of the water tables between Hevermündung and Husum (32 km):

Normal MThw	24 cm difference
Storm surge of 17.2.1962	70 cm difference
Storm surge of 3.1.1976	91 cm difference

Height differences of the water tables between Trischen and Meldorf (25 km):

Normal MThw	12 cm difference
Storm surge of 17.2.1962	78 cm difference
Storn surge of 3.1.1976	81 cm difference

A persistent problem is the tendency for storm surges to occur more frequently during certain periods. Henning (1897), for example, tried to discover why certain days during the winter months were especially susceptible to storm surges, but he came to no conclusive results. More recently Duphorn (1976) has mentioned the coincidence between storm surge frequency and global climatic changes. The relationship, however, appears to be rather complex. Seliger (1983) attempted various correlations between storm surge frequency and climatic factors, but likewise came to no general conclusion. Linke (1981: 782) like Duphorn concluded that the storm surge periods were related to changes in the atmospheric circulation pattern, and not directly correlated with the well-known historical glacier oscillations.

Results of the storm surges

Storm surges in former times were defined mainly by the resulting damage rather than by the measured height of the water table. Knowledge of historic storm surge levels is rather insufficient. In the reports of past storm surges losses of human lives are mainly recorded from the low-lying dyked mainland areas and very rarely from the barrier islands. Frequently houses were damaged, gardens and pasture were destroyed by salt water inundation, and the farmland was often covered by marine sands.

When marshlands are covered by marine sands not only is the existing vegetation destroyed, but the saline sand also prevents regeneration. For example, the aerial photograph in Fig. 12 shows the sand-covered area behind a dyke breach of 1962 in the Ostland area, on the Isle of Borkum. This photograph was taken in 1966 and demonstrates that the vegetation had still not regenerated four years later. Recent aerial photographs taken in 1984, however, show no traces of the 1962 dyke breach.

The most marked changes along the coastline resulting from storm surges have been intensive erosion of the island dunes and the Pleistocene core of the Isle of Sylt. To a lesser degree the cliffs of the mainland (Emmerlev cliff) and the leeward coasts of the islands (Morsum cliff on Sylt, 'Ual anj' cliff on Amrum) have also been eroded.

The small cliffs of the seaward margins of the Halligen Islands and of the mainland salt marshes are more severely eroded by light storm surges when the waves break in this zone. During severe storm surges the entire salt marsh area is inundated and the waves break further inland or at the foot of the dykes.

In the tidal flat areas the direct effects of storm surges are limited. The extensive widening of the hydraulic cross section of channels within the tidal flats keeps stream velocities relatively low and therefore only minor sediment transport is possible. Stronger redeposition occurs on the shoals flanking the tidal inlets. Rapid sea level fall immediately after the storm surge is often associated with extensive sediment removal and temporary overdeepening within the tidal inlets (Petersen & Rohde, 1977: 28).

The effects of storm surges on dykes are caused by a number of different processes. When waves break at the dyke base water can push up to the crest of the dyke and cause erosion of the landward side. This erosion is increased if the core of the dyke becomes saturated during high water levels and seepage occurs. In addition, waves breaking on the dyke slope can cause considerable damage through compressive thrust and cavitation.

Figure 12: Aerial photograph of the eastern part of the Isle of Borkum. The sand-covered area resulting from the dyke breach of 17.2.1962 in the Ostland is clearly visible; the vegetation has not yet regenerated. Aerial photograph of the Niedersächsisches Landesverwaltungsamt, Landesvermessung, of 17.4.1974, Bildflug Ostfriesische Inseln (1086), Str.1/015, freigegeben durch NLVA - Abt. Landesvermessung -, Hannover, unter Nr. 34/74/1086.

Figure 13: Old dyke on Borkum with deep ponds on the leeward side which are the remnants of former dyke breaches. According to Dekker (1892: 94), this dyke was breached in a number of places during the storm surge of 1825 (1984).

When waves collide with vertical walls extreme forces are released. In this way three tetrapods from the foot protection of the Westerland seawall (Isle of Sylt), each weighing 6 tons, were transported several metres by waves during the storm surge of 17.2.1962 (Petersen & Rohde, 1977: 27). Modern dykes therefore generally have a gentler profile to reduce these forces.

Dyke breaches usually cause considerable erosion. During the storm surge of 17.2.1962 in Hamburg scour hollows 11 m in depth were formed (Kolb, 1962: map 8). During the storm surge of 1.2.1953 in the Netherlands, basins of up to 38 m depth were even formed (Gierloff-Emden, 1954: 17). As in earlier times, there is no satisfactory way of filling such deep hollows and new dykes are normally erected seaward of the eroded holes. Today, the ponds found behind old dykes often indicate the positions of former dyke breaches (fig. 13).

Land losses during the Middle Ages

Reports of storm surges have been documented since the end of the 10th century. These early sources, mainly from the Netherlands and from Friesland, usually only refer to the date of the event, and often only the year. In most cases the chroniclers had not witnessed the floods themselves, and only knew of them through hearsay. Consequently, such information tends to be inaccurate. There are no first-hand reports on storm surges predating the 15th century (Petersen & Rohde, 1977).

Geological and archaeological investigations give hints about the age of the early great land losses. The Zuider Zee seems to have formed between 1000 and 1300 A.D. (van der Heide, 1955; Hallewas, 1984: 301), and the Middelzee near Leeuwaarden formed between the 2nd and 10th century. It was already completely reclaimed again by the early 16th century (Halbertsma, 1955).

In the Wadden Sea area the oldest mentioned storm surges occurred in 1010, 1020, 1041, 1075 and 1094 (Arends, 1833), followed by the inundations of 1102 and 1114. One of the first major floods to cause considerable damage following the construction of the dykes, which started about 1000, was the Julianen Storm Surge of 17.2.1164 (Petersen & Rohde, 1977). A number of catastrophic floods followed, leading to transgressions in the Wadden Sea coastal area. The Lucia Flood of 14.12.1287 initiated the formation of the Dollart, and probably also the Jade Bay (Reinhardt, 1979: 27-42, *op. cit.* in Brandt, 1981: 82). According to contemporary reports of 1338 the Eider estuary was then widened to its present funnel shape (Petersen & Rohde, 1977).

The major storm surges of the Middle Ages did not immediately cause the inundation of vast areas of land as commonly assumed. Although during catastrophic floods the weak dykes were breached in many places, the erosion of the consequently unprotected marshes was initially minimal. Where gaps in the dykes could not be repaired, large forelands appeared which, like the present salt marshes, were flooded only during storm surges and lay dry for most of the year. These were only dissected by a number of tidal creeks (Homeier, 1979 a).

The major and partially enduring land losses resulting in the formation of the Jade Bay, the Dollart and the Zuider Zee did not occur as the result of one large event but gradually, through many smaller stages. These major land losses were caused mainly by the lack of a technical infrastructure capable of protecting the vast forelands before the destructive effects of subsequent storm surges in the following decades. Land reclamation was possible in a number of cases but only through projects that lasted over the centuries (Ehlers & Mensching, 1983).

The greatest land losses resulted from the storm surge of 16.1.1362 (Second Marcellus Flood), known as the 'Große Mandränke' (Great Drowning Disaster). Jade Bay, Dollart, Harle Bay and Ley Bay were enlarged, and in Nordfriesland vast areas of land were lost including the legendary Rungholt, east of the present island of Pellworm.

Several severe storm floods in the 14th century (1380, 1387, 1391, 1393 and 1395) occurred at least partly during the late spring and early summer (end of April to June). This means that in these years the dykes were not yet fully restored and unable to resist even lesser summer storm surges (Petersen & Rohde, 1977).

A period of dyke reconstruction

The storm surges of the 15th century caused less damage than those of the 14th century. It may be that only the relatively safe areas were left unscathed by the preceding catastrophies. It was at this time that the reclamation of the inundated land began.

In Nordfriesland, the Großer Kohldammer Koog, Bargumer Koog and Langenhorner Alter Koog were dyked in the 15th century (Prange, 1978: fig. 1). The Hattstedter Alter Koog was dyked in 1478 (Prange, 1968: 632), and the adjoining Brekklumer, Bredstedter and Bordelumer Koog dyked not later than the early 16th century. During the flood of 26.9.1509 the Dollart reached its greatest extension (Petersen & Rohde, 1977), and the Jade Bay reached its maximum extension after the storm surges of 1509, 1510 and 1511 (Brandt, 1981: 78); after these events reclamation also began here. More land was already reclaimed in the Harle Bight at this time than before the catastrophic flood of 1362 (Homeier, 1979).

During the 'All Saints Day Flood' (1.-6.11.1570) in Ostfriesland only about 13 km^2 of land was lost. The first flood water level records date from this period. The flood level of the 1532 storm flood was marked in the church at Klixbüll (Schleswig-Holstein) at NN + 4.16 m, and the flood heights of 1570 are recorded at the church of Suurhusen near Emden (NN + 4.40 – 4.50 m) and in Dangast on the Jade Bay (NN + 4.41 m) (Petersen & Rohde, 1977).

Land reclamation continued into the early 17th century. There was even a plan to reclaim Rungholt Bay on Strand island by constructing a dam from Pellworm via Südfall to the present Isle of Nordstrand (Higelke, 1982).

The 'Second Mandränke'

On October 11th, 1634 another catastrophic flood occurred in Nordfriesland (Arends, 1833, Vol.2: 144). This severe storm flood followed a series of storm surges in the early 17th century that weakened the dykes and resulted in the destruction of the major part of the horseshoe-shaped island of Strand and a number of the Halligen Islands. These were not reclaimable (Karff, 1968).

The series of major land losses along the coast of the Wadden Sea ended with the severe storm surge of 1634. The destruction of the 'Weiße Klippe' (White Crag) on Helgoland on 1.11.1711 and the subsequent partition of the 'Dune' from the main island nine years later were due to special circumstances on the island.

Later damaging floods

The next major damaging flood after 1634 occurred at Christmas 1717. In northern Germany and the Netherlands numerous dykes were breached and vast areas of land were flooded. The newly dyked Dieksander Koog (Dithmarschen) had to be abandoned. The 1751 flood saw the loss of the Christianskoog on Nordstrand (Prügel, 1942: 19); these were the only notable land losses during the 18th century. Severe storm surges occurred during 1824 and on 3./4.2.1825 but only on the Halligen salt marsh margin was there any major erosion.

The storm surges of the late 19th and early 20th century (1855, 1906, 1911) caused considerable local damage; the settlement of Wangeroog had to be moved from the western end of the island to the centre following the 1855 flood. In most cases, however, catastrophic consequences were avoidable. During the storm surges of 18. and 27.10.1936 several dykes were breached, including Trendermarschkoog on Nordstrand and Neufelder Koog in Dithmarschen, and there was severe erosion of the island coasts (Prügel, 1942: 70). During the storm surge of 9./10.2.1949 extremely high water levels were registered along the western coast of Schleswig-Holstein (Petersen & Rohde, 1977: 60), but there was only slight damage.

Only the severe storm surges of recent years have reminded the public of the potential hazards of the coastal areas. On 31.1./1.2.1953 143,000 ha land were flooded and almost 2,000 people drowned in the Netherlands, and on 16./17.2.1962 a storm surge cost 340 lives in North Germany.

Consequences

The present configuration of the coastline is the result of a number of mutually-interactive processes. The factors leading to the formation of the coastline as discussed in sections 2.1 to 2.4 above may be summarized as follows:

1. When the rising Postglacial sea level reached the present coastal area at about 7,500 B.P. (Jelgersma et al., 1979), it encountered a landscape sculptured by Pleistocene glaciers. The depositional landforms consisted of moraines and sandur plains, the submerging of which led to the formation of an irregular coastline, subdivided by the Elbe, Weser, Ems and Rhine estuaries.

2. Almost a thousand years later, around 6,700 B.P., the coastal barrier system developed. This resulted in the formation of a smoothed shoreline, and to the cutting-off of lagoonal areas and large back barrier tidal flats (van Straaten, 1965 b: 77).

3. Major parts of the North Sea coast, flooded during the Postglacial transgression, emerged after the transgressive hemicycle. This development was promoted by human interference. Man has considerably reduced the marine influence through the dyking of large areas since the Middle Ages. This has led to changes in the size of the catch-

ment areas of most of the tidal inlets and these have to be taken into consideration when interpreting the coastal evolution of the area.

4. Vast parts of the coastal area were subsequently affected by storm surges and considerably re-shaped. The most notable changes were the formation of the Zuider Zee (before the first dykes were built), the Lauwerszee, the Dollart, the Harle Bight, the Jade Bay and the destruction of Nordfriesland by the two flood surge disasters of the 'Mandränken'. Marine embayments formed with great influence on tidal processes in adjacent areas. Lasting alterations were created only in areas where deep channels and basins were eroded (Dollart, Jade Bay, Nordfriesland).

5. Both the continuing transgression and changing coastline configuration led to changes in the tidal range which subsequently influenced the morphodynamics of the coastal area. The fossil beach barrier systems of Dithmarschen and Eiderstedt are but two examples of this.

6. In the course of its development the barrier island coast has been continuously reshaped, always adjusting to the changing natural conditions. Such adjustment is active today and will continue into the future.

CHAPTER 3

BARRIER ISLAND DEVELOPMENT

Barrier coasts similar to those of the North Sea are found in various parts of the world. The East Coast of the United States is particularly comparable and has been the subject of intensive investigations since the beginning of the century. However, research in the Wadden Sea area has developed largely independently. As in the American literature (cf. Leatherman, 1982: 4), three different basic concepts of barrier island genesis have been considered for the North Sea coast:

1. Drowning of coastal dune ridges.
2. Formation of spits, which were later breached by the sea.
3. Island development from emerging shoals.

Haage (1899: 51) believed that the barrier islands had been formed by drowned dune ridges, as a consequence of coastal subsidence. This concept was also favoured by Wahnschaffe (1901: 249).

Gripp (1944: 25, 1956, 1964) promoted the idea that the chain of islands along the North Sea coast had developed from a chain of spits. Coast parallel sand transport would have created the barrier. The sand was assumed to have migrated into the inner parts of the German Bight through the Channel and along the Dutch coast, with a second transport path extending from the Isle of Sylt southwards. This interpretation, however, is no longer valid.

Within the investigation area, major spit formation is only observed on the Isle of Sylt. There is an imminent danger that the existing spits will be breached, both being heavily eroded on their western sides. Whether such a breach during a storm tide would actually lead to a separation of the spit from the island, or whether the newly formed gap would be rapidly closed again by coast-parallel sand transport remains an open question.

Ordemann (1912: 46) appears to have been the first to argue that the islands had formed as a result of emerging sand banks. Behrmann (1921) also favoured this concept.

In the course of his geological mapping of the East Frisian Islands, Keilhack (1925: 5) came to the same conclusion that the island chain had been built up by sediment supply from the sea. From the formation of the Isle of Memmert, which had taken place 'under our eyes', he invoked the following model of barrier island formation:

1. Sand supply from the sea leads to the formation of sand banks near the coast.
2. Increasing sand supply causes the shoals to rise until they finally emerge during low tide.
3. If accumulation continues, they eventually remain dry even during ordinary floods.
4. Under the influence of the wind, parts of the former shoal finally grow above spring tide level, forming an island.

Keilhack (1925: 4) explained the sand supply by invoking sediment transport under tidal influence from the easily erodible Tertiary sands eastwards along the coasts of Belgium and northern France. At first Wildvang (1929: 88) accepted this interpretation. However, after van Veen (1936) had demonstrated that at least recent sediment movement through the Channel had been insignificant, Wildvang assumed that the sands originated from the bottom of the North Sea (Wildvang, 1938: 186).

This interpretation is widely accepted today (Sindowski, 1973; Jelgersma & Ente, 1977: 32). Hageman (1969) assumes that barriers existed in the coastal area during the entire Holocene, these being progressively shifted landwards during phases of transgression.

Information about the development of the barrier islands is only preserved where the coastline has not undergone major retreat, such as in parts of Holland and in southern Schleswig-Holstein. In spite of the considerable differences in the regional development of barrier formation the results from both these areas permit a general reconstruction of their development.

During the first millenia of the Holocene sea level rose very rapidly. During this time the horizontal transgression rate was the controlling factor in shaping the coastline. Pratje (1951) assumed a medium transgressive rate of 150 m per year, but Bäsemann (1979: 116), basing his results on more recent data, calculated even faster rates (267 m per year). It seems reasonable to conclude that during such a rapid landward displacement of the coastline there was not enough time for the formation of a barrier system.

Formation of the barrier in Holland

This situation changed as soon as the sea reached the present coastline and the transgression ceased. This event represents the inception of the coastal barrier. This was followed by beach ridge formation along the coast of the Netherlands. Lagoonal sediments and tidal deposits of great thickness accumulated behind the protective barrier. The so-called 'Old Dunes' were rapidly formed on the beach ridges which, partly covered by younger dunes, can be traced along the west coast of Holland from the south of Den Haag to the north of Alkmaar.

Development of the beach ridges was a two-stage process. The first stage, stabilization of the shore, started around 6700 B.P. The second stage, shore regression, was initiated around 4100 B.P. (van Straaten, 1965).

Along this preserved section 80 km in extent, the older beach ridge system was breached in three places. Two of the gaps were connected to former courses of the Rhein-Maas or Oude Rijn river systems; the third gap (at Bergen) was a real tidal inlet. The coastline at that time had a configuration considerably different from today; the present smooth outline resulted from later erosion. The formation of the tidal inlet at this specific point was not a random feature, the gap in the dune ridges coinciding well with a depression in the Pleistocene subsurface (Urstromtal of the Vecht) (Jelgersma, 1983).

During the Dunkerque-0-Transgression (3700 – 3000 B.P.) the continuous coastal barrier was breached in several places. Whereas in Holland and in the western parts of the present Wadden Sea only minor tidal inlets with small drainage areas were formed, in Groningen vast areas were covered with marine clay. These areas were settled as a result relatively early because they were free of swamps in contrast to the other coastal areas (Roeleveld, 1971, 1974; Jelgersma *et al.*, 1979).

During the Dunkerque-1-Transgression (about 2300 B.P.) the freshwater lake, Lake Flevo, was formed in the area of the present IJsselmeer, linked in the west and north with the sea. This lake was later extended to form the Zuider Zee by the marine inundations of the early Middle Ages (Jelgersma *et al.*, 1979).

With the onset of Roman settlement at about 2000 B.P. the main phase of 'Old Dune' development came to an end (Jelgersma *et al.*, 1979). Minor aeolian sedimentation continued, however, until the Middle Ages.

The formation of the 'Young Dunes' started on a large scale in the 12th century. This process was intensified during the 15th and 16th centuries, when large quantities of sand became available for aeolian redeposition from eroding shore sectors. These 'Young Dunes' closed the mouth of the Oude Rijn at Katwijk and the mouth of the Utrecht-Vecht river between Egmond and Bergen (Jelgersma *et al.*, 1979).

Whereas the 'Old Dunes' formed mainly elongated, relatively low dune ridges on the surface of the old beach ridges, the 'Young Dunes' are often represented by huge parabolic forms, which have migrated landward from the coastline for several kilometres. The vast quantities of sand represented by the 'Young Dunes' suggests a preceding phase of strong coastal erosion. This phase may have been accompanied by a steepening of the underwater relief seawards of the barrier (Roep, 1984).

The 'Young Dunes' did not form simultaneously. Radiocarbon dates show that dunes first formed in the west and only in the late 15th century started to develop on the eastern West Frisian Islands (after de Jong, 1984: 274):

Terschelling	1220 – 1290 A.D.
Ameland West	1040 – 1290 A.D.
Ameland Centre	1290 – 1390 A.D.
Schiermonnikoog	1430 – 1480 A.D.

Historical records demonstrate that the main phase of dune formation on the Dutch mainland coast ceased during the 17th century (Zagwijn, 1984: 267). On the islands, however, dune formation has been active up to the present day.

Formation of the barrier in Schleswig-Holstein

Transgression first effected northern Dithmarschen around 5000 B.P. During the transgression marine sediments up to 15 m thick were deposited locally. North-south striking beach ridges of gravel, coarse sand and shell fragments accumulated early during this phase, and these ridges strongly influenced subsequent local sedimentation. As a result of the protective influence of the barrier mainly fine-grained material was deposited in this zone (Hummel & Cordes, 1969: 106).

Coarse clastic sedimentation continued during

the Sub-Boreal (around 1800 B.P.) after an interruption during the Atlantic. The beach ridge system was extended southwards and finally developed into a long barrier system ('Lundener Nehrung'). These barriers were relatively narrow ridges which grew to a height of 1–2.5 m above NN. On the surface of the beach ridges small dunes accumulated. As a result of the subsequent regression, the barrier system became inactive, and during the following Dunkerque-1-Transgression, until about 2000 B.P., the barrier was no longer influenced by marine processes (Hummel & Cordes, 1969).

The 'Lundener Nehrung' is not probably a spit in the classical sense in that no abrasion area, which might represent a sediment source, has so far been identified. Gripp (1964: 295) suggested that the barrier was constructed by a marine current from the north, and that the abraded Pleistocene areas were now covered by younger sediments on the Eiderstedt peninsula; no such areas have, however, been found. One argument against this hypothesis is that a continuous spit across the mouth of the River Eider seems rather unlikely. Hummel & Cordes (1969: 107) concluded that the 'Lundener Nehrung' is the result of redeposition of marine sediments by shoaling waves during the rise in sea level (cf. Shepard, 1963: 187). The 'Lundener Nehrung' therefore seems to be an equivalent of the Dutch beach ridge system on which the 'Old Dunes' were formed.

This early barrier system, known only from a few parts of the Wadden Sea area, became inactive during the regression phase between the Calais and Dunkerque Transgressions. The subsequent transgression led to a rapid reshaping of the coastline, resulting in the formation of a new barrier system. During this process some parts of the old barrier were eroded, whilst other sections became the foci for renewed deposition and were accordingly conserved.

The dunes on the North Frisian Islands are equivalent to the 'Young Dunes' in the Netherlands. On Amrum the dunes overlie artifacts and settlements dating from the Bronze Age to the Viking Era. Dune formation also seems to have ceased here during the 17th century.

It is not necessary to assume that the early barrier system was continuous. Linke (1979, 1982), on the basis of palynological and diatom data, shows that the regressive sediments in the Cuxhaven - Arensch area were not deposited under the protection of a coastal barrier, but that the area was continuously connected to the open sea. This type of discontinuous barrier very closely resembles the present situation where the barrier between Jade Bay and the Eiderstedt peninsula is at best only fragmentarily developed.

Sediment balance

A more detailed examination of the total sediment balance is of interest in relation to the reconstruction of the Holocene development of the coastal areas. Evaluation of a number of profiles through the coastal area of Niedersachsen (Streif, 1975), in addition to areal investigations, have demonstrated that during the Calais Transgression the sediment thickness in the coastal area increased at roughly the same rate as the rise in sea level. In the area of the Emden-West map sheet of the Geological Map 1:25,000 (Barckhausen & Streif, 1978: 32), during the period 5200 to 1600 B.P., about 12.7 cm of sediment accumulated per century. During the Dunkerque Transgression, on the other hand, the sedimentation rate increased dramatically. Although the sea level has risen at most only another metre up to the present day, during the period before dyking began in the 11th century, about 42.4 cm of sediment were deposited every hundred years (Barckhausen, Streif & Vinken, 1979: 59).

During the course of the Holocene there was a concomitant increase in the sand content of the sediments. The transgressive overlap of the sandy marine facies on the clay- and siltrich deposits of the lagoonal facies is therefore explained by the continuing Holocene transgression (Barckhausen, Streif & Vinken, 1979: 60). This process was described as 'bulldozing' by Hageman (1969: 385).

The positive sediment budget of the Wadden Sea area has been confirmed by recent investigations in the Netherlands. In order to estimate the potential for sand dredging in the Wadden Sea, the sediment balance of the area has been calculated. About 3 million m^3 sand is assumed to be supplied annually by the sea in the Dutch part of the Wadden Sea alone (Rijkswaterstaat, 1981).

Sand movement

Veenstra (1982, 1984) has shown that the grain-size parameters of the beach sands from the coastal barriers of East and North Frisia make it unlikely that continuous longshore sand transport occurs. In fig. 14 the median values of 320 grain-size analyses are shown. If there was any sediment movement along the coast a gradual decrease in grain-size from West to East should be expected theoreti-

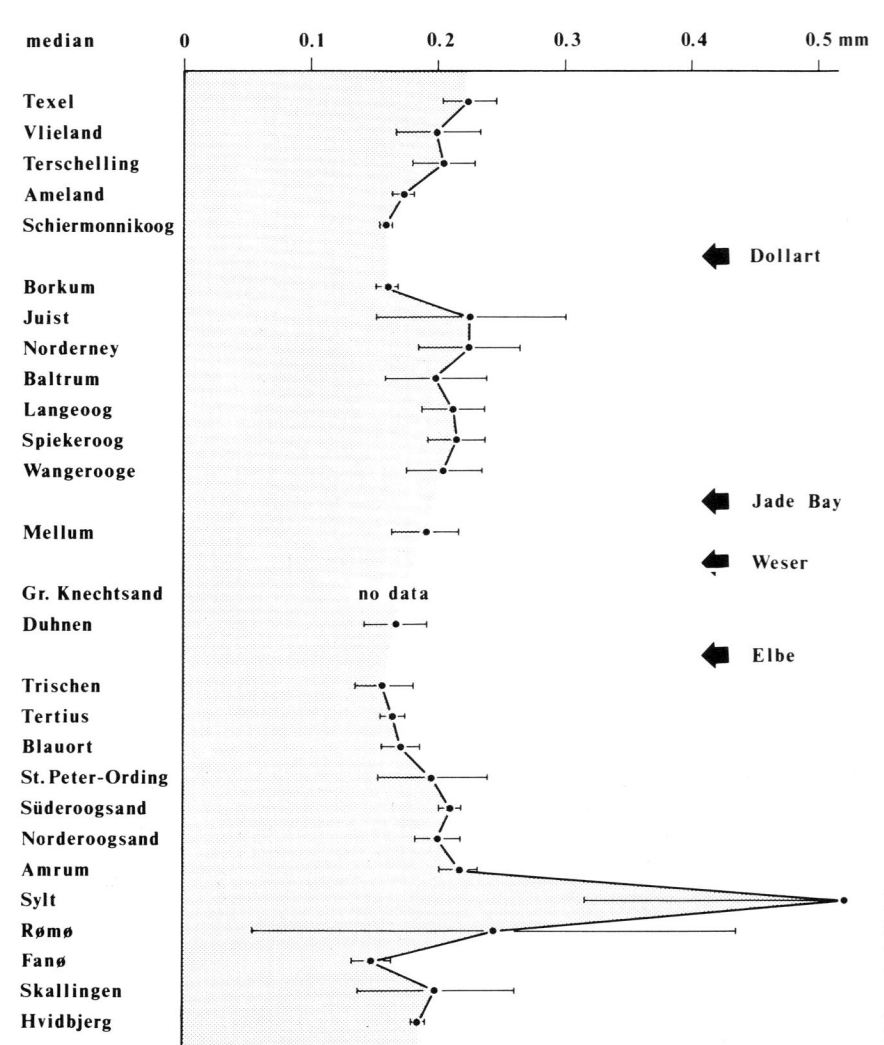

Figure 14: Median values of grain-size analyses from the beaches of the barrier islands (after Veenstra & Winkelmolen, 1976; Veenstra, 1982, 1984). The samples were taken at low water mark. The horizontal bars indicate the variability.

cally. In reality, there is a dramatic increase in grain-size on the Isle of Juist. This is most easily explained by a supply of fresh material from the Pleistocene rise on the sea floor northwest of Juist in the Borkumriffgrund area.

Veenstra's analyses are important in another sense (fig. 15). Particularly low median values are consistently found where the coast is either directly eroded (3), or where, in the course of bar migration, sand is transported from one island to another (1). On the other hand, where the bars come into contact with the beach of the juxtaposed island especially high median values are found (2). This suggests that the slightly coarser fractions seem to be more mobile than the finer ones.

Considerable variations in the grain-size distribution are also found along the North Frisian coast (fig. 14). Here it is mainly the influence of the Pleistocene core of the Isle of Sylt that is significant. In the St.Peter–Süderoogsand area there is the possible additional influence of the totally abraded Hever Pleistocene core (Dittmer, 1952: 149; Hoffmann, 1982: 250) to which the spits of the Eiderstedt Peninsula may have been attached.

These results suggest that the barrier system of the south-eastern North Sea was not built up successively by coast-parallel sand transport from the margins, but that it was constructed almost simultaneously along its whole length by landward directed sand supply. Differences in the shape of the barrier (beach ridge or island formation) appear to be largely the result of differences in pre-existing relief, exposure and tidal range.

These results seem to directly contradict the

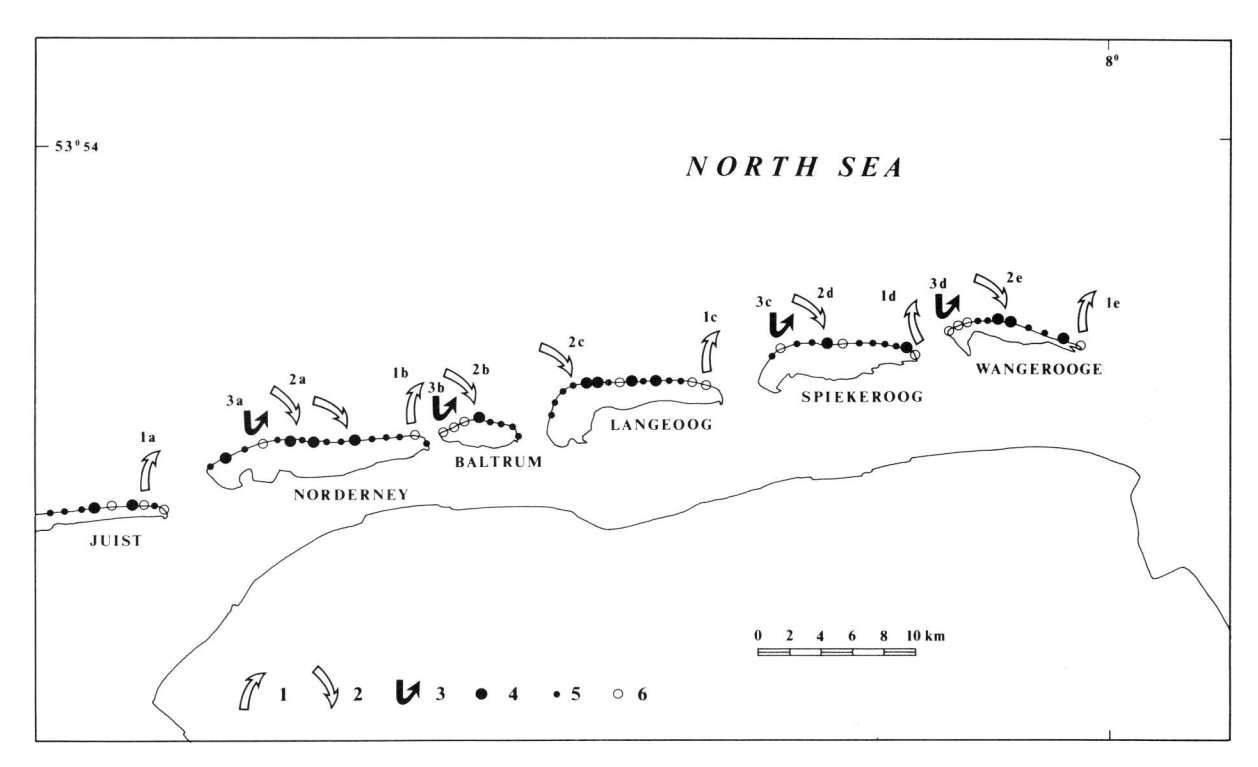

Figure 15: Differences in the grain-size distribution along the beaches of the Ostfriesische Inseln (East Frisian Islands); samples from low water mark. The analyses are from Veenstra (1982). 1 = seaward sediment transport at the western margins of the tidal inlets; a = Norderneyer Seegat, b = Wichter Ee, c = Otzumer Balje, d = Harle, e = Blaue Balje; 2 = landward sediment transport at the eastern ends of the shoal arcs; a = Norderneyer Seegat, b = Wichter Ee, c = Accumer Ee, d = Otzumer Balje, e = Harle; 3 = seaward sediment transport in areas of strong coastal erosion; a = Weiße Düne (White Dune), Norderney, b = western end of Baltrum, c = northwestern end of Spiekeroog, d = western end of Wangerooge; 4 = noticably high median value; 5 = normal median value; 6 = noticably low median value.

traditional concept of coast-parallel sand transport, approximately parallel with tidal current direction as deduced from morphological analyses and map evaluation. Both hypotheses can, however, be combined.

It is generally accepted that there is lateral sediment transport along the island chains, i.e. in the case of the West and East Frisian Islands from west to east. The displacement of the bars of the ebb-tidal deltas is directed in the same direction, and the sediment bodies on the island foreshores also move from west to east. There can thus be no doubt that some sand transport along the coast of the Wadden Sea is directed into the inner German Bight. In the area between Jade and Eider, however, into which sand transport is directed from both sides, no related major deposition has taken place. Some of the sediment therefore appears to be withdrawn from circulation by deposition on the tidal flats behind the islands.

In addition, the channels of the Ems and Jade, as well as those of the Weser and Elbe, are kept artificially overdeepened by continuous dredging. This prevents significant cross-channel sediment transport. In theory this should lead to a significant sediment deficit leewards of the overdeepened channels, but this is not the case.

The coast-parallel sand transport is not only retarded by the deep navigation channels, but also by the tidal inlets. As tidal current velocities in the inlets are considerably higher than along the island coasts, strong sediment displacement in landward as well as in the seaward direction may occur. The foreshore and beach seem to serve only as transport paths from one inlet to another where strong sediment displacement normal to the coast can take place. In these areas sand may be withdrawn from the coast-parallel transport system and deposited on the tidal flats behind the islands, or additional sand can be supplied from the sea. The intensity of this coast-normal sand transport determines the degree to which sediment transport

Figure 16: Sand migration model, assuming coast-parallel sand movement and prevailing landward transport. Open arrows: direction of sand transport; black arrows: direction and strength of the tidal currents.

from one island to another can take place.

The results of Veenstra's investigations have demonstrated that the coast-normal sand transport plays an important role in the sediment budget of the Wadden Sea. The landward-directed component may be the dominant one. This concept is illustrated in fig. 16.

Climatic influence

Van Straaten (1961: 380) was the first to demonstrate a relationship between coastline changes and climatic influence. A phase of intense erosion along the Dutch mainland coast between Camperduin and Scheveningen between 1865 and 1900 coincides well with a precipitation maximum, an increase of the mean annual temperatures as well as a marked increase of westerly and southwesterly winds.

The period between 1850 and 1900 was also a period of striking morphological changes on the barrier islands, characterized by excessive sand supply and island growth:

1. During this period on *Texel* the low water line

was displaced about 200 m seaward, giving rise to the formation of several new coast-parallel dune ridges (Elorche, 1982).

2. Considerable accretion began in the east of *Vlieland* after 1853, permitting the short-termed dyking of the Kooremansvallei (Joustra, 1971).

3. A period of considerable westward growth was initiated on *Terschelling* following the merger with the Noordvaarder (Joustra, 1971). Dune formation on the hitherto barren eastern Boschplaat began about 1866.

4. The dune core of *Juist* extended 2,600 m to the east between 1840 and 1900, and after 1880 a new line of coastal dunes was formed seaward of the old dune core.

5. Under the protection of recently formed dunes, salt marsh began to develop on *Memmert* in 1891.

6. The dune cores on *Norderney* and *Baltrum* began to extend rapidly eastwards after 1860 (Homeier, 1980).

7. The Ostplate of *Spiekeroog* started to extend rapidly eastwards after about 1865 (Homeier, 1961).

8. The dune core of *Wangerooge* extended eastwards after 1880 (Homeier, 1961).

9. Salt marsh formation on *Mellum* began after 1870.

10. Vegetation became established on *Trischen* in 1854; after 1880 dunes formed on the island (Wohlenberg, 1950).

11. The Kniepsand began to merge with *Amrum*.

12. Similarly the Havsand and Juvre Sand merged with *Rømø*.

13. *Fanø* started to extend northwards, with the formation of new dune ridges.

14. Dune formation began shortly before 1870 on *Skallingen*, leading to salt marsh development on the previously barren sand flat.

Some of these morphological changes may have been caused by local circumstances. However, the striking coincidence of so many positive shoreline changes makes it very likely that some common process may have caused them. The period under discussion marks the end of the 'Little Ice Age' (about 1450–1850), characterized by a marked retreat of the Alpine glaciers after about 1855 (Bradley, 1985: 240) and increased drift ice around Iceland (Lamb, 1982: 243).

The formation of the youngest barrier islands

The formation of the large barrier islands occurred during Prehistoric times, and, as a result, the conditions of their formation remain unclear. In the inner German Bight however, the more recent formation of new barrier islands provides more potential evidence.

The Isle of Trischen, for example, did not exist in the 16th century. A sandbank emerged on its site only at the beginning of the 17th century, but it showed little development during the succeeding centuries. The evolution of this bank, the so-called Buschsand, into the Isle of Trischen has been documented by Wohlenberg (1950) and Lang (1975). The first permanent vegetation, protected by the supratidal sand bank, developed around 1854, long before the onset of dune formation around 1880 (Wohlenberg, 1950: 166). As a result of the failure of the dyking of the small island (the dyke erected around the Trischenkoog between 1922 and 1925 was finally destroyed by the storm tide of 16.10.1942) Wohlenberg called his article 'Formation and Destruction of the Isle of Trischen'. However, only the Trischenkoog itself was destroyed; the island continues to exist and is being displaced landwards at almost the same rate as in the 19th century (fig. 17).

This applies also to other young barrier islands. The displacement of Scharhörn from 1935 to 1970 has been demonstrated by Göhren (1970: fig. 48), and the displacement of Mellum by Wunderlich (1979).

Fig. 18 shows that Trischen displays the typical subdivision of a barrier island, with a low dune core in the centre allowing salt marsh growth on the lagoonal side. The rectangular pattern of the drainage system is inherited from the reclamation works undertaken during the dyking of the Trischenkoog. Flood spits are found both to the north and south of the dune core. The spits are submerged during higher tides. The resultant washover processes lead to west-east sand transport across the island, whereas during normal tides coast-parallel sediment transport occurs on the foreshore. This is directed towards the north in the northern part and towards the south in the southern part of the island.

The migration rate of Trischen is about 30 m per year; this rate has remained largely constant over the last 100 years. The mobility of Trischen, Scharhörn and Mellum is similar to that of the Außensände (Outer Sands) seawards of the Halligen as well as of the sand banks of Tertius, Blauort and Großer Knechtsand (cf. Taubert, 1982).

The young barrier islands show much higher migration rates than the older, comparatively more stable islands (fig. 19). Luck (1975) has demon-

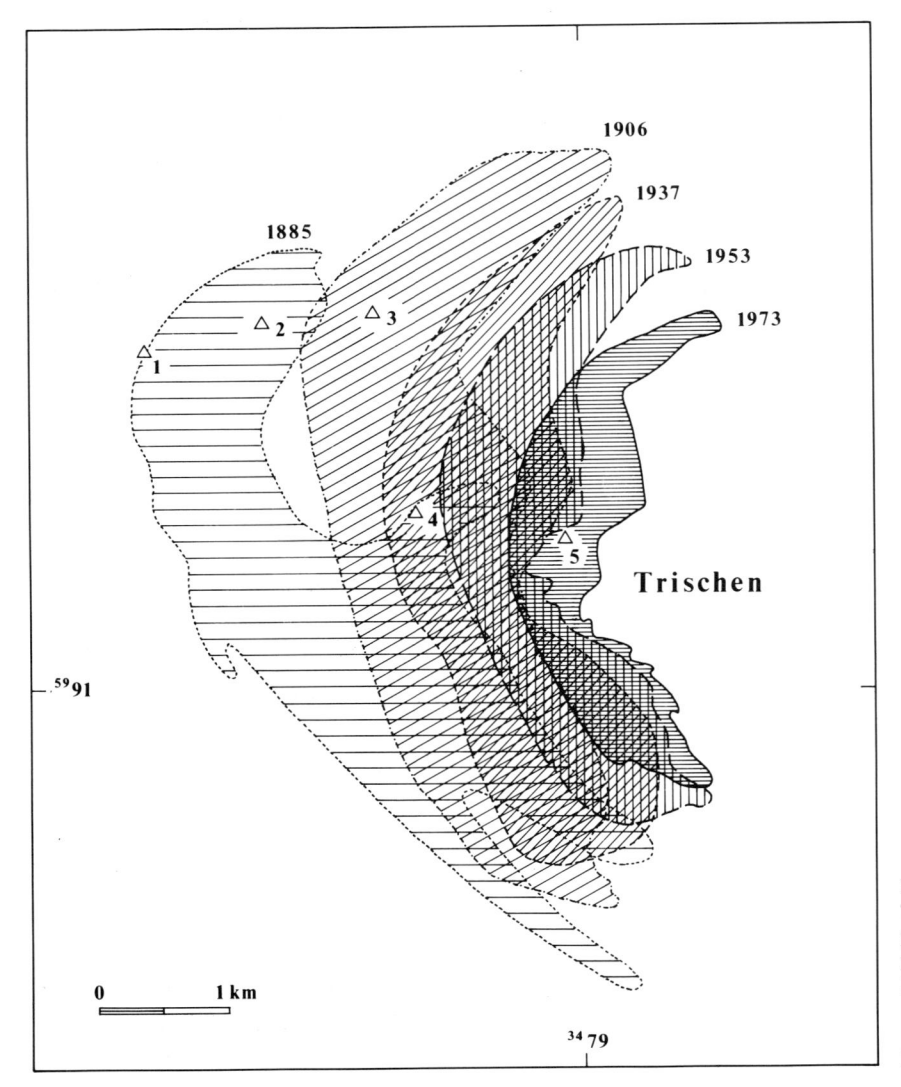

Figure 17: Shift of the Isle of Trischen from 1885 to 1973. The points 1–5 indicate the different positions of the Trischen beacon: 1 = 1884–1890; 2 = 1890–1911; 3 = 1912–1924; 4 = 1924–1948; 5 = 1951–today (after Wohlenberg, 1950: 176; Wieland, 1972: 131).

strated that barrier islands need to maintain a certain size in order to survive as viable independent members of the island chain. The fate of the Isle of Buise, which was eroded during the 18th century and the degradation of Baltrum, halted only by massive coastal protection structures, illustrate this process. The young barrier islands in the inner German Bight therefore migrate eastwards until they either are eroded or merge with the mainland. Historical examples can be cited for both cases (cf. chapter 7).

Although the young islands may not constitute permanent members of the barrier island chain, such as those of the West- and East Frisian Islands, they must be regarded as integral parts of the barrier system, because they separate the extensive tidal flats of the inner German Bight from the open sea.

From his investigations in the East Frisian Islands, Luck (1975: 25) concluded that under the existing morphological conditions, with chains of barrier islands, no new tidal inlets or islands can be formed. Comparison with the USA demonstrates, however, that new islands and inlets will be formed whenever there are strong enough impulses such as extreme and repeated storm surges.

The young barrier islands demonstrate that submarine bar aggradation (the de Beaumont-Johnson theory; de Beaumont, 1845; Johnson, 1919: 348) is the most likely process to have led to

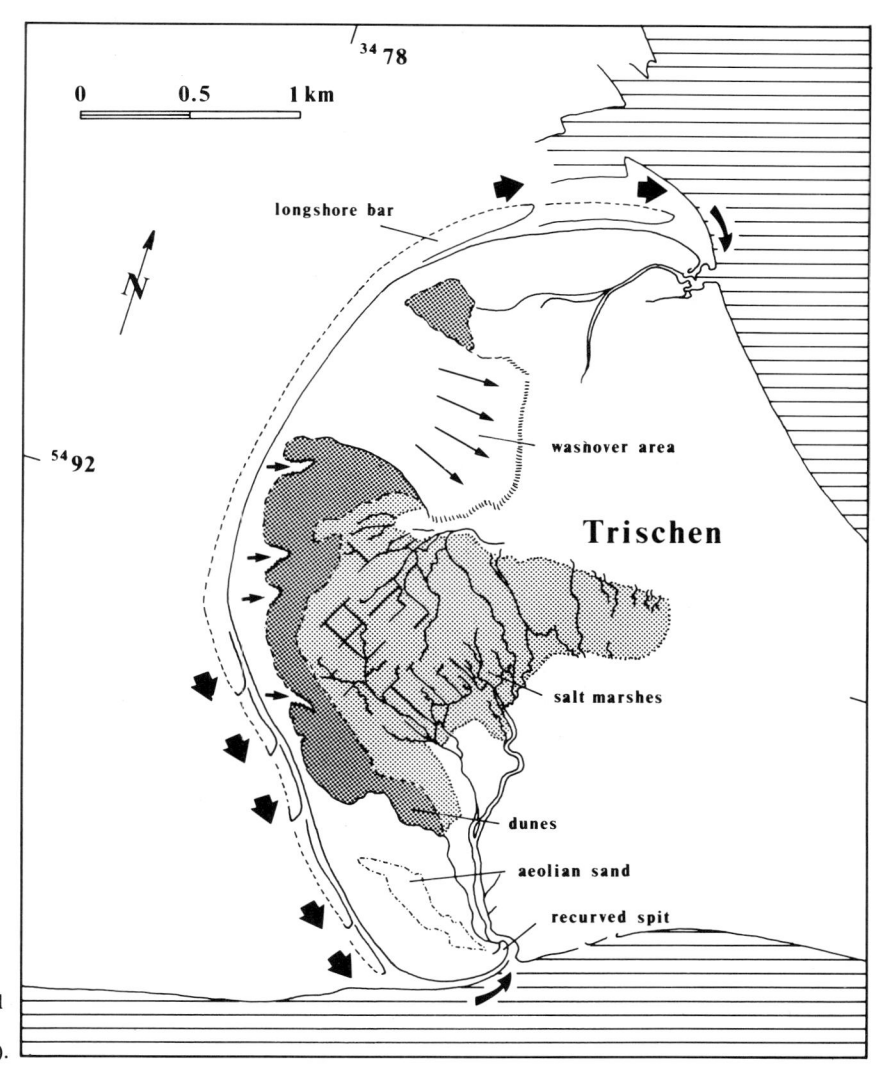

Figure 18: Position and shape of the Isle of Trischen based on aerial photographs of the Landesvermessungsamt Schleswig-Holstein (1980).

the formation of the large barrier islands in the past. According to this theory, landward directed sand transport leads to the formation of a submarine sand bank seaward of the coast which finally emerges above sea level. Otvos (1970, 1979, 1981, 1985) was able to demonstrate this method of island formation citing several islands along the Gulf Coast of the southern USA.

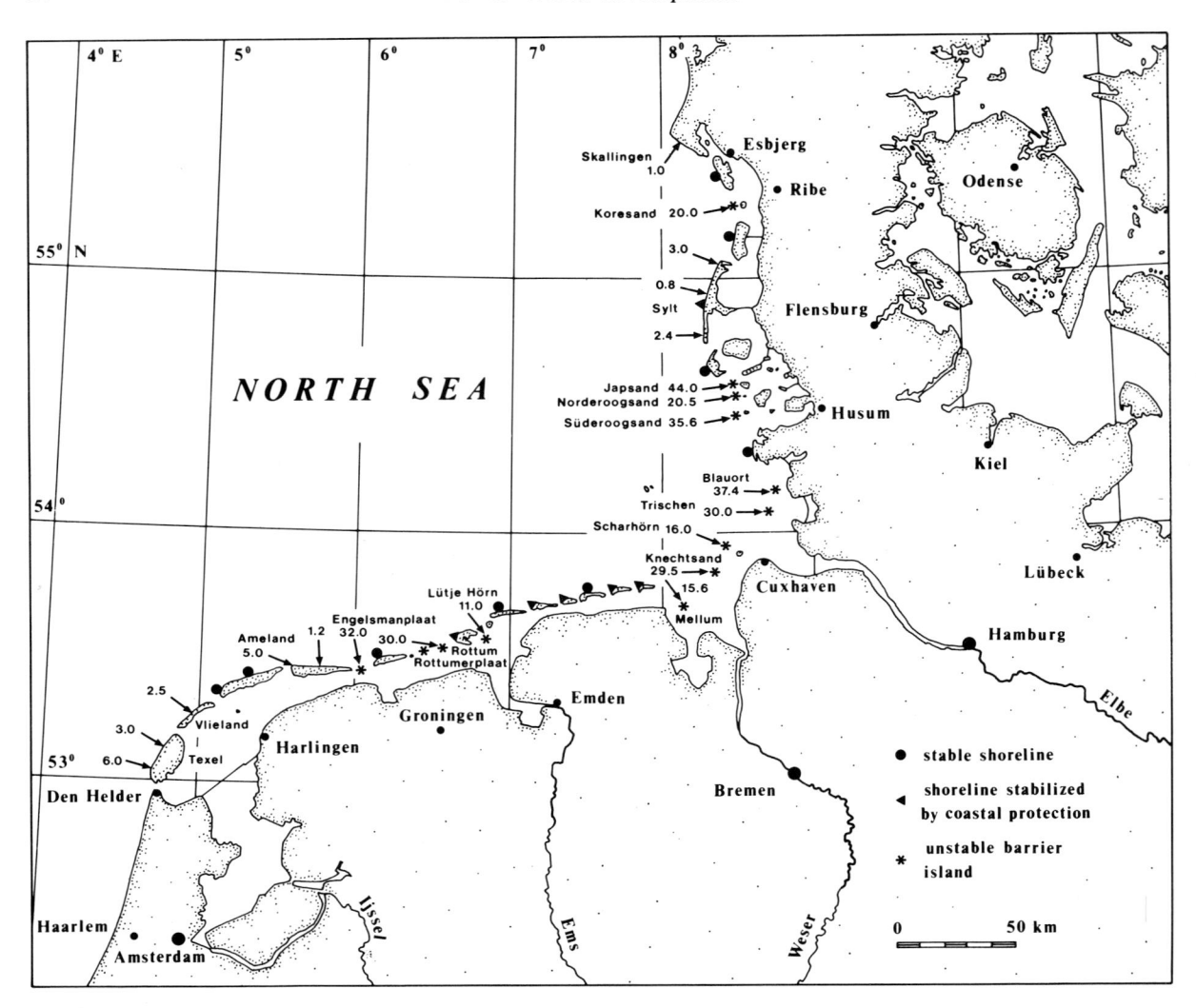

Figure 19: Historical shoreline changes along the Wadden Sea coast. Data from the last 100 years have been evaluated; the arrows indicate the direction of coastal retreat; the erosion rates refer to the actual situation with a number of islands almost totally stabilized by coastal protection measures.

CHAPTER 4

RECENT GEOMORPHOLOGICAL PROCESSES

4.1 CURRENTS

Geomorphological processes in the Wadden Sea are almost exclusively associated with flowing water. Current measurements conducted over the whole Wadden Sea area have given an impression of the nature and intensity of these water movements. Extensive measurement programmes have been conducted, such as those associated with the planning of the Neuwerk deep-water port in the Wadden Sea between the Elbe and Weser estuaries (Göhren, 1969; Koch & Luck, 1973).

Tidal currents

Ebb and flood currents in the open North Sea flow in almost diametrically opposed direction. The flood current flows towards the east and southeast into the inner German Bight, whereas the ebb current is directed towards the west and northwest with current velocities increasing towards the inner German Bight. The highest values are always recorded in the narrow tidal inlets and tidal streams of the Wadden Sea (fig. 20).

According to Hjulström (1935: 298) a current velocity of 20 cm/s is required to erode fine sand of 0.2 mm grain-size. At velocities as low as 50 cm/s fine gravel of more than 2 mm grain diameter can be entrained. At 150 cm/s clasts of 15 cm diameter can be eroded and transported, but such coarse material occurs only in a very few places within the investigation area, principally where Pleistocene strata are eroded.

The current velocity on the tidal flats rarely exceeds 50 cm/s. As the tidal flats are mainly composed of fine to medium sand, this velocity is sufficient to transport the available material. In the tidal creeks the current velocity peaks are about 100 cm/s; in small tidal inlets such as the Wichter Ee between Norderney and Baltrum currents in excess of 150 cm/s occur frequently (Koch & Niemeyer, 1980).

It is clear, therefore, that on the tidal flats the ebb current has higher velocities than the flood current; this contrast is not found, however, in the tidal inlets. Here, current velocities vary depending on whether the measurements are taken in the ebb or in the flood channel of the inlet. In the ebb channels the ebb current reaches maximum speeds, and *vice versa* in the flood channels (fig. 21).

Residual currents

Excluding the influence of the wind or a limiting water depth, ebb and flood currents are approximately of the same strength in the open sea. If the

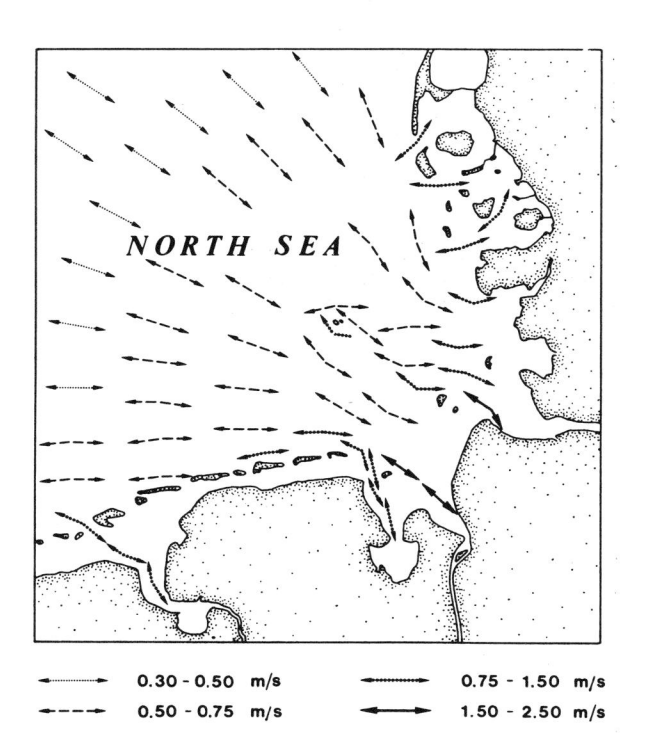

Figure 20: Direction and strength of the tidal currents in the German Bight; the flood current is directed towards east to southeast, the ebb current towards west to northwest (after Reineck, 1978 a).

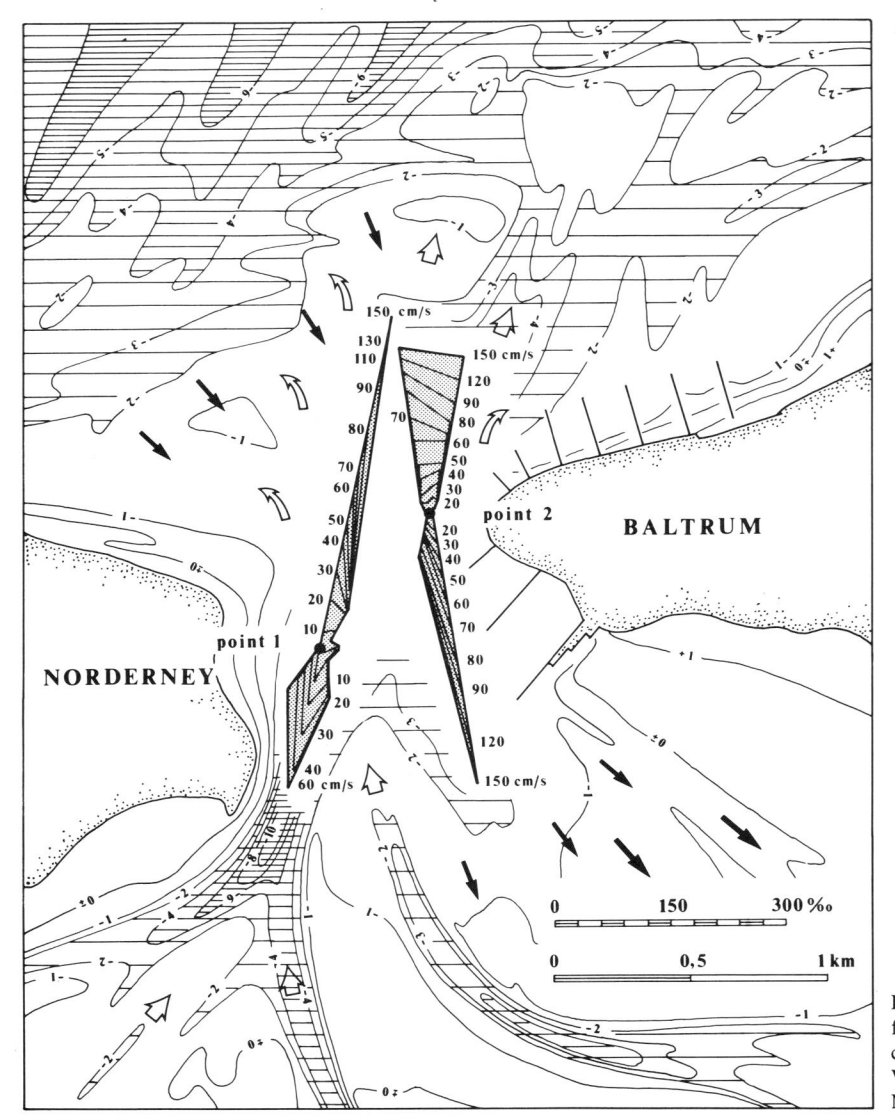

Figure 21: Frequency of ebb and flood current velocities in the tidal cycle (in parts per thousand), in the Wichter Ee tidal inlet (after Koch & Niemeyer, 1980).

movement vectors measured at one point over one tide are connected to a vector train then a symmetrical ellipse will result.

Göhren (1969 a) has demonstrated that these conditions do not operate in the shallow waters of the Wadden Sea. Water movement can only take place at or near high tide because of the limiting water depth, and accordingly the movement vectors measured at one point rather resemble a half ellipse. If the water movements at one point are measured over a number of tides, a garland-like vector train results; this is orientated parallel to the current direction near high tide. The resulting current is called the residual current (Göhren, 1974: 8) and can be used to calculate the most likely sediment transport direction (Göhren, 1979).

Drift current

In addition to the tidal currents, currents dependant on the influence of the wind become important in the coastal areas. These so-called drift currents (Göhren, 1968) are controlled by water depth as well as wind speed and become more effective as the water depth decreases. Even moderate wind speeds result in drift currents faster than tidal currents over the tidal flats. This is shown by the measurements of Koch & Niemeyer (1980) from the Baltrum tidal flat water divide (figs. 22 a, b and c).

The drift currents move considerable water masses across the tidal flats, which again contribute to sediment dispersal. Göhren (1968: 250 f.)

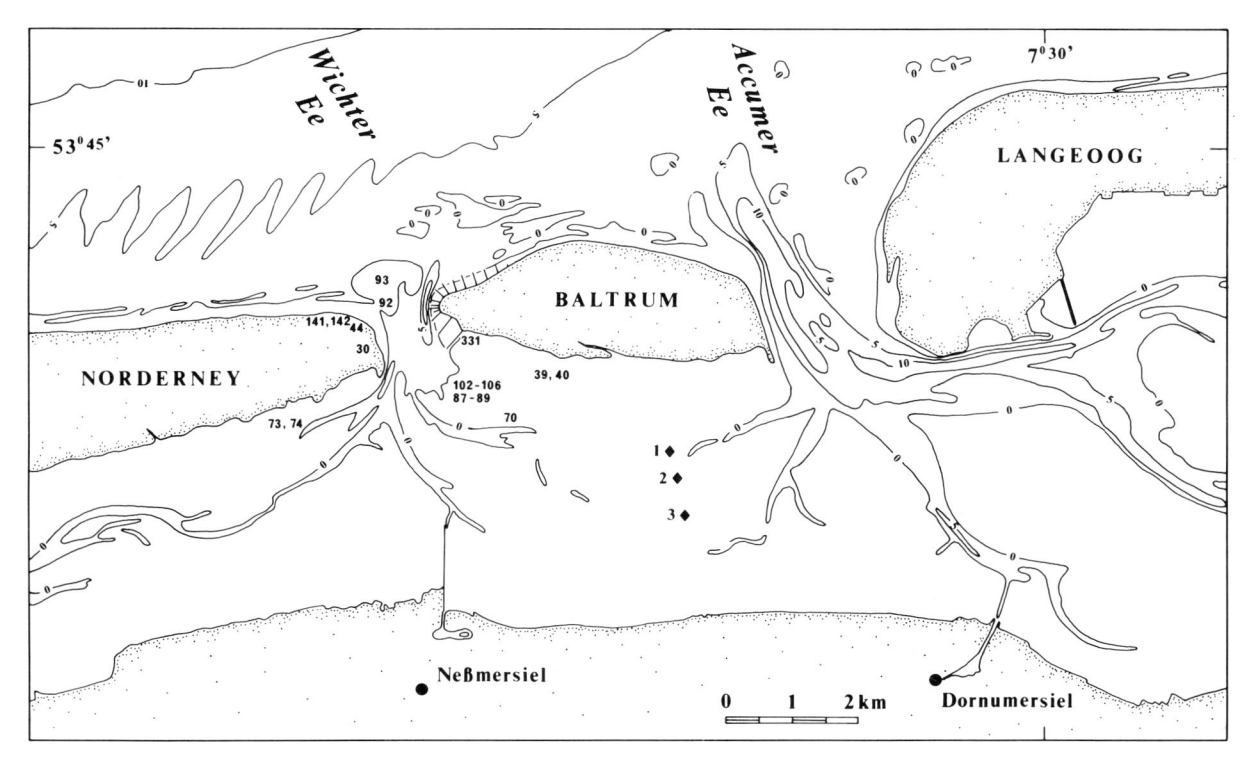

Figure 22 a: Location map of the Baltrum area, showing where the photographs and plates of small-scale features were taken. 1, 2, 3 = locations of current measurements (figs. 22 b and c).

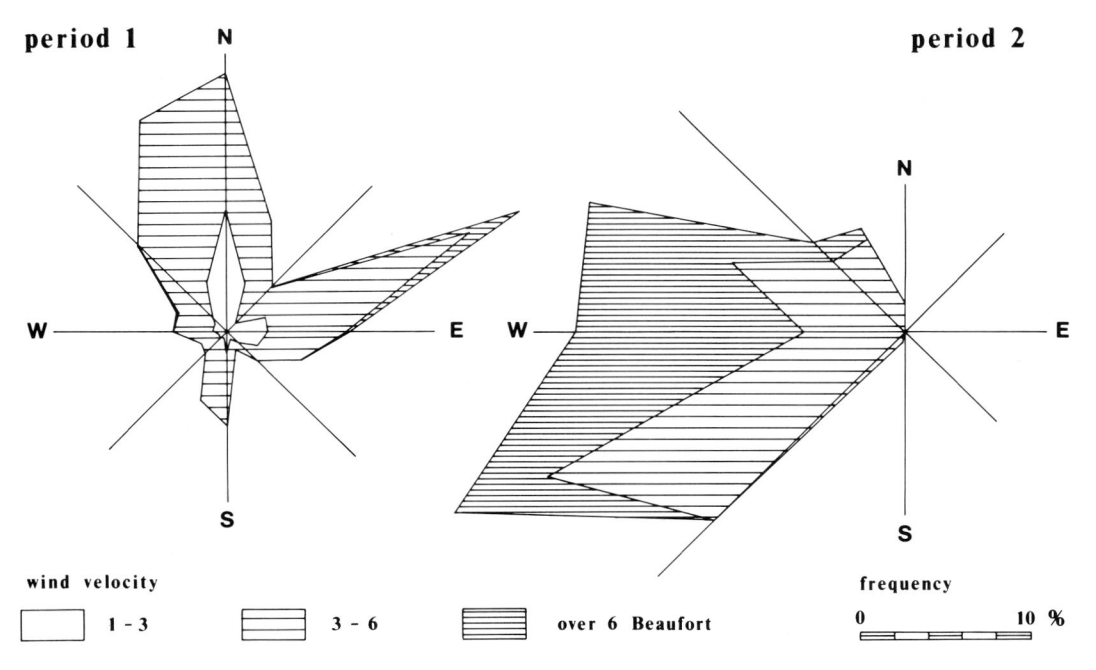

Figure 22 b: Wind directions and velocities during the current measurements in the Baltrum tidal flat water divide area (after Koch & Niemeyer, 1980).

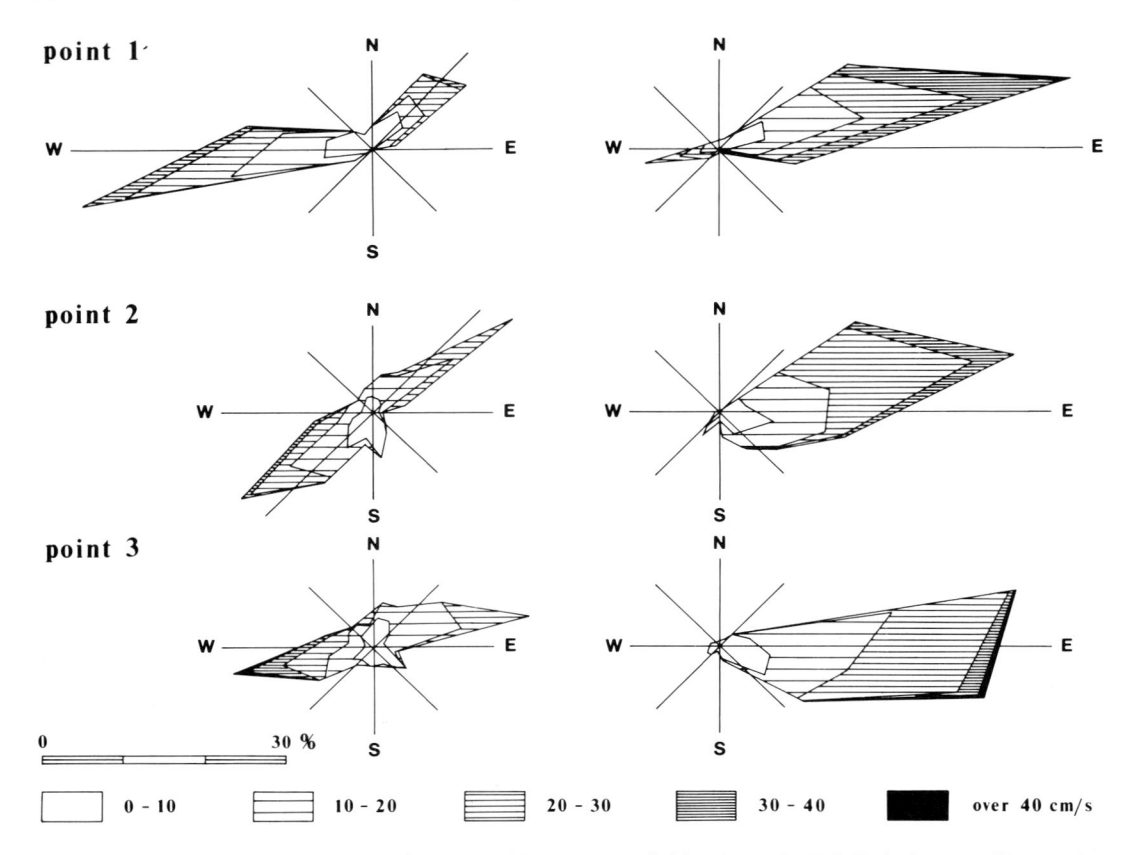

Figure 22 c: Current measurements in the Baltrum tidal flat water divide area; series 1 (left) during prevailing northeasterly winds; series 2 (right) during prevailing westerly winds (after Koch & Niemeyer, 1980).

has demonstrated that sediment redistribution in the Wadden Sea by drift currents during strong wind weather conditions is much higher than redistribution by normal tidal currents. More detailed investigations would be necessary to quantify this difference more precisely.

Göhren (1969 b) demonstrated how wind-induced deviations of the tidal currents affect sediment transport at the bottom using fluorescent tracers in the Neuwerk tidal flats. All measurements taken during strong wind weather conditions demonstrate a strong correlation between wind and sediment transport directions. A characteristic example is shown in fig. 23.

The direction of the drift current is also controlled by the position and configuration of the tidal creeks. In deep tidal creeks the wind influence is minimal when compared with the tidal flats, and enables the water which is driven over the tidal flats to concentrate and eventually run off against the wind direction (fig. 24).

Current measurements

Current measurements using automatic devices have normally been restricted to selected localities and to limited periods of time because of the heavy costs incurred. It is therefore impossible to gain a spatially or temporally complete picture of the current patterns.

Our present knowledge of current movements in the Wadden Sea is based mainly on data gathered by automatic current measurement devices. These devices have the drawback that measurements are always taken at a fixed height above the seabed. As a result both currents at the bottom as well as at the surface have been neglected. For example, during the measurements in the Wichter Ee tidal inlet (fig. 21) the current meters were installed 1.4 m above the bottom. On the tidal flats (fig. 22) the measurements were made 0.35 m above the bottom. The currents immediately above the bottom, which are critical for sediment transport, have

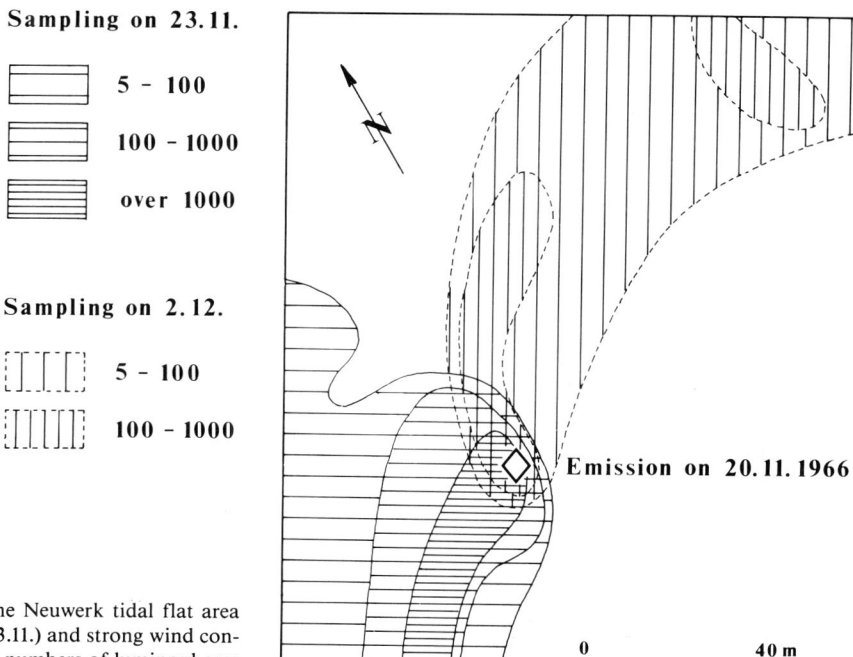

Figure 23: Sediment migration in the Neuwerk tidal flat area during calm weather (sampling on 23.11.) and strong wind conditions (sampling on 2.12.); shading: numbers of luminophores per sample (after Göhren, 1969 b).

therefore not been measured. This means that on the tidal flats the conditions immediately after the onset of flooding, as well as during the final run-off, have not been recorded.

Determination of currents from aerial photographs

Fig. 25 shows the currents in the Lister Tief drainage area during high tide, drawn from aerial photographs. These show current directions both directly, in the form of stream lines, and indirectly, as stripes of foam. Comparison with the contour lines of the topographical 1:25,000 Wattkarte demonstrates that in general the currents follow the major topographical features. However, during sufficiently high tides they tend to shorten their routes and produce a larger curve radius.

Aerial photographs can also be used successfully to record another phenomenon. Shortly after low tide, during the onset of the flood current, a considerable body of water remains on the tidal flats that rise about 2 m above sea level. This water is not dammed by the incoming flood current but continues to run off seawards beside the flood current.

Amongst other indicators, the stripes of foam which commonly form on the boundaries between differently directed currents are especially helpful in this context (fig. 26). Another aid to the separation of tidal currents is differential transparency; the water of the ebb current is normally more sediment-laden and therefore more turbid than the flood current.

Where water bodies of different densities (salinity, sediment content) and current direction meet, vortices form. This is also the case at the ebb and flood current boundary. Seawards of the ebb-tidal deltas, the sediment-rich water of the ebb current mixes with marine water containing less sediment, and large vortices develop (fig. 27), which, under favourable conditions, can be seen in aerial photographs.

Since aerial photographs are normally taken during a single flight, it is impossible in most cases to measure current speeds directly. The time interval between exposures is only a few seconds, which is far too short to interpret current-induced changes. In exceptional cases, however, where two parallel rows of photographs have been taken, it is possible to measure the tidal currents.

Fig. 28 shows the situation in the inner part of the tidal inlet between Borkum and Juist shortly after low tide. Interpretation of the photographs, taken at 16 minute intervals, suggests that shortly after the onset of the flood current there is a coexistence of the currents. The ebb current in this case has a velocity of almost 20 cm/s; the velocity of

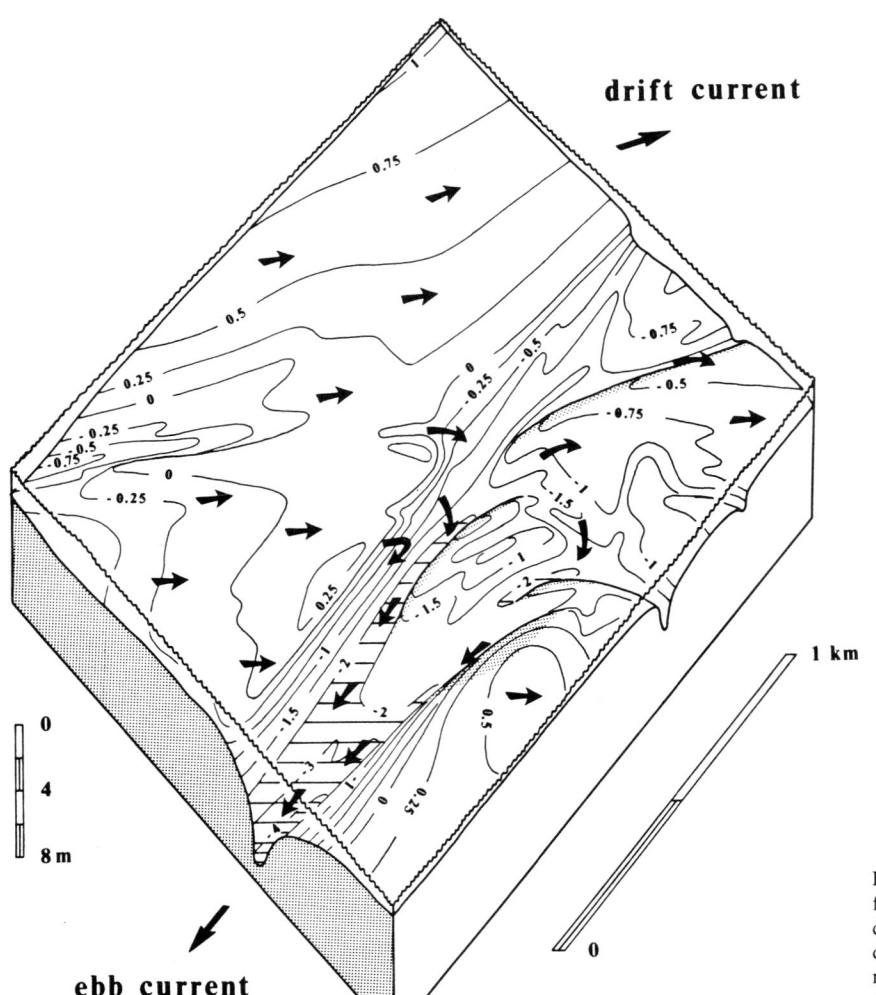

Figure 24: Drift currents in the tidal flat area as a function of water depth; in the deeper tidal creeks the drifting water can eventually flow at right angles or directly opposite to the wind direction. Tidal flat area south of Baltrum.

the flood current is somewhere between 10 and 15 cm/s.

Determination of currents from satellite imagery

A visual impression of the current patterns in the whole Wadden Sea at a given time can be gained through satellite imagery. Satellite images cannot show small scale details such as foam stripes. However, the winter drift ice cover can be used as a current indicator.

The satellite image of 4.1.1979 (fig. 29 a) is the only cloud-free image currently available, which shows the Wadden Sea covered by drift ice. Because of the seasonally low angle of the sun in the original picture even comparatively small landforms, such as dykes and dunes, can be distinguished. When the image was taken there was a slight wind from the NNE. This can be seen from the orientation of the smoke trail over 20 km long from the Hemmingstedt oil refinery. The wind has caused a partial ice jam on the tidal flat side of the islands.

The drift ice movement is more wind-dependent than is the water movement. When the image was taken, however, there was only a slight wind, and therefore the ice distribution can be regarded as almost exclusively controlled by the tidal currents. On 4.1.79 low water in Cuxhaven was at 11.10 a.m., in Westerland (Sylt) at 11.39 a.m. and in Norderney (Riffgat) at 9.45 a.m. The image therefore shows the situation shortly before, during, and shortly after low tide.

The dry tidal flats are hardly distinguishable on the image because of the strong contrast in brightness between the snow and ice, and the dark sub-

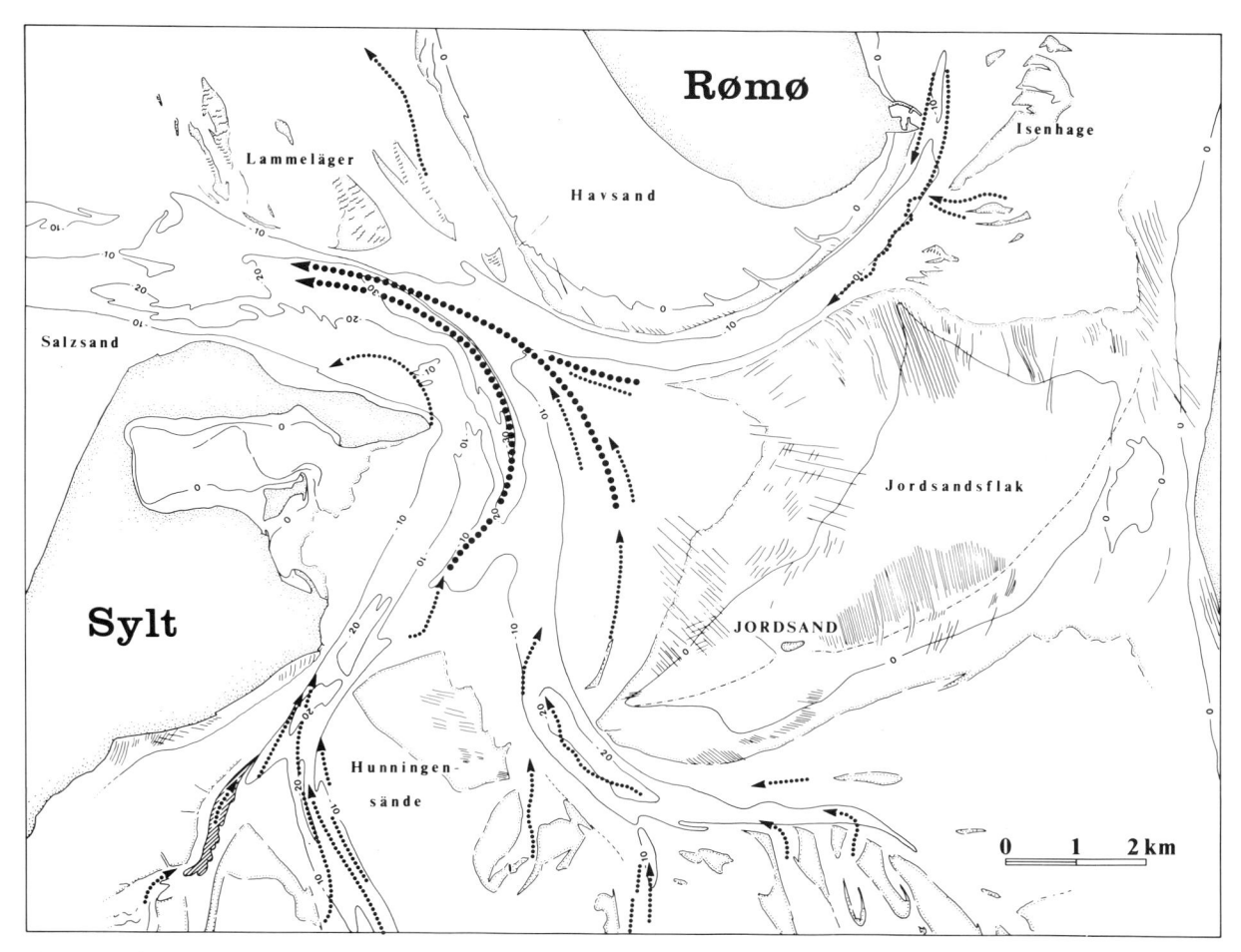

Figure 25: Tidal flats in the Lister Tief drainage area. The superficial currents do not directly follow the channel-bed relief, but curves with larger radii. Sources: Küstenkarte 1:25,000, Sheet 0916 K List and Sheet 1016 K Kampen in addition to aerial photographs of the Geodætisk Institut, Copenhagen, Flight No. 7609 (1976).

Figure 26: Stripes of foam marking the boundary line between currents of different velocity or direction (Outer Jade, 1981).

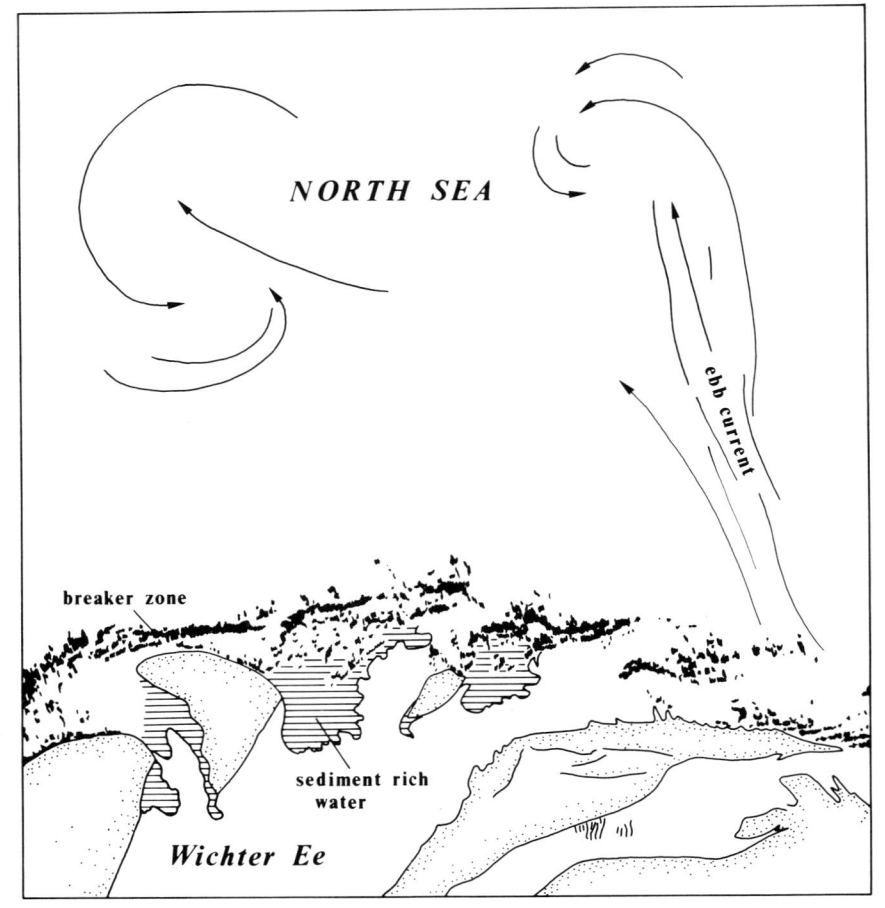

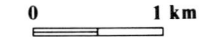

Figure 27: Diagram showing the formation of vortices with a diameter of 1–3 km seaward of the Wichter Ee inlet during the mixing of the sediment-rich ebb currents with the sediment-poor waters of the North Sea. Drawing from an aerial photograph of the Niedersächsisches Landesverwaltungsamt, Landesvermessung, Bildflug Ostfriesische Küste (419) of 29.5.1966.

stratum. The flats are punctuated with ice floes of different sizes. Most of the drift ice has been removed by water runoff into the creeks and tidal inlets. During this process numerous ice floes become rafted on the banks of the creeks so that in many places the courses of the small creeks are characterized by garlands of ice floes. This pattern is well developed on the tidal flats between Cuxhaven and Eiderstedt.

The ice floes are mainly formed during high tide on the calmer areas of the Wadden Sea. In the major tidal streams during each ebb the ice floes are transported seawards. Consequently, during low tide, the major drift ice concentrations are found far out at sea whereas the inner parts of the tidal streams are mainly ice-free (Weser estuary, Eider, Vortrapptief).

The transportation of drift-ice seawards extends approximately as far as the -10 m submarine contour. When this depth is reached the flood current transports the ice floes back onto the tidal flats.

On the satellite image (fig. 29 a) the beginning of this transition of current direction can be seen seawards of Amrum.

In the Wadden Sea area between the Elbe and Weser estuaries the boundary between the subtidal and intratidal areas is marked roughly by the -2 m submarine contour. Fig. 29 b demonstrates that in the subtidal areas seawards of this boundary the courses of the tidal currents, under calm weather conditions, are largely determined by the underwater topography, which is in turn fashioned by the tidal currents. This feedback relationship can be traced down to a depth of -10 m.

4.2 WIND

The formation of currents in the Wadden Sea is dependant not only on the tides but also on the wind. The wind therefore plays a significant part in the shaping of the tidal flats and islands. Al-

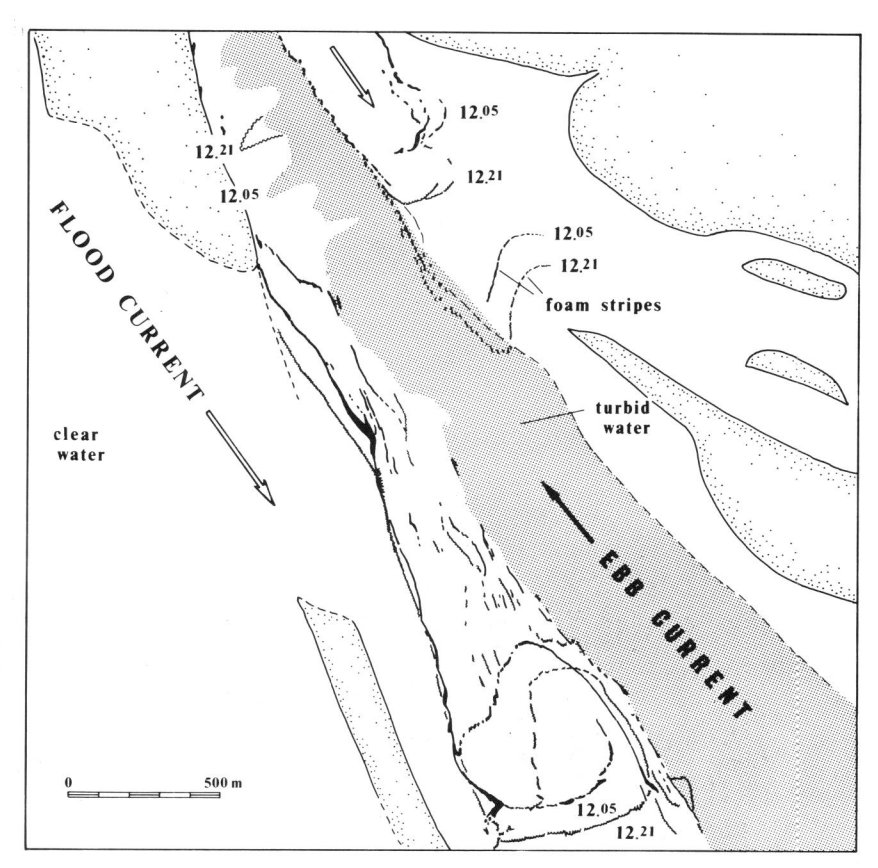

Figure 28: Ebb and flood current between Borkum and Juist. From two aerial photographs, taken at an interval of 16 minutes, the sediment clouds and foam stripes permit measurement of the true water movement. Drawing from aerial photographs of the Niedersächsisches Landesverwaltungsamt, Landesvermessung, Bildflug Ostfriesische Küste (419) of 29.5.1966.

though the tide-induced currents are primarily controlled by the configuration of the tidal creeks the wind is not subject to this limitation and can act areally and unidirectionally over large parts of the Wadden Sea.

The prevailing wind direction strongly influences the intensity and direction of sediment movement. Two aerial photographs of the eastern end of Norderney (taken on 2.8. and 10.9.1982) demonstrate this (figs. 30 a and b). The first photograph was taken after a period of relatively calm weather which was only occasionally interrupted by strong winds from the N and NE. This caused increased sand transport around the eastern end of Norderney, enhancing the recurved spit. It also led to the formation of a broad sand tongue that extended far to the west at the outer margin of the salt marsh (fig. 30 a).

During the next month calm summer weather dominated, only occasionally interrupted by periods of strong wind from the W to SW. As a consequence, the sand at the margin of the salt marsh was partially eroded and the recurved spit facing the Wichter Ee tidal inlet was significantly lowered. Simultaneously the beach ridges on the seaward side of Norderney were further developed (fig. 30 b).

The differences between the two photographs are exaggerated by the fact that the first photograph was taken close to neap tide whereas the second was taken at spring tide. Water level on the second photograph is about 0.5 – 1.0 m lower than that on the first, and as a consequence the beach and tidal flat areas appear to be drier on the former.

On the inner parts of the island the situation is the reverse. Here the second photograph shows a comparatively high water level in the runnels and ditches, and large ponds of water in the washover areas (cf. section 5.3.4). This is a consequence of the spring tide being about 0.75 m higher than the neap tide. During spring tide, the inner parts of the island are flooded at high tide and, because of the poor drainage conditions, they remain rather wet at low tide.

The moulding of the beach and dune areas by the wind depends heavily on the moisture content of the sand. The aeolian morphodynamics of the islands will be dealt with in section 5.3.3.

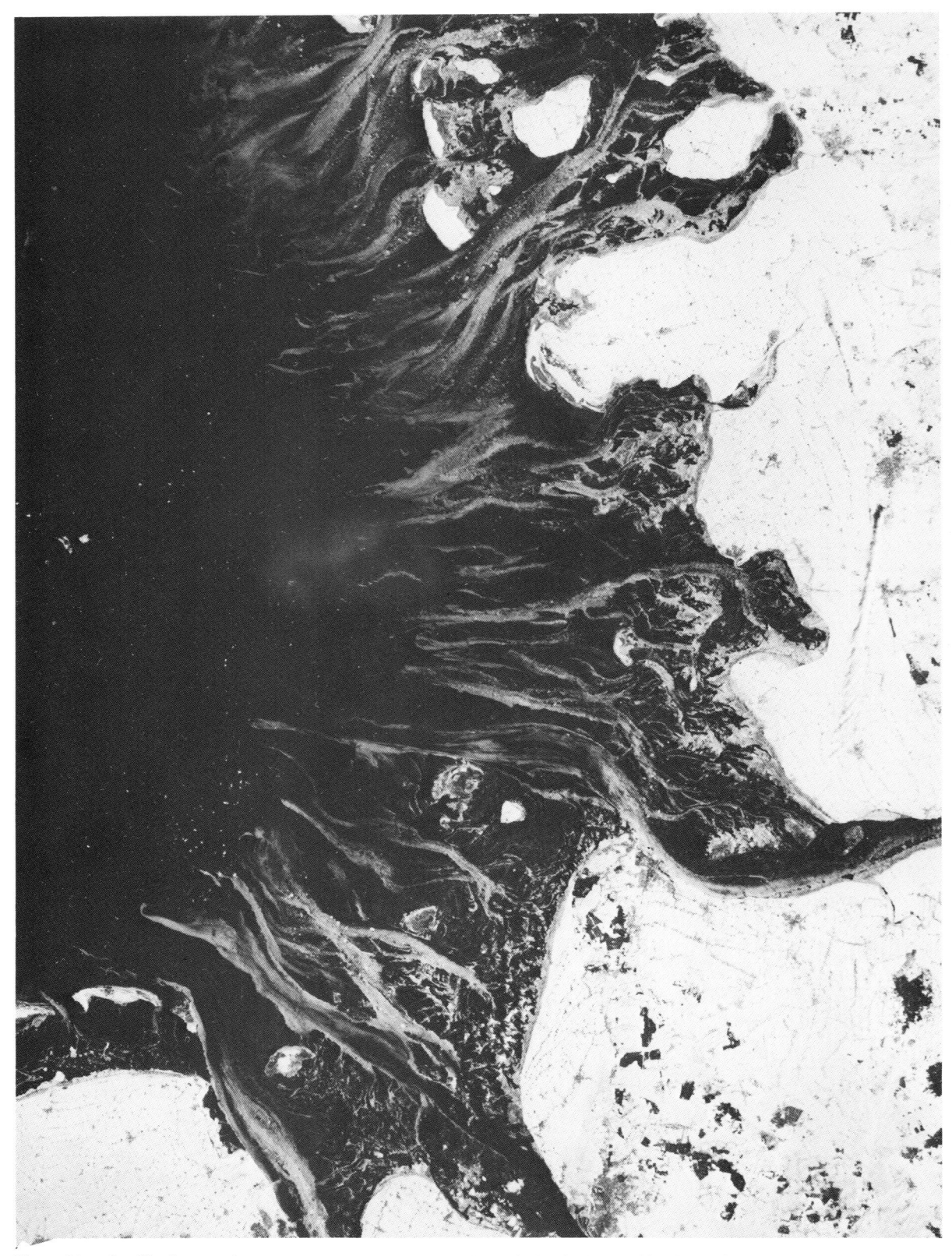

Figure 29 a: Satellite image taken on 4.1.1979, at 11.43 h; showing drift ice in the Wadden Sea. The current-induced drift of the ice floes is clearly visible (Image: Landsat 2, 212-22).

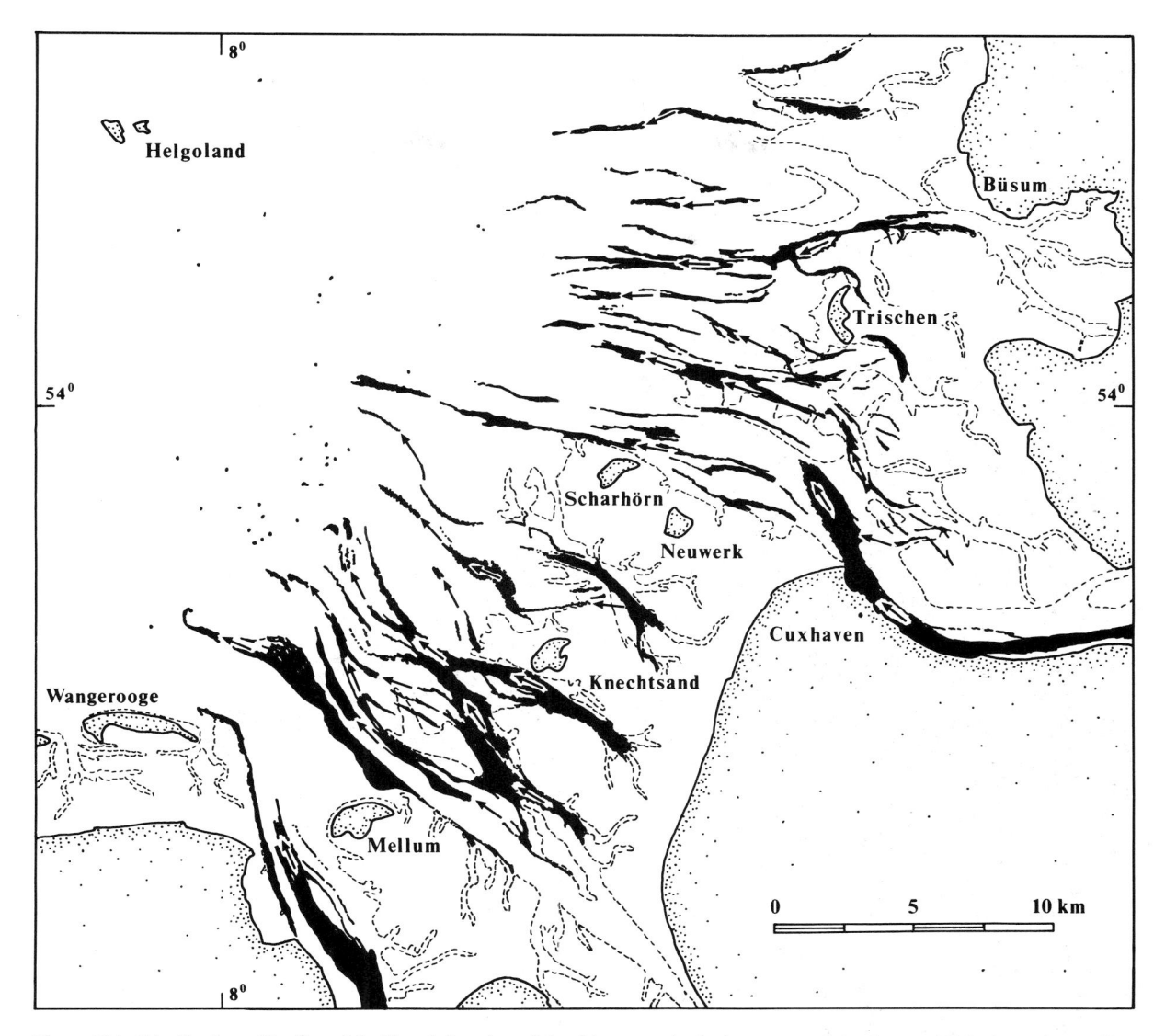

Figure 29 b: Distribution of ice floes (black) and direction of the ebb current in the inner part of the German Bight from the satellite image taken on 4.1.1979 (fig. 29 a). Stippled line: 2 m isobath.

Figure 30 a: Aerial photograph of the eastern end of Norderney, taken on 2.8.82. After some periods of strong winds from N and NE, the recurved spit at the eastern end of the island is more strongly developed. Aerial photograph of Fa. N. Rüpke, Hamburg, of 2.8.1982, Bildflug Wichter Ee, Str.2/150, freigegeben durch Luftamt Hamburg unter Nr. 818/82.

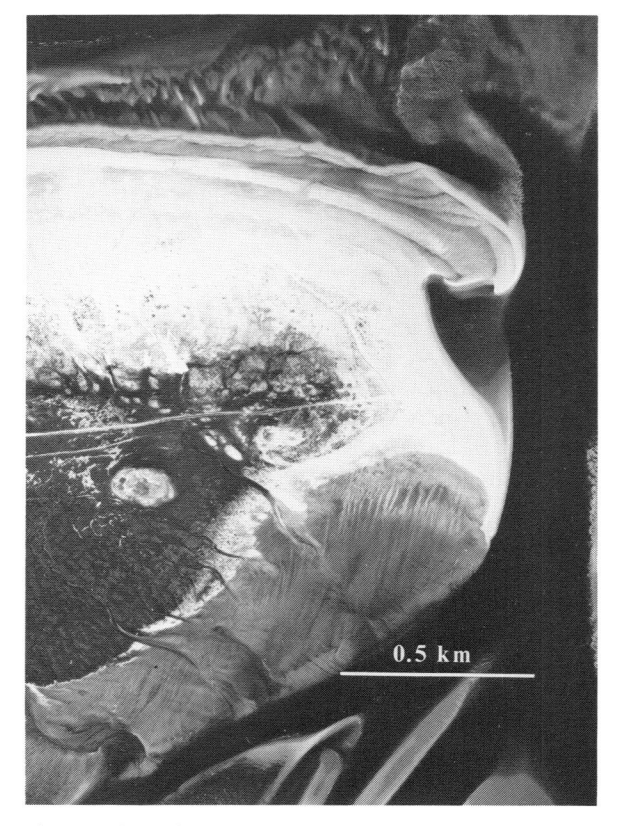

Figure 30 b: Aerial photograph of the eastern end of Norderney, taken on 10.9.82. After strong winds from the W and SW, much sediment has been removed from the recurved spit. The beach ridges are more strongly developed. Aerial photograph of Fa. N. Rüpke, Hamburg, of 10.9.1982, Bildflug Wichter Ee, Str.1/43, freigegeben durch Luftamt Hamburg unter Nr. 818/82.

onshore winds. The wind drives the water film across the sand flats at relatively high speeds causing the entrainment of the saturated sand surface resulting in the formation of long sand tongues (excellent examples are given by Gripp & Martens, 1963). Long subparallel ribbons covering the tidal flats may form if the process lasts for a longer time period. Small ripple marks are removed by the smoothing effect of this sand/water film. Considerable expanses of the flats become covered with garland-like patterns if the process persists for long enough (fig. 31).

The restructuring of the small ripples begins at the point where the water becomes too shallow for normal ripple migration. At first, ripple crests are flattened (Klein, 1977: 33), and at the same time small sand tongues, which drift in the wind direction, are formed (fig. 32 a). As this process continues the sand tongues become more elongated, cross the adjoining ripple trough and contact the

Micro-relief

The influence of wind on the formation of drift currents was dealt with in section 4.1. In this section the role of wind-driven shallow water on the beaches and tidal flats is considered. The resulting landforms have not yet been fully described. Whereas in deep water the forms created by drift currents resemble those created by other currents (i.e. ripples, comet marks etc.), the process assemblage in shallow water is completely different.

Large parts of the surface, and especially those areas which are not juxtaposed to drains, remain covered by a shallow film of water during most of the low tide. This water film is only a few centimetres thick but, as a result of the shallow gradient, it is unable to run off.

This residual water can contribute considerably to the sculpturing of the small-scale relief on the tidal flats and the beach, especially when driven by

Figure 31: After wind-current remodelling a small ripple field at St.Peter-Ording has become covered by garland-like connected sand ribbons (1981).

next sand tongue (fig. 32 b). Finally, the former rippled surface becomes completely smoothed, and the various sand tongues grow together to form a single sand plain (fig. 32 c).

In places where larger obstacles hinder the flow of the shallow water, small recurved spits are formed (fig. 33). Ripple crests lying above the water surface can be completely remodelled. This process may continue until the original ripple form can no longer be distinguished (fig. 34).

The flattening of the ripple crests occurs by sand movement in the upper parts of the ripples only. Photographs of sand tongues and recurved sand ribbons in a small ripple field seawards of Mellum (figs. 35 and 36) demonstrate that the ripple troughs are barely filled with sediment whilst at the same time the crests are lowered and broadened; sand transport is largely confined to the sand tongues. It may be that only the upper, freely mobile, water layer is propelled by the wind whereas the deeper water trapped between the ripple crests remains almost static.

The formation of new small ripple fields may occur during periods of strong wind, enabling a thicker water film to be driven across the sand flats. Fig. 37 shows that the resulting sand patches are orientated parallel to the wind direction. Allen (1984 a: 443) presents similar structures from the Norfolk coast formed by wave action.

The forms mentioned above were observed on sand flats and in shallow beach areas. Wind-sculpturing is not restricted to these specific areas, however, and also affects parts of the muddy flats. Fig. 38 shows the successive remodelling of a small ripple field in a slightly muddy sand flat near Arensch. The ripple crests are at first supplemented by spurs which are then elongated in the wind direction (fig. 38 a). As the influence of the wind-driven shallow water becomes stronger the spurs become more distinct and start to replace the ripple marks (fig. 38 b) until the whole flat is finally covered by wind-parallel stripes (fig. 38 c).

Until recently it was not realized that even the major sand banks and shallows on the tidal flats could be considerably remodelled by wind-driven water, and landforms created which resemble beach barchans (fig. 39). The wind influence ceases when these forms emerge during the course of the ebb tide. Because the remaining water patches provide very low potential fetch distances, the remaining water runs off towards the nearest drain following the natural surface gradient which may be at right angles to the dominant wind direction (fig. 40).

Wind streaks on the tidal flats

Strong winds create wind-parallel streaks on the tidal flats which are nicely displayed on aerial photographs. The photograph, fig. 41, was taken southeast of Schiermonnikoog during a southeasterly wind. The wind-driven water film and the parallel streaks are clearly visible; the streaks represent flow lines. The distance between the in-

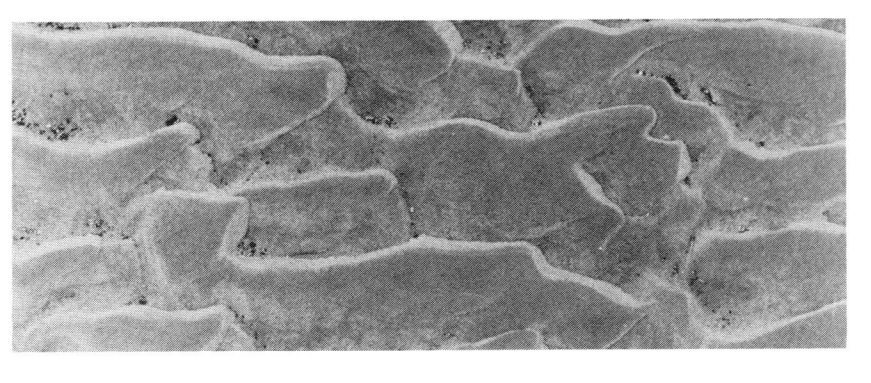

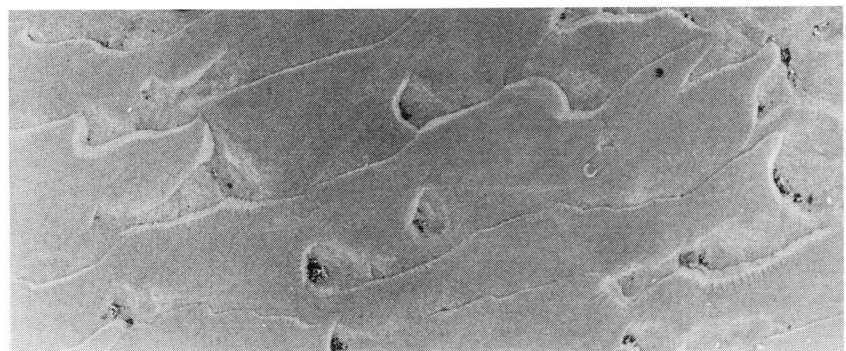

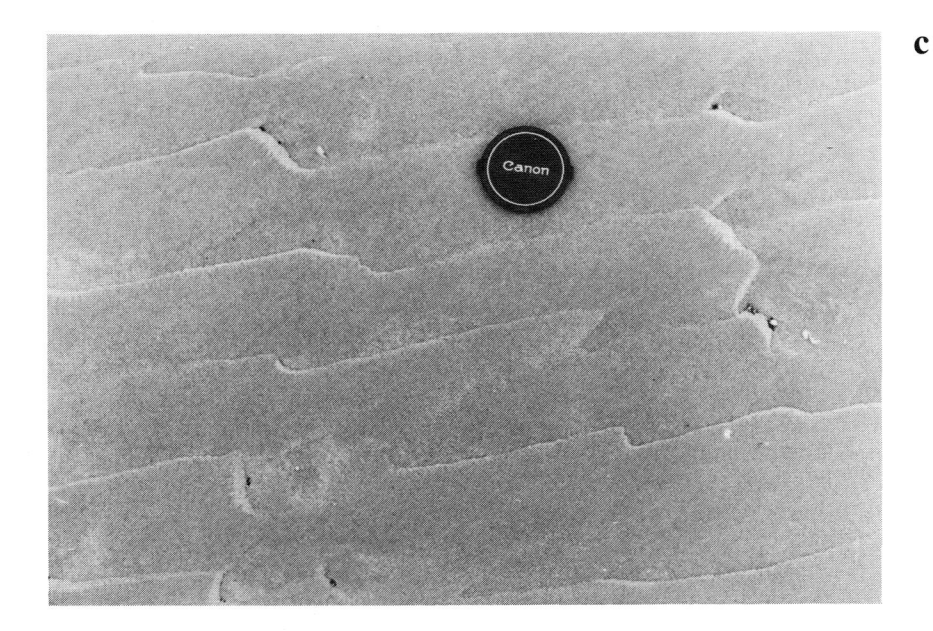

Figure 32: Progressive flattening of a small ripple field (Amrum, Kniepsand) by wind-driven shallow water (1981).

dividual streaks is around 1 m. The windward slopes of the tidal creeks are always drier than the leeward slopes. Where the water film meets a creek draining parallel to the wind direction, the water is drawn away leaving a small, sharply delineated dry area behind the creek.

4.3 WAVES

The effect of waves may extend to considerable depths in the water column, thus influencing the sediment-water interface. Where sediment is reworked by the waves, bioturbation patterns are de-

Figure 33: Wind-current related recurved spits connected to partly flooded megaripples in a runnel on Kniepsand, Amrum (1981).

Figure 34: Given enough time for reshaping, the wind-driven water film can completely change the small-scale morphology. Levelled higher parts of a small ripple field at St.Peter-Ording with two recurved sand ribbons (1981).

Figure 35: Sand tongues and recurved sand ribbons in a small ripple field seawards of Mellum, which was reshaped by a wind-driven film of shallow water (1981).

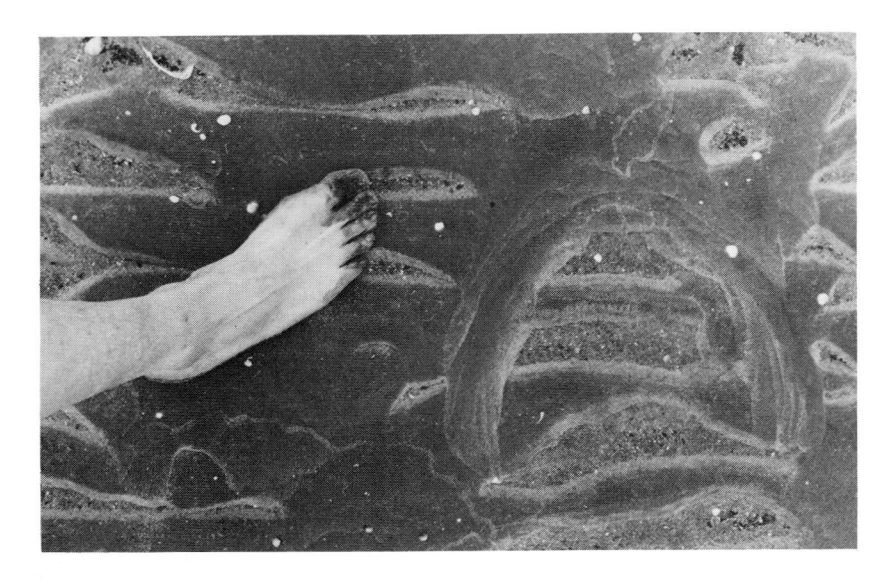

Figure 36: Detail of fig. 35: Sand tongues and recurved sand ribbons on Mellum (1981).

Figure 37: Formation of a small ripple field by sand, which was drifting in a film of shallow water. The remaining sand patches are striking parallel to the wind direction (Ameland, 1981).

Figure 39: Barchan-like forms on the Baltrumer Inselwatt, created by wind and shallow water (1980).

stroyed. This enabled Aigner & Reineck (1983) to demonstrate in a series of undisturbed sediment cores taken off Norderney, that the wave effects in summer reach down to a water depth of 3–5 m and in winter to a depth of 6–12 m.

Refraction

Decreasing water depth leads to increasing deformation of the wave paths. As water depth decreases to a value of less than half of the deep water wave length, wave refraction starts to reorientate the wave trains. Originally orientated at an angle to the shoreline, wave refraction causes wave crests to align parallel with the coastal depth contours. This gives the observer on the beach the impression that the waves are always moving at right angles to the coast.

Figure 38: Progressive reshaping of a small ripple field on a slightly muddy sand flat at Arensch (Niedersachsen). The ripple crests are first supplemented by small leeward spurs (fig. 38 a), which successively get longer (fig. 38 b) until finally the ripple structure is completely destroyed and replaced by wind-current related sand ribbons, formed in shallow water (fig. 38 c; 1983).

50　　　　　　　　　　　　*Recent geomorphological processes*

The situation is different on the tidal flats. Waves normally break at the outer margin of the tidal flats so that strong swells cannot actually develop on the flats (Siefert, 1969: 56; 1971: 21; 1974: 18). At the mouths of the tidal creeks, however, refraction leads to the reorientation of the wave direction into the creek, and, characterized by a permanent loss of energy, waves can extend deep into the tidal creeks and contribute to sediment movement.

The intrusion of relatively long-period waves into the tidal creeks under favourable conditions enables the discrimination of the tidal creeks on aerial photographs taken at high tide (Siefert, 1969: 56). This discrimination is even better on SEASAT radar images (fig. 42) because here the disturbing influence of light reflection and changing light conditions (Wieczorek, 1982: 41) are avoided.

SEASAT images only exist from the western-

Figure 40: Barchan-like form on the Baltrumer Inselwatt. The wind-parallel streaks on the crest clearly indicate the way, in which the 'barchan' was formed (1980).

Figure 41: Aerial photograph of the tidal flats SE of Schiermonnikoog showing the wind-driven film of shallow water and wind streaks on the flats. Aerial photograph K.L.M. Aerocarto Holland, flight 5-15-29 for Rijkswaterstaat, 2.10.1979.

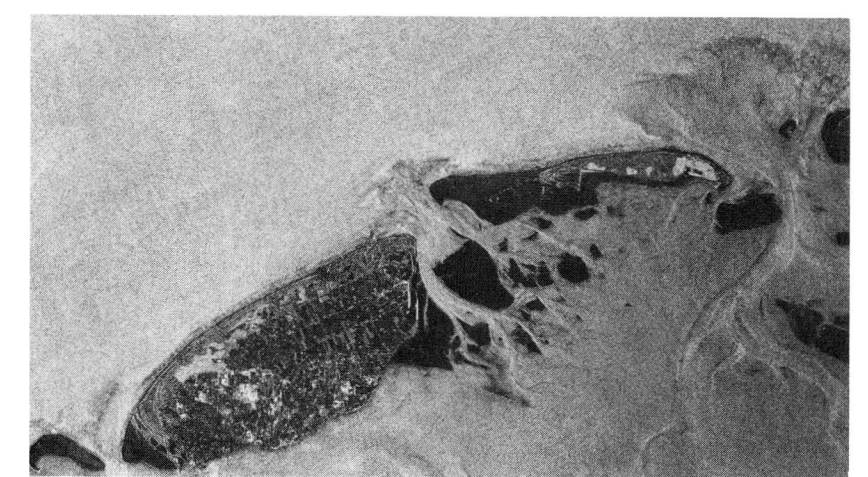

Figure 42: SEASAT satellite image of the Eijerlandse Gat and Vlie tidal inlets (Netherlands). On the left is the Isle of Texel, and on the right Vlieland. The black areas in the Wadden Sea represent emerged tidal flats. The picture was taken close to high tide. The underwater relief is reflected by the wave patterns. SEASAT SAR image S 0891 N 05313 E 00450, 28.8.1978, processed by DFVLR.

Figure 43: Aerial photograph showing part of the Norderneyer Seegat ebb-tidal delta during high tide. The diffraction of the waves leeward of the shoals is clearly visible. Aerial photograph of the Niedersächsisches Landesverwaltungsamt, Landesvermessung, of 29.5.1966, Bildflug Ostfriesische Küste (419), Str.2/267, freigegeben durch Regierungspräsident Münster/W. unter Nr. 2316/66).

Figure 44: Ripple troughs spared by the levelling of the surf at Norderney beach (1982).

Figure 45: Outcropping megaripple bedding on the beach at Spiekeroog visible after the upper sand layer was removed by deflation (1984).

most part of the Wadden Sea (Texel–Ameland). The few images available, however, give an impression of the enormous possibilities of such a system, which, in contrast to optical images, can provide high quality pictures even during periods of dense cloud cover and at night.

Diffraction

The diffraction of waves takes place leewards of man-made obstacles such as breakwaters or groynes (Komar, 1976: 113) and natural obstacles such as the shoals of a tidal delta. After overtaking the obstacle wave energy diffuses at right angles to the original wave orientation (Siefert, 1969: 69) leading to the formation of semi-circular wave crests. This is exemplified by the concentric waves landward of the ebb-tidal delta of the Norderneyer Seegat (fig. 43).

Surf

In shallow water wave speed decreases because of increasing bottom friction, whereas the forward-directed movement of water particles on the wave crest accelerates. As soon as their velocity exceeds the speed of the wave, the wave breaks. Further landward new, smaller, translation waves are formed which are able to break again later but with less force than the original waves. In this way several surf zones can exist, one behind the other.

During the breaking process a considerable amount of wave energy is expended. Part of this energy is transferred to the ground, whilst another part of it is exchanged into coast-parallel surf currents by slightly oblique incoming waves. In addition, another part of the energy is used to generate surf congestion and rip currents. A major part of the energy, however, is consumed in the formation of a water-air mixture in the breaking wave (Führböter, 1971).

The greatest part of this energy loss occurs precisely at the breaking point of the wave. This effectively means that at this point the transporting capacity of the wave is suddenly reduced resulting in the deposition of most of the sediment load (Führböter, 1971: 41). This leads to the accumulation of foreshore sand ridges in the surf zone.

Levelling by the surf

Reineck's (1963: 28) investigations have demonstrated that sand ridges on the foreshore essentially consist of megaripple bedding. These ripples are not normally visible at low tide as their ridges are levelled by the surf as soon as they emerge. However, small remnants of ripple troughs are occasionally found on the otherwise level surface of the beach (fig. 44). The underlying megaripple crossbedding only becomes visible where the sand cover created by the breaking waves is removed from the beach surface by deflation (fig. 45). The bedding within the ridges is orientated almost parallel to the coast (cf. Reineck, 1961 a).

The surface of the shoals of the tidal deltas are also levelled by the breaking waves. The degree of levelling varies according to relative exposure of the site. In the outer parts of the deltas, in the crescent bar region (cf. section 5.2.1), the whole surface is usually flattened. Aerial photographs reveal that those parts of the crescent bar region which are not exposed to the breaking waves, because they remain submerged even at low water, are completely covered with megaripples.

Further landward, on the outermost major shoals of the tidal delta, the influence of the surf is restricted to the seaward and consequently most exposed parts of the shoals. Here the surf creates

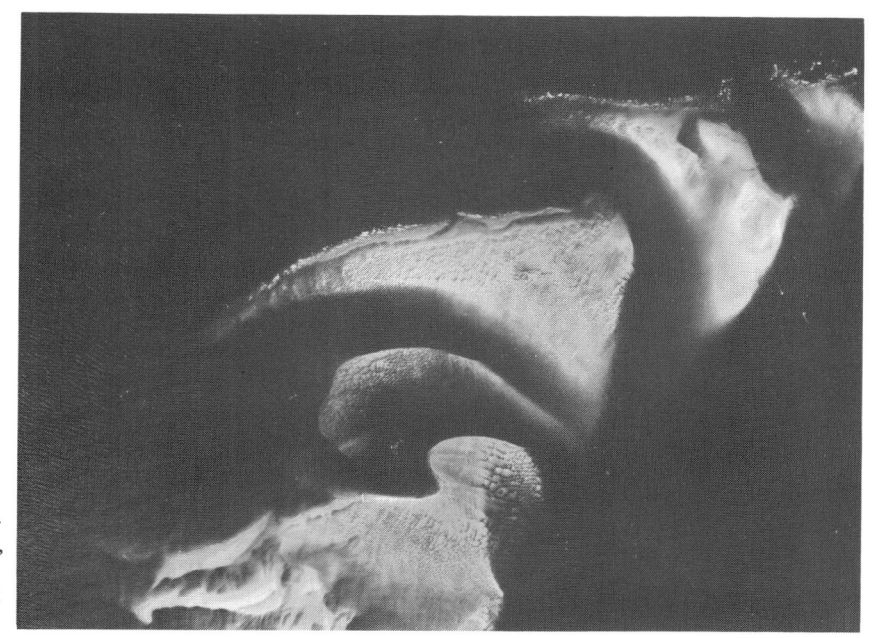

Figure 46 a: Part of the ebb-tidal delta of the Wichter Ee tidal inlet covered by megaripples (Aerial photograph of Fa. N. Rüpke, Hamburg, of 2.8.1982, Bildflug Wichter Ee, Str.1/145, freigegeben durch Luftamt Hamburg unter Nr. 818/82).

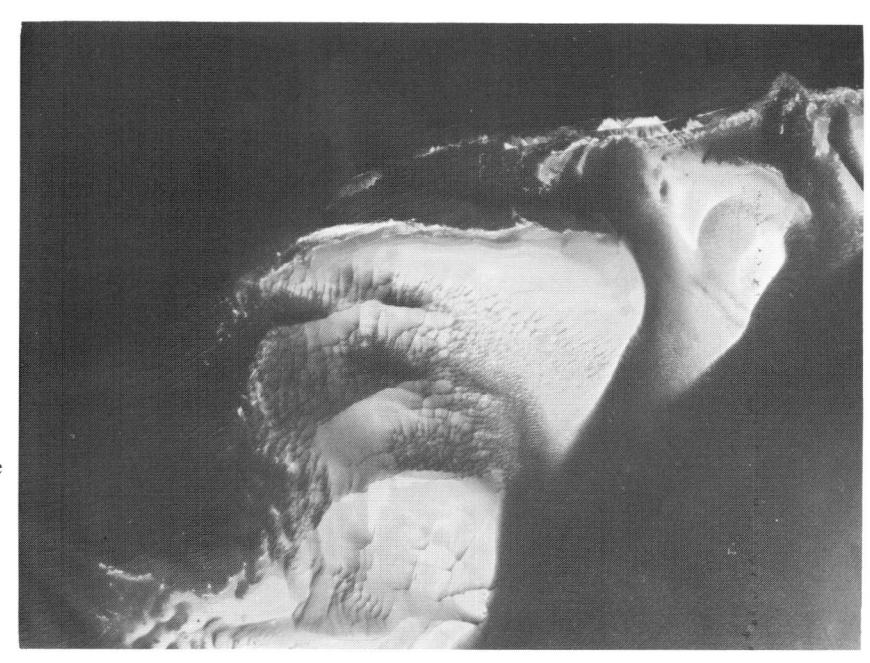

Figure 46 b: The same area as shown in fig. 46 a during a period of stronger wind from the NW. The shoals have been partially levelled by the surf. Only leeward megaripples were preserved (Aerial photograph of Fa. N. Rüpke, Hamburg, of 10.9.1982, Bildflug Wichter Ee, Str.2/69, freigegeben durch Luftamt Hamburg unter Nr. 818/82).

perfectly flat surfaces whereas further leeward the ripple forms are preserved (figs. 46 a and b).

The influence of waves on the mainland shore

Wave height depends on water depth and fetch. The influence of waves on the coast is therefore strongest where they are able to reach the coast unbroken. As a consequence the strongest effects of the surf can be expected on the seaward side of the islands and tidal deltas. Surf cannot be generated between the islands and the coast because of the small water depth.

The salt marshes of the mainland are most exposed to wave action in the areas behind the tidal inlets, where, due to the relatively long fetch and great water depth, higher waves can generate. The tendency of the salt marshes to grow seaward is

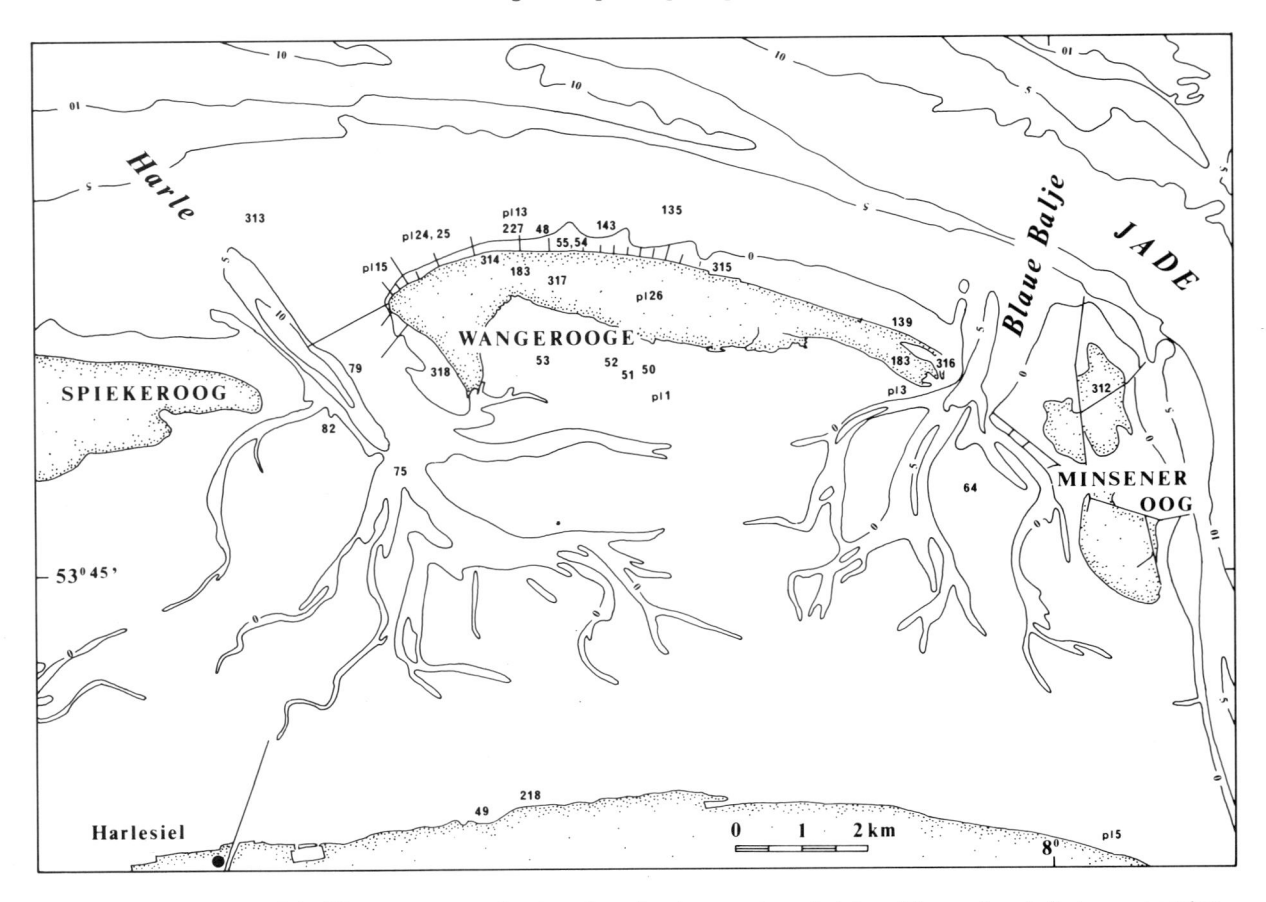

Figure 47: Location map of the Wangerooge area, showing where the photographs and plates of the small-scale features were taken.

thus minimal in these areas; this is clearly reflected by the configuration of the salt marsh margins (Niemeyer, 1977; 1979).

4.4 ICE

Because sea water contains 35 parts per thousand of salt, its freezing point is considerably lower than fresh water. A cooling to -1.9° C is required to freeze sea water. In the shallow areas of the Wadden Sea where there is no exchange with deep warmer water freezing starts at temperatures as high as a little below -1.0° C (Venzke, 1985: 254).

In the area investigated, ice plays a minor role as an agent of geomorphological change because of the relatively mild winters experienced. If sea ice is able to form, however, the whole surface of the beach and tidal flats can become covered by cryogenic forms very rapidly. Although ice creates a number of characteristic features, few traces of its existence will be preserved in the sediment record since the ice invariably degrades after a few days or weeks.

A number of ice features have been described in literature; the articles by Reineck (1956), Gripp (1963), Dionne (1968, 1969), Park (1974: 155), Reineck & Singh (1975: 54, 69 and 72) and Wunderlich (1978) are representative. During the relatively severe winters of 1978/79 and 1979/80 the author observed the effects of ice on the North Sea coast (for locations of the examples see fig. 47). A few of the features described have not previously been mentioned in the sedimentological literature.

Three different types of ice can be found on the tidal flats:

1. Floating ice blocks, driven by the wind and tidal currents, which shape the sediments from the outside.

2. Ice-impregnated frozen ground, the internal structure of which is changed.

3. Ground ice, that forms on the tidal flats;

Figure 48: Frozen sand on the beach starting to thaw. The frozen layers are undercut by the waves. The thawing process initiates from circular hollows which finally lead to the degradation of the whole layer (1979).

when uplifted by tides, it removes frozen-on sediment from the tidal flat surface.

Frozen ground

In spite of the relatively mild climate frost phenomena are observed almost every winter along the North Sea coast. Wangerooge has, on average, 52 days with sub-zero temperatures annually. When the water in small puddles on the beach freezes, the imprints of the ice crystals may be preserved on muddy surfaces after thawing (Pettijohn & Potter, 1964: plate 96 a; Reineck & Singh, 1975: 54). In severe winters, however, the frost penetrates much deeper.

During strong winters not only do ice blocks form in the sea but the uppermost layers of sediment on the beach and tidal flats also become frozen. The flats freeze despite being covered by the sea twice daily. The sands forming the concave banks of meandering tidal creeks have been observed to freeze to a depth of 5 cm. Where this frozen layer was being undercut by erosion, fragments of frozen sand (appr. 20 cm × 50 cm) fell into the creek.

The penetration depth of freezing in the sediments depends mainly on the pore volume and the resulting water content. Therefore, the frozen layer normally consists of a number of more or less well frozen strata, each depending on its specific grain-size characteristics.

Thawing usually takes place under a sediment cover of several centimetres to decimetres. It extends downwards from the surface. Frozen sediment layers, directly exposed through wave erosion in the shore, indicate that thawing is initiated from small hollows a few centimetres in diameter; they are circular in form and have a very sharp margin. These hollows expand laterally destroying the frozen layer, which, if undercut by the waves, breaks down and finally thaws completely (fig. 48).

These thaw holes have been observed at many localities on Wangerooge. They can also be found along the coast of the salt marshes south of the barrier islands. Although on the seaward facing beach of the island they were developed in medium sand, along the mainland coast they have been observed to form in a layer of muddy silt to fine sand (fig. 49).

The origin of the thaw holes is unclear. They may possibly be related to the repeated wetting of the sediment by the incoming waves. Salt water persists during falling tide in natural depressions on the sand flat surface. Thawing intensifies and further extends the hollows. Waves further enlarge them by removing loose sand grains and allowing deeper water penetration. The development of the holes may therefore represent a positive feedback mechanism.

Unfrozen salt-water lenses which are found within a frozen sediment layer composed of pure or nearly pure fresh-water ice may, on the other hand, play a role. During the freezing of salt water only fresh water is frozen at first, the salt remain-

Figure 49: Frozen layer at the outer margin of the salt marsh beginning to thaw. Undercutting and hollow formation are similar to features created on the beach (cf. fig. 48) (1979).

Figure 50: Tidal flat area south of Wangerooge striated by ice block dragging (1979).

ing in solution. The differentiation may result in pockets of unfrozen salt water remaining within the frozen sediment layer.

Drift ice

Drift ice may produce typical casts in bottom sediments under shallow water. The preservation of these casts depends on the dynamic conditions in a given area. In tidal flats behind the islands the casts may endure through several tidal cycles.

Drift ice forms mainly in Wadden Sea nearshore areas because these shallow water regions cool most rapidly. The drift ice created is propelled by wind and tidal currents. With the rising tide the ice blocks move towards the coast and onto the tidal flats. Reaching shallow water, the drift ice blocks become stranded. Their irregular bottom surface may etch grooves into the underlying bottom deposits (Pettijohn & Potter, 1964; Pettijohn,

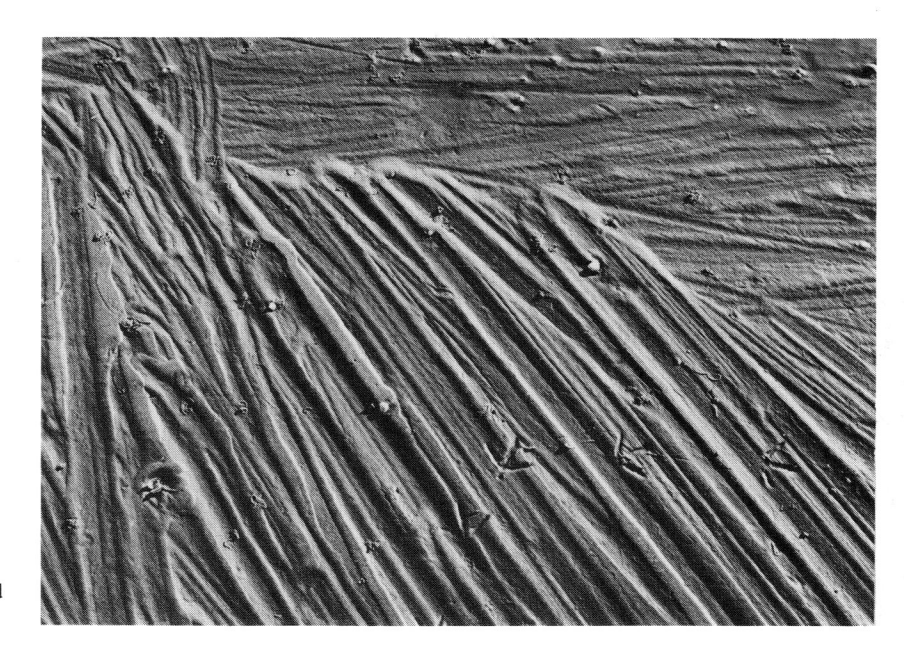

Figure 51: Several generations of grooves reflecting changes in wind direction (1979).

Figure 52: 'Thrust moraine' (7 cm high) created by a shoaling ice block (1979).

1975). Scars more than 90 m in length have been observed by the author.

Eventually the entire base of the ice block makes contact with the ground, and, as the ice remains in motion, the surface of the tidal flat becomes striated. After some metres or tens of metres the ice block finally strands and ceases to move.

Drift-ice striae may cover the entire tidal flat surface in rare instances. Fig. 50 shows a striated surface south of Wangerooge formed in this way, with fig. 51 showing detail from the same area; several generations of striae can be formed with changes in wind direction.

The normally straight and elongated striae may be deformed to a sinuous pattern under the influence of waves (Wunderlich, 1968; Reineck, 1976). Under calm weather conditions however, the straight lines are by far the dominant features. Sudden zig-zag-shaped changes in the directions of striae may be attributed to the collision of ice

Figure 53: Melt-out depressions after stranded ice blocks that sank into muddy tidal flats south of Wangerooge (1979).

blocks rather than to wave influences. The normally minimal wave heights of the Wadden Sea are even further reduced under drift-ice cover.

The process of stranding in some cases leads to the formation of pushed ridges like small 'end moraines' to the front of the ice blocks. Dionne (1969: 15) has described similar forms from the St. Lawrence Estuary. The small thrust ridges in fig. 52 have a maximum height of 7 cm, the whole arc having a diameter of about 30 cm. Much larger forms have been described from the beaches of Arctic coasts (Nichols, 1953; Ellenberg & Hirakawa, 1982).

Alternately no such end moraines may be formed, the ice blocks simply sinking in the bottom at the end of their trajectory (fig. 53). Large ice blocks with a thickness of 2 to 3 m have been observed to exert a considerable pressure on the scarcely consolidated sediments of the tidal flats. The blocks sink several centimetres into the sediment resulting in characteristic load casts or stationary tool marks as described by Reineck (1976: 198). Wunderlich (1973) demonstrated that the underlying layers of sand and mud are deformed down to a depth of about 20 cm.

Ice blocks do not only strand on the tidal flats but also on the beach where they influence erosional processes. On the seaward end of the ice blocks semi-circular current crescents are eroded by the waves (fig. 54). Reineck (1976) mentions that these bedforms can also be observed on sandy flats behind the barrier islands. Such current crescents are created upstream of all obstacles on beaches or in streams of shallow water (cf. Pettijohn & Potter, 1964). Only in the case of ice blocks however, does the obstacle generally disappear more rapidly than the eroded hole.

During severe winters ice blocks accumulate as a strand on the beach at high tide level. If the next tide rises higher than the preceding one the ice blocks are propelled further landward and packed one upon another.

If the strand of accumulated ice blocks is carried in this way to such a high position that the following tides can no longer reach it, it becomes subject to aeolian processes. Aeolian sand deposition

Figure 54: Current crescent formed upstream of a stranded ice block on the beach (1979).

Biogenic processes

Ground ice

On the tidal flats of the North Sea coast a form of ground ice develops when the ground becomes frozen at low tide during clear frosty nights. When these areas are flooded during the rising tide, the ground becomes covered with an ice blanket which may increase in thickness during subsequent tides. During higher water levels the ice may be uplifted by the tide, the uppermost sediment layer remaining frozen to the bottom of the ice (Ausschuß 'Küstenschutzwerke', 1981: 64). A similar effect may be caused by the freezing of ice blocks onto the underlying frozen sediment; in both cases the result is sediment transport by drift ice.

Most of the marks and bedforms described above are not preserved for long periods of time as they are rapidly destroyed by the influence of wind and waves. In certain protected situations, however, such forms may persist in the sediment record and thus document aspects of the recent environmental conditions.

4.5 BIOGENIC PROCESSES

Vegetation and animals are an important factor in shaping the landforms on the tidal flats and islands. As the influence of the vegetation on salt marsh growth will be dealt with in section 5.3.5, and on dune formation in section 5.3.3, this section is restricted to a description of the biogenic processes on the tidal flats.

The finer suspended sediment particles of the Wadden Sea would stay in suspension, were they not extracted from the water and deposited as pseudofaeces by bivalves such as the cockle (*Cerastoderma edule*; earlier: *Cardium edule*), the edible mussel (*Mytilus edulis*) and the sand gaper (*Mya arenaria*).

Large colonies of edible mussels (*Mytilus edulis*) occur often on tidal flat surfaces. They form an irregular micro-relief easily identifiable on aerial photographs (see section 5.1.1). Whereas the other bivalves burrow into the sediment, the edible mussel grows on the tidal flat surface (fig. 56). However, it is also found to a maximum water depth of 20–25 m. Individuals are connected with each other by byssus threads so that beds several layers thick can develop. These mussel beds can reach a maximum height of about 60 cm.

The edible mussels are important mud accumulators. By binding particulate matter in their pseudofaeces they are able to produce a mud layer 60

Figure 55: Thawing ice blocks on the beach partly covered by aeolian sand (1979).

takes place mainly in the lee of the obstacles so that finally the whole accumulation of ice blocks may become sand-covered. With the thaw of the ice the thin sand cover collapses and a row of craters is formed on the beach where the ice blocks were formerly situated (fig. 55). Where the ice fragments are extremely small the resultant structures may resemble cavernous beach sand (see section 5.3.2) although the conditions leading to their formation are completely different. Greene (1970: 422) describes similar forms from the arctic coasts of Alaska.

Figure 56: *Mytilus* colony on muddy tidal flats south of Ameland (1981).

cm thick within two years (200–400 mm³ faeces per day per individual; Niemeyer, 1972: 69). In the Dutch part of the Wadden Sea alone, the annual production of largely inorganic mud by edible mussels was estimated at one million tons (Verwey, 1981 a: 115).

The cockles and sand gapers live at shallow depths on the tidal flat floor and contribute to the mud accumulation. They catch their food via a syphon through which water and suspended matter is transported from the tidal flat surface. As the cockle is unable to digest coarse sediment particles it only contributes in a minor way to the mud accumulation (Verwey, 1981 b: 116). Thiel, Grossmann & Spychala (1984: 280) report that the permanent mud accretion by the cockle is small, and that if there is any current activity it may be absent altogether.

Apart from the bivalve species mentioned above, the lugworm (*Arenicola marina*) is also of importance. The lugworm lives in a U-shaped burrow at a depth of 30–40 cm below the surface and eats sand. Sand is supplied at one end of the tube where a small crater forms whilst at the other end the digested sand is released. The faecal casts of the lugworm determine the micromorphology of large areas of the tidal flats (fig. 57). The lugworm promotes better ventilation of the Wadden Sea bottom and also causes strong bioturbation (Thiel, Grossmann & Spychala, 1984: 289).

The edible mussel, cockle, sand gaper and lugworm together represent about three quarters of the total weight of the benthic biomass over 1 mm in size in the Wadden Sea area (Beukema, 1976: 136).

After deposition, the mud is further consolidated by diatoms. Whenever these diatoms are covered by fresh sediment, they migrate upwards towards the tidal flat surface in their search for light for photosynthesis. During this process they deposit a mucous layer which hardens the sea floor and protects the tidal flat surface against erosion (cf. Wunderlich, 1979; Reineck, 1979 b; Führböter, 1983). Where the resultant cohesion is degraded by waves, drifting ice blocks or by sea birds, for instance, small pools are created bounded by up to 5 cm high scarplets (fig. 58).

The cohesiveness of the tidal flat surface can be so effective that it influences the development of the drainage pattern. In one case a dendritic drainage network was observed at the margin of a small tidal creek on slightly muddy tidal flats south of Vlieland (fig. 59), which differs markedly from the normal drainage patterns.

The migration of megaripples is also influenced by micro-organisms. Allen & Friend (1976) assumed that the observed difference in boundary conditions for the start of sediment movement before and after neap tides were partly caused by stabilization of the sediment by algae. De Boer (1981) has confirmed this by experimentation. The sterilized section of a megaripple was completely eroded after two tides, whilst the remainder of the ripple remained intact.

Figure 57: Faecal casts of the lugworm (*Arenicola marina*) at Ameland (1981).

Figure 58: Pools on the tidal flats south of Wangerooge, formed where the superficial diatom mat was destroyed (1977)

Local destruction of the tidal flat surface is caused by food-searching sea birds. The herring gull (*Larus argentatus*), for instance, sits on the tidal flat surface and moves slowly backwards in its search for food causing furrow-like disturbances of the diatom mats (fig. 60). The trampling moulds of shelduck (*Tadorna tadorna*) and black-headed gull (*Larus ridibundus*) have similar effects (fig. 61).

The influence on the sedimentary conditions by the large numbers of sea birds should not be underestimated (numbers after Goethe, 1983 a, b, c).

Numbers of pairs of selected sea birds occurring in the Wadden Sea

Herring gull	66 000
Black-headed gull	91 000
Shelduck (during moult)	30 000

Figure 59: Dendritic drainage pattern at the margin of a shallow tidal creek, south of Vlieland. This has formed as a result of the cohesion of the upper sediment layer caused by diatom mats (1981).

Figure 60: Resting mould of herring gull (*Larus argentatus*) on the tidal flats at Mellum (1981).

These animals are much more likely to cause the numerous pools on the tidal flat surfaces than wave action, which is very limited leeward of the islands. The footsteps of tourists may have similar effects (fig. 62).

Figure 61: Trampling moulds of shelduck (*Tadorna tadorna*) or black-headed gull (*Larus ridibundus*) on the tidal flats at Ameland (1981).

Figure 62: Trampling of the tidal flats between Neßmersiel and Baltrum by tourists. The footsteps damage the superficial diatom mats (1982).

CHAPTER 5

MORPHODYNAMIC UNITS

5.1 TIDAL FLATS

5.1.1 Subdivision of the tidal flats

The tidal flats in the area under investigation consist mainly of sand. Mud areas are mostly limited to the margins of tidal creeks and to the mainland coastal areas. In the Wangerooge area a number of samples were taken for grain-size analysis. Dr. Figge of the Deutsches Hydrographisches Institut (DHI) kindly provided the results of additional grain-size analyses. Selected examples are presented in fig. 63.

Wadden Sea sediments are characterized by a dominance of fine sand with variable amounts of medium sand. Coarser fractions only occur in overdeepened sections of the tidal inlets (samples 342, 430 and 1860) and at the bottoms of the tidal streams (samples 314 and 595) which erode into the gravel-bearing strata of the subcropping Pleistocene.

In the sand flat areas (samples 11, 13, 2975, 2977 and 2979; plates 1 and 2) clay and silt content is about 1 %. True mud flats have a clay and silt content of over 50 % (Figge *et al.*, 1980: 192). Samples 12, 14 and 15 therefore belong to the so-called

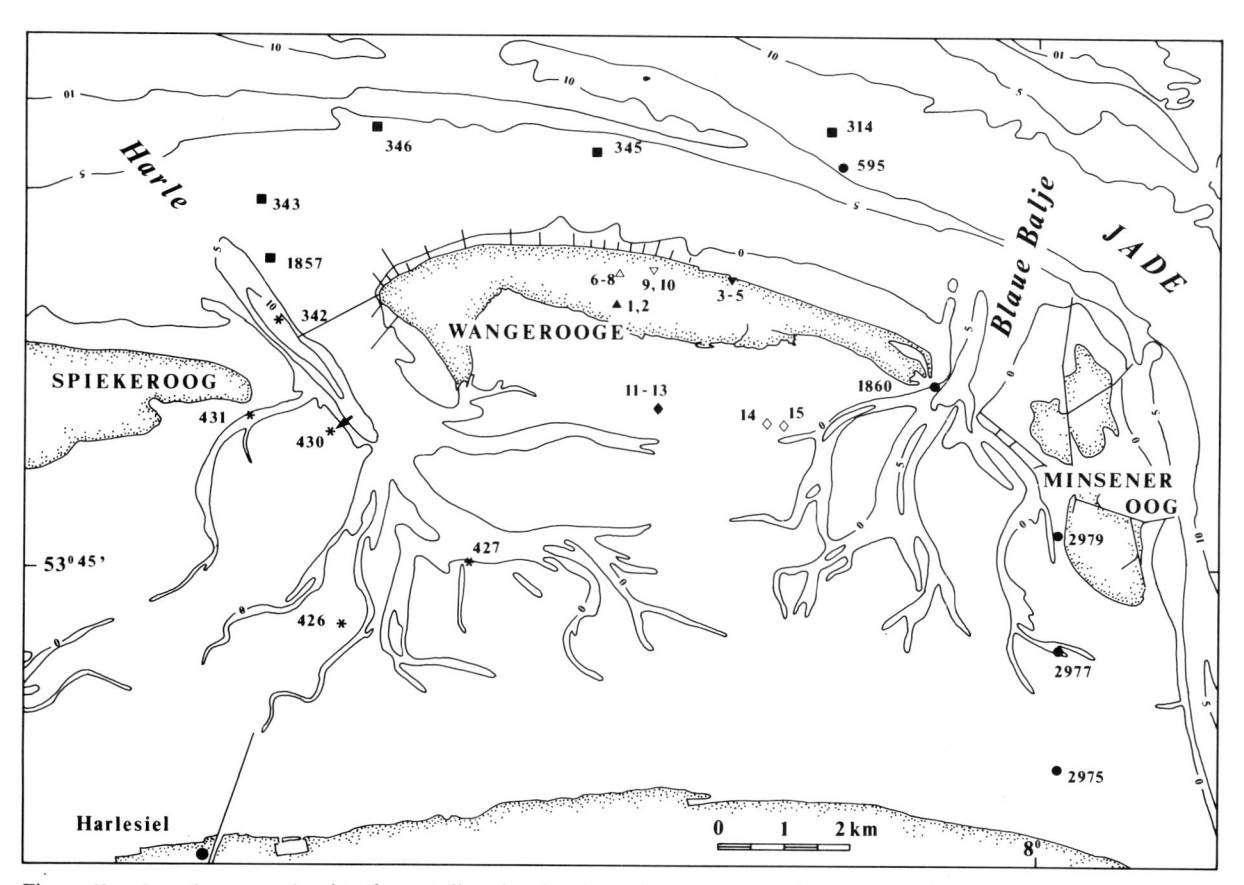

Figure 63 a: Location map, showing the sampling sites for the grain-size analyses shown in figs. 63 b–g.

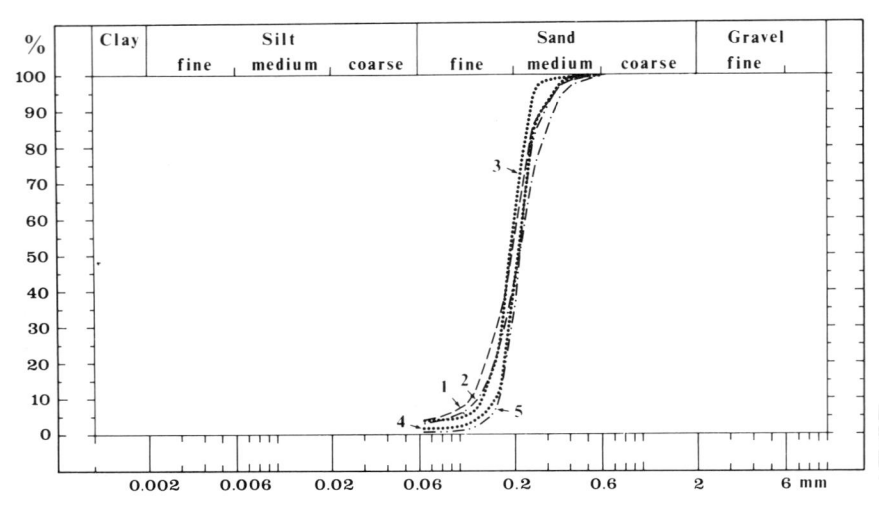

Figure 63 b: Grain-size analyses of sediment samples from the Wangerooge area. For locations see fig. 63 a.

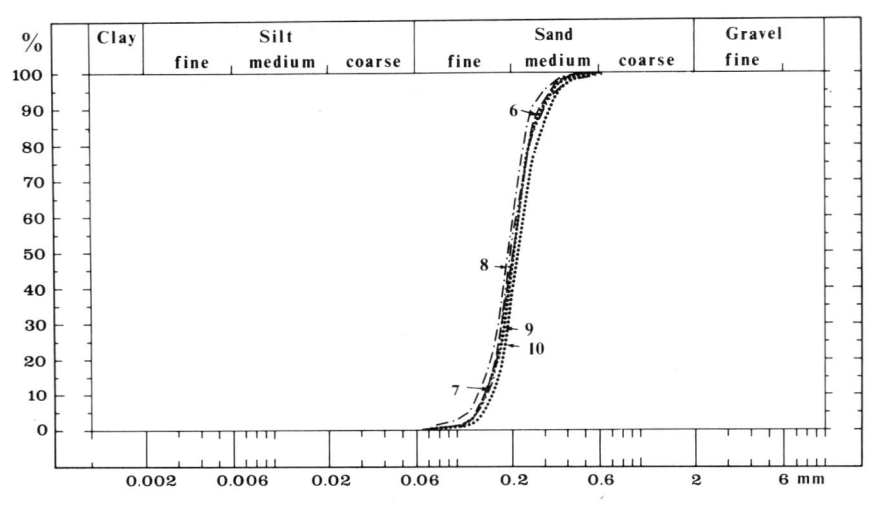

Figure 63 c: Grain-size analyses of sediment samples from the Wangerooge area. For locations see fig. 63 a.

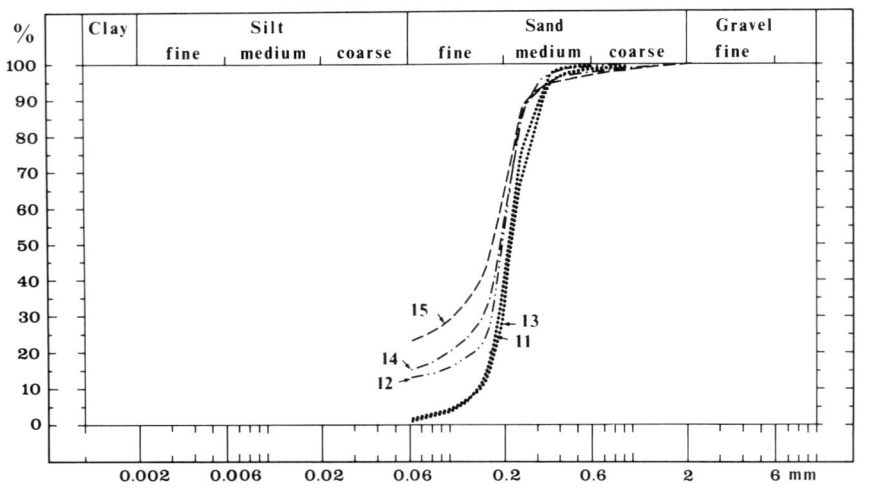

Figure 63 d: Grain-size analyses of sediment samples from the Wangerooge area. For locations see fig. 63 a.

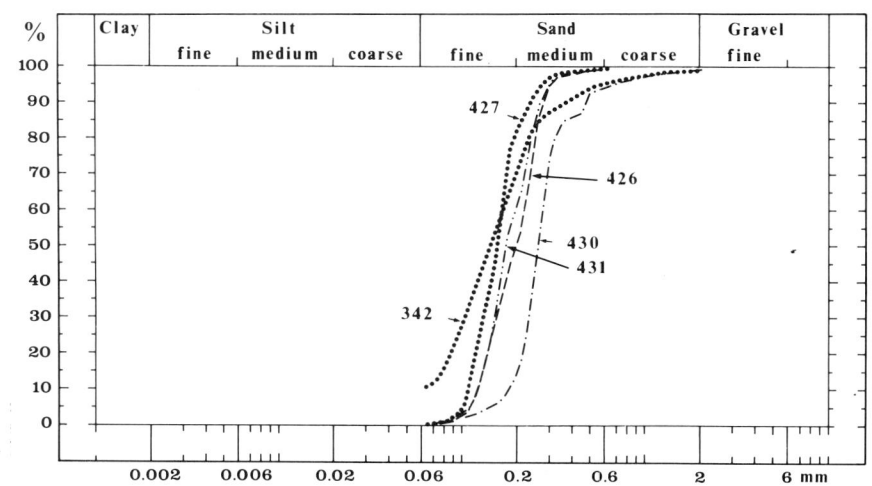

Figure 63 e: Grain-size analyses of sediment samples from the Wangerooge area. For locations see fig. 63 a.

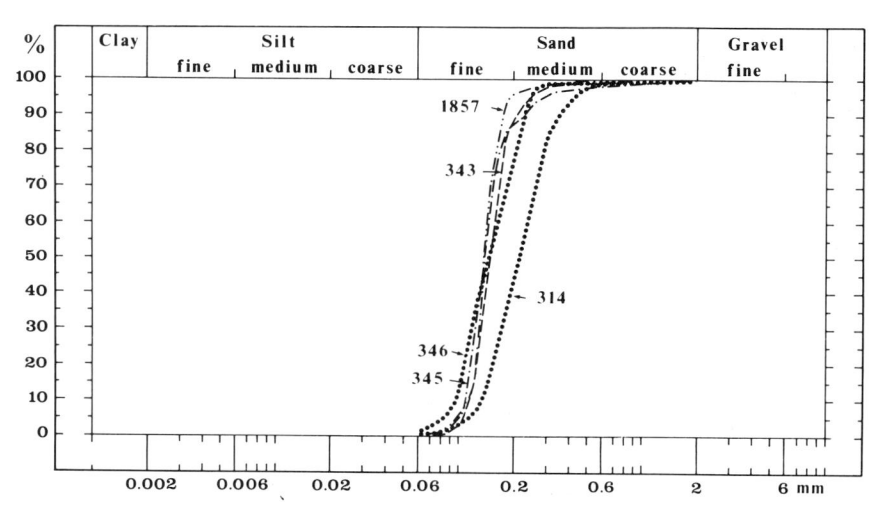

Figure 63 f: Grain-size analyses of sediment samples from the Wangerooge area. For locations see fig. 63 a.

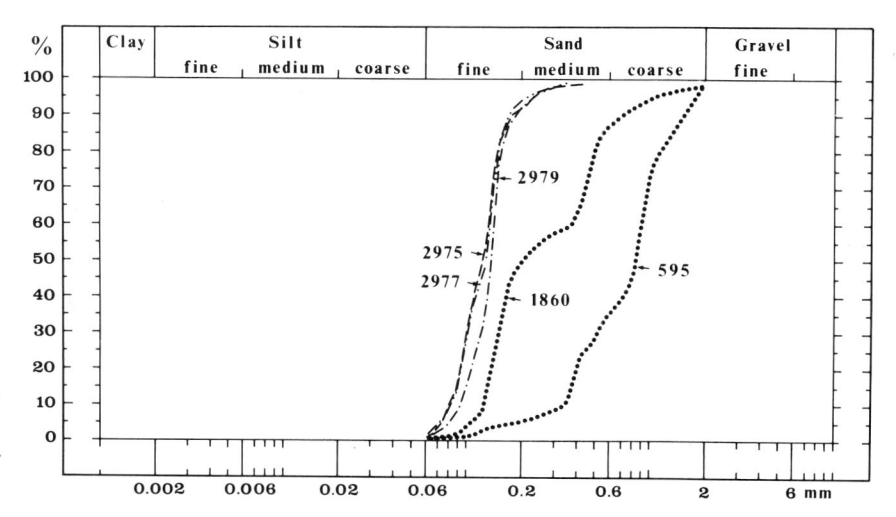

Figure 63 g: Grain-size analyses of sediment samples from the Wangerooge area. For locations see fig. 63 a.

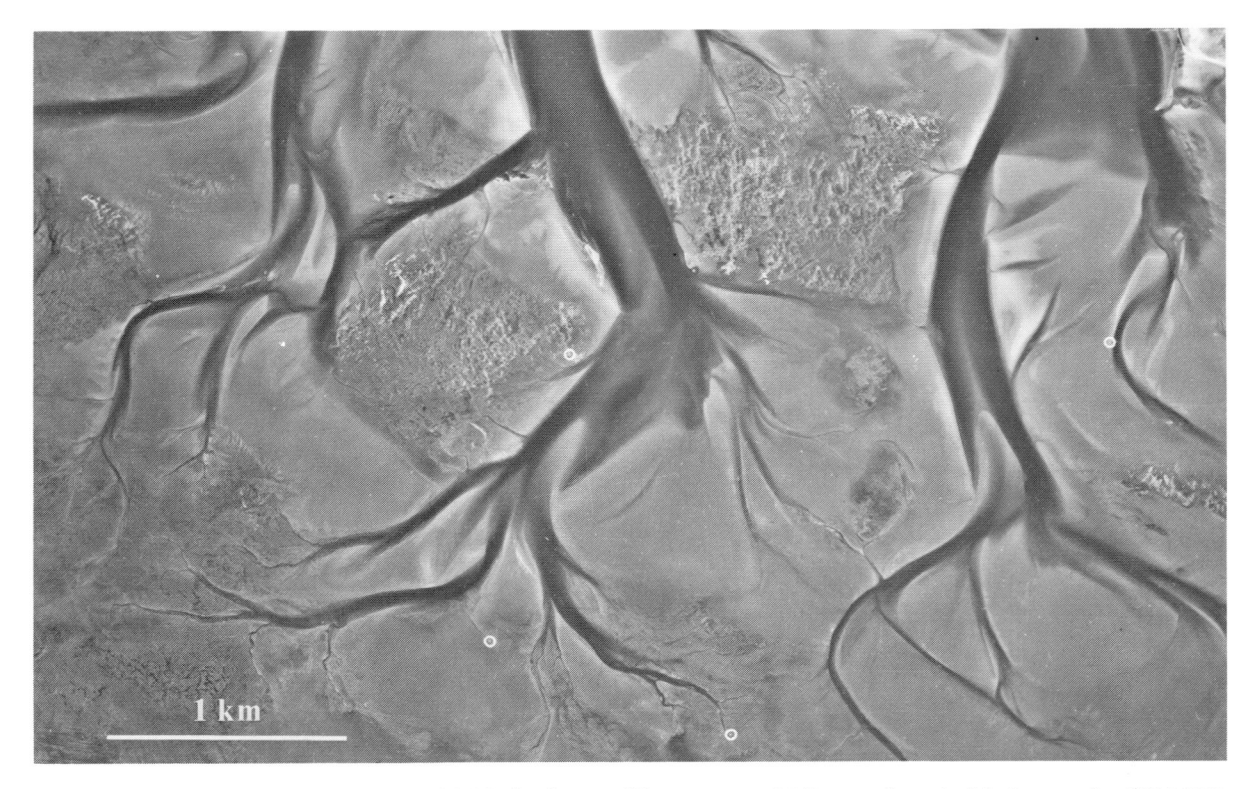

Figure 64: Aerial photograph of mud flats with *Mytilus* between Wangerooge and Minsener Oog. Aerial photograph of 17.4.1974, Bildflug Ostfriesische Inseln (1086), Str.4/083, freigegeben durch NLVA – Landesvermessung -, Hannover unter Nr. 34/74/1086.

'mixed' flats (10–50 % clay and silt). These samples have a bimodal grain-size distribution; the normal dominant mode in the fine sand fraction is complemented by a subdominant peak in the clay fraction which may be of equal magnitude.

The mud flat sediments are characterized by a high organic content. In thin section (plate 3) it becomes obvious that whilst the sand fraction consists mainly of quartz grains, the clay and silt fractions consist almost exclusively of organic matter. In addition, carbonate particles also contribute to the fine fraction. These consist partly of ground shell fragments and partly of possibly biochemically precipitated carbonate (Kamps, 1963: 14).

Exposure to, or protection from, marine erosion plays a decisive role in mud flat development (van Straaten & Kuenen, 1957; Postma, 1961). Linke (1939) observed that grain-size decreases towards mean highwater level (i.e. towards the land). Consequently mud flats are often found within, or directly adjoining, zones of sedimentation at the coast (plates 4 and 5). This is caused by the decreasing intensity of water movement and the increasing influence of organisms on the sediment composition. This has been largely confirmed by recent investigations (Reineck, 1978 a: 54). In addition, halophytic vegetation (e.g. *Salicornia*) helps to reduce the currents.

The mud flats in most cases have a slightly irregular surface caused by large colonies of edible mussels (*Mytilus edulis*) which act as mud accumulators (cf. section 4.5). The mussel beds can be seen as dark rises above the tidal flat surface from a great distance. They form hummocky, often net-like structures easily detectable on aerial photographs (fig. 64).

The major mud flats are found on the lee side of the higher Wadden ridges and in protected positions along tidal creek margins (but always at some distance from the area of influence of the nearest tidal inlet), where the deposits are sheltered from being remobilized by the strong westerly winds.

No sharp boundary, but a gradual transition exists between mud flats and sand flats. In addition, temporary changes in sedimentation caused by longer periods of deviating wind directions or velocities, for instance, can lead to stratified sand/mud deposits. It has been observed that in the summer mainly finer particles are deposited whereas during the winter coarser sediments pre-

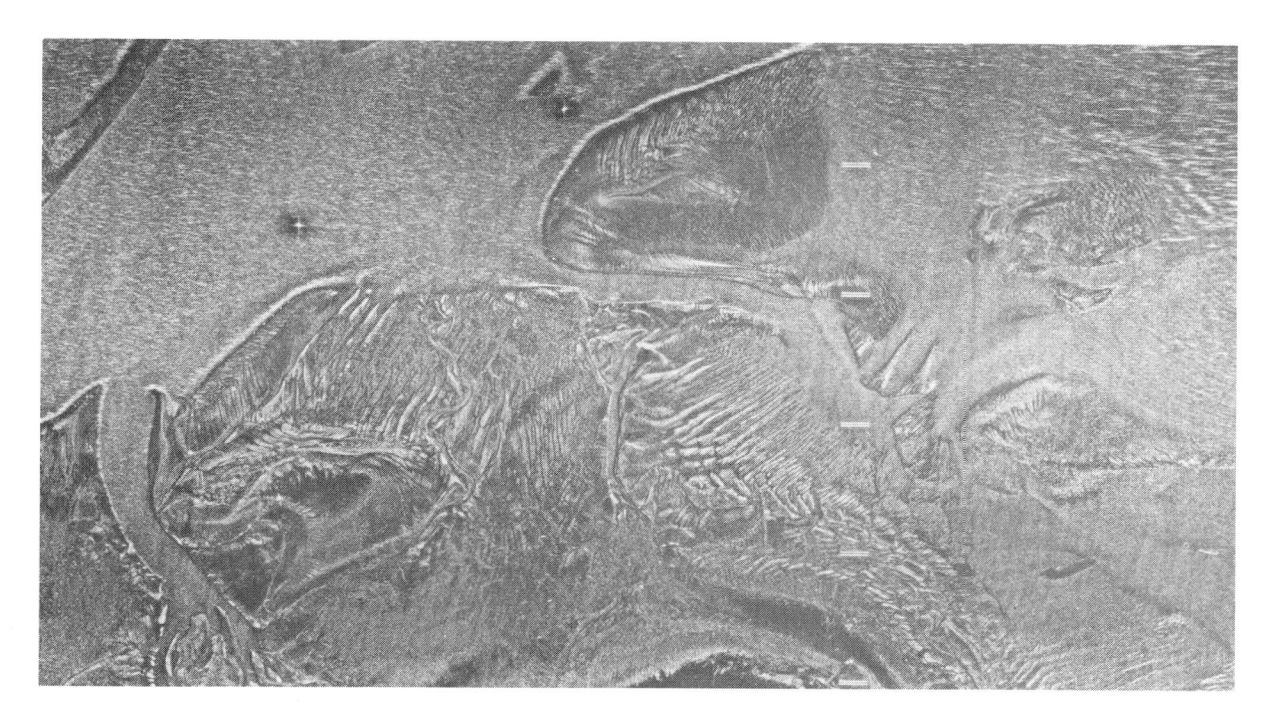

Figure 65: Drainage network in the Unterer Wittsand sand flat area, north of the Weser estuary. The radar image demonstrates the straight creek courses; side branches are scarce (Freigabe Luftamt Hamburg unter Nr. 715/84).

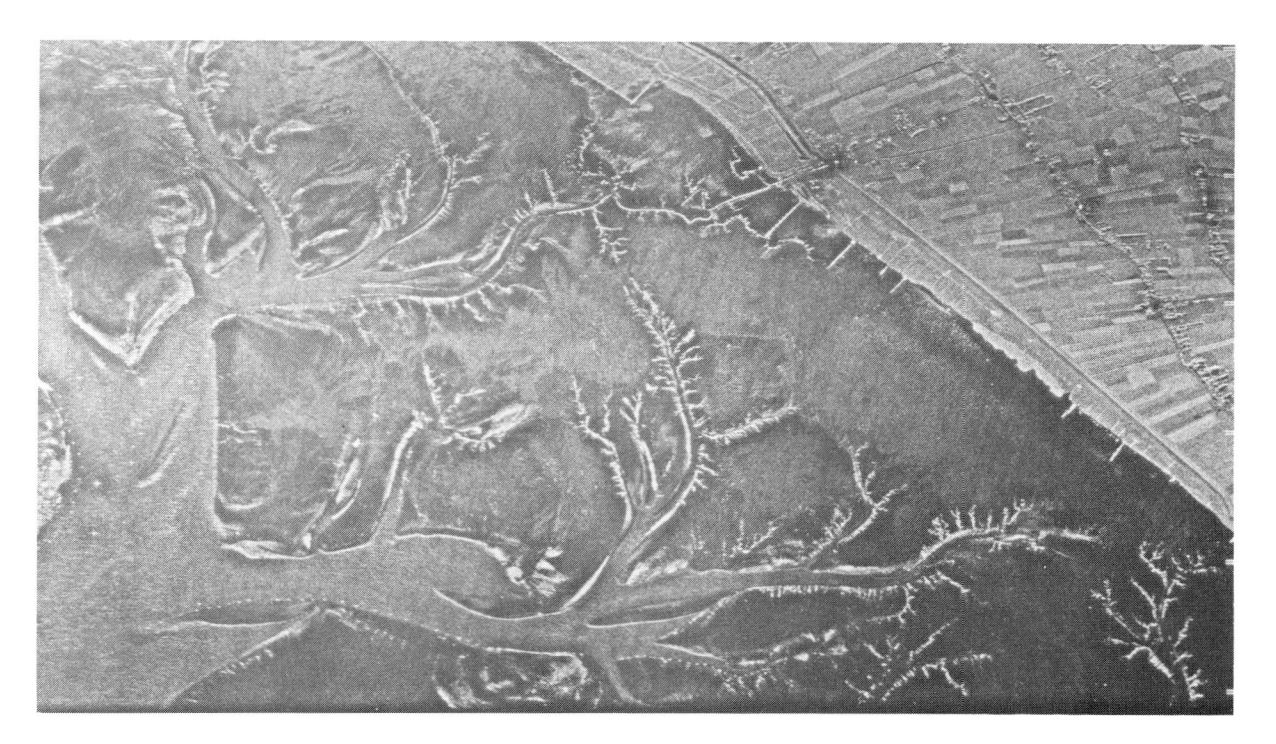

Figure 66: Drainage network in the mud flat area west of Cappel-Neufeld, north of the Weser estuary. The radar image demonstrates that the creeks are more densely spaced and have many filigree-like branches (Freigabe Luftamt Hamburg unter Nr. 715/84).

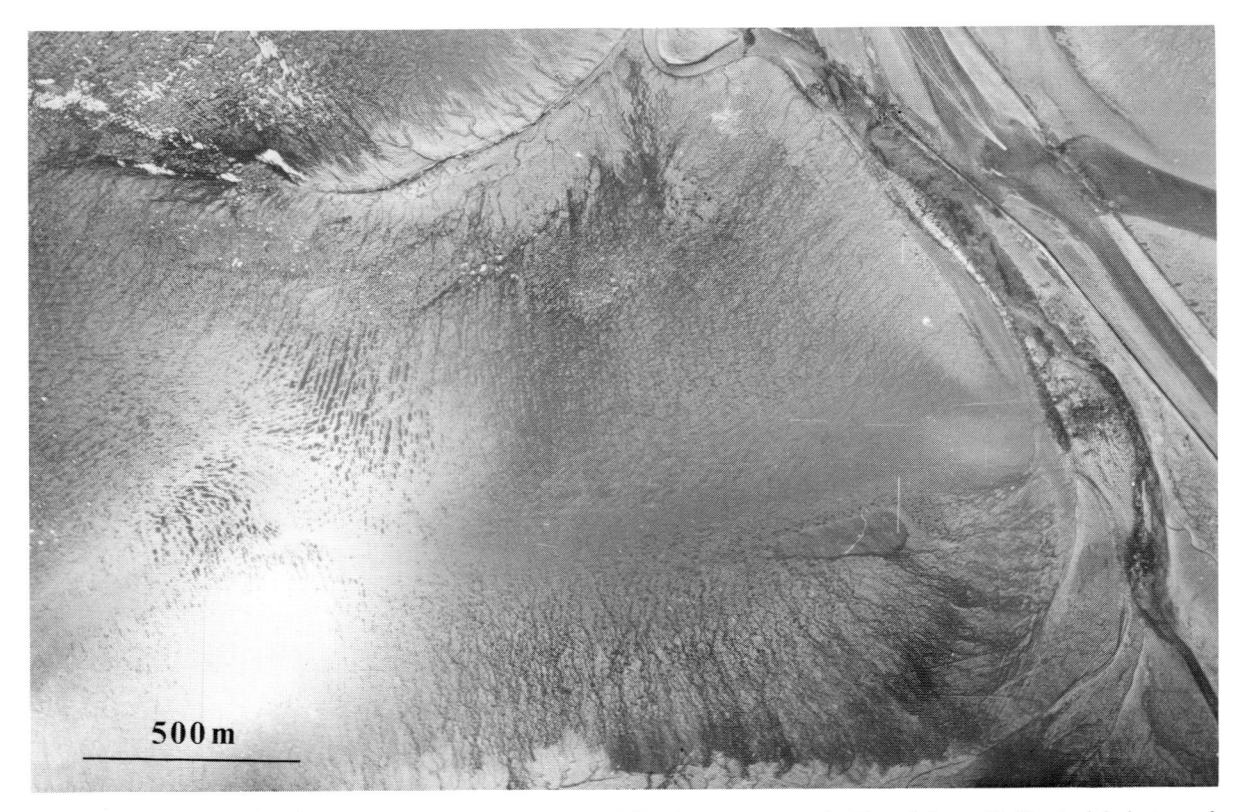

Figure 67: Micro-relief on the sand flats off Bensersiel. The sand flat slopes are covered with radial runoff rills. Aerial photograph of the Niedersächsisches Landesverwaltungsamt, Landesvermessung, of 29.5.1966, Bildflug Ostfriesische Küste (419), Str.3/231, freigegeben durch Regierungspräsident Münster/W. unter Nr. 2316/66.

vail (Ragutzki, 1979: 132). The typical interbedding of sand and silt is easily detectable in thin sections (plate 3). The mud areas on the tidal flats rarely exceed 1 m in thickness. Usually the mud only forms a top layer a few decimetres thick overlying dominantly sandy older sediments.

The major tidal creeks which transect the tidal flat areas, have relatively straight courses making them clearly distinguishable from mainland streams of the same size. A number of minor creeks which drain the major flats are confluent with these major creeks. The creek densities in sand flat and mud flat areas are very different. Relatively few, rarely branching, creeks are present in the sand flat areas, whereas the mud flats are drained by a dense creek network (Gast, Köster & Runte, 1984: 203). The creeks in the sand flat areas are mostly short and straight, whereas the creeks of the mud flats meander and have many branches (figs. 65 and 66).

Micro-relief

The sand flats have relatively smooth surfaces which appear light-coloured on aerial photographs. However, a more detailed examination reveals that the surface is not absolutely plane. Not only is it covered by small ripples but it is also subdivided into numerous small-scale form elements. Meyn mentioned these small forms as early as 1876 (p. 669). He noted that 'because of their low elevation they may easily escape the eye of the observer'.

The micro-relief is detectable on aerial photographs only under favourable conditions, in particular when the lower parts of the sand flat surface are still covered by a water film which reflects the sun. Aerial photographs of the tidal flats at Bensersiel (figs. 67 and 68) demonstrate that the form assemblage of small rills and ribbons owe their origin to different processes.

One major factor in sculpturing the sand flats is gravity. Following the gradient of the buckle-shaped sand flats, the water runs off radially towards the adjoining tidal creeks during the ebb. Slight meandering of these radial rills creates a slightly irregular surface pattern.

Other processes prevail on the level higher areas

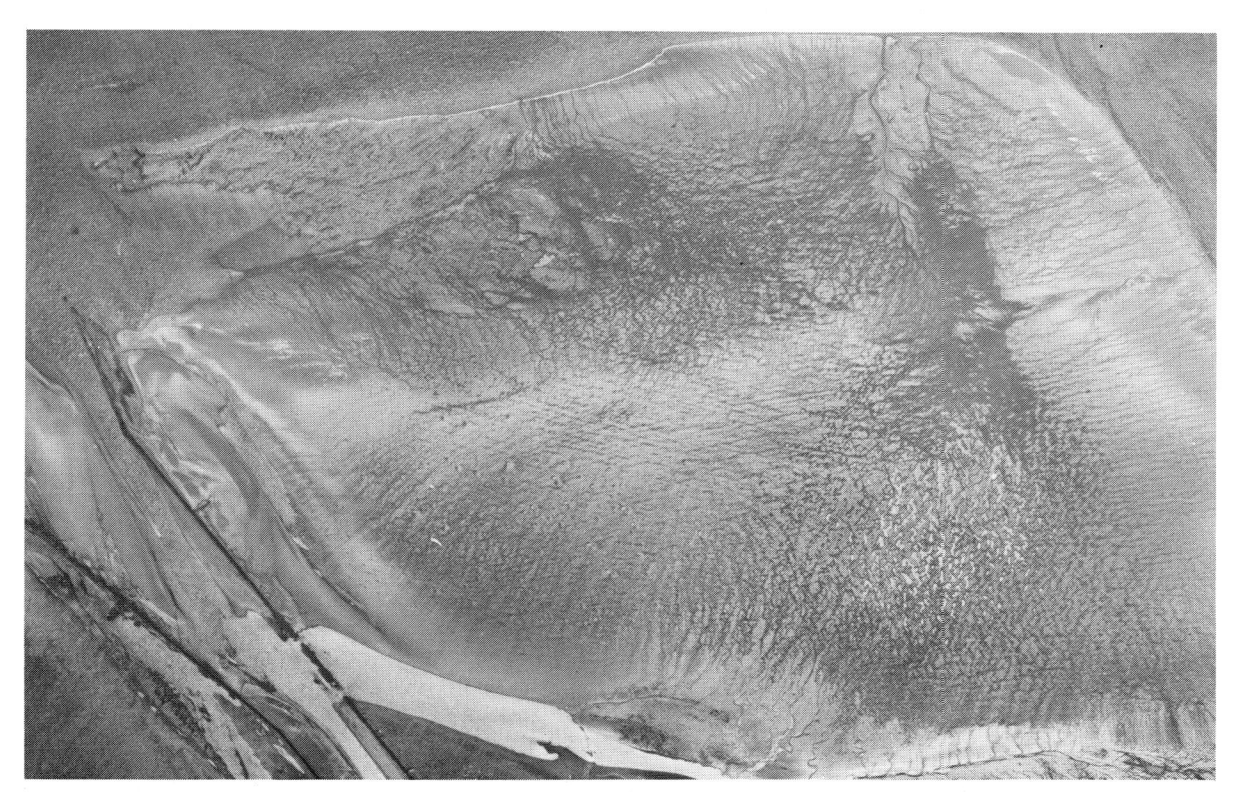

Figure 68: Micro-relief on the sand flats off Bensersiel. As well as the gravity-controlled runoff-rills, on the higher parts of the flats there are also two systems of sand ribbons parallel to the dominant wind directions. These were formed by drift currents during two consecutive low tides. Aerial photograph of the Niedersächsisches Landesverwaltungsamt, Landesvermessung, of 29.5.1966, Bildflug Ostfriesische Küste (419), Str.3/229, freigegeben durch Regierungspräsident Münster/W. unter Nr. 2316/66.

Figure 69: Current stripes on the tidal flats near Arensch. The stripes are roughly parallel to the orientation of the creek. Weakly developed megaripples are crossed at right angles. The distribution of sand and shell fragments in the right foreground indicates that the stripes are partly erosional and partly accumulative in origin (1983)

of the sandy flats. These areas are sculptured mainly by the wind. Drift currents create parallel-striking sand ribbons which may even survive one or more high tides under favourable conditions. The photographs (figs. 67 and 68) show two superimposed wind ribbon fields, differing in direction by about 20°, representing the wind directions of two low tides.

A number of small landforms are useful in the interpretation of local current conditions. These include the densely spaced stream stripes (fig. 69) which occur particularly on the slopes of shallow

72 *Morphodynamic units*

Figure 70: A comet mark (sand tail), several metres in length, in the lee of a stone on the tidal flats of Baltrum (1980).

Figure 71: Comet marks of a length of several tens of metres on both sides of the former Ameland Dam, which was destroyed by the sea. The marks are partly ebb- and partly flood orientated (1981).

tidal creeks. Reineck (personal communication) believes these features to be a mixture of erosional and depositional forms the origins of which are poorly understood.

Behind obstacles on the tidal flats (e.g. stones, mussel beds) elongated sand or mud tails are often constructed (fig. 70). These 'comet marks' reach considerable dimensions where larger obstacles, such as the abandoned Ameland Dam, are regularly overflown by strong currents (fig. 71).

Plate 1: Sand flats south of Wangerooge, covered with small ripples (1977).
Plate 2: Thin section of sand flat sample from south of Wangerooge. Note mollusc shell fragments.
Plate 3: Thin section of mud flat sample from southeast of Wangerooge. Note thin, well stratified sand layers.
Plate 4: Mud flats southeast of Wangerooge with footprints (1981).

1	2
3	4

PLATES 1-4

PLATES 5-7

Comet marks and similar sand ribbons are even more pronounced where strong currents occur in the subtidal area. Belderson, Johnson & Kenyon (1982: 46) demonstrate the sonar signatures of such marks. The length of comet marks in the lee of shipwrecks, for instance, can exceed 1 km. The orientation of the marks reflects the dominant current and sediment transport directions (cf. Werner & Newton, 1975).

Borrow pits

Sand has been repeatedly dredged from the tidal flat areas for beach replenishment or for dyke construction. It was initially thought that the borrow pits would soon refill. Evidence from the West Frisian Islands (Terschelling), from Norddeich and from Pellworm (fig. 72), however, demonstrates that this is not the case. These pits can be dangerous for inexperienced tidal flat tourists.

5.1.2 Landform changes

The elevation of the tidal flat surfaces are subject to continual change. There have been repeated attempts to quantify this change by measuring height differences over a period of time (e.g. Knop, 1963). The sediment budget measured in this way provides an interesting insight into the mass balance of the tidal flats, but does not explain the morphodynamic processes operating. In order to do this it is necessary to investigate the landforms of the tidal flats and their changes through time.

Sand waves

Aerial photographs of the Wadden Sea always show a distinct pattern of sub-parallel stripes on the tidal flats landward of the barrier islands, but on the ground these stripes are hard to detect. Around the East Frisian Islands the most significant of these stripes strike in a NE-SW direction. The stripes form ridges about 1-2 decimetres in height on the tidal flats. The ridges dry earlier than the intervening troughs and thus appear as bright lines on the aerial photographs (fig. 73).

Figure 72: A borrow pit in the tidal flats W of Pellworm. Borrow pits fill with sediments very slowly. Erosion pools are visible at the groyne heads. Scale: ca. 1:20,000 (Aerial photograph of 1980, freigegeben durch das Landesvermessungsamt Schleswig-Holstein unter Nr. SH 510/80, Bildflug Husum/Brunsbüttel, Str.4/889, vervielfältigt mit Genehmigung des Landesvermessungsamtes Schleswig-Holstein vom 3.12.85).

Gierloff-Emden (1961: 42) interpreted these features as runoff channels formed by the ebb current. Reineck (1978 b: 73) rather neutrally refers to them as 'sand waves' without interpreting their genesis. Newton & Werner (1969: 14, figs. 17-19) describe similar landforms from the Scharhörn tidal flat area. Asymmetry and internal structure demonstrate that the forms are migratory in character; they constitute 'transport' forms.

The presence of sand waves along the lagoonal island shores indicates that the tidal flats participate in the general coast-parallel sediment trans-

Plate 5: Mud flats with single tussock of *Spartina* at Minsen (1979).
Plate 6: *Suaeda maritima* on Baltrum (1984).
Plate 7: *Artemisia maritima* (light grey) on the elevated, sandy levée of a salt-marsh creek on Baltrum; in the foreground *Atriplex littoralis* (1984).

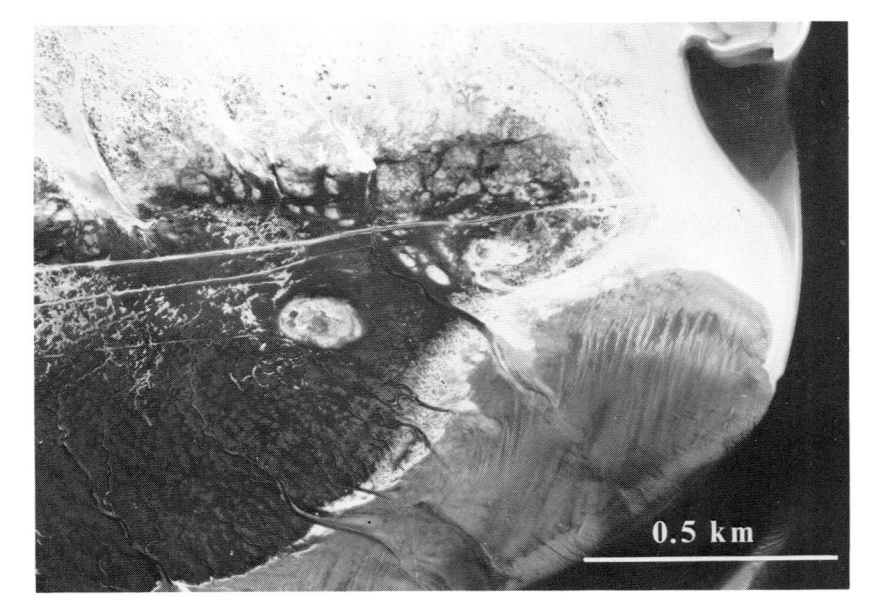

Figure 73: Sand waves on the tidal flats south of Norderney (Norderneyer Inselwatt). The stripes are dissected by a number of small tidal creeks which drain the island marshes. Sediment fans are always located on the eastern banks of the creeks suggesting an easterly direction of sand movement. Aerial photograph of Fa. N. Rüpke, Hamburg, of 10.9.1982, Bildflug Wichter Ee, Str.1/43, freigegeben durch Luftamt Hamburg unter Nr. 818/82.

Figure 74: Sand waves on the Norderneyer Inselwatt, as seen from the ground (1982).

port. The eastward directed movement of sediment along the East Frisian Islands can be seen in the aerial photographs (fig. 73). The sediment removed from the island marshes by small tidal creeks always accumulates on the eastern side in the form of small, half-developed, sediment fans. The sand waves off Norderney have a slightly asymmetrical profile as described by Newton & Werner (1969). The sandwave slopes face southeast, indicating southeastward sand movement (fig. 74).

Shifting of tidal creeks and flats

Major shifts in the positions of Wadden Sea tidal creeks and flats are usually the result of either the reduction or expansion of the drainage basins of the tidal inlets; this may be the result of dyking, migration of the water divides, or migrations of the tidal inlet mouth.

Jessen (1922: 163) found that the tidal inlet axes along the North Sea coast have a tendency to shift in a clockwise direction. He interpreted this migration as an adaptation to changing tidal conditions. In most cases, however, the observed rotation of the channel axes can be attributed to shifting of the water divides, which is largely a consequence of drift current action.

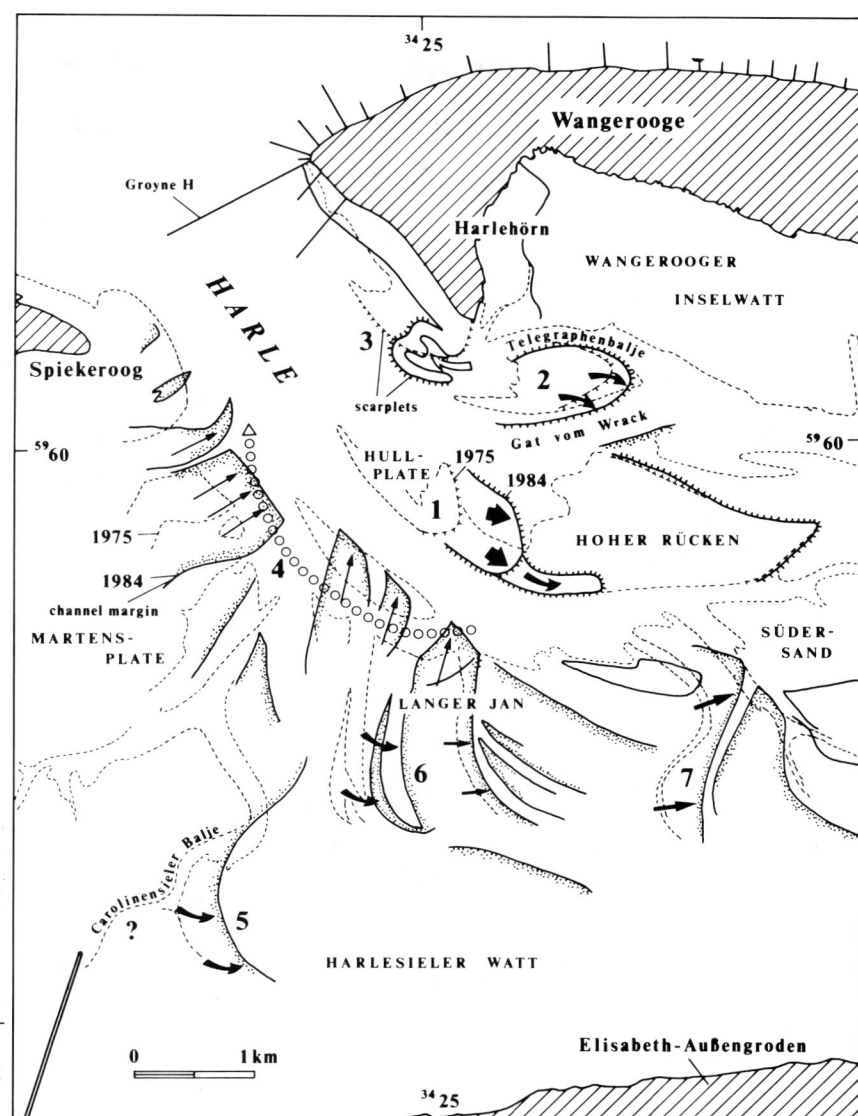

Figure 75: Morphological changes in the tidal flat area southwest of Wangerooge, between 1975 and 1984. Dotted line: former course of a main tidal creek. Sources: Engel, Mauser & Stibig, 1984; Aerial photographs of the Niedersächsisches Landesverwaltungsamt, Landesvermessung, of 17.4.1974 (Bildflug Ostfriesische Inseln); Küstenkarte 1:25,000, Sheet 2212 K Wangerooge; Geomorphologische Karte 1:25,000, Sheet 10 Wangerooge.

Example 1: Wangerooger Watt

Considerable changes have taken place in the tidal flat area between Wangerooge and the mainland over the last few years. These have been caused mainly by the sagging of the westernmost section of Groyne H, which is over a kilometre in length, at the western end of Wangerooge. Its ineffectiveness resulted in an enlarged cross section of the Harle Inlet and in an eastward shift of the inlet axis.

Comparison of aerial photographs taken in 1974 with the Landsat 5 TM image of 1984 (plate 32; cf. Engel, Mauser & Stibig, 1984) reveals the extent of the changes (fig. 75). A number of technical differences militate against direct comparison, however:

1. The different scale and resolution of the aerial photographs (1:28,000) compared with the satellite images (smaller than 1:250,000) permit comparisons of only the larger forms.

2. The aerial photographs were taken at low tide whereas the satellite image was recorded at half tide when major parts of the flats were still submerged. Only the positions of the steep edges of bars and channels can therefore be regarded as equivalent; these forms only are shown in fig. 75.

Despite these problems, the satellite image can

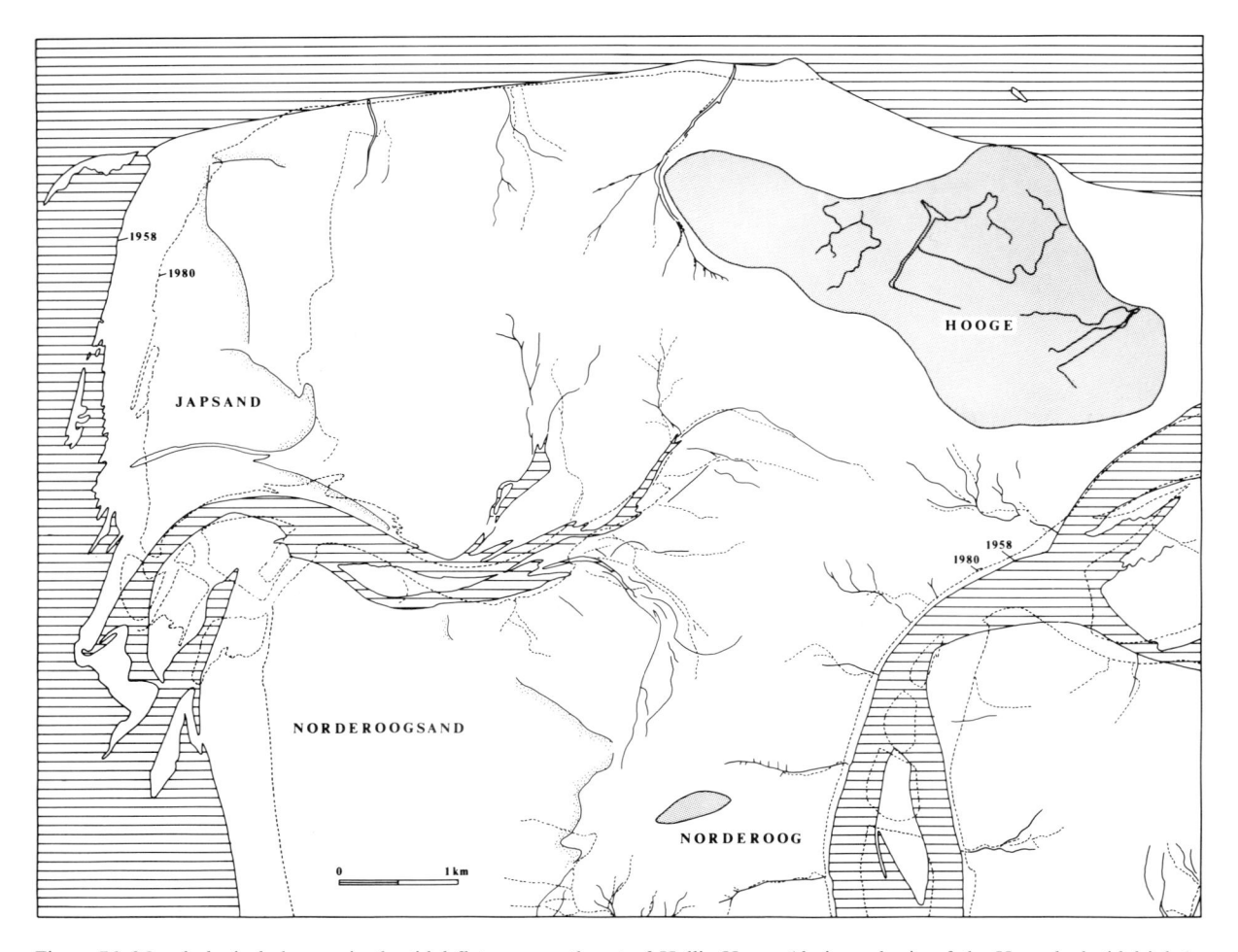

Figure 76: Morphological changes in the tidal flat area southwest of Hallig Hooge (drainage basin of the Hoogeloch tidal inlet) between 1958 and 1980. Sources: König, 1972, fig.18; Aerial photographs of the Landesvermessungsamt Schleswig-Holstein of 1980; Küstenkarte 1:25,000, Sheet 1416 K Hooge.

be used to identify the major changes of the Harle Inlet morphology:

1. During the last ten years the Hullplate bar has migrated 600 m towards the SE (1). It built a spur of about the same length that extended eastward over the Hoher Rücken flat. However, the enclosing boundary of the Hoher Rücken remained essentially in the same position.

2. Over the same period the Muschelbank bar north of the Hoher Rücken has migrated about 200 m towards the SE.

3. The bar SW of the Harlehörn peninsula (3), which in 1974 was part of the flood delta of the Harle, now forms part of the ebb delta of the Telegraphenbalje.

4. The Breite Legde (4) has merged with the adjoining tidal creek to the north forming a single main creek; the Eversand bar has disappeared. In the process the mainland flats (Festlandswatt) north of Harlesiel have extended over 500 m towards the NE.

5. One branch of the Carolinensieler Balje creek (5), the creeks west and east of the Langer Jan bar (6) and the creeks at the western margin of the Südersand flat (7) have moved markedly towards the east.

All these changes suggest a recent easterly extension of the Harle drainage basin. The adjoining drainage basin of the Blaue Balje inlet appears to have been restricted by sand replenishment south of Minsener Oog (fig. 63 a).

Example 2: Tidal flats southwest of Hooge

Despite changes in the length of tidal creeks in response to tidal inlet drainage area changes, the

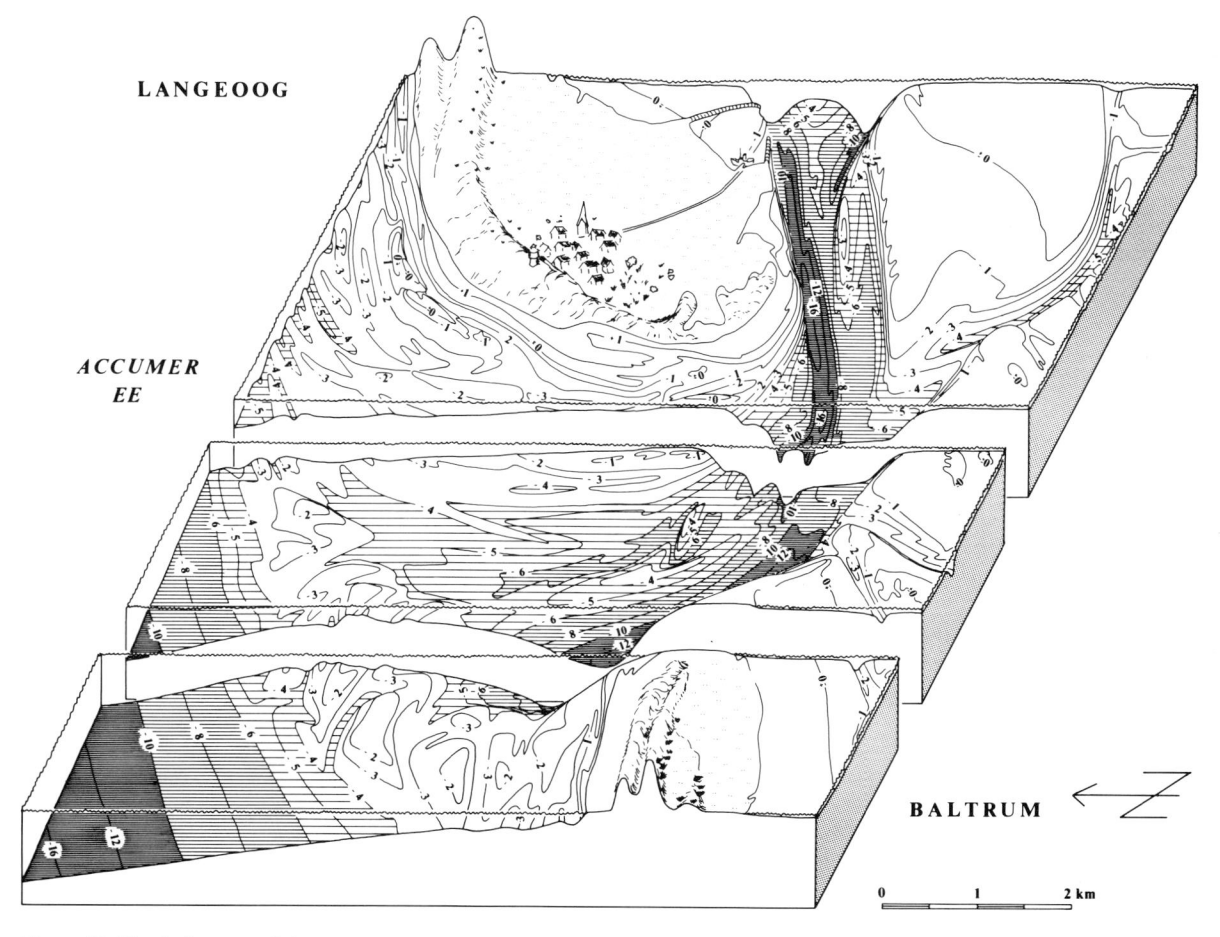

Figure 77: Block diagram of the Accumer Ee tidal inlet between Baltrum and Langeoog; illustrating the strongly overdeepened tidal channels and high lying ebb-delta (Sources: Küstenkarte 1:25,000, Sheet 2210 K Langeoog and Sheet 2310 K Accumersiel).

general configuration of the landforms remained the same. Fig. 76 shows the changes which have occurred in the tidal flat area southwest of Hallig Hooge. The 1958 reconstruction is based on aerial photographs published by König (1972), whilst the 1980 reconstruction is based on aerial photographs of the Landesvermessungsamt Schleswig-Holstein.

The drainage area of the Hoogeloch Inlet has decreased considerably as a result of the landward migration of the outer sands of Japsand and Norderoogsand (Taubert, 1982). The changes in the tidal creek pattern, however, are minimal. After 22 years, the position of even the smallest tidal creeks remains unchanged.

Measurements taken in the Wangerooge tidal flat area (Ehlers & Mensching, 1982) have demonstrated that minor creeks alter their position by a few decimetres between tidal cycles but the changes are reversible. Over long time periods, the tidal creek positions displayed no significant shifts.

Consequences

Small landforms on the tidal flats (sand waves and bars) undergo rapid replacement and reshaping; in some cases these may even occur within one tidal cycle. The larger landforms (flats and major tidal creeks) undergo only minor alterations. Thus the general drainage pattern can be maintained for many decades.

Major changes only occur where the size of the drainage area is altered or where the courses of tidal streams are altered for other reasons, such as through human interference. Most morphodynamic changes are controlled by the behaviour of the tidal inlets (see section 5.2).

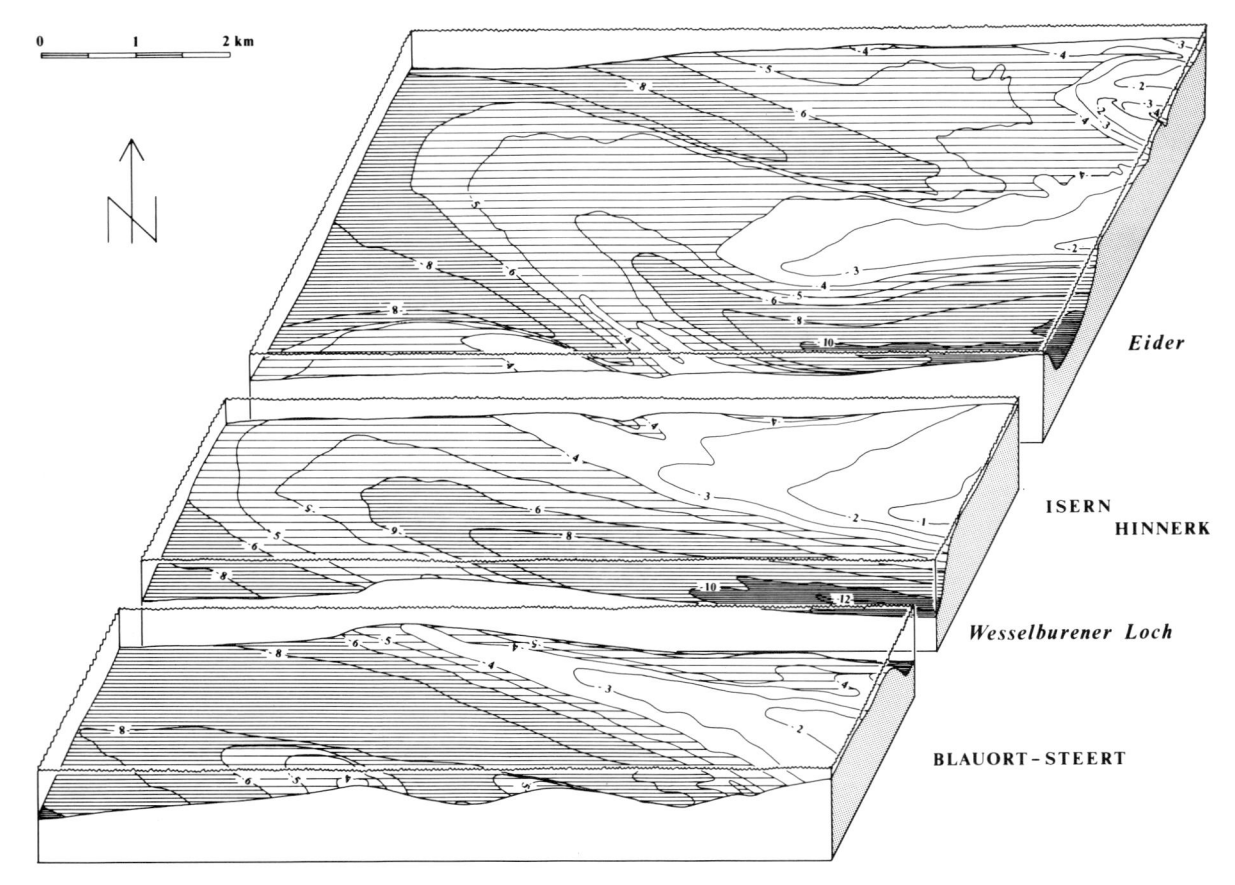

Figure 78: Block diagram of the mouths of the Eider and Wesselburener Loch tidal streams south of Eiderstedt; instead of the intertidal ebb-delta shoal arc, permanently water-covered low bars have developed (Sources: Küstenkarte 1:25,000, Sheet 1716 K Außeneider and Sheet 1816 K Norder- und Süderpiep).

5.2 TIDAL INLETS

5.2.1 SUBDIVISION OF THE TIDAL INLETS

Tidal inlets between the barrier islands are influenced twice daily, both by flood and by ebb currents. The large water volumes that are forced through these inlets during short periods cause a strong current to develop, the so-called 'tidal jet', which has considerable erosive power. The resulting, overdeepened channel segments occasionally are eroded to depths much greater than the adjoining sea floor (fig. 77).

Shoal arcs, linking adjoining islands, are found at the mouths of the tidal inlets. Comparison of aerial photographs taken at different times reveals that the shoal arcs are not stable features but undergo rapid changes (cf. section 5.2.3).

The tidal flats between Jade Bay and Eider Estuary, which are not barred from the sea by major islands, show a slightly different arrangement. The overdeepened tidal channels are also found here (e.g. Wesselburener Loch and Eider in fig. 78), but, instead of the limiting shoal arc at the seaward ends, only relatively low-lying bars occur. These bars also undergo rapid modification (cf. section 5.2.4).

Ebb delta

The discharge of a tidal current into the sea is morphodynamically comparable to the discharge of a major river. As a result of its high velocity the ebb current transports large volumes of sediment out through the tidal inlet. At the inlet mouth this sediment load is quickly deposited as the current velocity rapidly decreases. The sand accumulation created in this way is called the 'tidal delta' (van Veen, 1950). As the tidal delta is formed by the ebb current it can also be called the 'ebb delta' (fig. 79).

The ebb delta of a tidal inlet is not only shaped

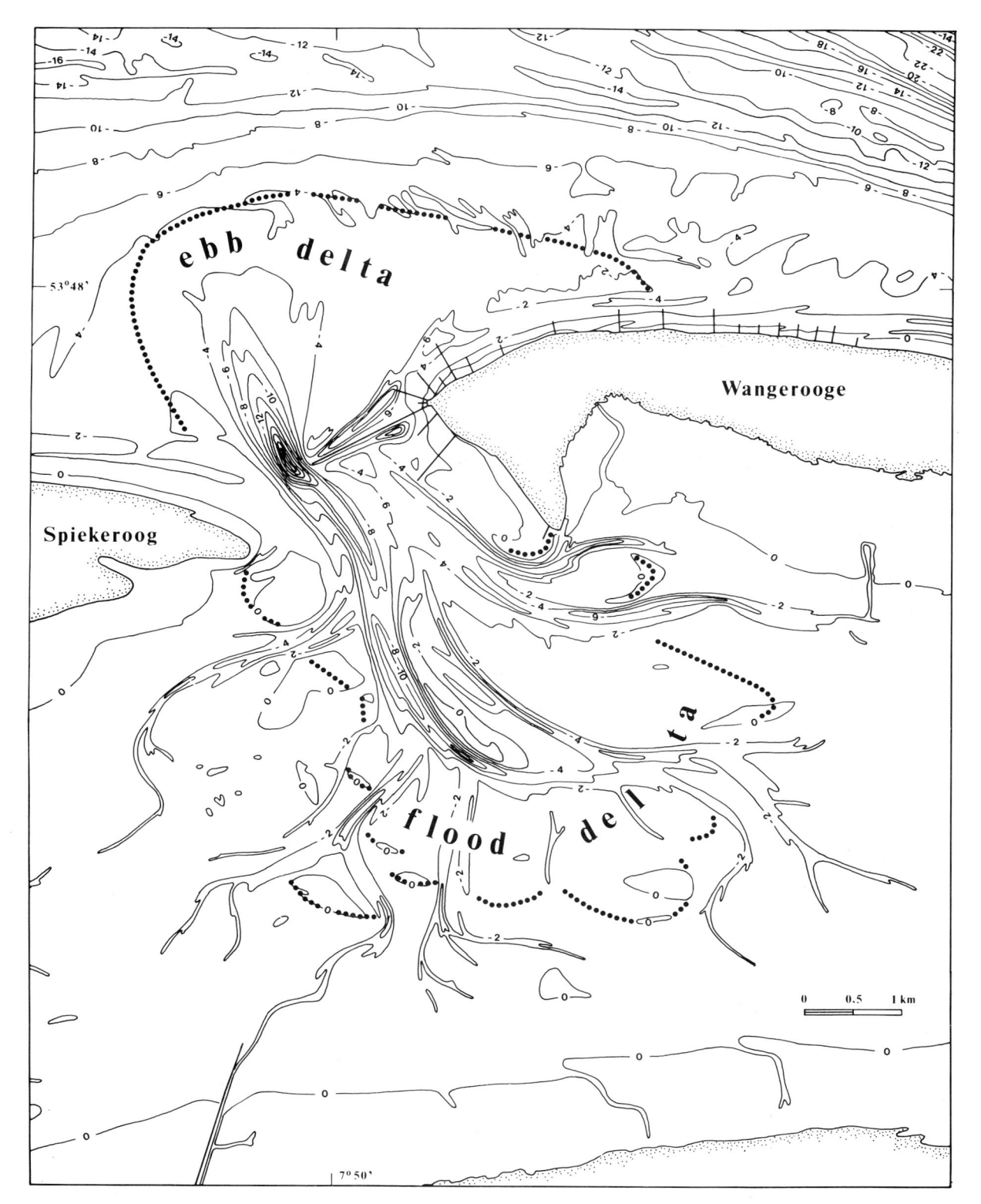

Figure 79: The ebb delta and flood delta of the Harle tidal inlet (From: Ehlers & Mensching, 1983).

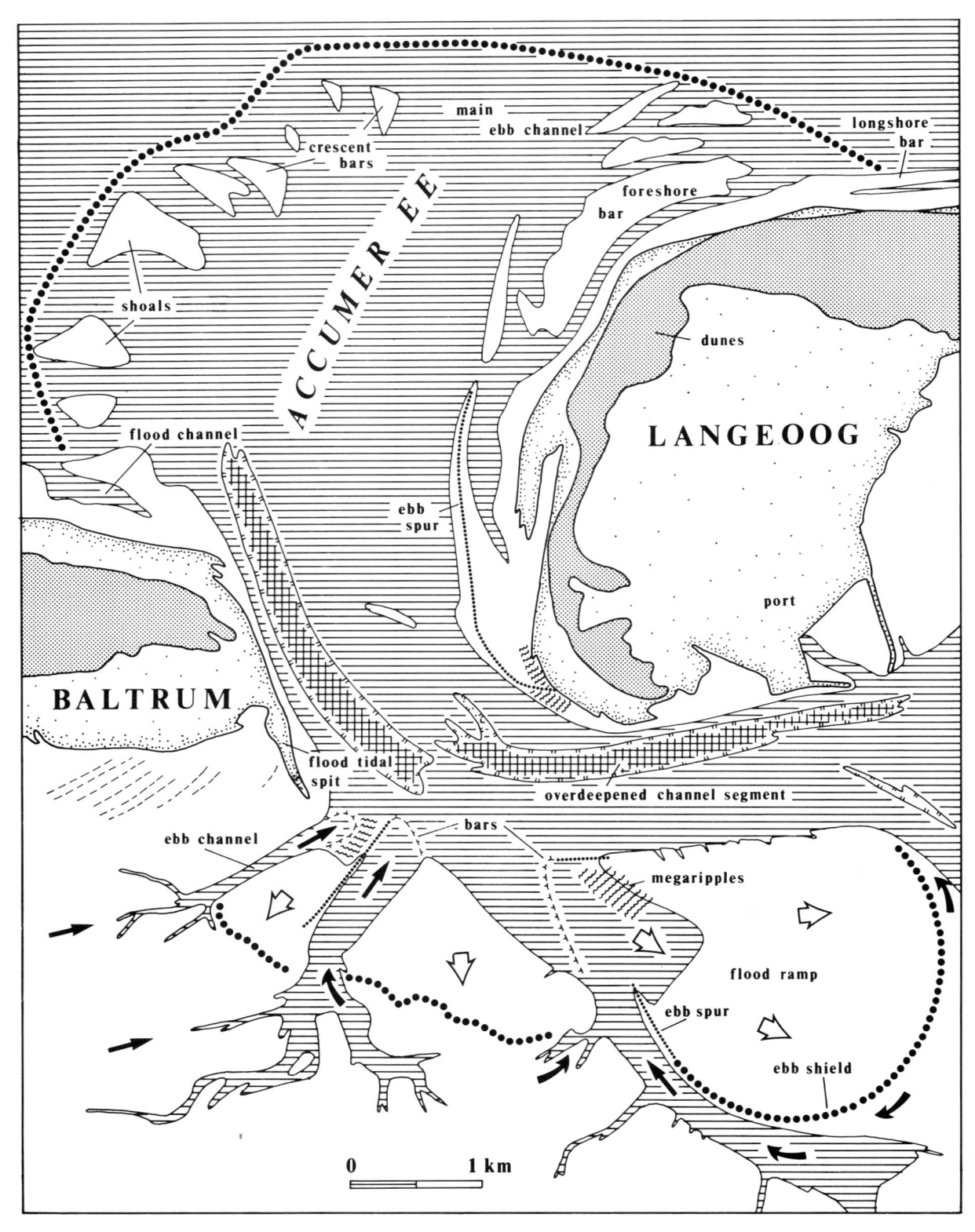

Figure 80: Landforms of a tidal inlet exemplified by the Accumer Ee between Baltrum and Langeoog. The arrows indicate current directions in the flood delta area during the rising tide. Open arrows: flood current, dark arrows: flow direction of remaining water draining from the higher tidal flats. (Sources: Küstenkarte 1:25,000, Sheet 2210 K Langeoog and Sheet 2310 K Accumersiel; aerial photographs of the Niedersächsisches Landesverwaltungsamt, Landesvermessung, Bildflug Ostfriesische Küste (419) vom 29.5.1966).

by the ebb current but also by marine processes and sediment migration in the foreshore area. The extent to which the ebb current can prevail against the coast-parallel tidal current depends on the tidal range and the size of the drainage basin. Where the ebb current dominates a large ebb delta is formed, extending far out to sea; where surf and coast-parallel currents dominate only a small ebb delta can be sustained (Boothroyd, 1978).

A tidal inlet is composed of a number of different form elements (fig. 80). Within the ebb delta there is normally one main ebb channel, usually leading directly out to sea. Its direction is not necessarily normal to the coast; in the case of the East Frisian Islands it is often displaced eastwards by the coast-parallel component of the sand transport. The actual position of the main ebb channel may be of critical importance for the sediment budget of the adjoining island margins. The shoal arc of the ebb delta is found at the mouth of the inlet.

Large groups of swash bars become detached from the island ends at the updrift end of the shoal arc. Major marginal flood channels are often found at this point as well as at the downdrift end of the arc (cf. Hayes, 1975: 13; Boothroyd, 1978: 340). The shoals are smaller in the seaward parts of the arc. Small isolated crescent-shaped bars are often found here; these are commonly referred to as crescent bars. Downdrift of the main ebb channel largely elongated swash bars are found which slowly move towards the beach of the adjoining island.

The landform distribution is a product of the interacting forces of the ebb and flood currents. During flood tide, converging currents approach the inlet (sink flow); but during the falling tide, the ebb current, formerly concentrated in the inlet trunk, shoots out of the inlet (jet flow). The ebb current therefore has considerably more erosive power. Only a low lying shield can be formed at the main ebb-channel mouth instead of swash bars (Özsoy, 1977: 148).

Flood delta

The flood current also creates a tidal delta. As a result of the relatively high lying tidal flats of the Wadden Sea area the tidal inlet flood deltas are not prominent landforms. On aerial photographs, however, the brighter sandy shoals of the flood deltas can often be easily distinguished from the darker silty tidal flats. Moreover, the flood deltas have a specific form assemblage which helps to

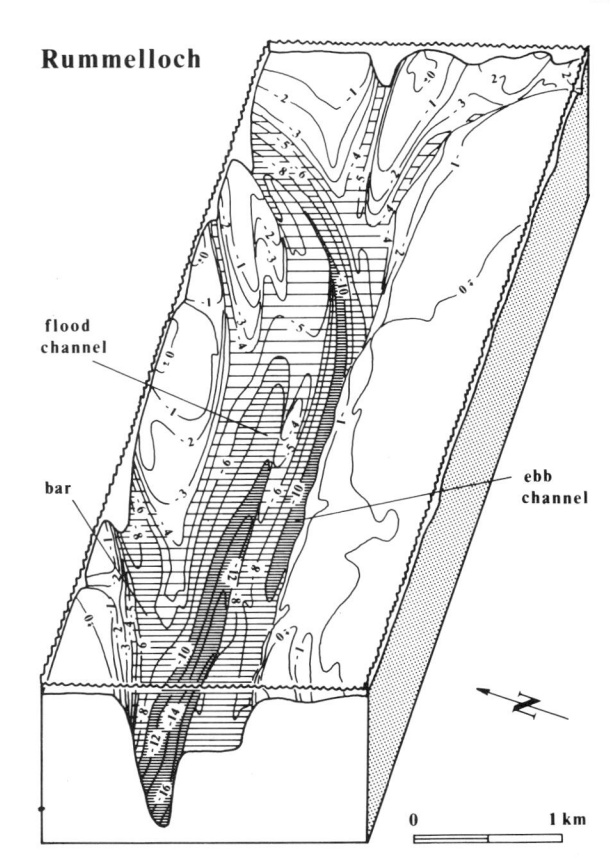

Figure 81: Ebb and flood channels in the Rummelloch area north of Pellworm. The channels are separated from each other by protective bars (Sources: Küstenkarte 1:25,000, Sheet 1416 K Hooge and Sheet 1516 K Süderoog).

distinguish them from the surrounding areas.

The morphodynamics of a flood delta differ considerably from those of an ebb delta. The influence of the surf, for instance, is much less. Since most parts of the flood deltas are emergent during low tide they provide an excellent opportunity to study the coexistence of ebb- and flood-orientated landforms.

Krümmel (1889: 138) distinguished between ebb- and flood-channels in the major inlets and estuaries as early as 1889. He pointed out that they were always shielded by inner and outer bars ('Binnenbarren' and 'Außenbarren'). Walther (1934: 151) noted that the deeper channels of the inlets were dominated by the ebb currents, whilst in the shallower channels a landward directed residual current was found.

Van Veen (1950) discovered that some of the channels and creeks in the tidal flat areas are dominantly controlled by the ebb current, whilst

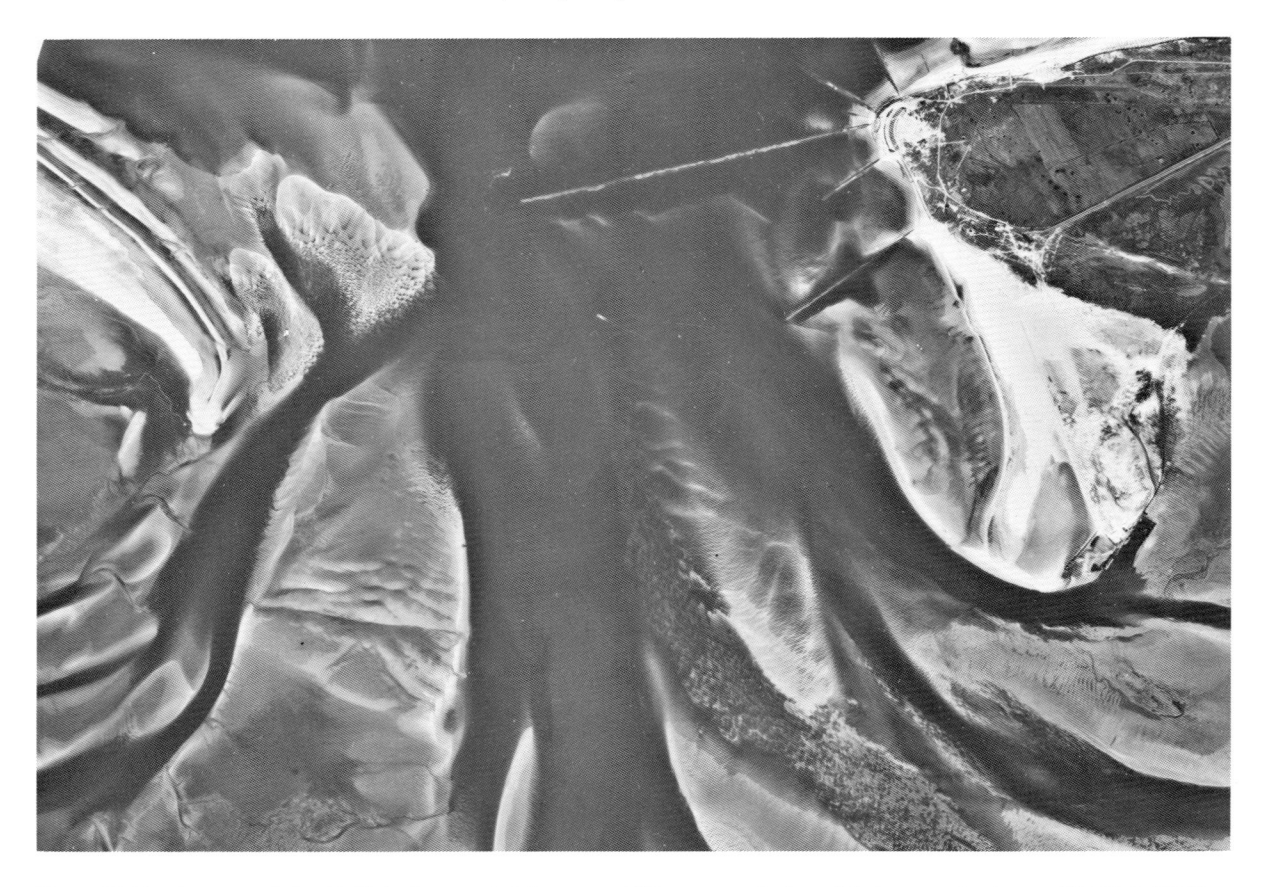

Figure 82: Development of the tidal inlet between Spiekeroog and Wangerooge. Left: aerial photograph taken on 27.4.1964 (Bildflug Wattaufnahme Spiekeroog, Wangerooge (225), Str.1/l98, freigegeben durch Luftamt Hamburg unter Nr. 700 418), right: aerial photograph taken on 17.4.1974 (Bildflug Ostfriesische Inseln (1086), Str.3/069, freigegeben durch NLVA,– Landesvermessung – , unter Nr. 34/74/1086).

others are controlled by the flood current (fig. 81). This is because even after the onset of the flood large quantities of water continue to run off from the higher tidal flats (cf. section 4.1). The channels through which this water drains at high velocity are consequently blocked to the rising tide and the flood current is forced to find alternative routeways.

The major channels predominantly are ebb channels. These are usually narrow and deep, and their mouths are blocked by seaward facing shallow bars or 'shield deltas' (van Veen, 1950). The channels used by the flood current are broader and open towards the sea. Their course is normally straighter than the ebb channels (Jakobsen, 1962, 1964).

The flood current creates broad flood ramps on the flood delta (fig. 80), which are protected against the ebb current by an ebb shield (cf. Hayes, 1975: 17). The flood ramps are often covered with megaripple fields. Elongated, levée-like ebb spurs, often several hundreds of metres in length, are found at the margins (fig. 80).

The areas which remain submerged at low tide are called subtidal, whereas those parts of the Wadden Sea which fall dry during the ebb are referred to as intertidal. Only the intertidal zone is readily accessible to direct field studies. Under favourable conditions aerial photographs permit a limited insight into the subtidal area.

Aerial photographs taken at low tide can only reveal the controlling processes to a certain extent. The last controlling process is always related to the ebb current, so the aerial photographs, in most cases, only reveal ebb-orientated megaripples. This has occasionally led to the assumption that the features controlled by the flood current are largely restricted to the tidal channel floors. This is clearly not the case, as will be demonstrated below (section 5.2.2).

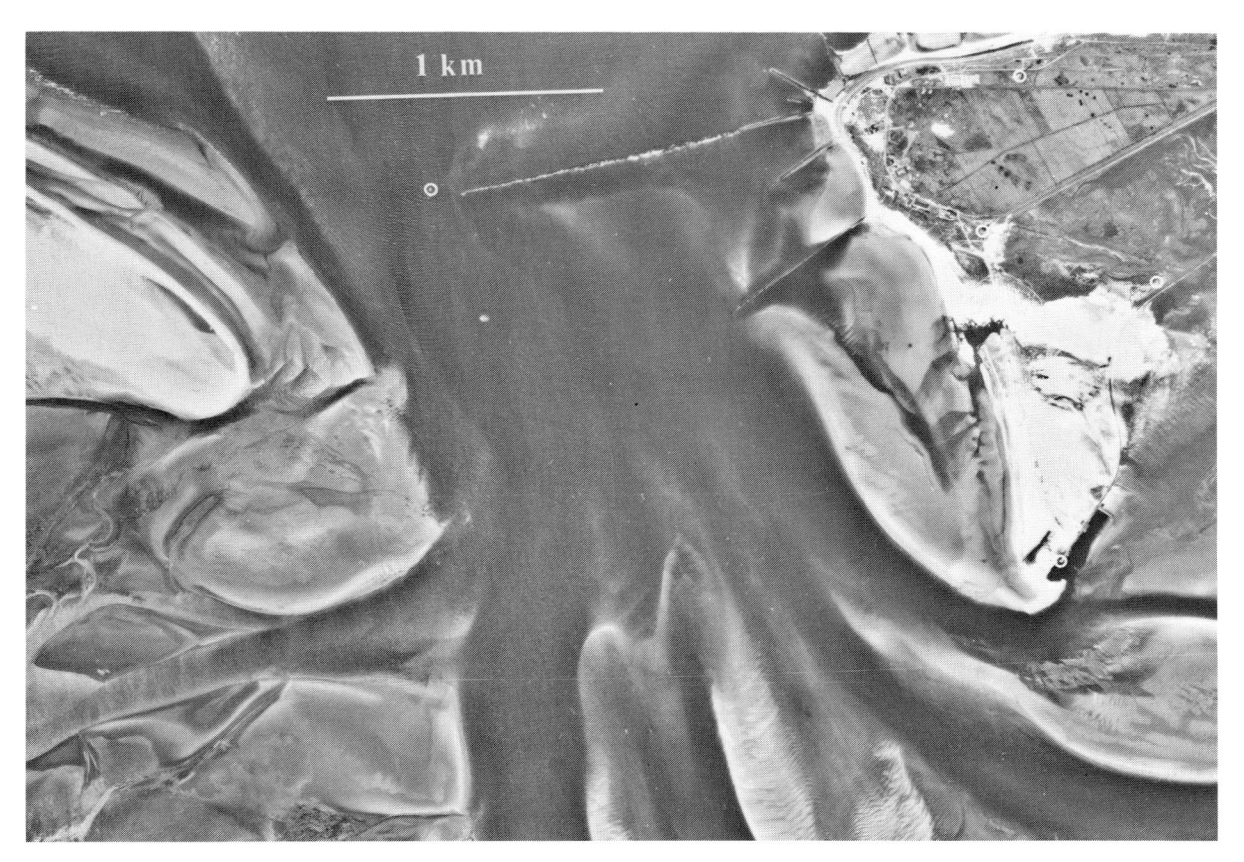

Figure 82 (cont.).

Inlet migration

The tidal inlets are morphodynamically the most active elements of the Wadden Sea. Their shift leads to a phenomenon which has been described in the popular literature as 'island migration'. The extent of inlet migration is very limited today as a result of coastal engineering countermeasures (Luck, 1975). Underwater groynes, which extend to over 1 km from the threatened island margins into the inlet, in addition to underwater revetments ensure that the deep channels are maintained some distance away from the island coast (cf. section 5.3.7).

Tidal inlet migration generally coincides with the direction of the littoral drift. However, there are exceptions. The Vlie Inlet between Vlieland and Terschelling migrates updrift towards Vlieland. The broad western beach of Terschelling (Noordvaarder) is the result of long term sand accumulation at that end of the island. Finley (1978) has pointed out that a surf-induced reversal of the sand transport direction can occur within the ebb delta which may ultimately lead to a change in inlet migration. Aubrey & Speer (1984) discuss several situations leading to updrift inlet migration.

The position of the main ebb and flood channels within an inlet is of prime importance in relation to short term erosion or deposition rates on either side of the inlet. This can be seen at the northern end of Texel, where accumulative phases alternate with periods of strong erosion depending on the position of the tidal channels within the Robbengat. It can also be seen at the western end of Langeoog, which is at present not threatened by the Accumer Ee (figs. 77 and 80).

Not only do the tidal inlets tend to migrate in the direction of the coast-parallel sand transport, but the watersheds between the drainage basins of the inlets are also preferentially shifted in the same direction. Exceptions to this rule are only found in cases such as the Wichter Ee between Norderney and Baltrum, where the inlet is too small to maintain its drainage area; in such cases it may be tapped from both sides.

Changes in the position of the drainage divide may cause a breach and the formation of a new water divide to the west. Old maps show that this

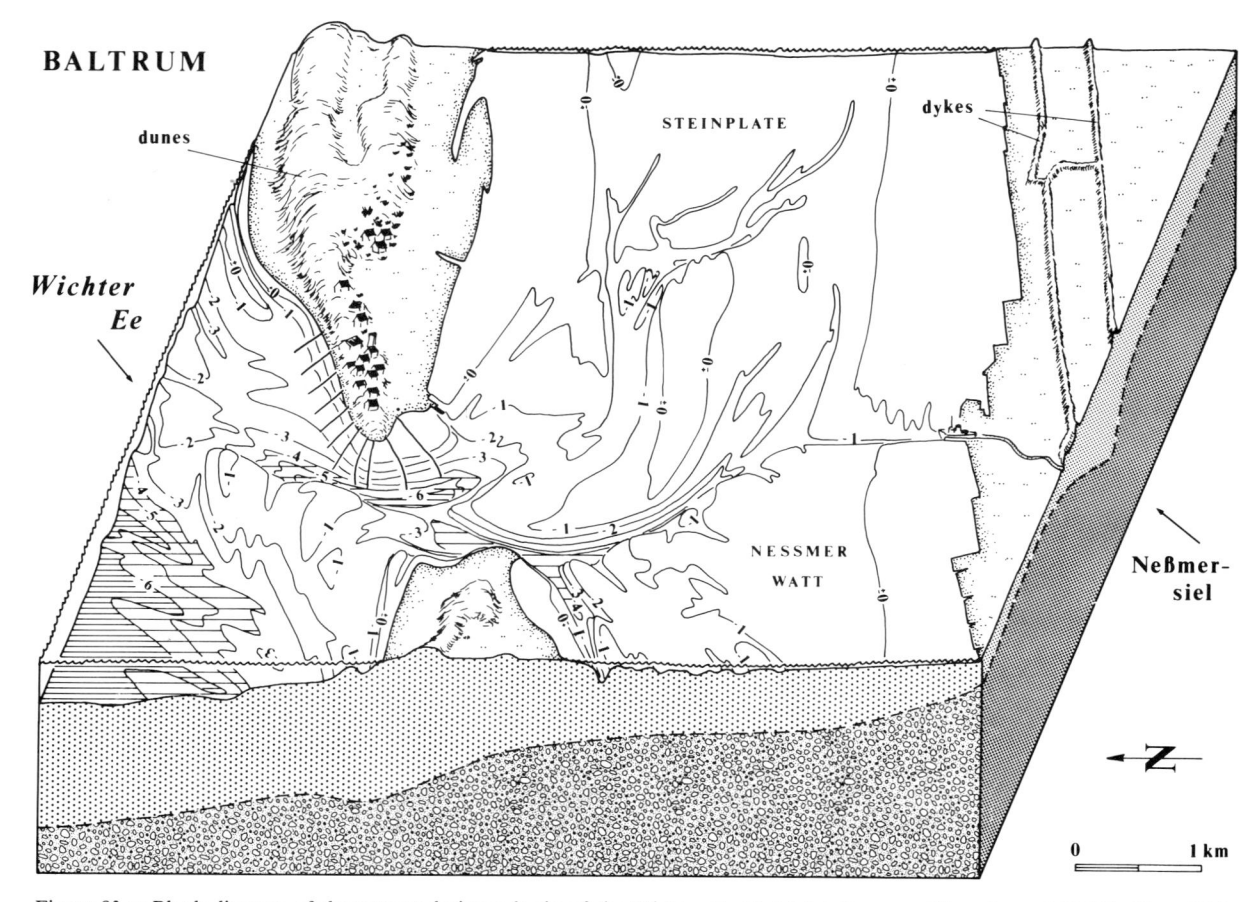

Figure 83 a: Block diagram of the present drainage basin of the Wichter Ee tidal inlet (Sources: Küstenkarte 1:25,000, Sheet 2210 K Langeoog and Sheet 2310 K Accumersiel).

seems to have occurred in the tidal flats off Schiermonnikoog. Formerly, the water divide ran from the island towards Friesland and the Lauwerszee drained through the Lauwers Inlet east of Schiermonnikoog. However, since the 14th century the water divide has extended from the island towards Groningen and the Lauwerszee has drained through the Zoutkamperlaag Inlet west of Schiermonnikoog (Haan, Ijbema & Reitsma, 1983: 10).

In some localities, for example between Juist and Norderney, there are two tidal inlets between two barrier islands instead of just one. In the case of Juist and Norderney this is due to the destruction of the Isle of Buise between Juist and Norderney in the 18th century.

Two tidal inlets formerly existed between Spiekeroog and Wangerooge; the Harle in the west and the Rote Balje in the east. They probably owed their existence to the large drainage basin, created by the formation of the Harle Bight, which reached its largest extension after the catastrophic flood of 1362. During the following centuries the drainage basin decreased in size through redyking. From 1362 to 1978 145.6 km² of land was reclaimed in the former Harle Bight area. At the same time the Otzumer Balje, the inlet west of Spiekeroog, extended its drainage basin further to the east, leading to a decline of the old Harle inlet. Later, the former Rote Balje was re-named Harle, and the old Harle acquired the name 'Alte Harle'. In 1963 it only constituted a type of micro-inlet west of the Harle (fig. 82) and by 1974 it had completely disappeared.

Primary tidal inlets

Pre-existing relief played a decisive role in the formation of tidal inlets as the sea transgressed during the Holocene. The present tidal inlets, therefore, largely follow the courses of buried channels in the pre-Holocene relief (cf. Jelgersma, 1983) which had already been active during the Eemian.

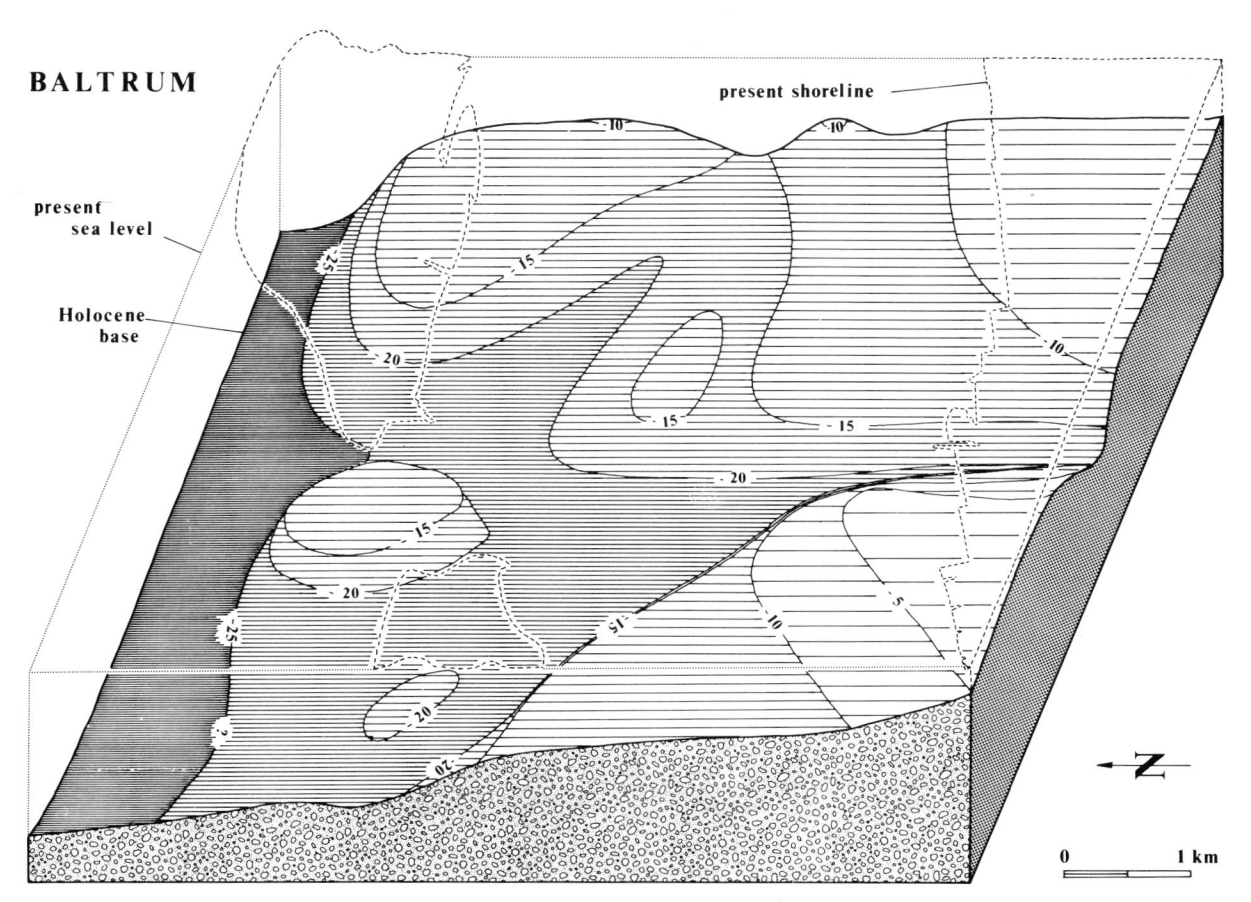

Figure 83 b: Block diagram of the drainage basin of the Wichter Ee at the beginning of the Holocene transgression (Source: Barckhausen, 1969: Plate III).

Fig. 83 a shows the present drainage basin of the Wichter Ee Inlet between Norderney and Baltrum, and fig. 83 b shows the situation at the beginning of the Holocene transgression (adapted from Barckhausen, 1969: plate III). The N-S trending channel at Neßmersiel is a former river valley which had existed during the Eemian. The original formation of the tidal inlets followed the trends of such valleys.

Major morphological changes have occurred in the area between Juist and Baltrum during the course of the Holocene, and it may therefore be accidental that the present Wichter Ee lies almost superimposed on its early Holocene predecessor. East of Baltrum the positions of the islands remained more or less stable. In this area geological mapping has revealed the extent of inlet migration during the course of the Holocene:

1. Continuous old tidal flat deposits under central Langeoog indicate that this area was never overrun by a tidal channel. Since its formation, the Accumer Ee tidal inlet has only migrated about 4.5 km to the east (Barckhausen, 1969: 257; 1970: 25).

2. The predecessor of the Otzumer Balje between Langeoog and Spiekeroog cannot be located with the same accuracy (Barckhausen, 1970: 25). It is most likely to have been situated under the eastern part of Langeoog. If the position given by Barckhausen (1970: plate 1) is correct, the inlet has migrated about 4 km to the east during the Holocene.

3. An Eemian river valley lies under Spiekeroog. At the beginning of the Holocene transgression this valley formed the predecessor of the present Harle Inlet (Sindowski, 1970: 29), which is now situated about 6 km to the east.

4. The old Jade channel was situated under Wangerooge (Sindowski, 1969: 17), the position of the present Jade is now about 6 km to the east.

Despite the lateral displacement of a number of these inlets, the majority have remained relatively stable. Some have enlarged their drainage basin

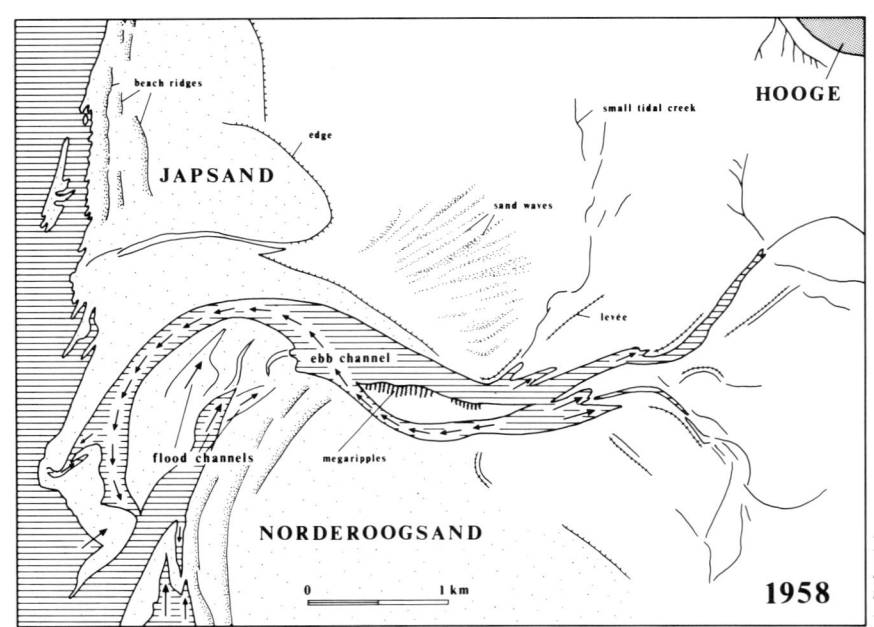

Figure 84 a: Landforms in the Hoogeloch area between Norderoogsand and Japsand, 1958 (Source: aerial photograph in König, 1972).

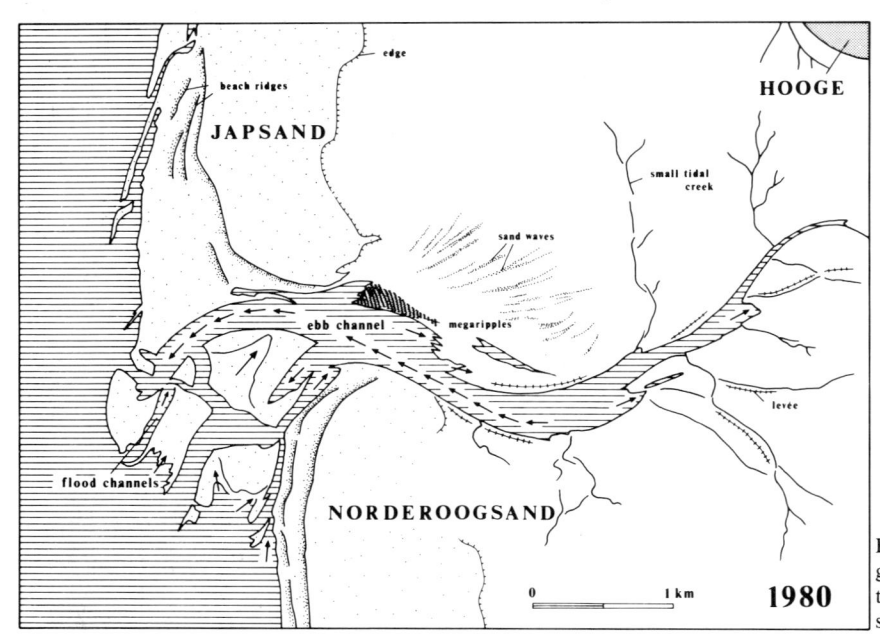

Figure 84 b: Landforms in the Hoogeloch area, 1980 (from aerial photographs of the Landesvermessungsamt Schleswig-Holstein).

through time and have consequently increased in size, e.g. the Harle. Others, such as the Wichter Ee, have lost parts of their drainage areas as a result of tide water diversion by adjacent inlets. The landform assemblage tends to remain relatively constant. The channel cross sections are controlled by the size of the drainage areas (Rodloff, 1970: 27).

Figs. 84 a and b show the changes between 1958-80 in the Hoogeloch area between Norderoogsand and Japsand. During this time, the drainage basin of the inlet was considerably reduced by the eastward migration of the Outer Sands. Nevertheless, the pattern of shoals and creeks and the configuration of bars at the inlet mouth, as well as the position of megaripple fields and sand waves have remained almost constant.

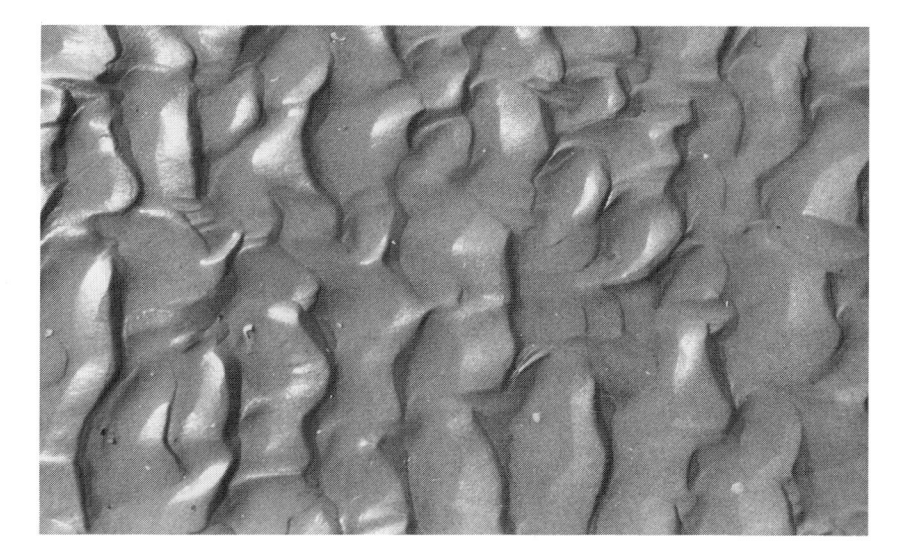

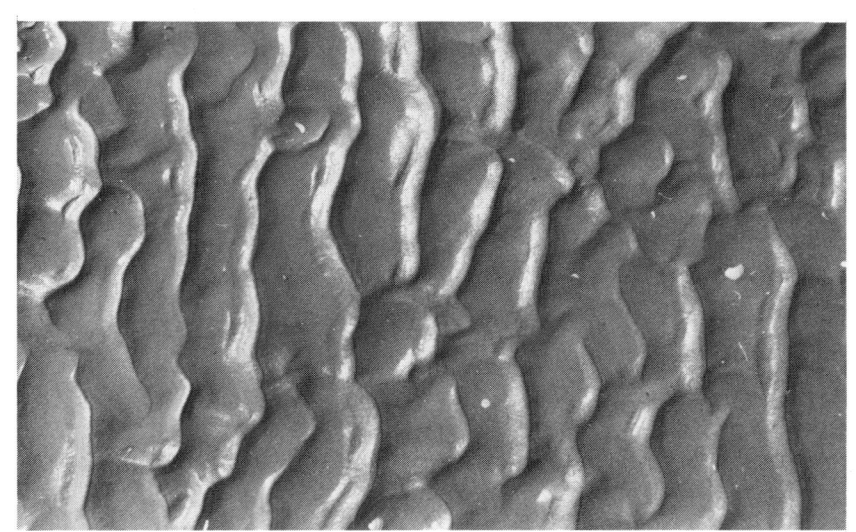

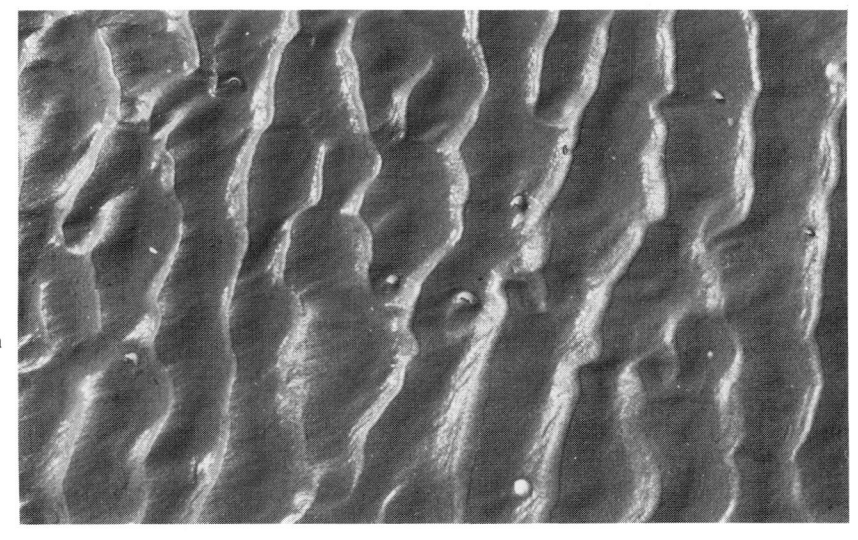

Figure 85: Small ripples in the ebb delta of the Slufter inlet on Texel, looking towards the south, direction of the ebb current: to the right. a: ebb-orientated small ripples; b: flood-orientated small ripples with strongly developed ebb crests; c: flood-orientated small ripples. Distances between the three photographs: 2 m (1985).

Figure 86 a: Pair of aerial photographs from the Wichter Ee tidal inlet, taken on 2.8. (left) and 10.9.1982 (right); ebb delta, western part. Aerial photographs of Fa. N. Rüpke, Hamburg, Bildflug Wichter Ee, freigegeben durch Luftamt Hamburg unter Nr. 812/82.

Secondary tidal inlets

Apart from the major, primary tidal inlets which were formed at the beginning of the Holocene transgression in the low-lying Pleistocene subsurface areas, a number of usually smaller, secondary inlets are found which were formed later. Secondary inlets may develop where the island dune barrier is breached during storm surges, and where the resulting washover processes are so intense that a channel may form that is sufficiently deep to initiate the formation of a tidal inlet. If this occurs the channel may be expanded and deepened during successive storm surges so that a new inlet becomes established.

In most cases these secondary inlets are rather small. As all inlets require a specific drainage basin the number of potential inlets along the North Sea coast is limited. Any new tidal inlet has to tap the drainage areas of existing inlets. This has suc-

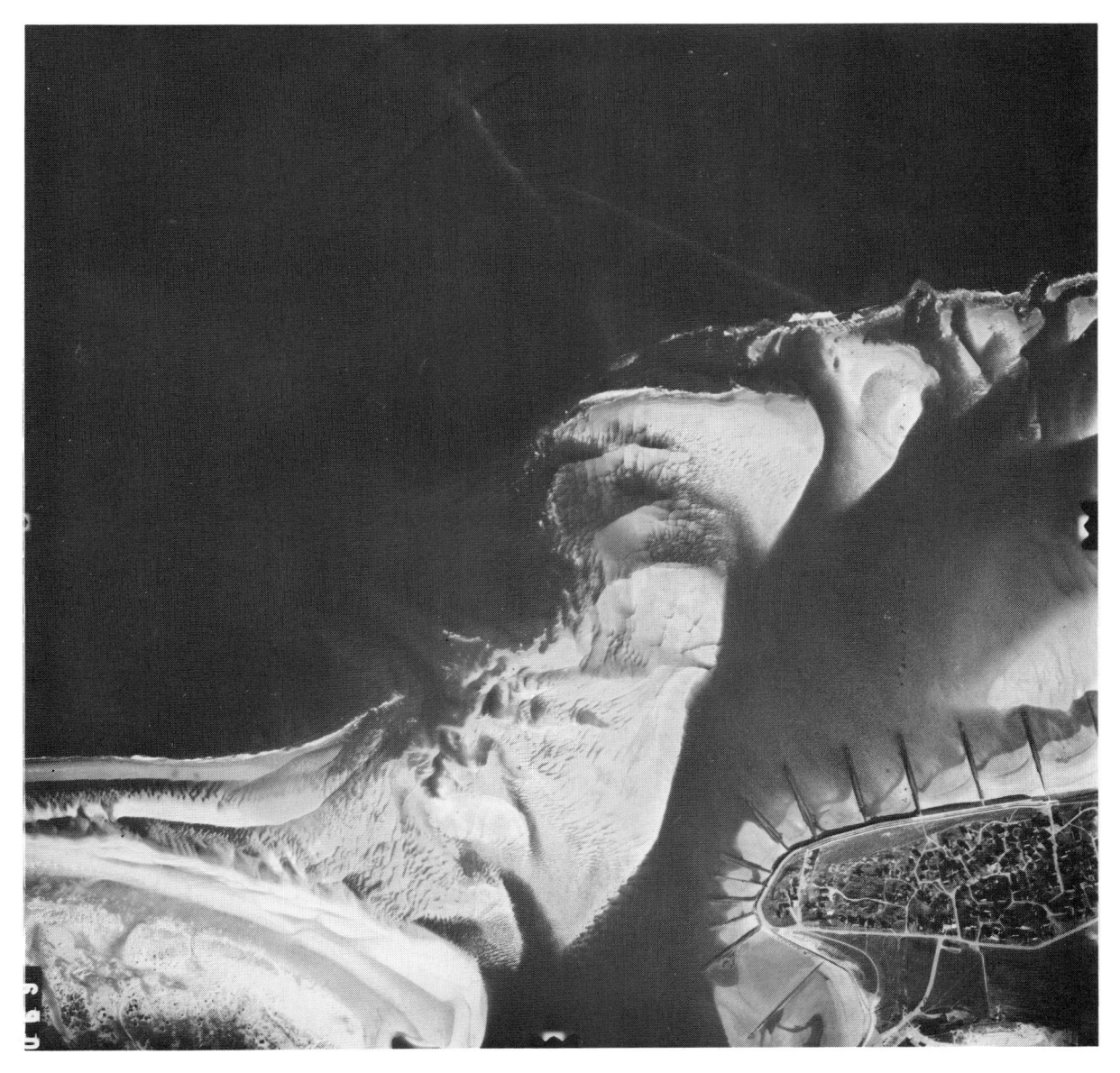

Figure 86a (cont.).

ceeded only in one instance, leading to the formation of the Marsdiep Inlet between Texel and Den Helder. The drainage basin of the Eijerlandse Gat between Texel and Eijerland was consequently so reduced that it ultimately disappeared.

In most areas, the present potential for tidal inlet formation is minimal because of dune- and coastal protection projects. New inlets may only form through the unprotected island ends. A small secondary tidal inlet existed from about 1966 until 1977 at the western end of Juist, but had disappeared by 1980.

In small tidal inlets such as the Slufter on Texel, current velocities and water depth are too low to allow the formation of megaripples. In such cases the ebb delta will be completely covered by fields of small ripples. As in the large tidal inlets, certain areas of the deltas are dominated by the ebb, and others by the flood current. Although normally all ripples are ebb-orientated at low tide, during periods of calm weather in the flood-dominated areas, the flood-orientated ripples may outlast the ebb. This was observed at the mouth of the Slufter on Texel in spring 1985 (fig. 85).

Fig. 85 a shows lingoid small ripples formed by the ebb current in close proximity to an ebb chan-

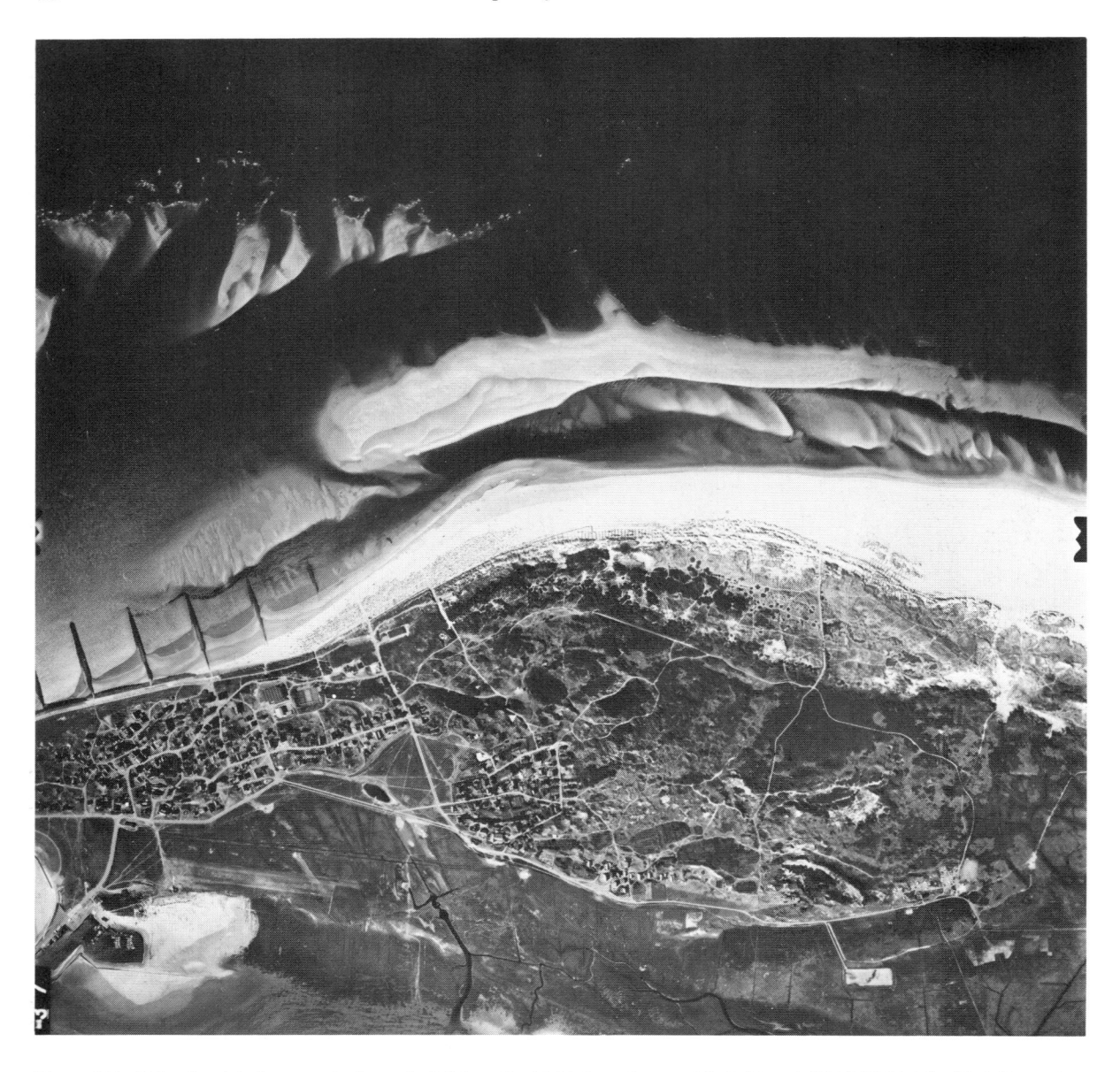

Figure 86 b: Pair of aerial photographs from the Wichter Ee tidal inlet, taken on 2.8. (left) and 10.9.1982 (right); ebb delta, eastern part. Aerial photographs of Fa. N. Rüpke, Hamburg, Bildflug Wichter Ee, freigegeben durch Luftamt Hamburg unter Nr. 812/82.

nel. About 2 m further landward the ebb current was no longer strong enough to completely rebuild the forms created by the flood current. Here only the ripple crests became re-orientated (fig. 85 b). Another 2 m further landward, at the margin of the ebb delta, well preserved flood-orientated small ripples were found (fig. 85 c) with only weak indications of crest re-orientation.

5.2.2 Migration of shoals

Much coastal research in recent years has focused on shoal and bar migration on ebb tidal deltas. Krüger (1911) was the first to observe the regular migration of the ebb delta shoals eastwards along the East Frisian coast. Gaye (1934: 297) and in particular Gaye & Walther (1935) were able to demonstrate the shoal migration pattern in the East Frisian tidal inlets. Homeier & Kramer (1957) thoroughly studied the form elements of an ebb delta (Norderneyer Seegat) and were able to distinguish between bedforms of three different orders of magnitude:

1. Shoal subunit (Plate); the smallest morpho-

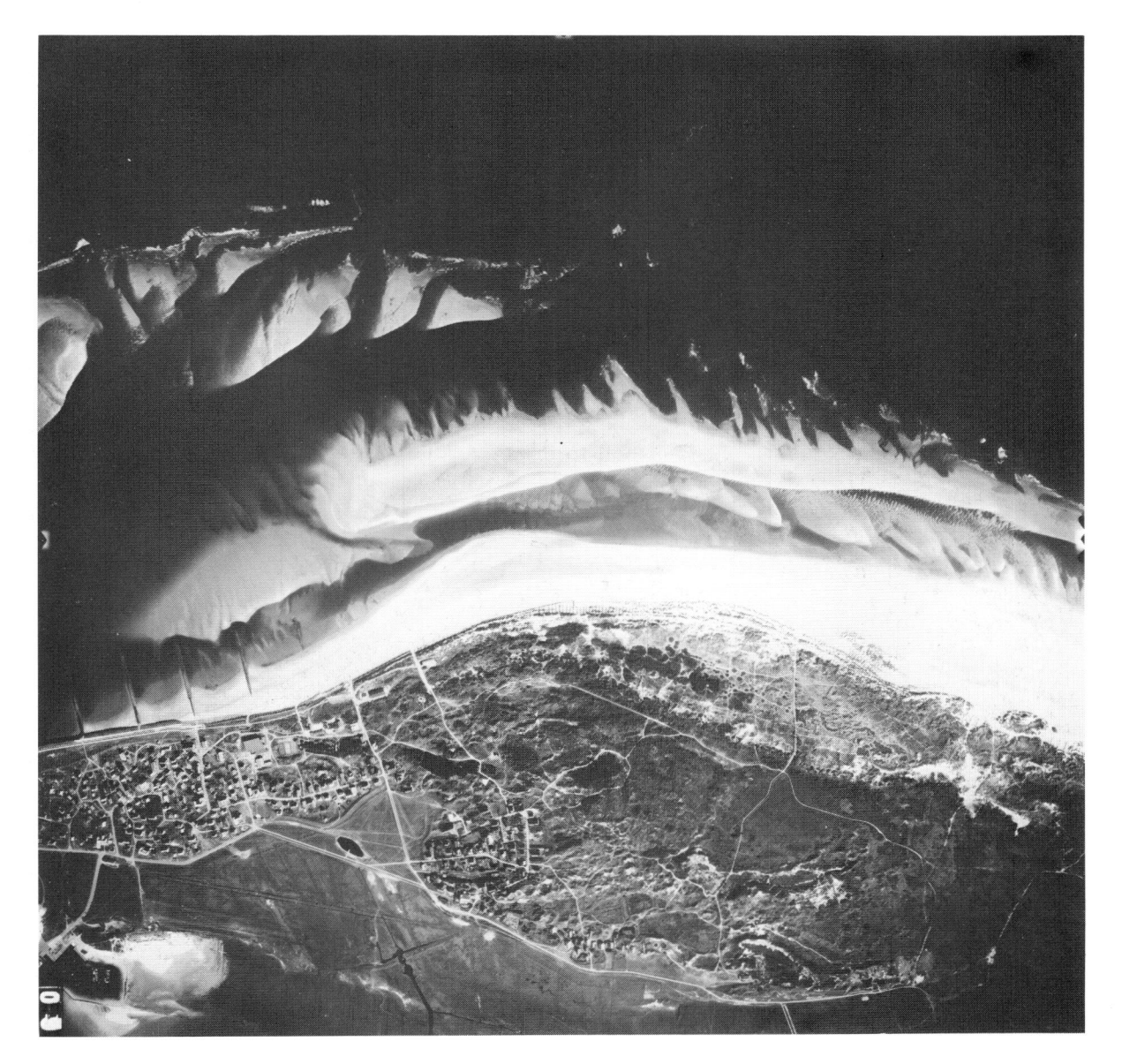

Figure 86b (cont.).

logical unit within the ebb delta, having a size of 200 to 300 m × 500 to 600 m with a small distance to the adjoining shoal subunit.

2. Shoal (Platengruppe); several shoal subunits, spaced closely together and forming a major morphological unit, separated from the ajoining shoal by deeper and broader channels.

3. Shoal arc (Riffbogen); all shoals of the ebb delta between two islands.

Through the examination of manuscript charts dating from 1926-57 Homeier & Kramer discovered that shoals in the East Frisian Islands migrate along the shoal arcs in an easterly direction. The migration rate measured was greatest in the apex of the arc and smallest where the shoals approached the beach of the downdrift island. The average migration rate measured was 405 m/year (±20 %) (Homeier & Kramer, 1957).

The 'Forschungsstelle Norderney' (Norderney Research Station) studied bar migration processes of several inlets, including the Accumer Ee ebb delta (Homeier & Luck, 1971) prior to investigating the entire East Frisian coast between Juist and Wangerooge (Luck & Witte, 1974). Aerial photograph surveys conducted over a number of consecutive years confirmed the previous results (cf. Nummedal & Penland, 1981).

The principles of shoal migration, originally ex-

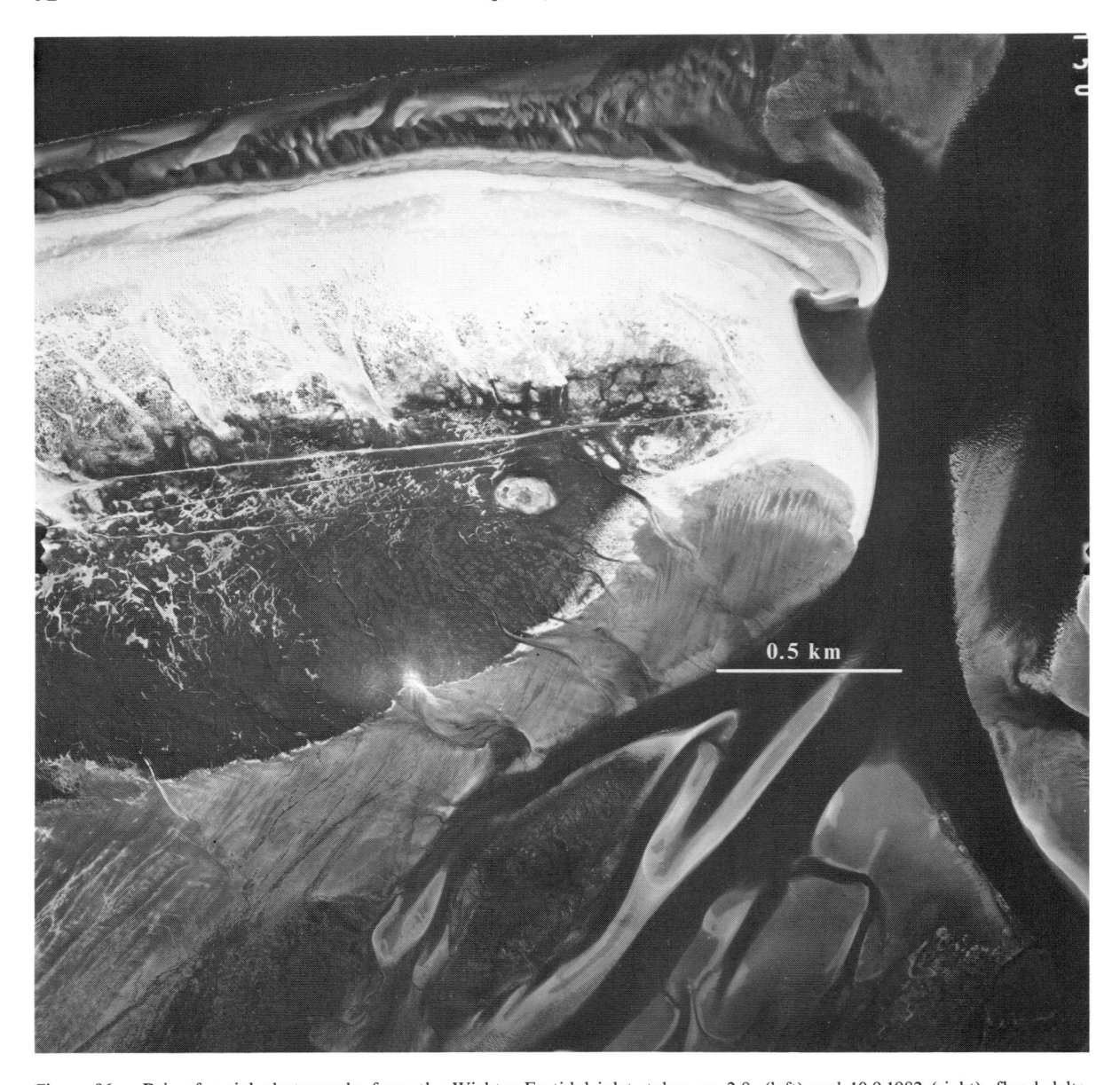

Figure 86 c: Pair of aerial photographs from the Wichter Ee tidal inlet, taken on 2.8. (left) and 10.9.1982 (right); flood delta, western part. Aerial photographs of Fa. N. Rüpke, Hamburg, Bildflug Wichter Ee, freigegeben durch Luftamt Hamburg unter Nr. 812/82.

emplified by Gaye & Walther (1935: 567) on the Norderneyer Seegat and Accumer Ee inlets, is much too simplified to represent the actual sand transport pattern. Based on the morphological investigations of van Veen (1950), Bruun & Gerritsen (1959: 94) developed a more sophisticated model. Using the example of the Eijerlandse Gat between Texel and Vlieland, they demonstrated the multiple backward and forward movements of sediment in the inlet, and also took into account the flood delta processes.

Since 1976 the Niedersächsisches Landesamt für Bodenforschung has conducted geological and sedimentological investigations in the Harle tidal inlet area between Spiekeroog and Wangerooge. Hanisch (1981), based on his own investigations, discarded the widely accepted bar-migration model and came to the conclusion that the sand did not migrate along the shoal arc, but more or less perpendicular to it. The flood current drives the sand deep into the inlet, and it is transported back to the sea by the ebb current. As both currents do not

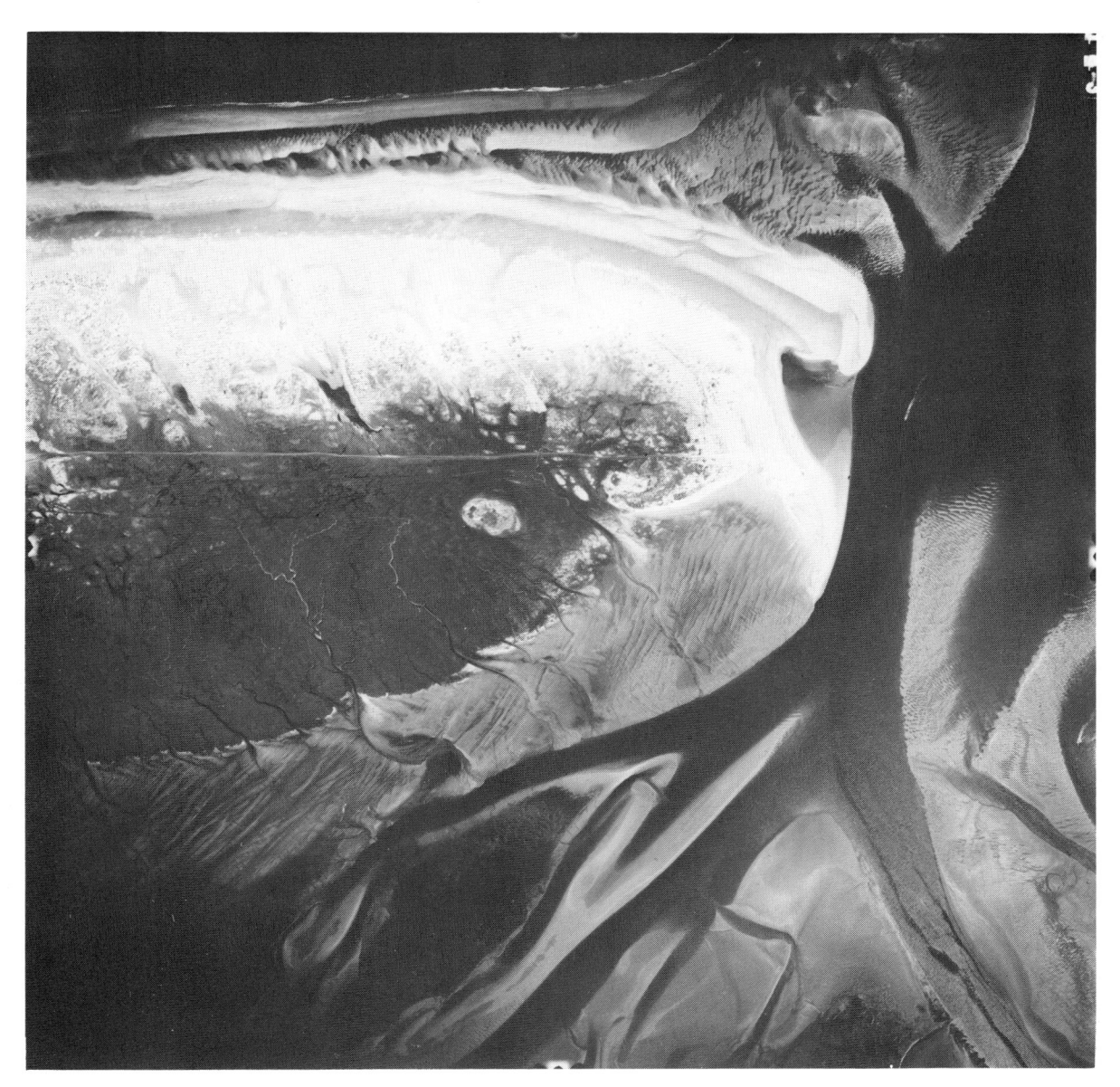

Figure 86c (cont.).

flow in precisely opposite directions sediment displacement leads to net sand transport along the shoal arc.

The study area

The Wichter Ee inlet was chosen for a closer examination of the morphodynamics of a tidal inlet. Situated between the East Frisian islands of Norderney and Baltrum, it has a drainage area of 25 km^2 (Homeier, 1980) and a medium tidal range of about 2.4 m. Maximum current velocities of over 200 cm/s have been measured in the inlet throat during winter storm tides (Koch & Niemeyer, 1980). Current directions and velocities are noted in fig. 22.

The Wichter Ee inlet is particularly appropriate for the study of sand migration because the shoals of both tidal deltas are relatively high and are generally exposed during low tide. This assists the acquisition of good aerial photographs and renders at least the marginal shoals of the deltas accessible from the adjacent islands. Two aerial photographic surveys were undertaken over a short interval in 1982 (on 2.8. and 10.9.) in order to de-

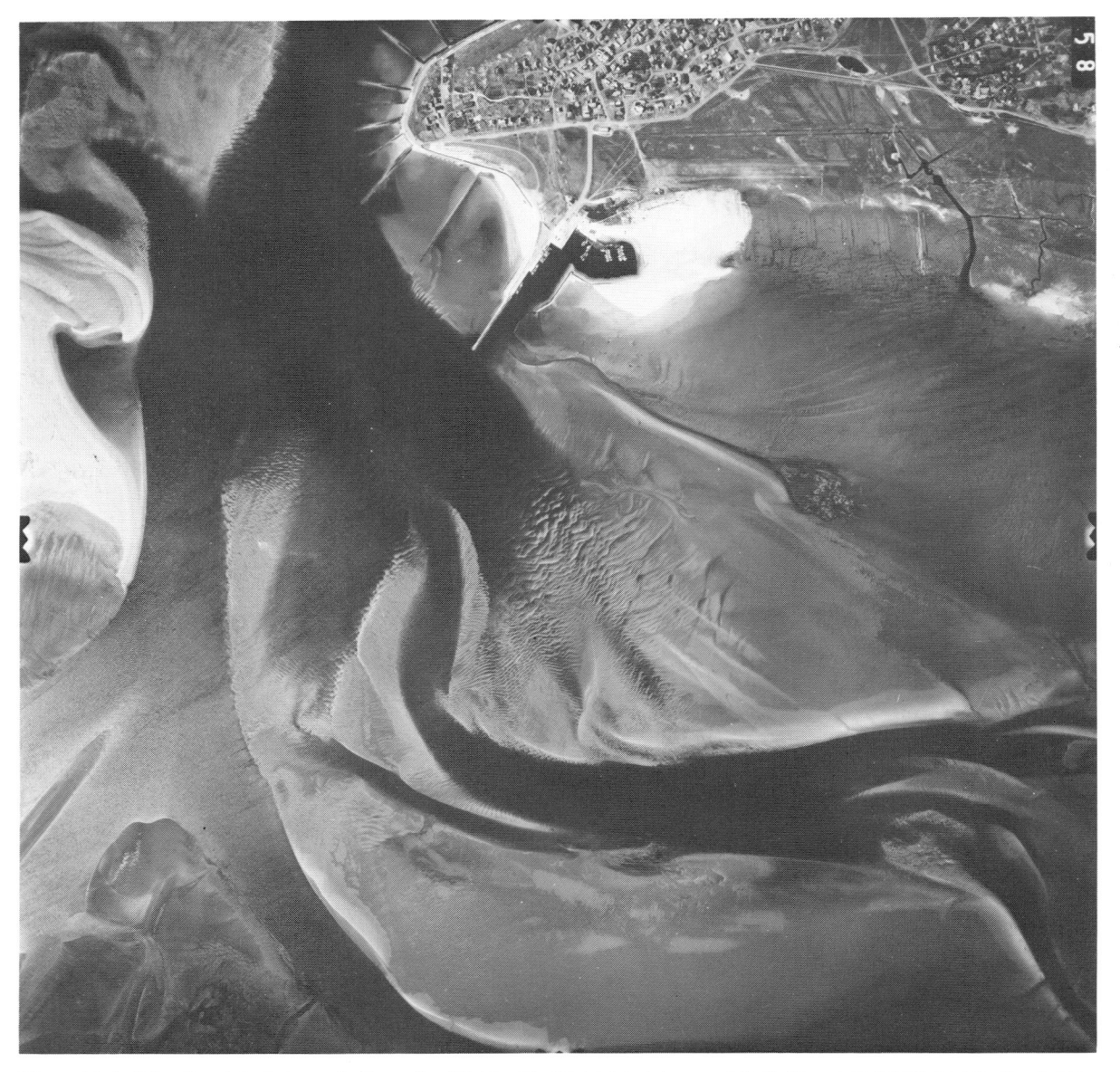

Figure 86 d: Pair of aerial photographs from the Wichter Ee tidal inlet, taken on 2.8. (left) and 10.9.1982 (right); flood delta, eastern part. Aerial photographs of Fa. N. Rüpke, Hamburg, Bildflug Wichter Ee, freigegeben durch Luftamt Hamburg unter Nr. 812/82.

tect short-term morphological changes (figs. 86 a-d).

Although previous investigations concentrated on the shoal arcs only, the flood deltas were also included in this survey. Assuming that the tidal deltas together with the inlet form one morphodynamic unit, the extent to which the shoals of the flood delta participate in the sediment transport processes of the tidal inlet should also be investigated.

Results of investigations

Flood delta

The flood delta of the Wichter Ee, like the ebb delta, is asymetrically shaped (figs. 86 c and d). In the west the adjoining Norderneyer Inselwatt flats are shielded against current influence from the Wichter Ee inlet by a flood tidal spit, several hundred metres in length. To the south the Ostbalje creek is flanked by sandy levées on both sides

Figure 86d (cont.).

whereas the adjoining shoal consists of a muddy core with a mussel bed surrounded by a sandy margin.

In the southeastern part of the flood tidal delta a sequence of ebb-flood-ebb channels occurs. This remained in an almost constant position during the investigation period. Only the outermost seaward spur of this system exhibits a small field of flood-orientated megaripples. The northernmost spit of the Neßmer Nacken consists of a sand flat about 500 m wide which is regarded as part of the flood delta. To the south it is clearly delineated by the mussel-covered mud flats.

Further to the east, the Neßmer Plate also forms part of the flood delta. It consists of a sand flat about 2 km long, with sand wave fields covering its northern part. This area and the adjoining Baltrumer Inselwatt in the northeast, form the inner part of the flood delta where the sediment transport rates are the highest. This is demonstrated by the sand wave fields on this large flood ramp.

The sand waves of the Neßmer Plate shoal and the western part of the Baltrumer Inselwatt exhibit a preferred flood-orientation (fig. 87). Ebb-orientated bedforms are found only locally at the

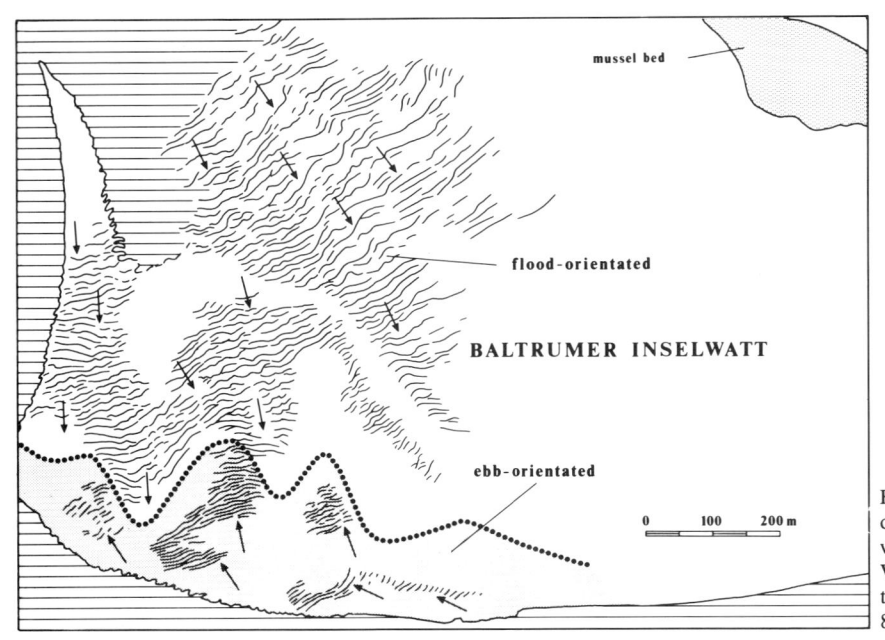

Figure 87: Ebb- and flood-orientated megaripples and sand waves in the eastern part of the Wichter Ee flood delta; lateral sand transport to the west (left); cf. fig. 89.

Figure 88: Ebb-orientated megaripples with spurs at the margin of the Baltrumer Balje (1980).

mouth of the Baltrumer Balje (figs. 87 and 88). The significant morphological changes are restricted to a small sector of the flood delta, indicated by comparison of the aerial photographs. A slight westward shift of the seaward end of the Baltrumer Balje and simultaneous minor reorientation of the adjoining megaripple and sand wave fields are the most significant changes. The shoal-like large sand waves on the southern part of the flood ramp migrate at a rate of 30–60 cm per tide towards the west, whereas the flood-orientated sand waves migrate by about 10–20 cm per tide towards the southeast (fig. 89) (cf. section 5.2.2).

The sand stripes in the southwestern Baltrumer Inselwatt (fig. 86 d) originate within the flood-orientated sand wave field. Their strike direction is not exactly perpendicular to that of the sand wave crests but, nevertheless, it forms an acute angle to it. These sand stripes remained stationary during the 39 day survey.

Ebb delta

Morphological changes are much greater on the ebb delta (Ehlers, 1984). The shoal arc between Norderney and Baltrum can be subdivided into three areas which differ in both morphology and morphodynamics (fig. 90):

1. The 'Othello Plate' (Othello Shoal Shield) at

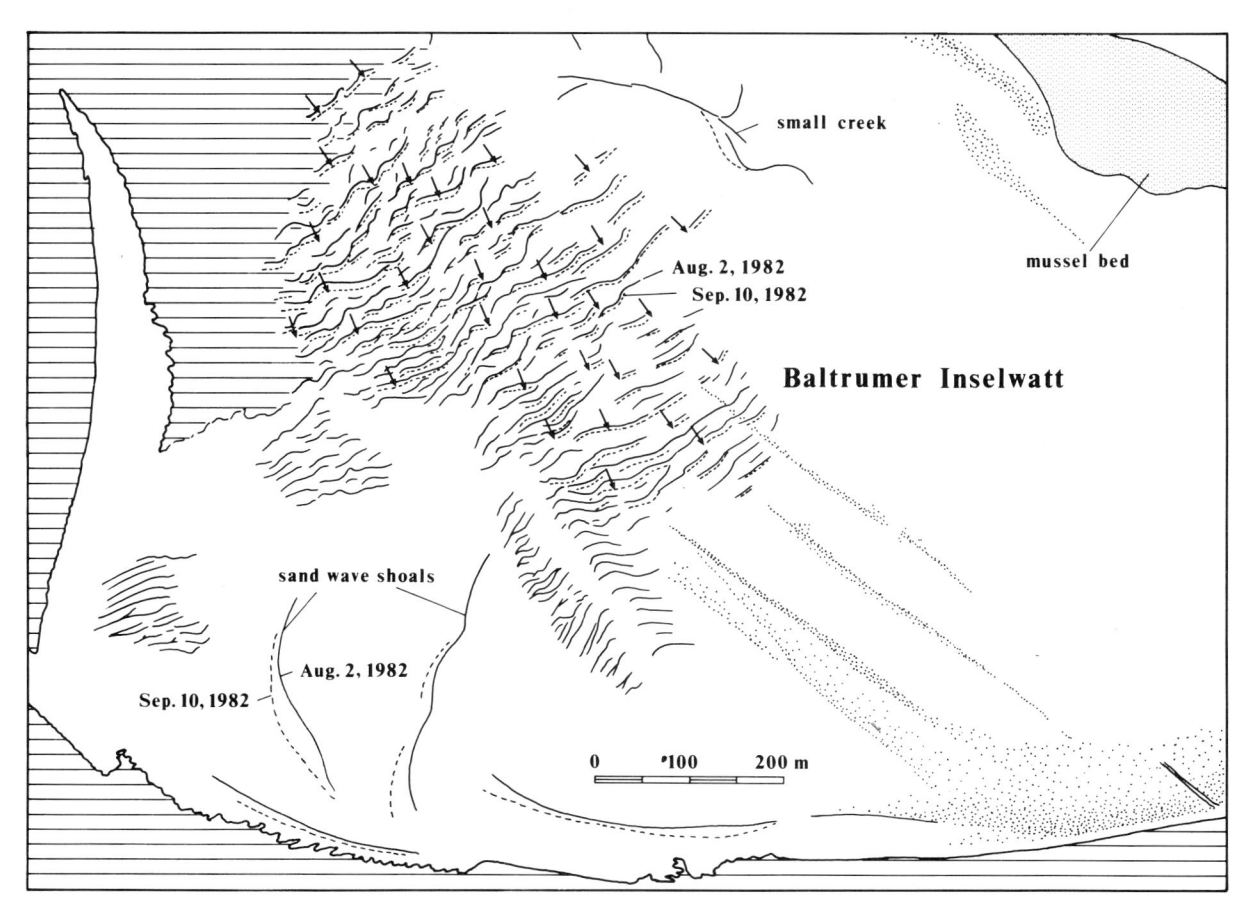

Figure 89: Sand wave migration in the eastern part of the Wichter Ee flood delta between 2.8.82 and 10.9.82 (interpretation of aerial photographs).

the eastern end of Norderney
2. The crescent bars at the apex of the shoal arc
3. The foreshore shoal on Baltrum.

An eastward directed current, produced by the flood current and the prevailing westerly winds, flows along the coast of Norderney. Coast-parallel sand transport occurs in the inter- and subtidal areas (fig. 91).

Sand transport occurs in two ways. Longshore ridges migrate both eastwards and towards the beach whilst sawtooth-like sand waves are transported eastwards in the runnel between the ridge and beach. Some sand is transported with the flood current around the eastern end of the island into the tidal inlet. This process, which also takes place on the other islands, leads to the formation of a series of recurved spits at their eastern ends ('Osterhook' on Baltrum and Langeoog).

The beach ridges also contribute to the sand transport. Fig. 91 shows that sand moved eastward along the beach and also up the beach foreshore during the study period (note the shift of the outward directed bend in the outer beach ridge from a to b). The upper beach ridge remained inactive during this period; it seems only to be moved during storm surges.

The western part of the Wichter Ee inlet is formed dominantly by the ebb current. A considerable part of the transported sand is entrained at this point by the tidal current and moved northwards or northwestwards on to the Othello Shoal Shield where extensive megaripple fields occur (fig. 92). The sand is concentrated into small shoal subunits which migrate longitudinally across the Othello Shoal Shield under the oscillating influence of the tidal currents (fig. 90). Comparison of the aerial photographs demonstrates that the shoal subunits migrate at a much higher rate than the major shoals (at about 2.5 m per day).

The Othello Shoal Shield does not always form

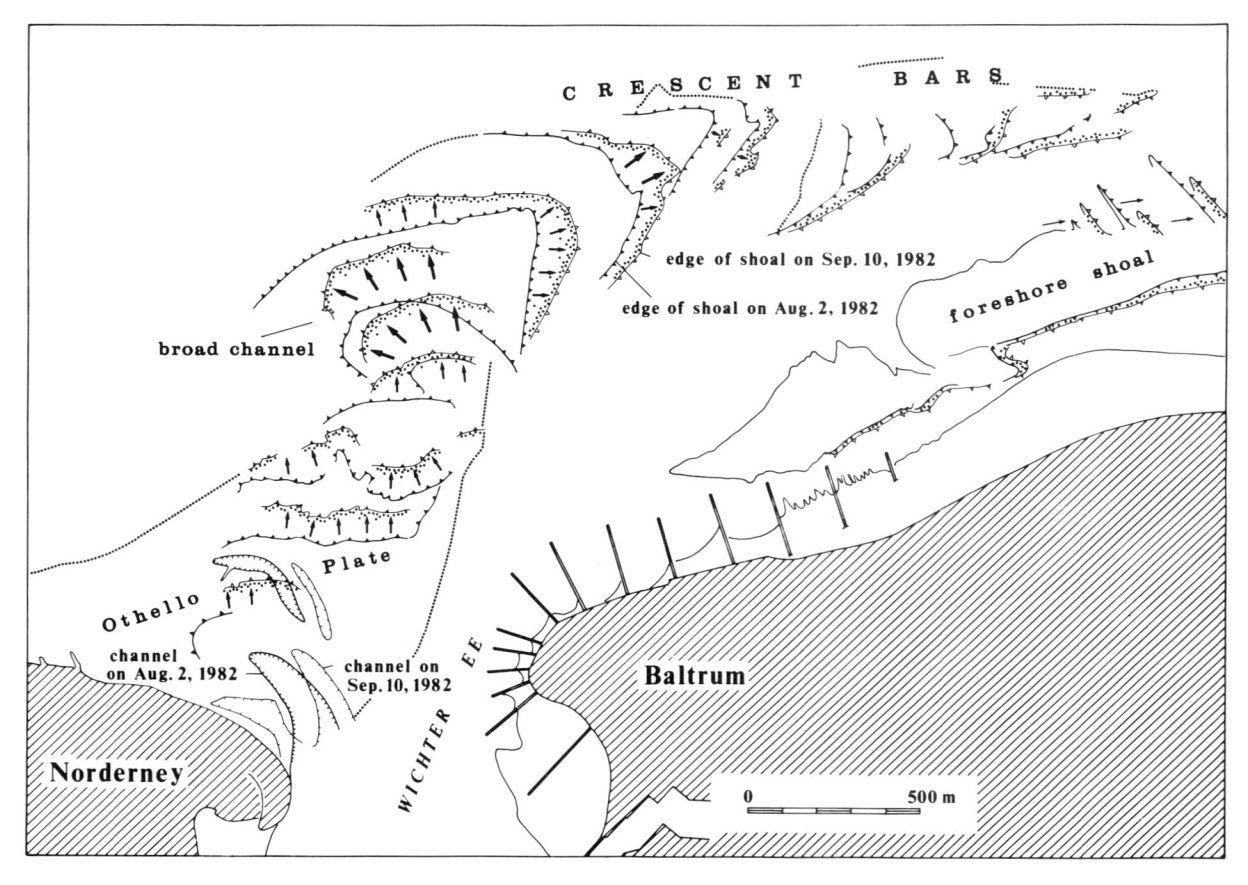

Figure 90: Migration of shoals on the Wichter Ee ebb delta between 2.8.82 and 10.9.82 (based on aerial photographs).

such an isolated and well-defined shoal as indicated on the aerial photographs taken in 1982. The shoal subunits are frequently much less closely connected (e.g. fig. 93). At any rate, the depressions between the shoal subunits are much shallower and smaller than the channels between the major shoals.

The channels between the shoal subunits are often either partly or completely closed at one end, indicating dominant formation by either the ebb or the flood current. In a few cases closed, trough-like features are found. Bars are always formed at the margins of the ebb delta where the tidal currents are no longer restricted to the channels but can diverge, resulting in lower current speeds and therefore lower transport capacities.

Concave as well as convex bedforms are shifted during sand movement. Fig. 91 shows how a short, broad ebb channel (I) and a closed depression (II) migrate northeast and eastwards. On 10.9., a new ebb channel formed close to the beach replacing the one migrating northwards.

The northern end of the Othello Shoal Shield is bordered by a broad channel which is used by both the ebb and flood currents. The channels between the adjoining crescent bars have a similar function. In particular, the broad first channel is crossed by migrating shoal subunits (fig. 90). As the transport capacity of the ebb current slightly predominates, movement towards the NNW is observed. The migration rate of the shoal subunits is about 4 m per day.

The distinctive shape of the crescent bars is a function of the oscillating ebb and flood currents (fig. 94; cf. Hanisch, 1981). As these do not flow directly opposite to each other, an easterly sediment transport results. This is especially striking in the eastern part of the crescent bar region where the shoals are smaller and the channels between them larger. Shoal subunits propelled either by the ebb or the flood currents follow each other at short intervals (fig. 94), producing a zip-like structure.

The migration pattern of the crescent bars is

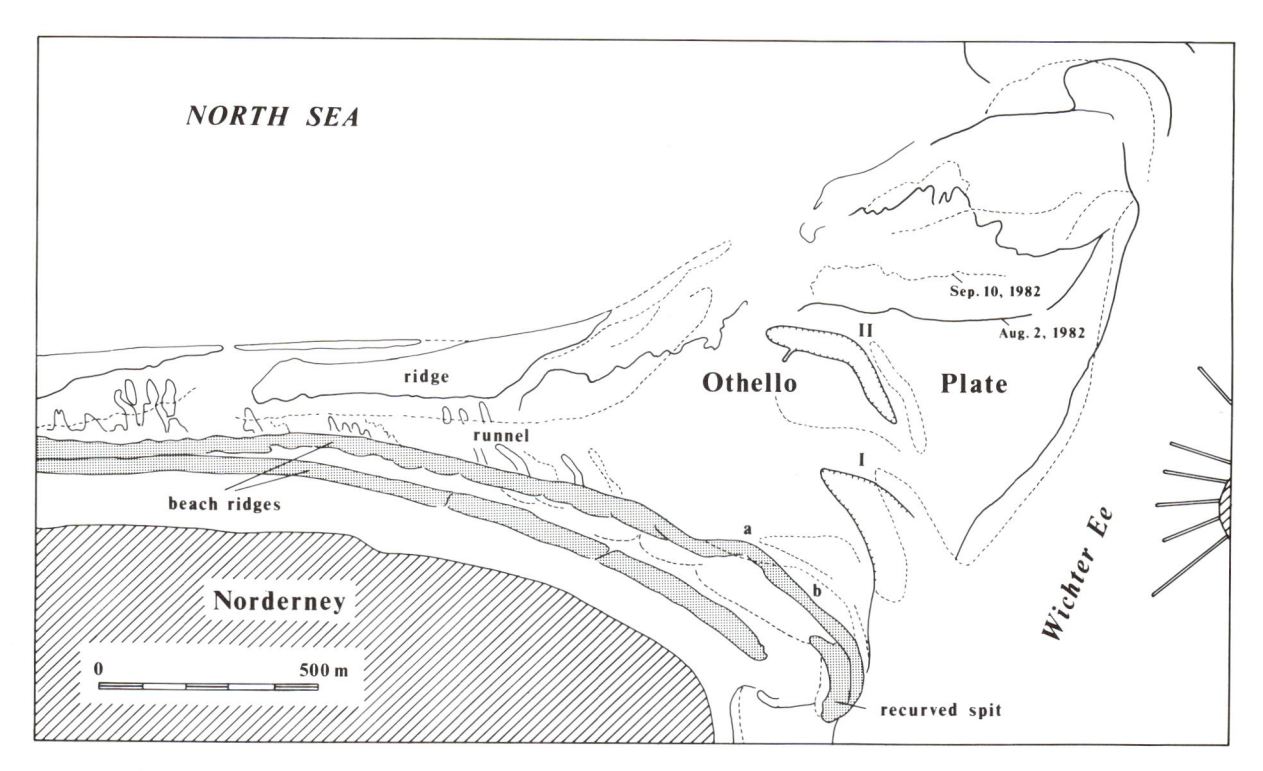

Figure 91: Migration of concave and convex bedforms at the Norderney beach and on the 'Othello Plate' (Othello Shoal Shield) between 2.8.82 and 10.9.82 (interpretation of aerial photographs).

Figure 92: View across the Othello Shoal Shield at the eastern end of Norderney towards Baltrum. Ebb-orientated megaripples in the foreground (height: 0.4 m, length: 3 m) (1982).

Morphodynamic units

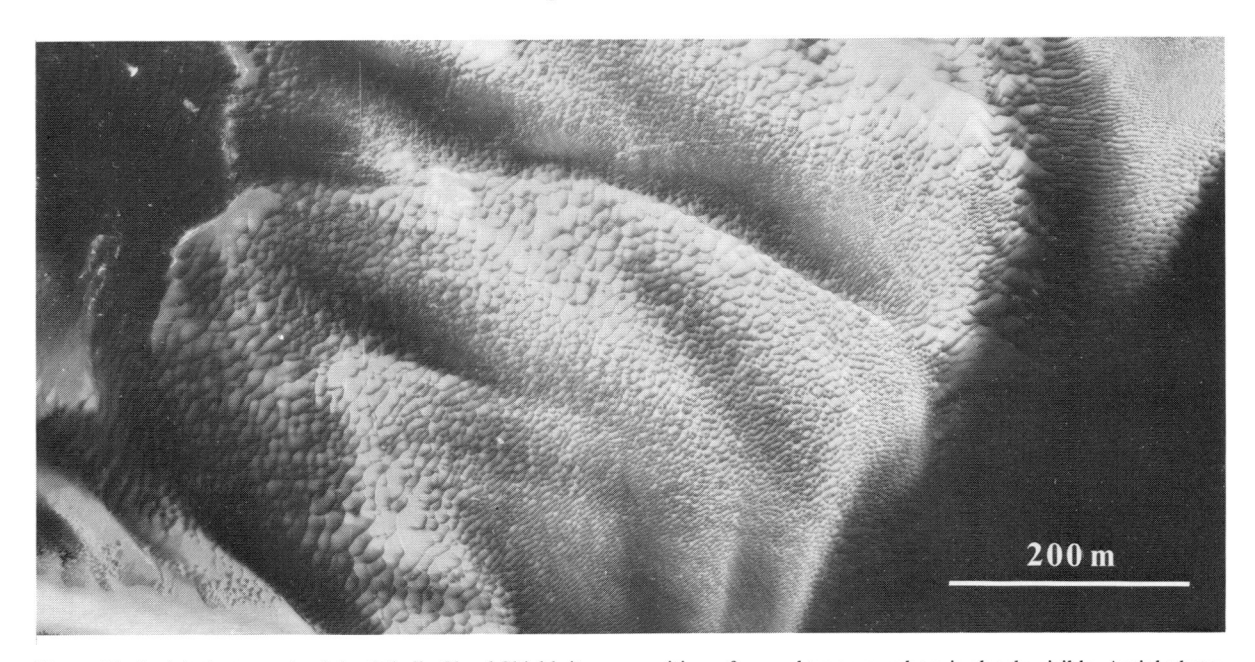

Figure 93: Aerial photograph of the Othello Shoal Shield; its composition of several transverse bars is clearly visible. Aerial photograph of the Niedersächsisches Landesverwaltungsamt, Landesvermessung, of 29.8.1977, Bildflug Norderney (1448), Str.1/4921, freigegeben durch NLVA−Abt. Landesvermessung - Hannover unter Nr. 78/77/1448.

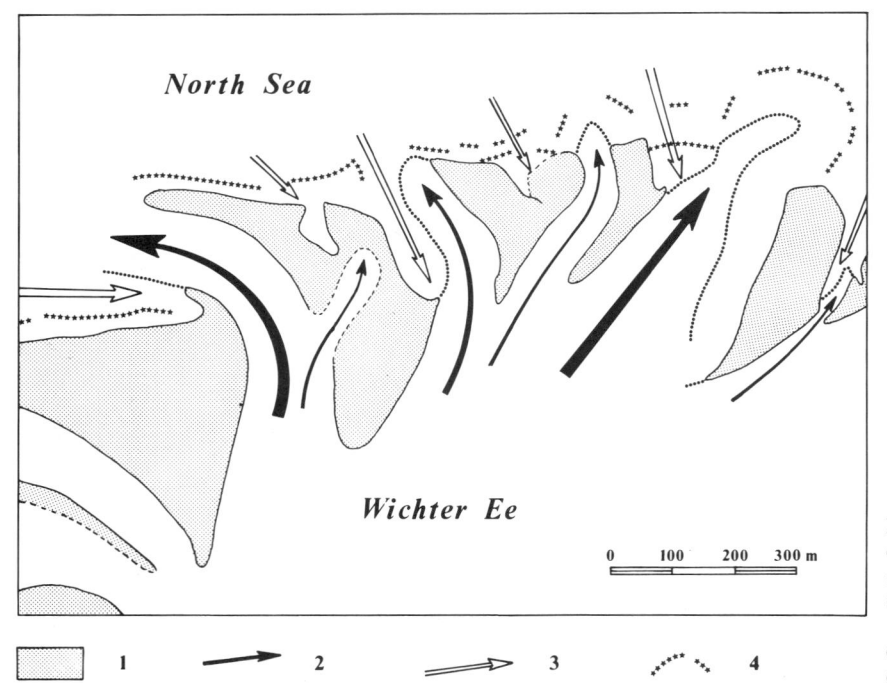

Figure 94: Formation of the crescent bars by interaction of ebb and flood currents. Situation on 2.8.1982 (drawn from aerial photograph). 1 = crescent bars, 2 = direction of ebb current, 3 = direction of flood current, 4 = breaker zone.

Tidal inlets

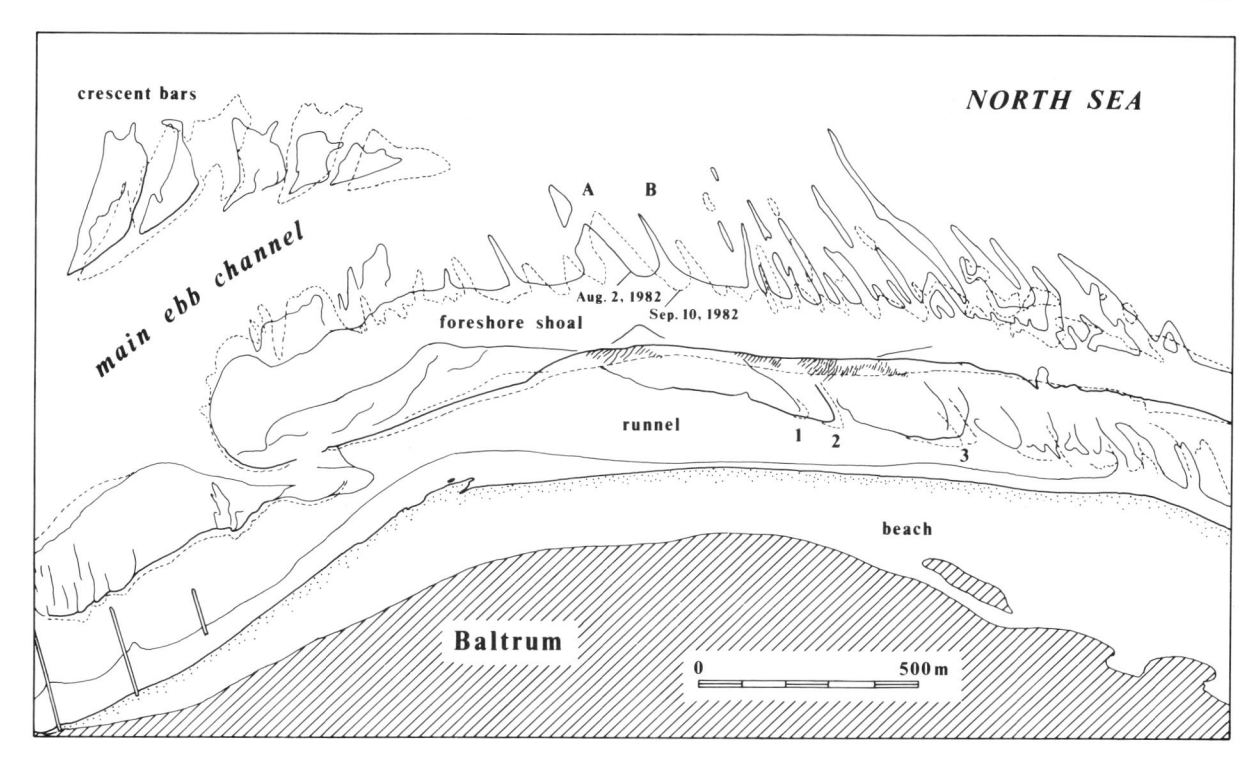

Figure 95: Migration of bedforms in the foreshore area of Baltrum between 2.8. and 10.9.1982 (interpretation of aerial photographs).

Figure 96: View from the foreshore shoal (area south of 'B' in fig.92) towards the western end of Baltrum. In the central area weakly developed megaripples migrate eastwards through the shore-parallel runnel (height: 0.3 m; length: 10 m) (1980).

fairly difficult to interpret from the aerial photographs. The two westernmost shoals migrated as a whole at rates of about 3 m per day, but the same could not be proven for the more easterly shoals. The number of shoals did not remain constant even though the interval between the surveys was only 39 days. During the first survey, of the 8 crescent bars, 3 were west of the central channel and 5 east of it, whereas 39 days later there were 4 shoals west of the channel and only 3 east of it. The position of the channel, however, remained virtually unchanged throughout (fig. 90).

The main ebb channel separates the crescent bars from the foreshore shoal of Baltrum. Material transport across this channel probably only occurs in suspension during storm surges as assumed by Hanisch (1981) for the Harle. Inward and outward sediment transport in the channel is clearly dominant during calm weather as experienced during the survey period. This is reflected in the behaviour of the sawtooth-like sand waves along the northern margin of the longshore ridge (fig. 95); a number of these could be identified on both sets of aerial photographs because of their characteristic shape (e.g. A and B). Their migration rate was about 1.5 m per day.

Part of the sediment migrates onto the foreshore shoal which slowly migrates beachwards during this process. The major replenishment of the beach, however, is supplied by megaripples and sand waves (figs. 95 and 96), which migrate from the ridge through the runnel towards the eastern beach of Baltrum (e.g. 1, 2, 3 in fig. 95). The migration rate of the sand waves is about 1.0 m per day. Although the foreshore shoal appears to move towards the central part of Baltrum, the sand approaching the beach usually reaches the island only at the very eastern extremity, where it soon comes under the influence of the Accumer Ee ebb delta transport system.

Megashoal migration

The migration of groups of shoals, as observed in the East Frisian ebb tidal deltas, constitutes only one form of sediment transport across an inlet, and cannot simply be transferred to all areas of the Wadden Sea. Variations include the very large shoals found in the Marsdiep, Vlie and Wester-Ems ebb deltas. These are all over 1 km^2 in size. Aerial photograph interpretation has shown that their migration rate is about 50 - 100 m/year. They will be referred to as 'megashoals' below.

Megashoals are only found in a few tidal inlets:

1. Marsdiep (between North Holland and Texel)
2. Vlie (between Vlieland and Terschelling)
3. Wester-Ems (between Rottum and Borkum)

In the Netherlands the Rijkswaterstaat has evaluated large-scale charts of the West Frisian tidal inlets covering a time-span of over 100 years (Joustra, 1971; de Reus, 1980, 1985; J.T. Postma, 1982; Elorche, 1983). This material forms a good basis for studies of tidal inlet morphodynamics. The development of the Marsdiep inlet between Texel and North Holland exemplifies the process of megashoal migration (fig. 97).

a. On the 1851 chart there was one dominant sand bank in the Marsdiep ebb delta, Onrust, differing from the other shoals in not being flooded during normal high tides. This megashoal was separated from the Isle of Texel by the over 10 m deep Noordergat channel.

b. 23 years later Onrust had migrated about 2 km to the east and thus approached the southern beach of Texel. Though the channel of the Noordergat had narrowed, it was still over 10 m deep.

c. 27 years later Onrust had migrated a further 2 km to the east and closely approached Texel. At the same time the southern beach of Texel had extended southwards. The Noordergat degenerated into a narrow, winding creek, little more than 5 m in depth. In the meantime the Noorderhaaks sand bank had considerably increased in size and expanded in the direction of the area vacated by Onrust. Thus far there had been no noticeable changes south of the main ebb channel of the Marsdiep.

d. The map of 1916/17 shows that Onrust had merged with the southern end of Texel.

e. By 1933 a new megashoal had formed on the Noorderhaaks, the Razende Bol. The configuration of shoals and channels closely resembles that of 1851.

f. By 1965/67 the portion of the Noorderhaaks which was emergent above high water level had grown considerably in size. Whilst this part of the megashoal remained in place, its eastern end approached the coast of Texel. The southern beach of Texel broadened, and the northern part of Noorderhaaks turned northeastwards. In this way the Molengat channel narrowed, causing deep scour.

g. By 1980/81 the Molengat had become narrower and deeper, its greatest depth increasing to over 20 m (fig. 98). The distance between the Razende Bol and the coast of Texel was reduced to a mere kilometre. At the same time the southern beach of Texel increased in breadth. It can be an-

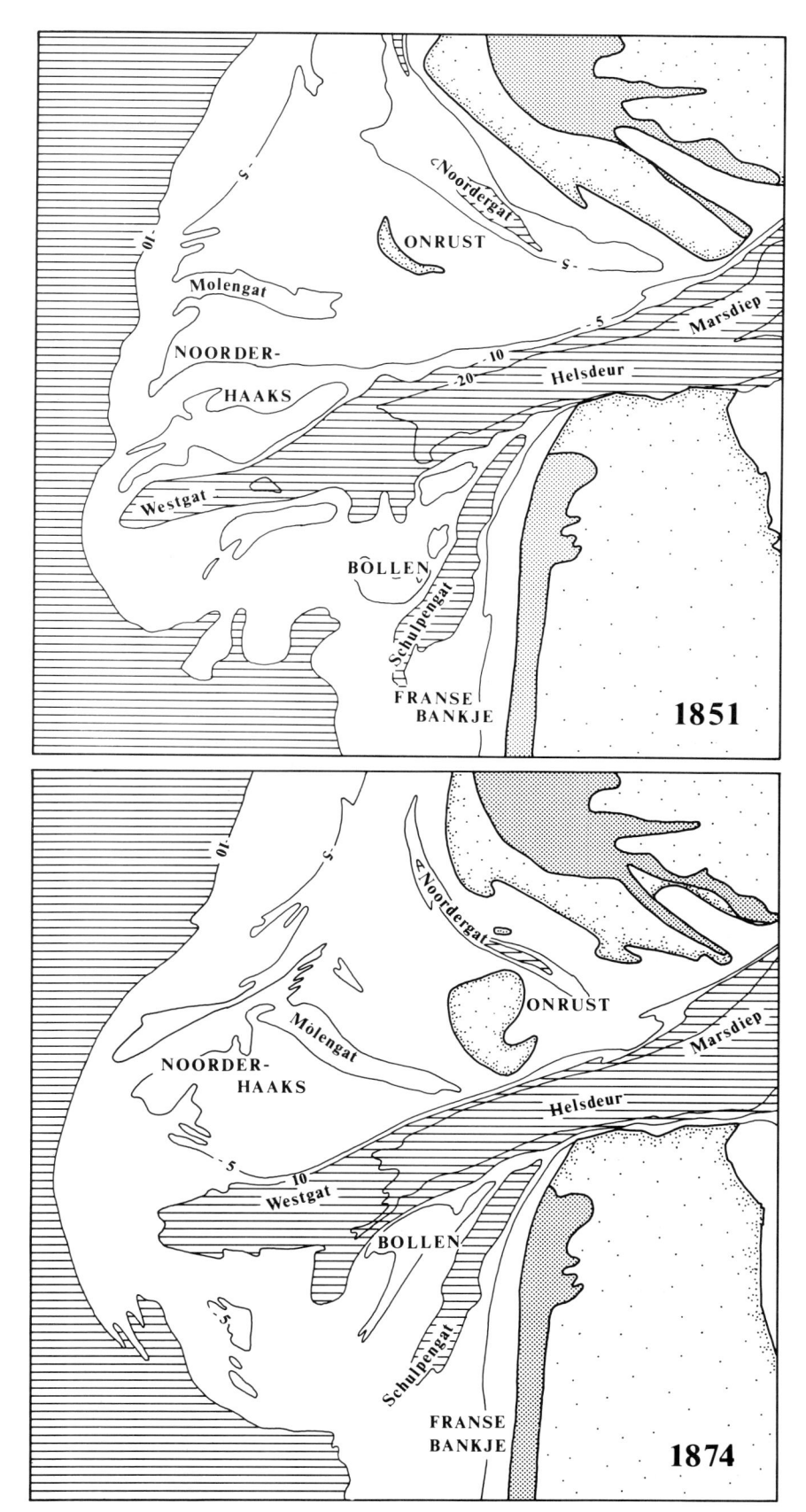

Figure 97: Development of the Marsdiep tidal inlet between Texel and North Holland from 1851 to the present. Sources: de Reus (1980, 1985).

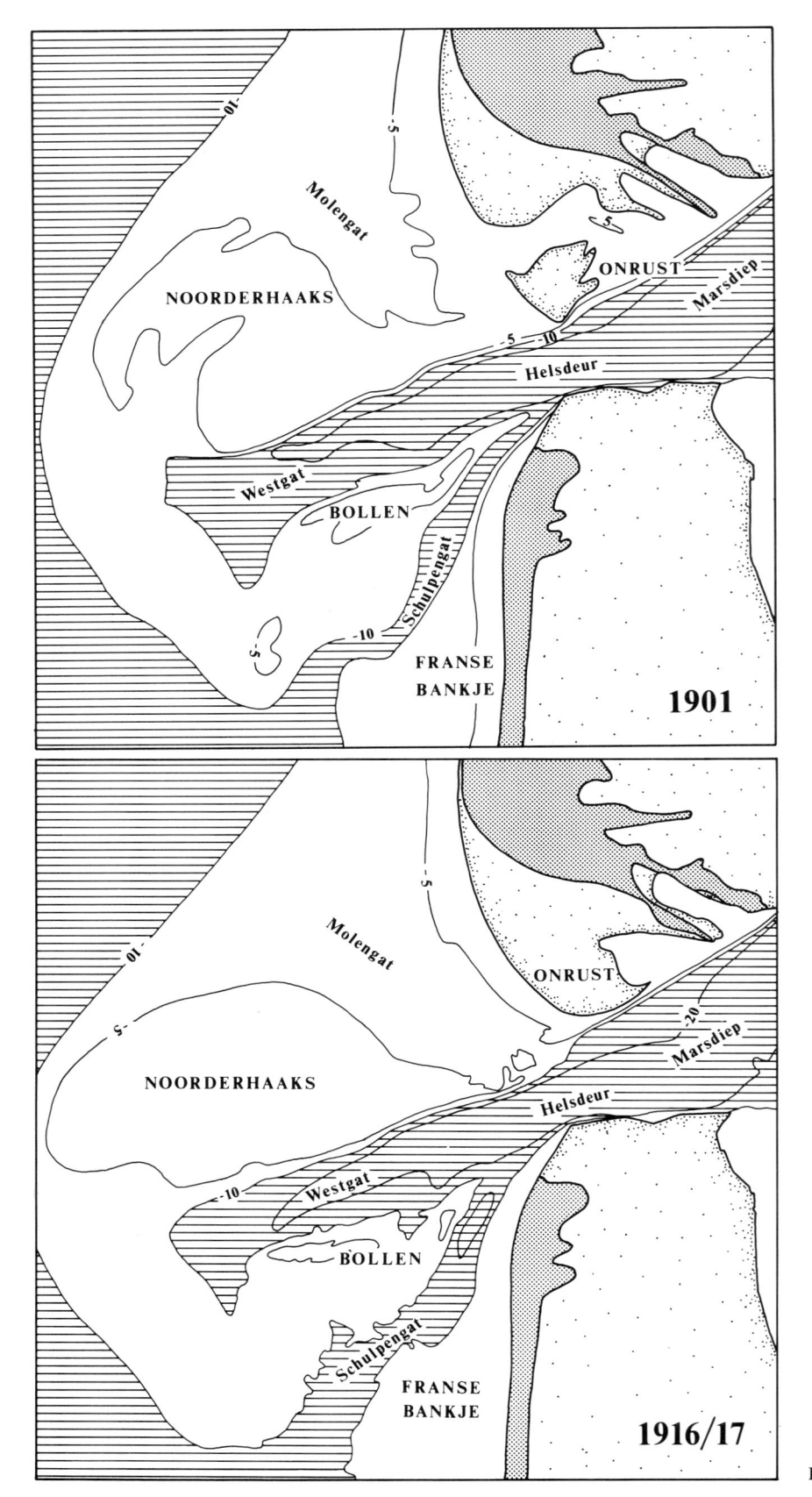

Figure 97 (cont.).

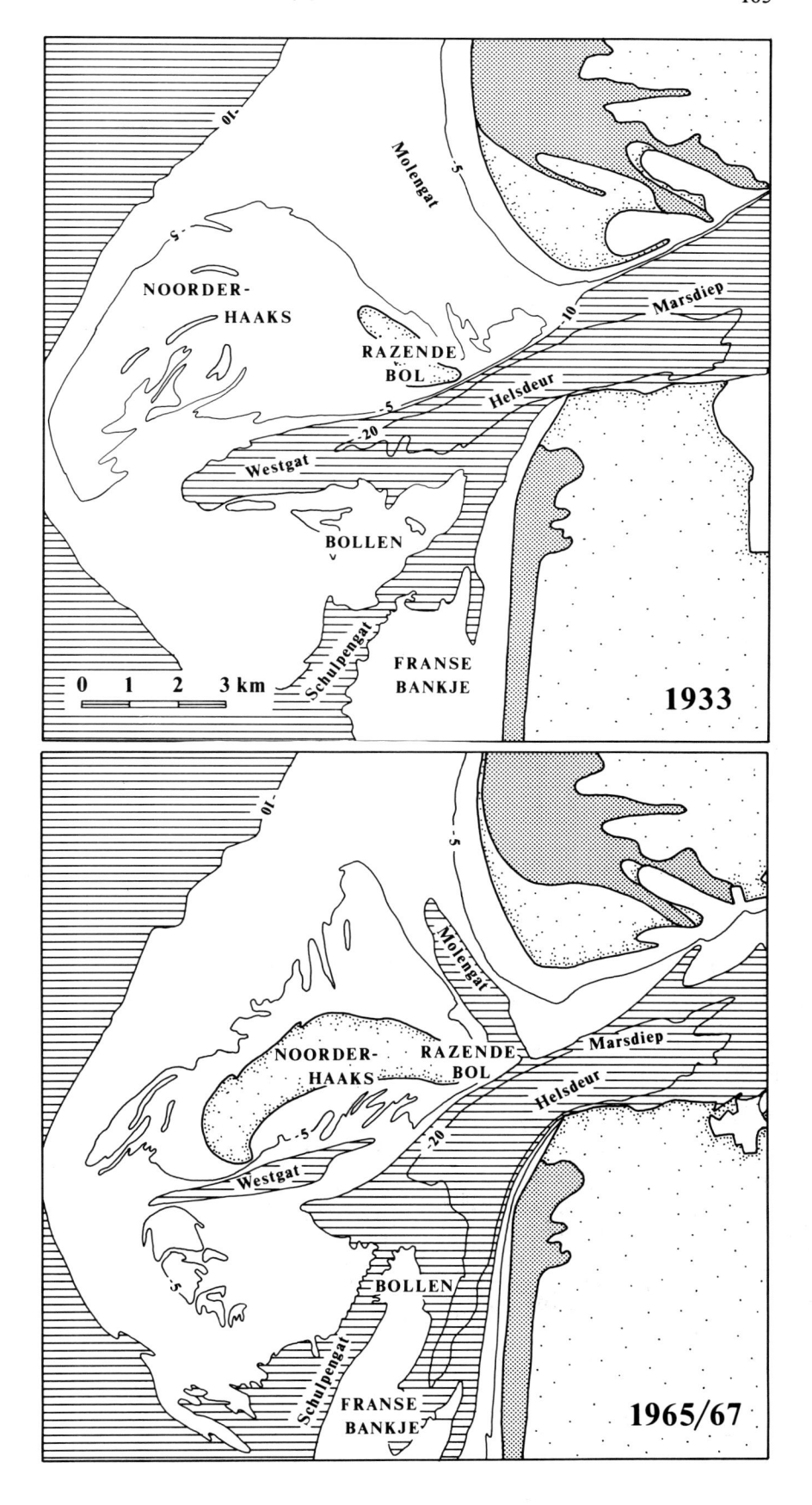

Figure 97 (cont.).

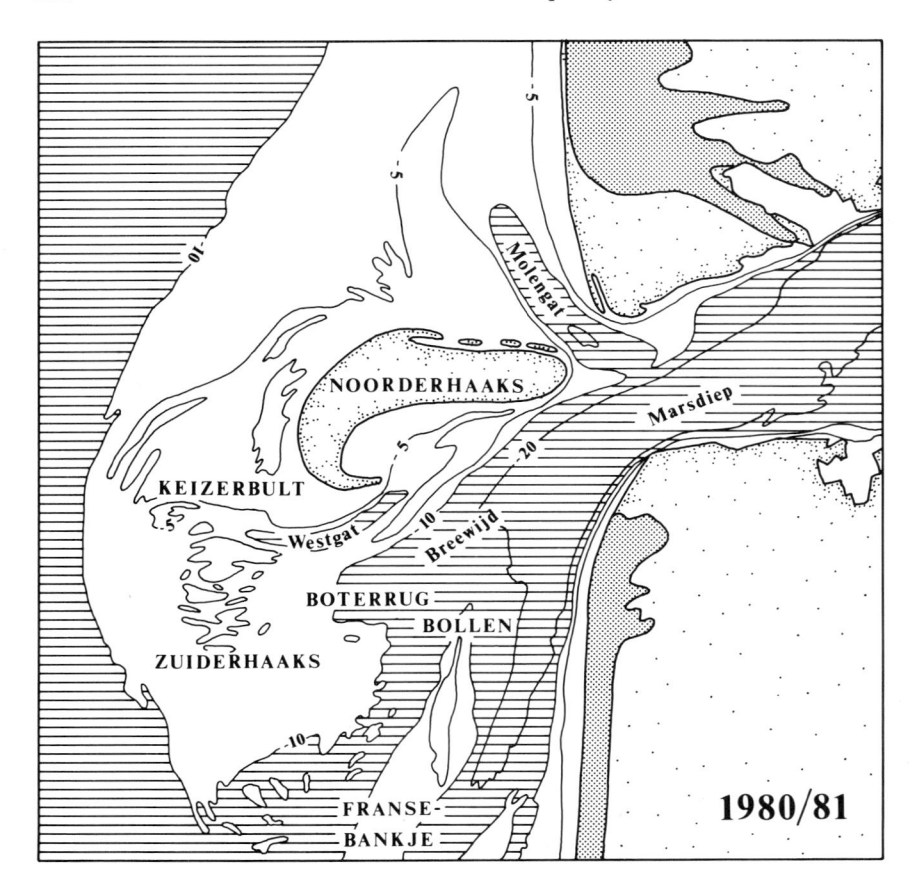

Figure 97 (cont.).

ticipated that the Razende Bol will merge with the island beach within a few decades.

If the migration of Onrust and Razende Bol can be taken as typical, the megashoal migration rate is about 4 km in 50 years, i.e. about 80 m/year.

The more detailed morphology recorded on the recent maps, since 1933, demonstrates that sand transport in the Marsdiep inlet is far more complicated than the slow migration of megashoals might suggest. Detailed investigations, however, are practically rather difficult because major areas of the ebb delta are permanently water-covered and thus not accessible for aerial photograph evaluation.

Although the migration of Onrust and Razende Bol forms just one aspect of the sand transport pattern in the Marsdiep inlet limited to the northern half of the ebb delta, it is of decisive importance for the coastal development of Texel. This fully justifies the attention concentrated on monitoring the migration of the major forms, and therefore the more rapid changes of the small-scale morphology have not been investigated.

The main ebb channel

The shoals of the ebb deltas migrate in the same direction as the coast-parallel sand transport, i.e. along the West and East Frisian coast from W to E. The channels in the ebb deltas follow this movement in most, but not in all cases.

If the rate of channel migration was equal to shoal migration, within a few years the main ebb channel would have reached a position at the eastern end of the delta directly in front of the downdrift island. This is not the case, however. In the Vlie and Marsdiep inlets the main ebb channels occur at the updrift ends of the deltas. The map series of the Marsdiep (fig. 97) demonstrates how the channel maintains its position despite shoal migration:

a. The Westgat formed the main ebb channel by as early as 1851. Its position remained very stable during the next few decades.

d. The Schulpengat became deeper around 1916/17. It continued to consist of two separate channel ends, the western end formed mainly by the flood current, the eastern by the ebb current.

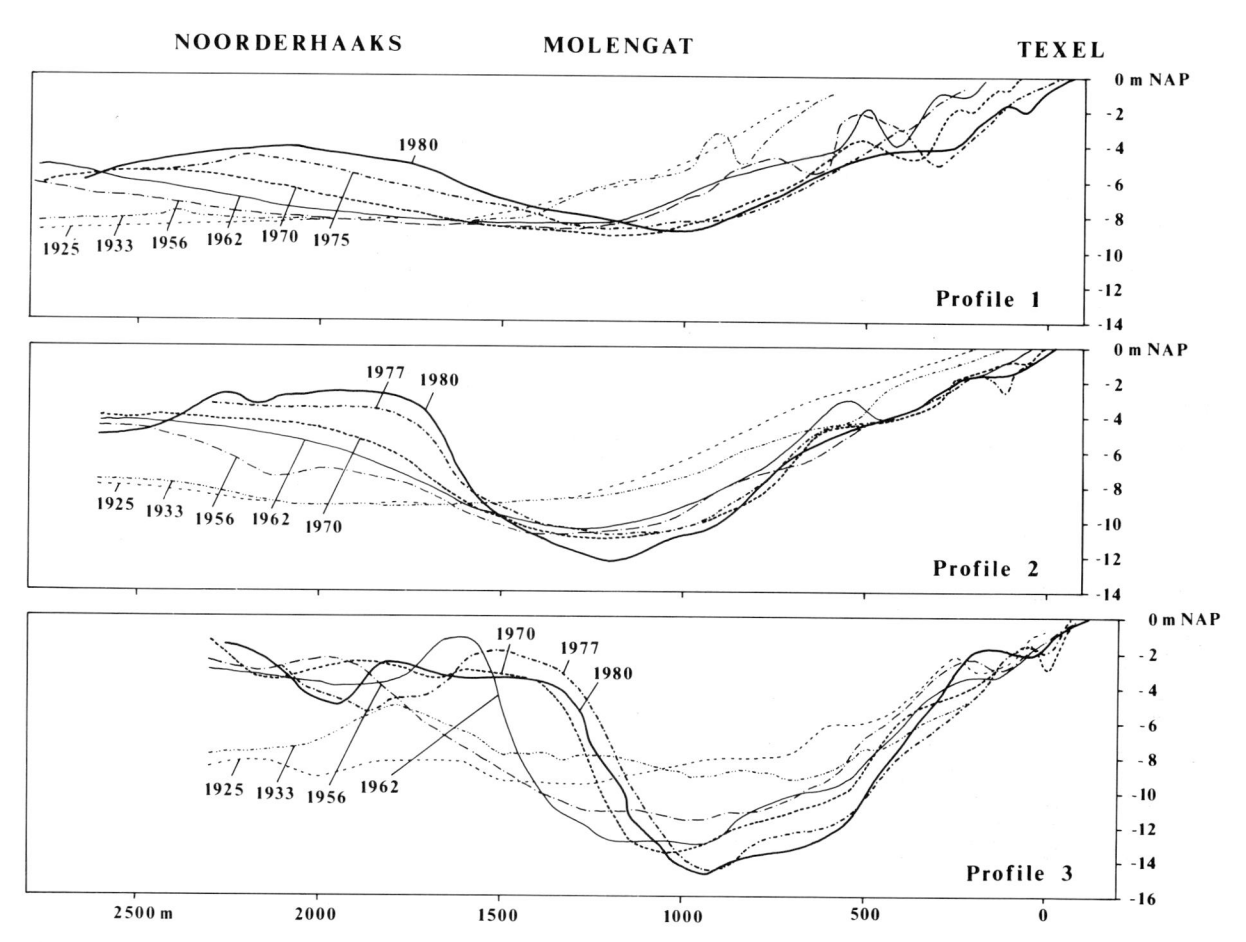

Figure 98: Cross sections showing the development of the Molengat channel from 1925 to 1980. The approaching Noorderhaaks megashoal has caused the channel to narrow and deepen. Source: de Reus (1980).

The connection between both ends, already visible on the 1901 map (fig. 97 c), intensified.

e. By 1933 the Westgat had decreased in depth but persisted as the main ebb channel. During this period a new channel formed close to the coast at Huisduinen, later called the 'Nieuwe Schulpengat'.

f. By 1965/67 the new channel had become the main ebb channel, the Westgat shrinking into a minor channel. The Schulpengat had simultaneously straightened its course and developed into the main flood channel. A section of the bar between the old Schulpengat channel and the Marsdiep was removed by dredging in 1965 between the Bollen and Botterrug shoals in order to form a new navigation channel to Den Helder.

g. At present the Nieuwe Schulpengat is clearly the new main ebb channel.

This recorded development of a new main ebb channel in close proximity to the Huisduinen coast has considerable morphodynamic consequences. In this extremely southeastern position the channel promotes erosion of the mainland shore. However, this situation will gradually improve as the channel has already started to migrate once again in a clockwise direction (cf. de Reus, 1980: 18).

The updrift shift of the main ebb channel was not a gradual process but occurred very rapidly within only a few decades.

This rapid switching of the main ebb channels is a comparatively rare event. For example, the investigations of the Forschungsstelle Norderney concerning the historical development of the East Frisian Islands show a dominant tendency for the main ebb channels to maintain their positions. No sudden changes have been recorded, only occasional movements of the channel axes over a few degrees, e.g. the Harle Inlet.

The deviating behaviour of the Marsdiep and Vlie inlets provokes the question whether human interference might have promoted the observed migrations. In both cases the strongest changes occurred after 1933, that is after the damming of the Zuider Zee, which considerably changed the tidal range and the drainage areas of the inlets.

The main ebb channels subdivide the ebb tidal deltas into two almost morphodynamically independent sections. This is exemplified by both the Wichter Ee inlet (section 5.2.2) and the Marsdiep (fig. 97). The marked development of the Noorderhaaks/Razende Bol complex in the centre of the ebb delta cannot be related to any detectable approach of major shoals from the south, but is simply an *in situ* recent formation in this shoal complex. Comparison with other inlets indicates that there are no instances of bar migration extending beyond the main ebb channel. The old bars degenerate updrift of the channel, and new bars develop on the downdrift sides.

5.2.3 MIGRATION OF THE DEEP BARS

While the tidal inlets of the West, East and North Frisian Islands all exhibit clearly developed ebb and flood deltas, another landform assemblage is encountered in the area of the large tidal streams between Eiderstedt and the Jade estuary. The seaward edge of this area is subdivided into a number of either ebb- or flood-orientated channels, separated from each other by deep bars (fig. 99). These deep bars form horseshoe-shaped sandbanks several kilometres in length. Their highest points mostly reach about −4 m below NN (NN = German ordnance datum) which means that they are between 2−3 m deeper than the shoals of the ebb deltas.

Old maps and charts indicate that these bars continuously change their positions. Early investigations in the Jade and Weser estuaries by Krüger (1911) and Poppen (1912) revealed the rapid morphological changes in the deep bar region. More recent investigations were made by Higelke (1978) along the west coast of Schleswig-Holstein. The sequence of bar migration there has been precisely monitored since about 1936, although the mechanism causing these changes is not yet understood.

The internal structure of the bars

Shallow seismic investigations in the area around the seaward end of the Hever inlet (north of the Eiderstedt Peninsula) have demonstrated that the sand bodies of the bars discordantly overlie the pre-Holocene land surface (Tietze, 1983). Horizontally bedded sands are found at the base of some of the bars, but the vast majority of the sediment bodies are composed of large cross-bedded sand bodies more than 10 m in height. The bedding planes slope gently towards the south.

Tietze (1983: 99) interpreted these delta structures as remnants of an old sand accumulation related to tidal inlet migration, or to the formation of the beach barriers of Dithmarschen (see chapter 3). The agreement between the dip angle in the cross-beds and the slope directions of the recent sand bars argue strongly against the interpretation of the sand bodies as relict. Thus they appear to be large tidal delta bars migrating slowly southwards. This direction conforms with the coast-parallel sand transport direction, as known through other studies. The bars therefore seem to be the morphodynamic equivalents of the ebb deltas of the tidal inlets.

Tietze's investigations provided the first opportunity for an examination of the internal structure of the bars. Although Ulrich (1973: map 7) represents the area as rich in mega- and giant ripples, megaripple bedding is not actually found in the internal composition of the bars; they are composed only of delta bed structures.

Bar migration

When evaluating charts for gemorphological comparison, it should be noted that chart contours are not related to a fixed sea level datum, but rather to the variable SKN (= mean spring tidal low water). Moreover, the charts are intended to serve as warning maps for navigation purposes, so that in areas with strong bottom relief, such as the megaripple-covered bottoms of tidal streams, only minimum water depths are given. A thorough discussion of these problems is provided by Higelke (1978: 24). Despite such problems, the 1:50,000 charts are accurate enough for use in the reconstruction of channel and bar migration.

As early as 1962, Bahr (p. 22) demonstrated that the seaward ends of the tidal streams between the Eider and Elbe used to migrate southwards. Higelke (1978: 117) provided a detailed reconstruction of the migration of the Norderpiep tidal

HEVER

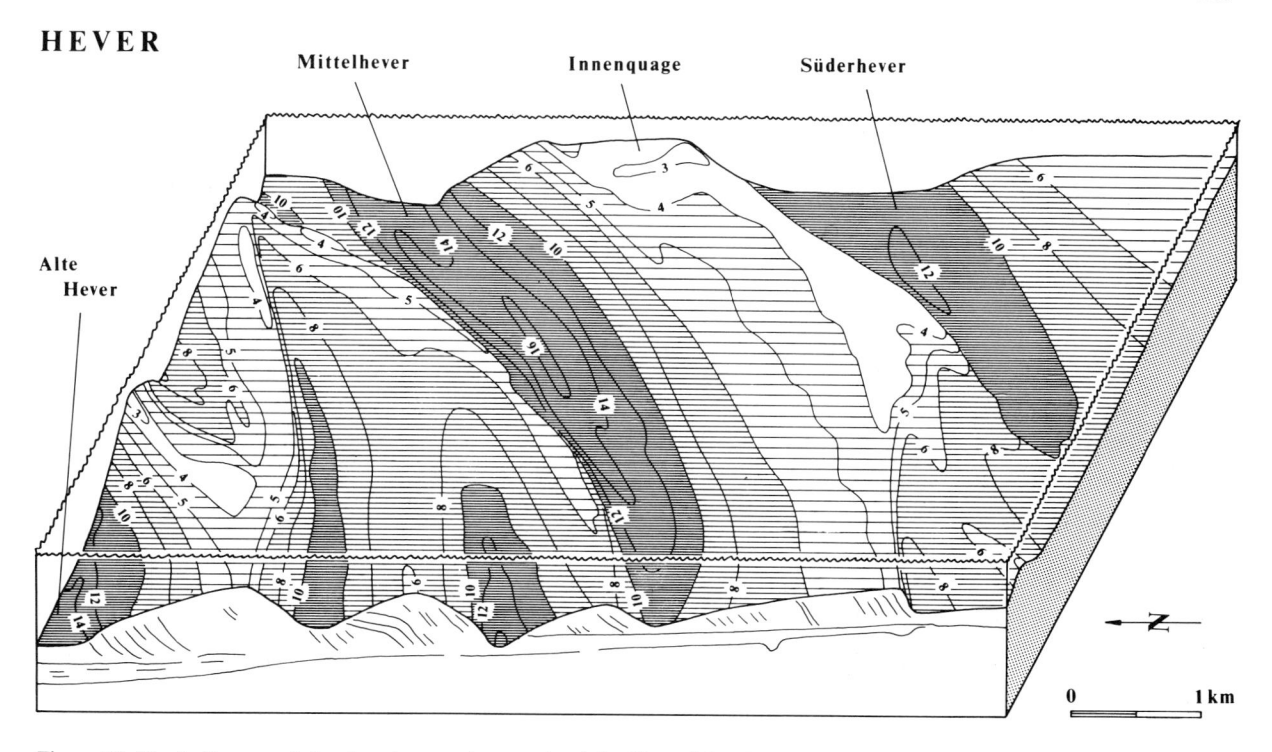

Figure 99: Block diagram of the deep bars at the mouth of the Hever inlet. Sources: Tietze, 1983 (fig. 15), Küstenkarte 1:25,000, Sheet 1616 K St.Peter.

stream. Profile maps drawn from data compiled by the DHI (Deutsches Hydrographisches Institut) show how the channel of the Norderpiep moved southward from 1947 to 1956. A small branch towards the northwest appears for the first time during 1959. This secondary channel developed into the new main channel in 1963/4 whilst the old Norderpiep migrated further to the south.

Whilst the outer extremities of the tidal streams move southwards, an opposite trend prevails further landward in some areas. In mouth of the Hever inlet this was not only shown through chart comparisons (Higelke, 1978: fig. 19) but also by the measured dip directions of the delta bedding within the bars. A similar development seems to have taken place in the area around Trischen (Higelke, 1978: figs. 15-29).

The deep bars in the mouth of the Hever inlet migrate at an average of about 20 m per year (fig. 100). In the Jade/Weser estuaries rates of about 20–50 m per year have been measured (see maps in Krüger, 1911: Atlas Bl.53 + 54; Popen, 1912: Plates 15-17; Göhren, 1965: 145). This is a very high rate for a process, which, rather than being compared with the bar migration of the Frisian Islands, has to be seen as the shifting of a whole tidal inlet. Not only is the eastward migration of the inlets of the East Frisian Islands much slower, but it was even slower before human interference began in the area. The western ends of the islands from Norderney to Wangerooge have been shifting eastwards at a rate of only 4 m/year since 1650. At the same time, the eastern island ends have been prograding at a rate of 11 m/year (Luck, 1975: 50). The difference between these values is explained by inlet narrowing over the centuries. The deep bars migrated 2–4 times faster than the average rate of inlet migration in East Frisia.

The interior structure of ebb delta shoals probably resembles the sedimentary structures of the deep bars. The megaripples which cover the shoals only form the uppermost metre of sediment. The ramp-like forms of most sub-shoals, as described in section 5.2.2, indicate that sand aggradation leads to seaward shoal progradation.

Conclusions

The process described above in section 5.2.2 has been summarized under the term 'bar migration'

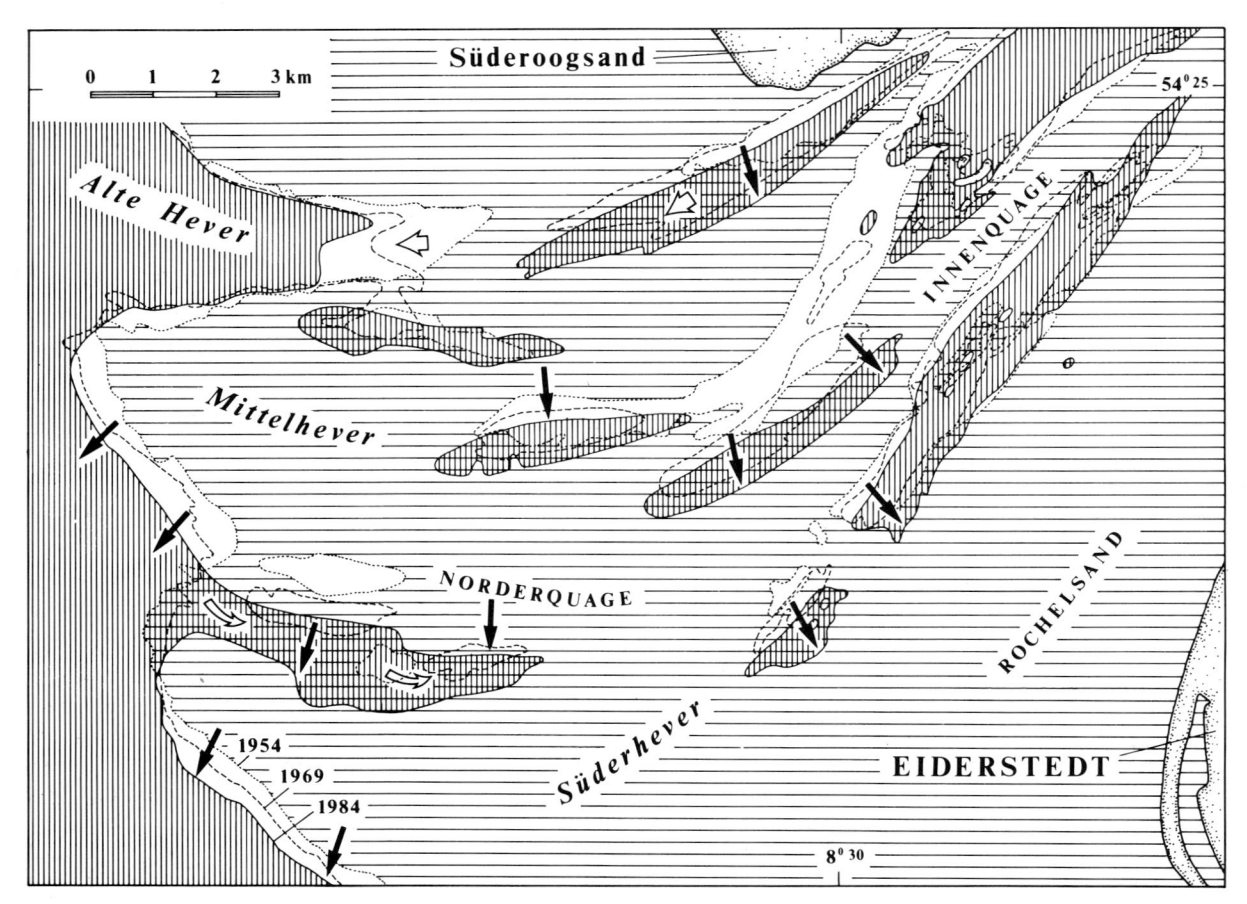

Figure 100: Migration of the deep bars at the mouth of the Hever inlet, west of Eiderstedt Peninsula. The 10 m bottom contours, related to SKN (= mean spring tidal low water) are shown for the years 1954 (dotted line), 1969 (striated line) and 1984 (solid line). The arrows indicate movement directions. Sources: Higelke, 1978 (fig. 19); Seekarte 1:50,000 Nr.106 Hever und Schmaltief, 3.Ausgabe 1984 IX.

(e.g. Luck & Witte, 1974). The existing measured migration rates, however, are only valid for the migration of shoals and not of shoal subunits, and then only in parts of the ebb deltas where it was possible to clearly re-identify the shoals. The migration of shoal subunits, where measured, occurs at a much faster rate. The calculated rates add up to about 1000 to 1500 m per year, compared to about 400 m for the shoals. The rate of migration is so rapid that it cannot be tracked by annual aerial photographs.

It must be remembered when evaluating the results that it is not the shoals, or shoal subunits, that are actually shifting, but the sand. The smaller the form elements observed the more their transport direction deviates from the shoal arc (cf. fig. 124). This will be discussed in greater detail in section 5.2.4.

In addition to the previously described shoals and groups of shoals there are large 'megashoals', over 1 km^2 in size, which migrate at a much slower rate than the groups of shoals. Amongst the existing megashoals are the Noorderhaaks/Razende Bol complex in the Marsdiep ebb delta, the Noordwestgronden in the Vlie ebb delta and the Hohes Riff in the Wester-Ems ebb delta. Their migration rates are about 50–100 m/year.

Within the shoal arcs of the ebb tidal deltas the main ebb channel forms a decisive boundary, separating the two parts of the ebb delta from each other. On both parts landforms are constructed and displaced independently. The shoals and groups of shoals do not migrate across the main ebb channels. Though the main ebb channel is often very stable, it may change its position by switching in an updrift direction.

The displacement of the main ebb channels is less dramatic in the area of the low bars, the height differences being smaller (cf. Higelke, 1978). Although the migration of the low bars resembles the shifting of complete tidal inlets, it occurs at a faster rate. The low bars migrate at an average rate of 20 m/year.

5.2.4 Megaripple migration

The Wadden Sea areas which are morphodynamically most active are covered with megaripples and sand waves. Ulrich (1973) and Visser (1978) have demonstrated that they cover the bottoms of the tidal inlets and deep tidal channels. Within the intertidal area megaripples are most frequently found on the ebb deltas and on the most exposed shoals of the flood deltas. Megaripple migration thus plays a decisive role in sediment movement in the Wadden Sea.

Observations on ripple movement date back to the 19th century. Hübbe (1861) pioneered the use of coloured sand in his investigations. He was able to demonstrate that ripples and sand waves on the tidal flats migrate in the current direction and that small ripples move much faster than the larger sand waves.

Laboratory experiments

Ripple and sand wave formation is related to both grain-size composition of the sediment and flow velocity. Laboratory tests at the Franzius-Institut in Hannover have produced the following results (Dillo, 1960: 179):

1. Current velocities below about 35 cm/s result in the formation of a planar bed.

2. Current velocities of about 35 – 90 cm/s (fine to medium sand, sample from Norderney) or 35 – 70 cm/s (medium sand, sample from Sylt) form small ripples. These migrate at speeds of 0.02 – 4.0 cm/s and 0.02 – 1.5 cm/s respectively.

3. At higher current velocities megaripples are formed. Ripple distance depends on the current velocity. The slowest migration rate of the megaripples is about 0.4 cm/s.

4. In fine sand current velocities of 100 cm/s or more do not lead to megaripple formation. Only small ripples are formed at higher velocities (over 50 cm/s). These migrate at a slower rate (0.03 – 3.0 cm/s) compared to ripples in coarser sediments.

The highest attainable current velocity in these experiments was about 110 cm/s.

Observations in the field

Until recently there were relatively few direct measurements of megaripple migration in the field. Hülsemann (1955: 370) investigated the megaripple fields on the beach of Mellum. He recorded strong changes between tides determined by wind direction and velocity.

Reineck (1961 a: 17) reported a 0.029 cm/s rate of megaripple migration on longshore bars on the Norderney foreshore. Assuming that migration during one tide can last about 4 hours, and that transport occurs only unidirectionally, a maximum migration rate of about 3.7 m per tide can be calculated. However, in inlets and on tidal deltas different, often opposing movement directions occur during ebb and flood, so that the actual rate is usually considerably lower.

Therefore it is understandable that in their studies of flood delta shoals in the mouth of the Oosterschelde estuary Nio & Siegenthaler (1979) observed very low migration rates. They measured a rate of a few cm per tide for flood-orientated megaripples. The migration rates were calculated from the sedimentary record. Still-water phases which occur at the turn of the ebb and flood tide are always accompanied by deposition of clay-rich layers (cf. Terwindt, 1981). The study also demonstrated that the most intense reworking always took place during spring tides.

Simon (1959: 34) observed over 9 days the migration of an ebb-orientated megaripple field on the Lühesand shoal in the Elbe. He found that ripple migration was controlled by moon phases as well as by wind direction and velocity. The maximum migration rates measured were 40 cm/tide. Part of the ripple field migrated in the ebb direction, whilst another part followed the flood current.

Allen & Friend (1979) observed an ebb-orientated megaripple field at Wells-next-the-Sea over one year. Megaripple migration was seen to occur almost exclusively during spring tides. These authors quote 60 cm/s as the critical velocity for megaripple migration, which corresponds well with the laboratory results reported by Dillo (1960).

Sand waves, which are larger than megaripples (spacing generally about 10 m), migrate at much lower rates. Holden (1980) made a series of measurements in the flood delta of the Essex Estuary in Massachussetts (tidal range: about 1.5 m). The migration rate of the sand waves was about 1200 cm in 100 days (Holden, 1980: fig. 27), i.e. approxi-

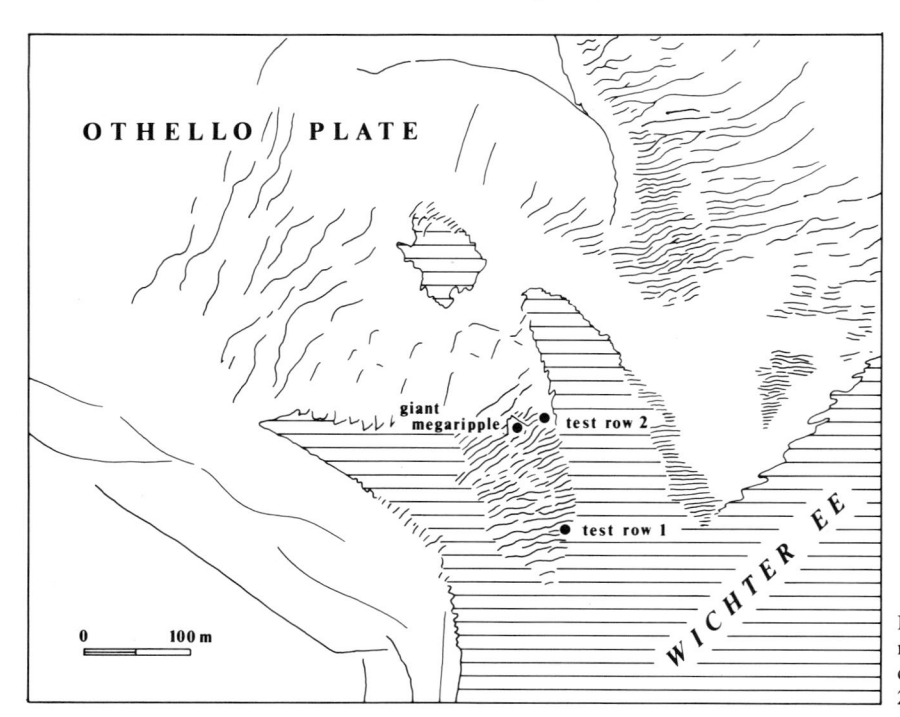

Figure 101: Position of the marked megaripple crests within the ebb delta of the Wichter Ee inlet; 29.9.–6.10.1984.

mately 6 cm per tide. The bedforms started to migrate at a current velocity of 47 cm/s with migration being strongest during spring tides.

Dalrymple, Knight & Middleton (1975) and Dalrymple (1984) measured migration rates of flood-orientated sand waves in the Bay of Fundy (Canada), reporting a long term migration rate of about 11 cm per tide.

In addition to these measurements on accessible flood deltas during low tide, the migration of bedforms has also been measured in the subtidal area, at the bottom of a tidal inlet. Ulrich (1979) measured a field of 'giant ripples' (sand waves) in the Lister Tief inlet between Sylt and Rømø. The test field comprised an area of mostly ebb-orientated forms in the southern part of the inlet and an area directly to the north of mostly flood-orientated forms. Sand wave height was about 2–11 m and spacing about 300 m maximum. The maximum current velocity measured was 130 cm/s. The migration rates for 3 m high bedforms were about 60 m per year, i.e. less than 10 cm per tide (cf. Ulrich & Pasenau, 1973).

Ulrich (1979: 338) investigated the tide-induced oscillation of sand waves in the Heppenser Fahrwasser channel of the Jade. He concluded that they were rarely reshaped completely during the turn of the tide. In most cases only the upper third of the ripples was altered by the reversal of the tidal current.

My own observations on the Wichter Ee tidal deltas led to the suspicion that the megaripple fields there were subject to more intense changes than those described above. Comparison of aerial photographs taken over an interval of 39 days revealed no similarities in megaripple distribution.

RESULTS OF THE PRESENT INVESTIGATIONS

Marking of ripple crest positions

Attempts to monitor ripple movement using an automatic camera installed in a waterproof case on the Wichter Ee ebb delta failed because of the high turbidity of the water. Similar problems were encountered by Luck (1979) when attempting to use an automatic underwater television camera.

In order to obtain information concerning tidal ripple migration, the positions of distinct megaripple crests were marked with tent pegs on the Othello Shoal Shield, bordering the eastern end of Norderney. The pegs were placed at 6 m intervals and were loosely connected by a line. It proved necessary to use 50 cm long pegs to ensure that most of them were still in position after one tide. This

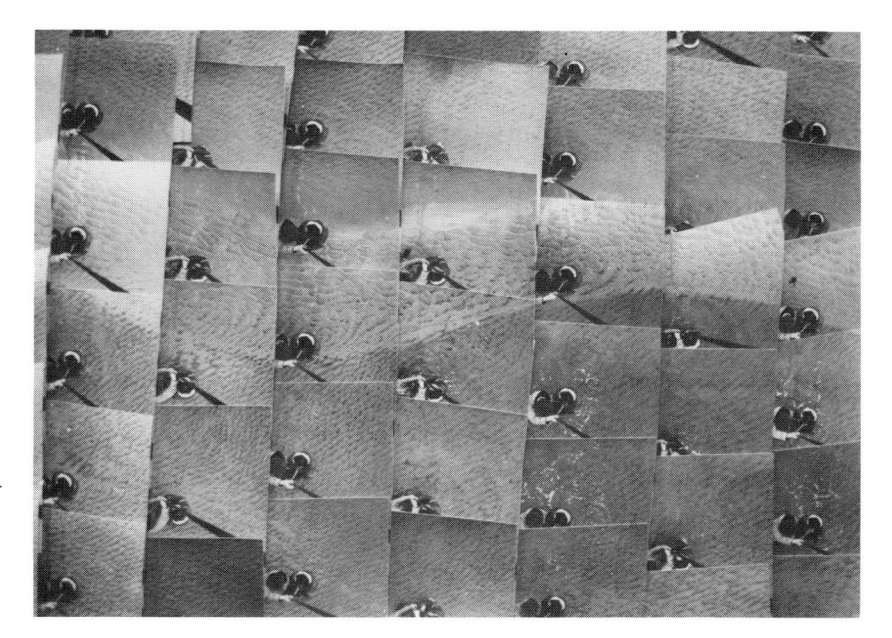

Figure 102: An example of an 'aerial photograph mosaic', composed of vertical photographs taken from 5 m height. Part of the investigated ripple field, flood delta of the Wichter Ee, 9th low tide (6.7.1984).

demonstrated that the depth of reworking was parallel to the maximum ripple height.

The results indicate that the form of a ripple is determined by its position on the shoal (top or flank) and by its migration rate. Test row 1 was located close to the Wichter Ee in an area where ebb-orientated megaripples migrate onto the first shoal of the Othello Shoal Shield (for location see fig. 101). The mean ripple length here was 5.9 m (min: 2.7 m; max 7.9 m), and the average ripple height 34 cm (min: 2 cm; max: 48 cm). Average migration rate was 3.20 m per tide in the direction of the ebb current (min: 2.05 m; max: 7.00 m).

The differences between minimum and maximum migration are partially the result of lateral movement of the ripple tongues and lobes during migration. Measurements were always taken at right angles to the strike direction of the ripple crests.

Test row 2 was situated 100 m north of test row 1, on the eastern flank of the shoal (fig. 101). Higher ebb current velocities were found here on top of the shoal, than in the adjoining channel. As a result, the migration rates measured in the southern part of the 36 m long test row (7 fixed points) were markedly higher than in the northern part. The following table shows the average ripple lengths, heights and migration rates for the 7 fixed points. Point 1 marks the northeastern end, point 7 the southwestern end of the row. The measurements covered an interval of 11 tides.

Point	Ripple length	Height	Migration rate
1	6.60 m	0.27 m	3.90 m
2	5.80 m	0.21 m	4.59 m
3	5.30 m	0.23 m	5.60 m
4	5.40 m	0.21 m	6.05 m
5	5.60 m	0.22 m	6.60 m
6	3.90 m	0.22 m	7.90 m
7	6.70 m	0.19 m	7.85 m

Twenty-five metres south of test row 2 an extremely large megaripple was found within the same ripple field, which was strikingly different in dimension from all of the surrounding ripples (average length: 31.9 m, average height: 0.22 m). The migration of this 'giant' megaripple was traced over a period of 9 tides. It had an average migration rate of 5.90 m per tide (min: 3.20 m, max: 8.00 m).

It may seem surprising that these values are so much greater than those recorded in other areas. Extrapolated over a year, annual migration rates of 1500–2000 m result. The difference can be attributed to the fact that in this study, for the first time, forms in the ebb delta of a tidal inlet were being measured where sediment movement rates are considerably higher than on the flood deltas. Only a small sector of the delta could be monitored, however, so these measurements alone should not be regarded as representative for the whole ebb delta.

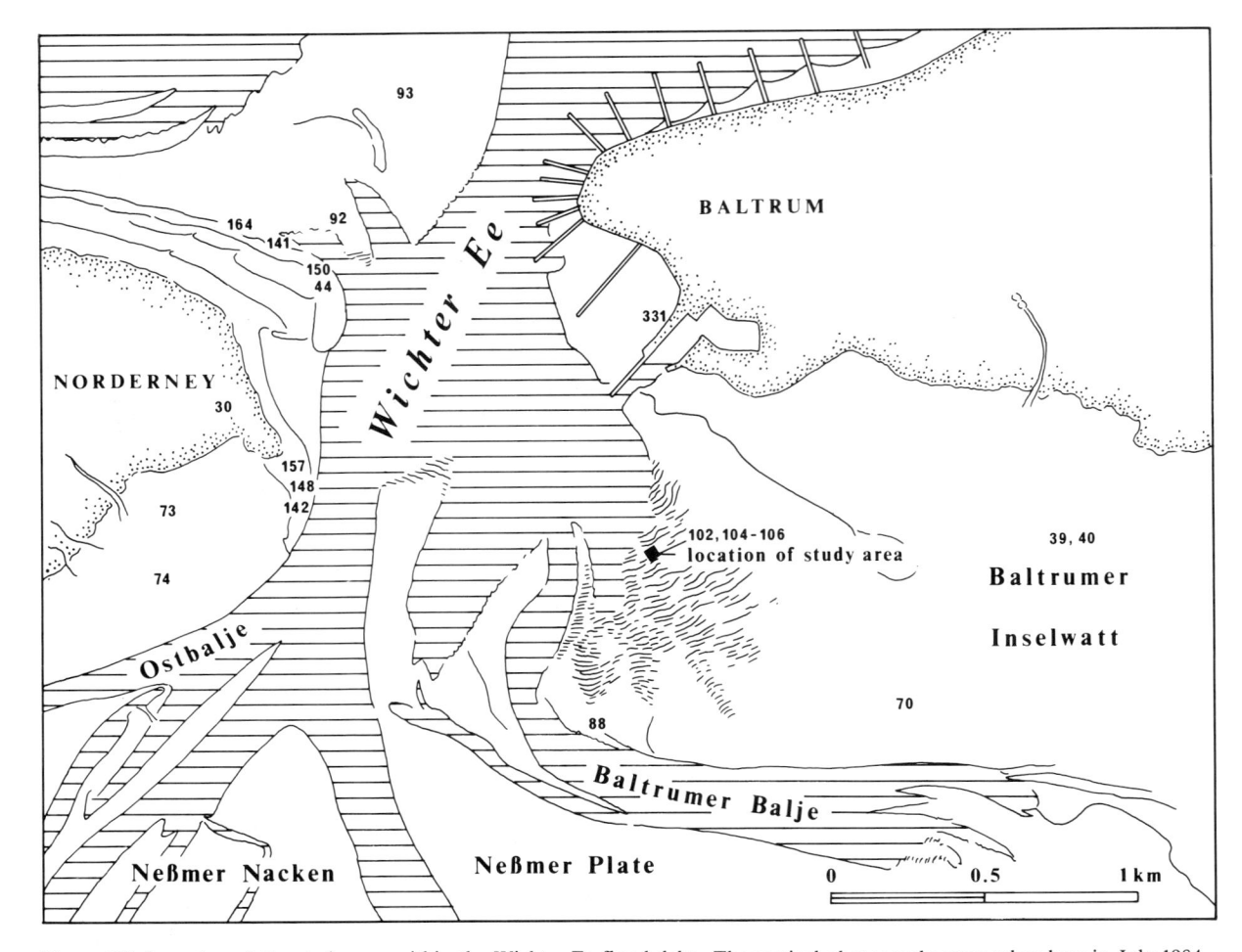

Figure 103: Location of the study area within the Wichter Ee flood delta. The vertical photographs were taken here in July 1984.

Vertical photographs

The marking of bedform crests in an area characterized by fast-moving sediments, with a few exceptions, cannot be totally objective. Often it is difficult to be certain that the measured crest is the same as that previously measured during the last low tide. It could easily be the adjoining, the third crest, or even the ripple could have moved in the opposite direction in spite of its ebb-orientation. An attempt was therefore made to substantiate the results by monitoring a marked megaripple field using vertical photographs taken from the ground (large scale 'aerial' photographs).

A camera equipped with a wide angle lens (24 mm) and a winder was mounted on a 5 m long steel rod. A vertical photograph taken from this height covers an area of about 4.20 × 6.00 m. A photograph was taken every 2 m in order to obtain sufficient overlap. During one low tide a maximum area of 3,000 m^2 can be covered by this method. The photographs can be quite easily combined to form an 'aerial photograph mosaic' (fig. 102).

Measurements on the flood delta

The sand wave fields of the flood delta of the Wichter Ee are easily accessible from the Isle of Baltrum. The area investigated formed part of a flood ramp of about 1.2 km^2. It is covered with flood-orientated sand waves (fig. 103). Field investigation demonstrated that the influence of the ebb current on this bedform field is minimal. Only the crests are vaguely reorientated by the ebb current (figs. 104 and 105).

Figures 104 and 105: Flood-orientated sand waves on the Wichter Ee flood delta. The flood ramp is shielded against alteration by the ebb current. Only the sand wave slopes are slightly flattened and the crests are reorientated by the ebb current (1980).

The flood current dominates because the flood ramp is well shielded against the Baltrumer Balje channel in the south and against the Baltrumer Inselwatt flats in the east so that any overflow by ebb currents can only occur during extreme high water.

The sand waves hardly changed their form and position during the investigation period (2.–6.7.1984). Only a slow migration in the direction of the flood current was recorded which reached a maximum of about 20 cm per tide, and on average considerably less (fig. 106). Marginal parts of the sand wave field even demonstrated a reverse tendency. Therefore it is not surprising that in this part of the flood delta an identical sand

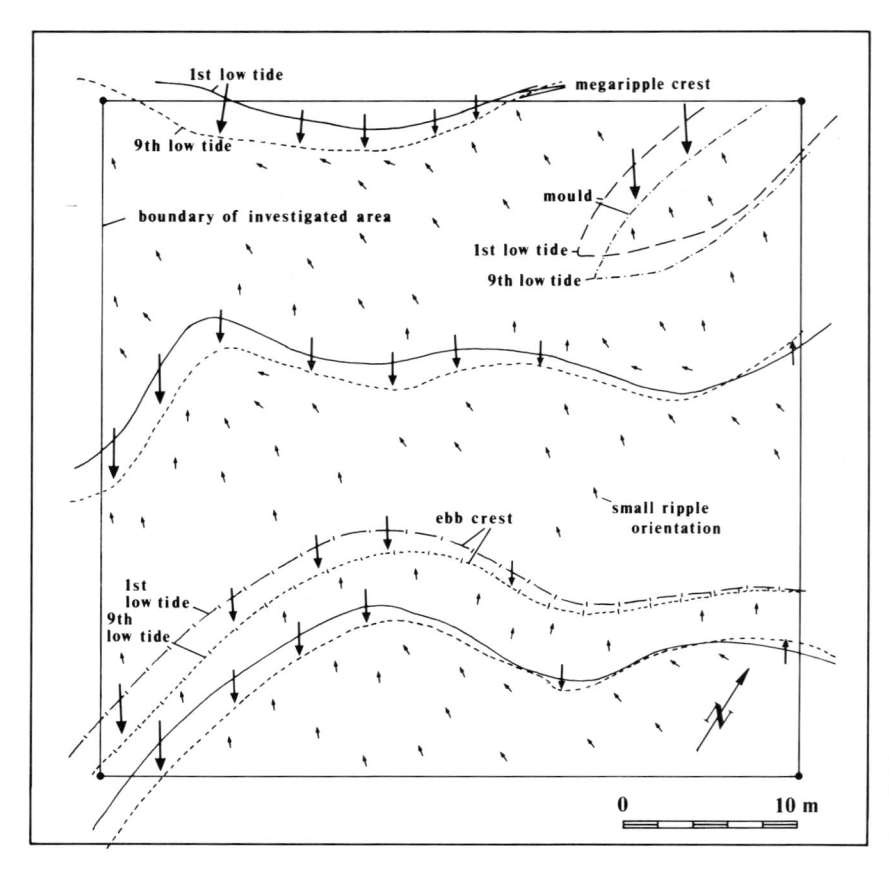

Figure 106: Sand wave migration in the Wichter Ee flood delta (July 1984) from the 1st to the 9th surveyed tidal cycle.

wave configuration occurred even after more than a month. On the aerial photographs of 1982, taken over a period of 39 days (cf. section 5.2.2), the same bedforms can be identified on both sets of pictures. The migration rate at that time was about 10 m at most, i.e. about 13 cm per tide.

Measurements on the ebb delta

The vertical photographs of the ebb delta yielded less conclusive results. The measurements covered the period 25.6.–1.7.1983 and were made on the first shoal of the Othello Shoal Shield bordering the eastern end of Norderney. An ebb-orientated megaripple field with crests striking NNE-SSW was surveyed. The ripple length ranged between 3-8 m.

The planned survey period was interrupted by strong westerly winds which prevented draining of the shoal at low tides. Before this interruption three sets of photographs were taken, and afterwards four sets. The photographs were taken during consecutive low tidal stages.

During the first three surveys the ripple field maintained the same pattern. Ripple lengths increased from SE to NW. This pattern changed after the storm, with shorter lengths being found in the NW (surveys 7, 8 and 9). During the tenth low tide the pattern became confused and the ripple crests from the preceding surveys could no longer be identified.

This study indicates that megaripple migration on the ebb delta occurs at a much faster rate than on the flood delta. The ripple pattern of a 30 × 100 m area often changes completely during a single tide. Only the ripple crests of the first and eighth low tide could with certainty be re-identified on the photographs of the following tides respectively. In both cases the ripple migration direction was opposite to their orientation, i.e. the bedforms migrated with the flood current (fig. 107).

This result differs from that obtained during the 1982 measurements. It also contradicts Hanisch's (1981) results which indicated ripple migration with the ebb current on the Harle ebb delta shoals.

In order to obtain clear results large parts of the ebb delta had to be investigated simultaneously. This was not possible with vertical photography from the ground, as indicated above.

The comparatively small changes demonstrated within the flood delta indicate that this part of the inlet contributes only insignificantly to the vast volume of transported sediment in the entire tidal inlet area.

Aerial photographs

Though it proved possible to follow sand wave migration within the Wichter Ee flood delta on aerial photographs spanning a period of over a month (see above), this interval turned out to be too large for examinination of the ebb delta. To obtain information about megaripple migration rates in this zone, vertical aerial photographs were taken from a light aircraft during 4 successive low tides, using a 6 × 6 camera.

Megaripple migration rates could not be measured directly; such a procedure would require many field assistants and an extensive battery of instruments. In addition, the study area is also extremely unaccessible. The shoals can only be reached by boat during calm weather. No fixed points could therefore be installed. Evaluation of the aerial photograph mosaics was therefore based on the following assumption; migration of the shoals occurs at a rate of approximately 1.10 m per day, i.e. about 0.65 m per tide. This rate is small in comparison with the migration rate of the megaripples. It therefore seems acceptable to regard the position of the shoals as 'constant' and measure only the migration of the ripples relative to the shoals. Where the shift is parallel to the shoal migration, the rates measured can be regarded as minimal and *vice versa*.

Surveys were run between 14.6. (evening low tide) to 18.6.1984 (morning low tide), during a period of strong northeasterly wind (Beaufort 7). High water was about 0.5 m higher than normal. Spring tide occurred on June 13th. Because of the low cloud ceiling the photographs had to be taken from an altitude of only 800 feet. It was therefore not possible to cover the whole breadth of the tidal delta during one transect, making several overflights necessary. An incomplete coverage with about 30 % of the area not covered, was the result and made it impossible to establish exact connections across the major tidal channels. These shortcomings were unavoidable because the coincidence

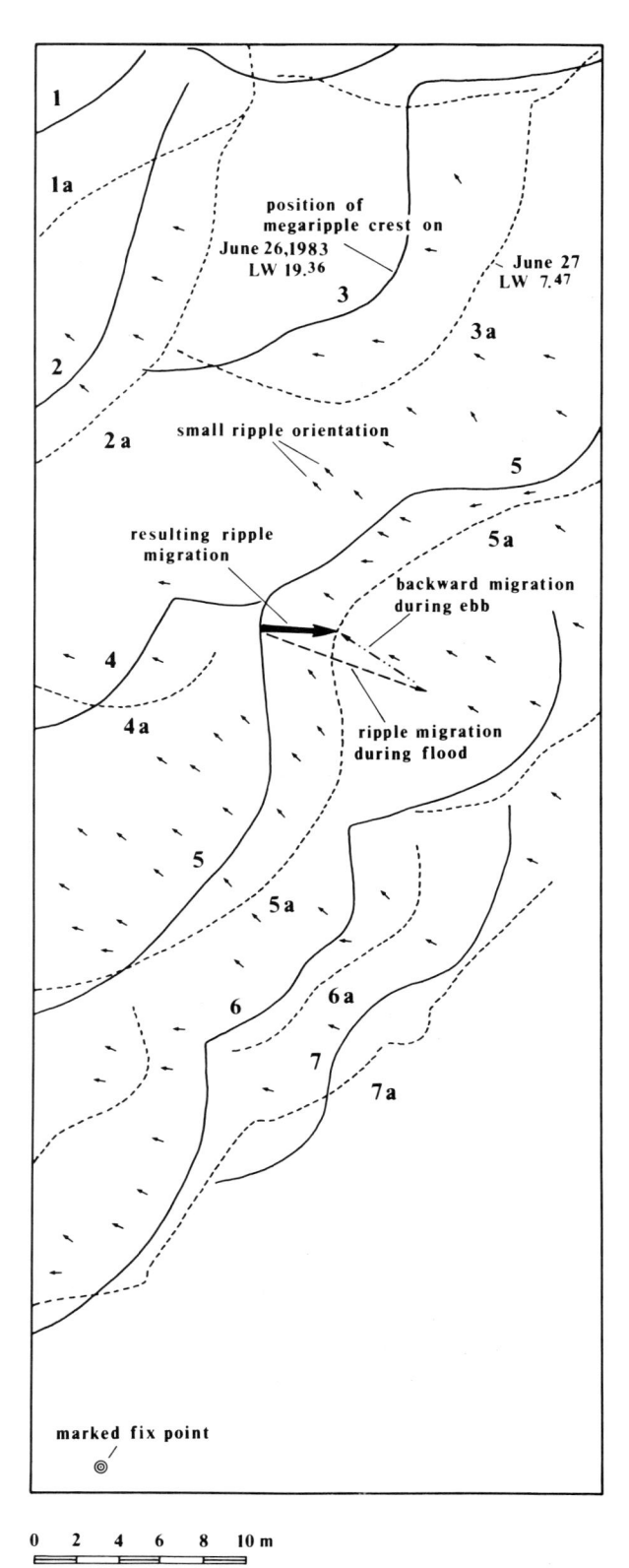

Figure 107: Megaripple migration in the Wichter Ee ebb delta (June 1983) from the 1st to the 2nd surveyed tidal cycle.

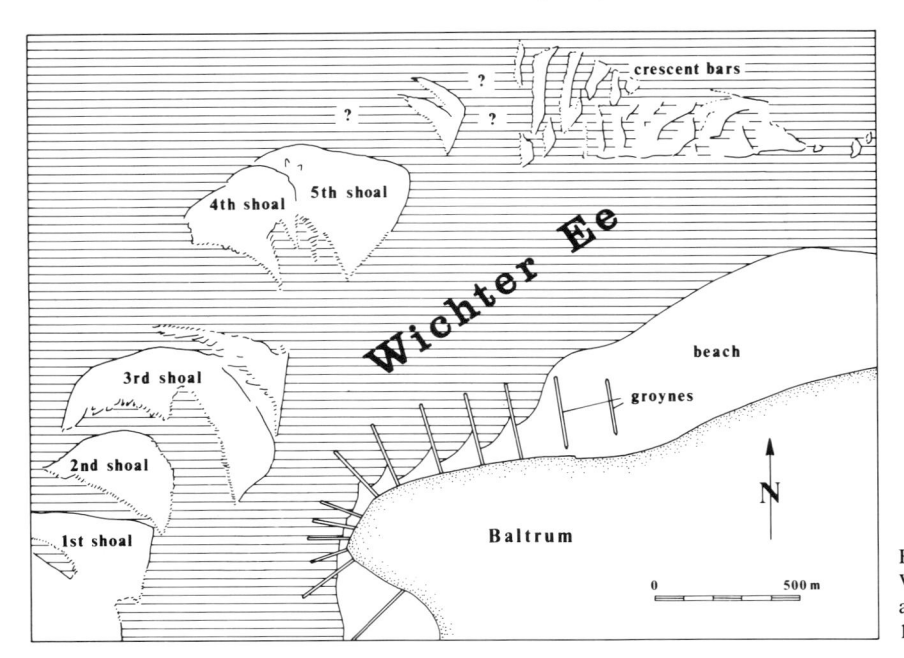

Figure 108: Location map of the Wichter Ee ebb delta during the aerial photograph survey in June 1984.

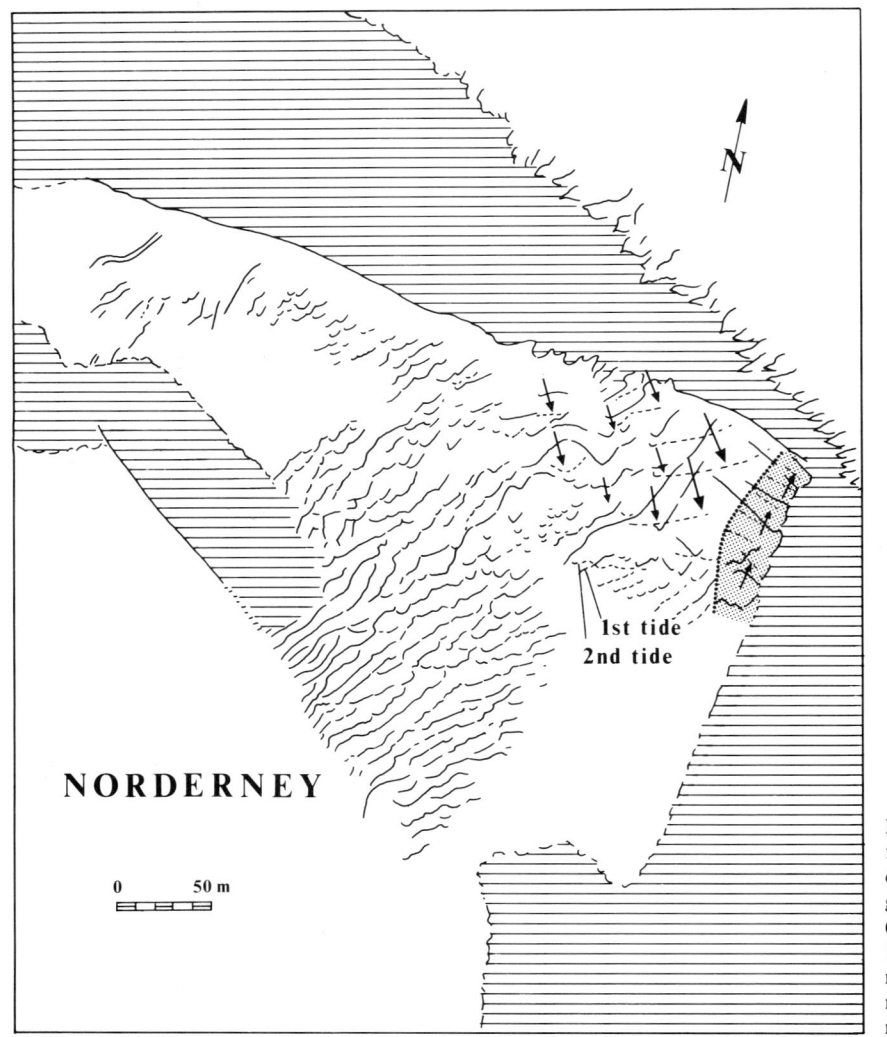

Figure 109: Ripple migration on the first shoal of the Wichter Ee ebb delta; comparison of aerial photographs of the 1st and 2nd surveys (evening of 14.6.84 and morning of 15.6.84). The direction of ripple migration is indicated by the arrows. The shaded area was formed mostly by the ebb current.

between low tides and daylight required allowed no delay in the surveys.

The surveys did, however, prove that changes within the ripple fields were enormous from one tide to the next. The position of ripple fields remained almost constant but it was often impossible to determine migration rates for individual ripples within the ripple fields. Many of the forms were not sufficiently characteristic to permit recognition after the modifications caused by the tide. Reliable results were obtained only where ripple length clearly exceeded migration rate. In this case a ripple length of about 10 m was normally sufficient.

The distribution pattern of the megaripples indicates that the intensity of sediment movement is variable within the ebb delta. At certain locations huge trapezoid megaripples occur with lengths often exceeding 25 m. These are indicative of high migration rates and, in contrast with the large megaripples found at the bottom of the channels, they are of low amplitude; their crests rarely exceed 10–20 cm. The ripple pattern indicates that the migration rate can vary from slow rates on the marginal slopes of the shoals to high rates actually on the shoals.

The observed bedform migration pattern can be illustrated with a few characteristic examples:

Example 1: Ripple migration at the northeast corner of the first (westernmost) shoal.

The northeastern corner of the first shoal of the ebb delta (adjacent to the eastern end of Norderney, fig. 108) was well covered during all four surveys. This part of the shoal is covered with ebb-orientated megaripples, with a ripple length of up to about 30 m. This area is therefore especially suitable for an investigation of the megaripple migration.

Comparison of the aerial photographs taken during the 1st and 2nd low tide reveals that the majority of the ripples migrated with the flood current, rather than with the ebb current (fig. 109). During this process the northeast-southwest trending crests of the largest ripples, in the northeastern part of the area dominated by the flood current, reorientated near to an E-W direction. The active part of the ripple field, which migrated with the ebb current, is shaded on the map. It is represented by only a small stripe at the eastern margin of the shoal.

From the 2nd to the 3rd low tides, the situation almost reversed (fig. 110); the area originally constructed by the flood current was remodelled by

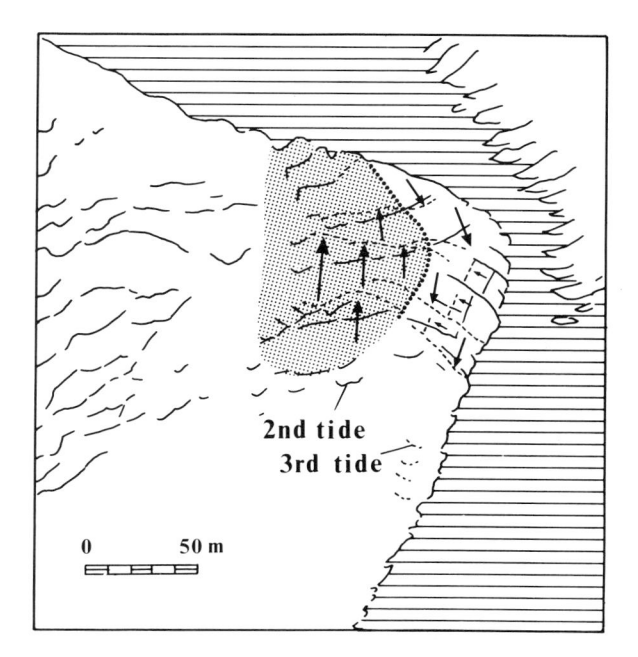

Figure 110: Ripple migration on the first shoal of the Wichter Ee ebb delta; comparison of aerial photographs of the 2nd and 3rd surveys (morning of 15.6.84 and evening of 15.6.84). The direction of ripple migration is indicated by the arrows. The shaded area was formed mostly by the ebb current.

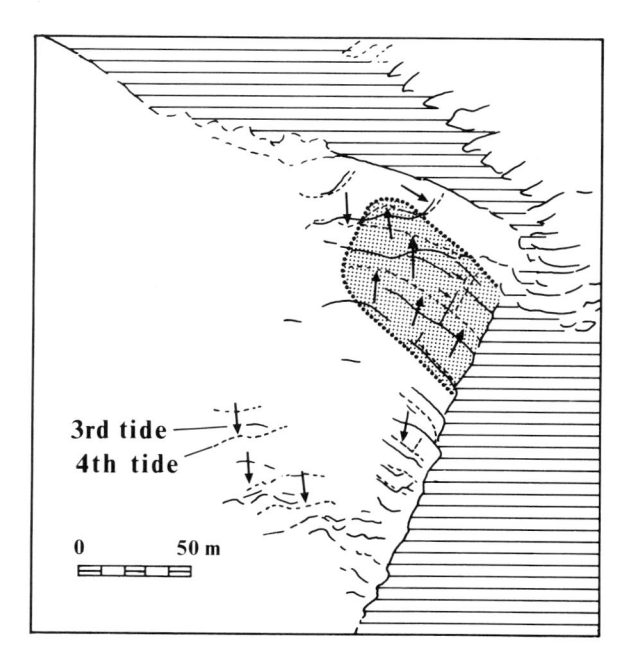

Figure 111: Ripple migration on the first shoal of the Wichter Ee ebb delta; comparison of aerial photographs of the 3rd and 4th surveys (evening of 15.6.84 and morning of 16.6.84). The direction of ripple migration is indicated by the arrows. The shaded area was formed mostly by the ebb current.

120   *Morphodynamic units*

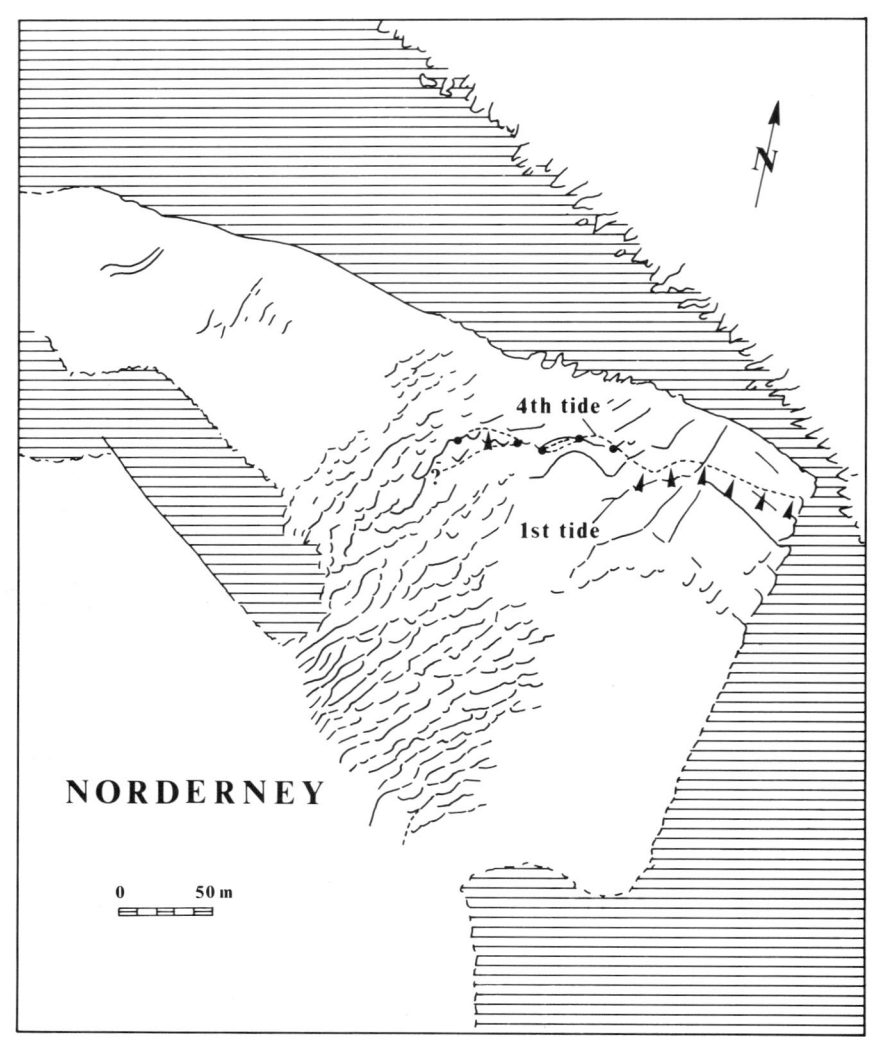

Figure 112: Migration of the transverse bar on the first shoal of the Wichter Ee ebb delta from 1st to 4th survey. Since the ebb and flood currents do not flow in precisely opposing directions, sediment is transported roughly parallel to the arcuate outline of the ebb tidal delta.

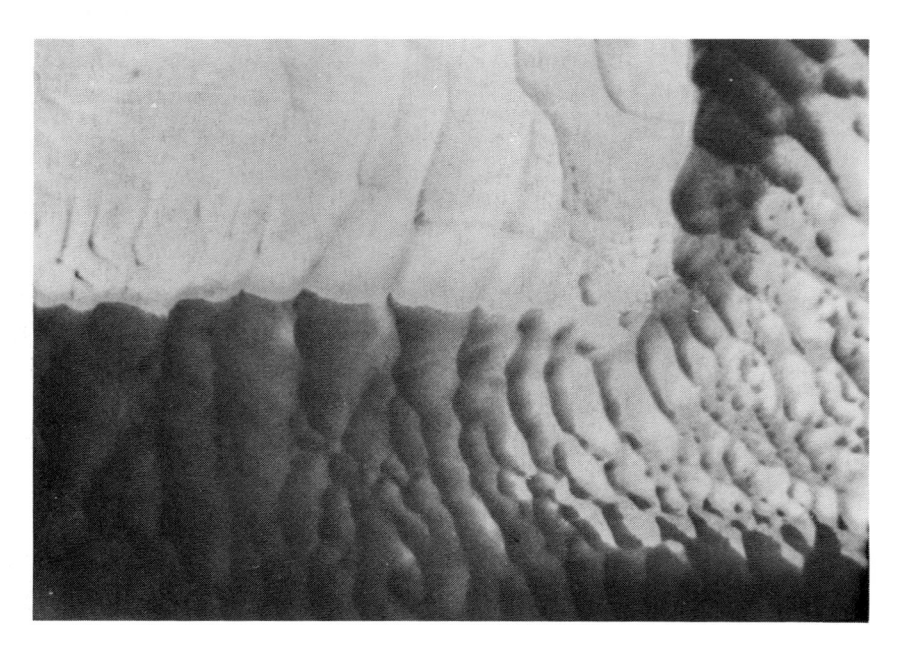

Figure 113: The NE corner of the first shoal of the Wichter Ee ebb delta during the 3rd survey (evening of 15.6.1984). Aerial photograph of F.Böker, freigegeben Brg. Nr. 5765/596.

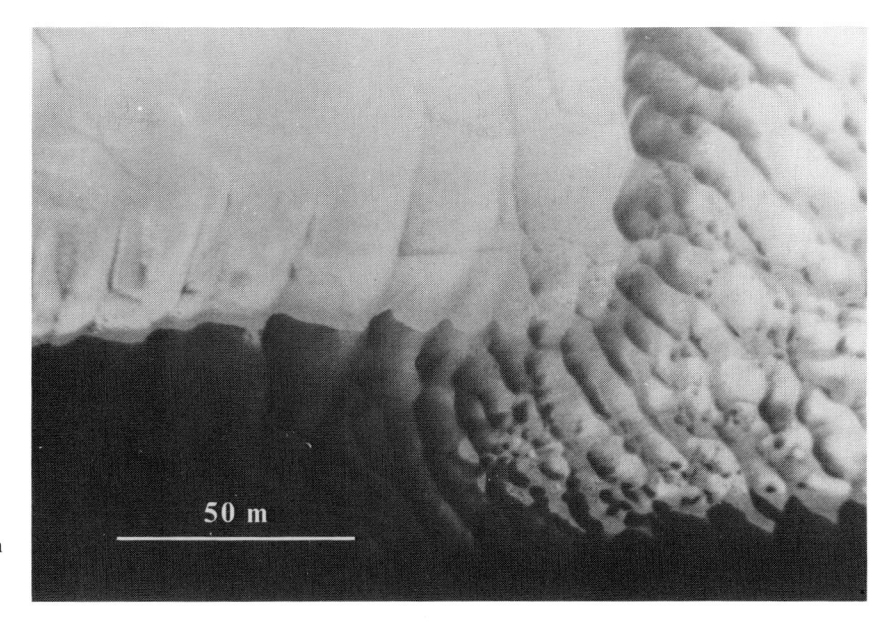

Figure 114: The NE corner of the first shoal of the Wichter Ee ebb delta during the 4th survey (morning of 16.6.1984. Aerial photograph of F.Böker, freigegeben Brg. Nr. 5765/597.

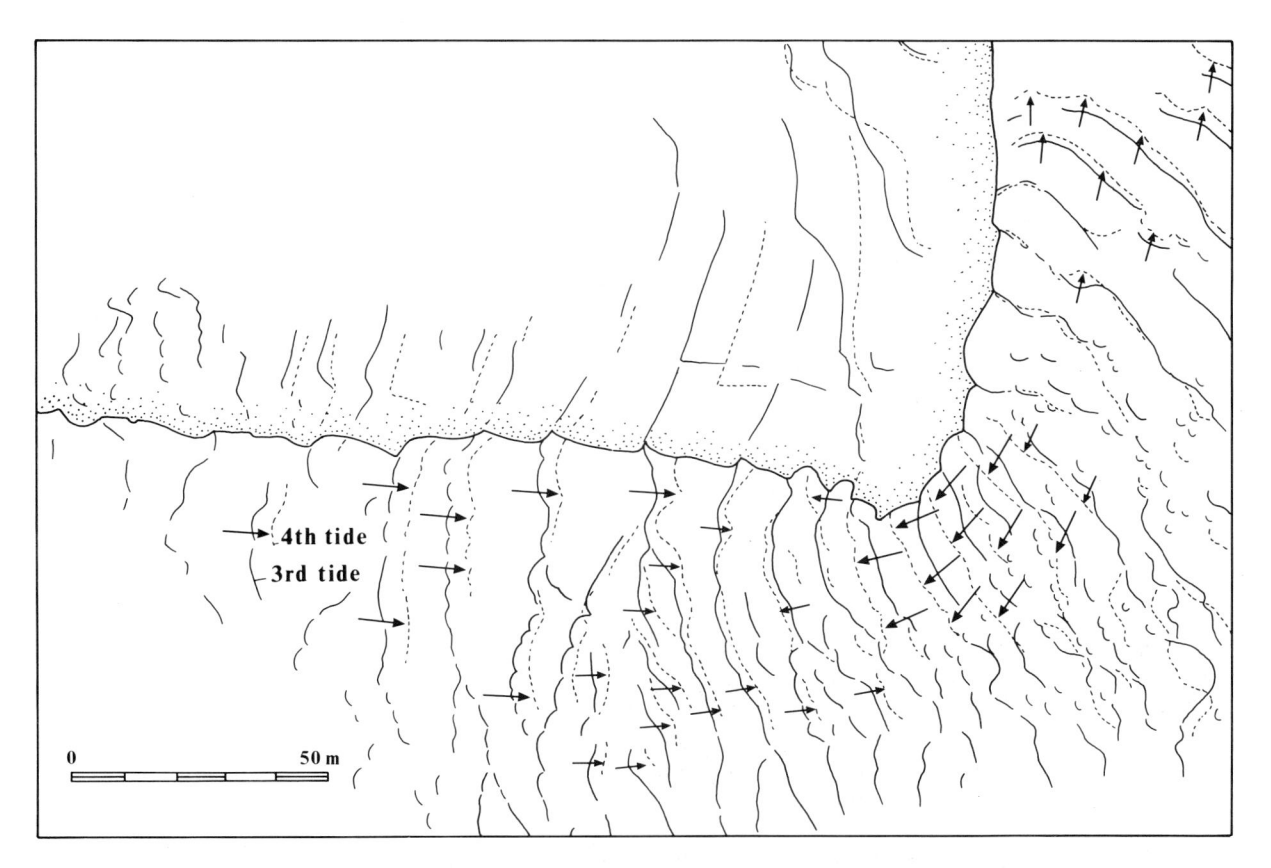

Figure 115: Ripple migration in the channel at the NE corner of the first shoal of the Wichter Ee ebb delta; comparison of aerial photographs of the 3rd and 4th surveys (evening of 15.6.84 and morning of 16.6.84). The direction of ripple migration is indicated by the arrows.

122   *Morphodynamic units*

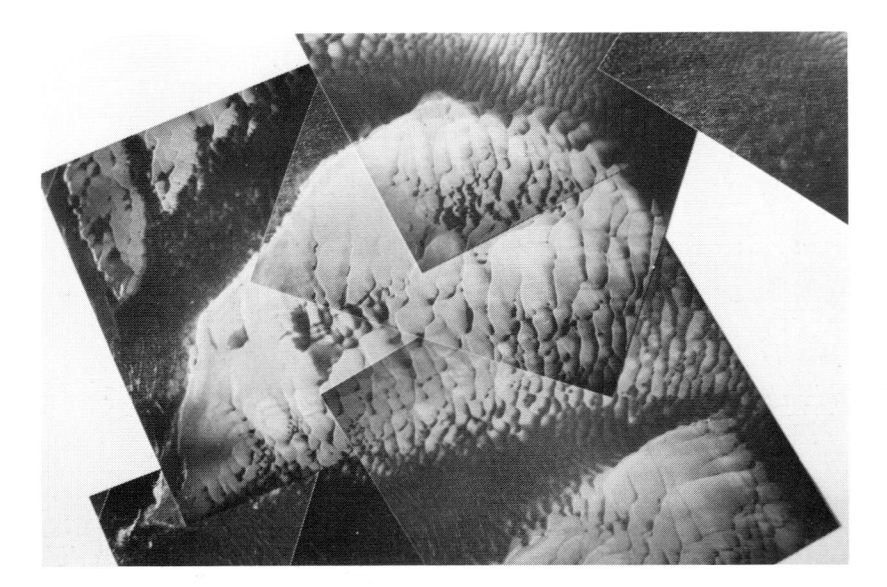

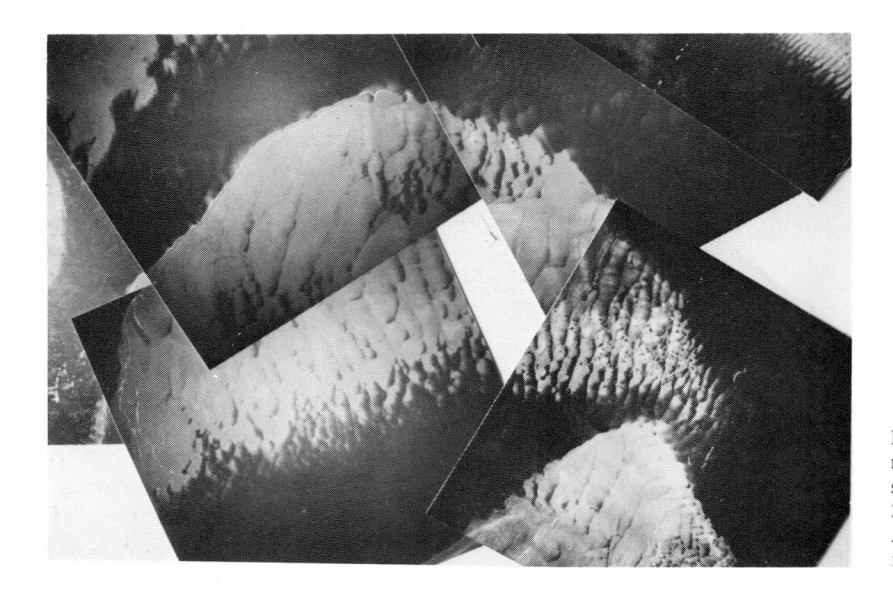

Figure 116: Aerial photograph mosaics of the 2nd shoal; (a) 1st survey (evening of 14.6.1984), (b) 2nd survey (morning of 15.6.1984). Aerial photographs of F.Böker, freigegeben Brg. Nr. 5765/598-600.

the ebb current, whilst the previously ebb-dominated section migrated with the flood current. With the exception of the largest ripples, the migration rates were roughly of the same order of magnitude as before (about 5–10 m).

Comparison of the aerial photographs from the 3rd and 4th low tides again demonstrates that some parts of the shoal were dominated by the flood current, and others by the ebb current (fig. 111). The individual forms permit no conclusions concerning their direction of migration. This can only be reconstructed by evaluation of the aerial photographs. The velocity of the ebb current is sufficiently high to modify all ripples into ebb-orientated forms before low tide.

Sand migrates along the shoal arc because the ebb and flood currents do not flow in precisely opposing directions. Flow directions form an oblique angle. This is obvious when the positions of the transverse bar (margin of soal subunit) are followed during the succession of tides (fig. 112). With the exception of a few small segments in

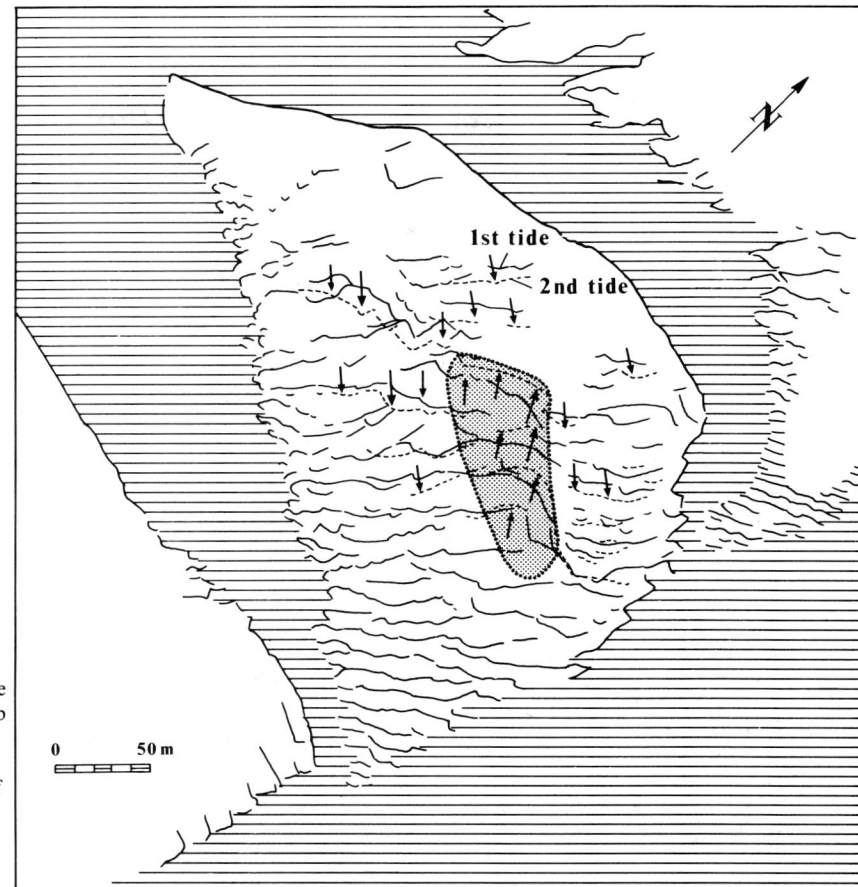

Figure 117: Ripple migration on the second shoal of the Wichter Ee ebb delta; comparison of aerial photographs of the 1st and 2nd surveys (evening of 14.6.84 and morning of 15.6.84). The direction of ripple migration is indicated by the arrows. The shaded area was formed mostly by the ebb current.

which no measurable shift has occurred, the transverse bar has moved in a northerly direction (cf. section 5.2.2).

Example 2: Ripple migration in the channels at the northeast corner of the first shoal (cf. map in fig. 108 and photographs figs. 113 and 114).

All the megaripples visible on these two aerial photographs are ebb-orientated. In particular, the shallow water megaripples at the margin of the Wichter Ee (lower part of the photographs), and in the channel between the 1st and 2nd shoal to the right, demonstrate characteristic forms which can be identified on both photographs. Their comparison shows that ripple sets in both sectors displayed different movements (fig. 115). In the marginal parts of the Wichter Ee the ripples migrated at a high speed towards the north (about 5 m/tide), whilst the ripples in the narrow portion of the channel between the two shoals (easternmost part of the first channel) have migrated about 3 m towards the southeast, i.e. with the flood current.

Where the lack of specific, recognizable bedforms prevents the identification of the ripple migration direction in the tidal channels, adjacent ripples on shoals may still provide valuable information. This is the case on the steep northern and eastern slopes of the first shoal. The photographs indicate that ripple migration rates on the shoal are smaller than in the channel, whilst the accelerating ripples at the margin of the Wichter Ee suggest higher migration rates on the shoal than in the channel.

Example 3: Ripple migration on the second shoal of the ebb delta.

The second shoal displays a characteristic ripple pattern that is easily identified on all aerial photograph mosaics (fig. 116). The low region in the northeastern part of the shoal is especially promi-

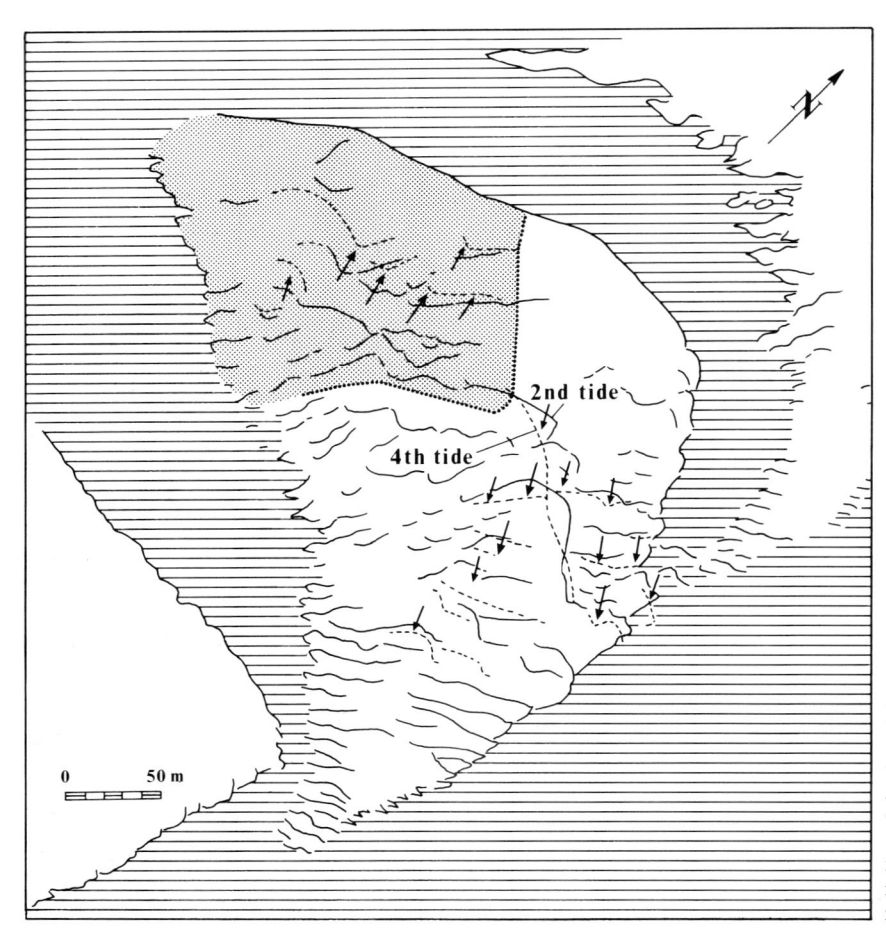

Figure 118: Ripple migration on the second shoal of the Wichter Ee ebb delta; comparison of aerial photographs of the 2nd and 4th surveys (morning of 15.6.84 and morning of 16.6.84). The direction of ripple migration is indicated by the arrows. The shaded area was formed mostly by the ebb current.

nent; it remains water-covered at low tide. Megaripples over 20 m in length and about 10−20 cm high occur here.

The most striking ripple crests are orientated in a SW-NE direction which, together with the ebb-orientation of the bedforms, suggests sand transport towards the northwest. Comparison of the aerial photographs of the 1st and 2nd low tides reveal, however, that during this period at least, sand transport by the flood current prevailed, in a south-easterly direction (fig. 117). Only in a small area in the central part of the shoal did ripples migrate in the direction of the ebb current.

As the photograph mosaic of the third low tide only covers a small part of the second shoal, fig. 118 shows a comparison of the photographs of the 2nd and 4th surveys. This again demonstrates that most of the second shoal was moulded by the flood current, whilst a minor portion in the northeast corner demonstrates ripple migration with the ebb current. The maximum ripple migration per tide was about 5−7 m.

As on the first shoal, the migration of the transverse bar was followed throughout the whole period of investigation (fig. 119). The total observed distance of migration ranged between 0−10 m, with maximum values slightly exceeding 3 m per tide. Sand moved nearly parallel with the seaward boundary of the shoal arc.

Example 4: Ripple migration in the channel at the northeast corner of the second shoal.

The photographs taken during the 1st and 2nd survey provide an opportunity to examine ripple migration in the channel between the 2nd and 3rd shoals (fig. 120). The ripples in the deeper parts of the channel migrated about 2−3 m in the same direction as the ebb current. Those in the shallower area at the channel mouth did not participate in this movement, similar to those in the channel be-

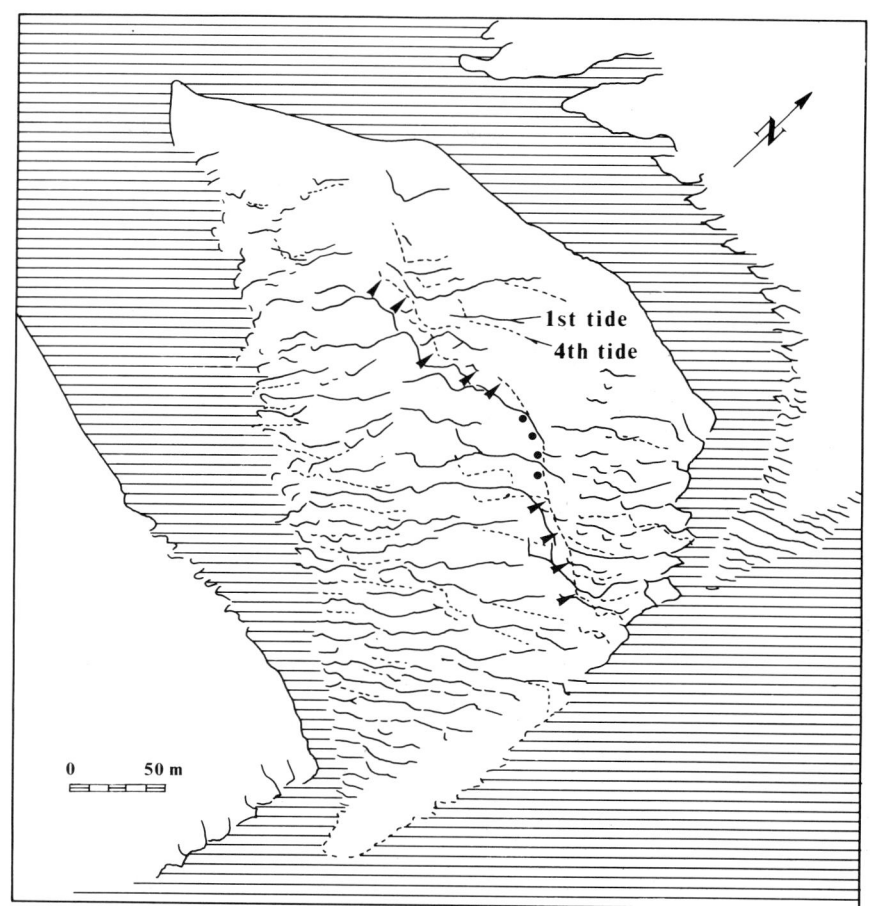

Figure 119: Migration of the transverse bar on the second shoal between the 1st survey (evening of 14.6.1984) and 4th survey (morning of 16.6.1984).

tween the 1st and 2nd shoals. No clear migration in either direction was observed.

Example 5: Ripple migration in the channel between and on the southern slopes of the 4th and 5th shoals.

Comparison of the aerial photographs of the 3rd and 4th surveys (fig. 121) reveal a clear pattern of megaripple migration in the same direction as the ebb current flow. This movement is especially pronounced in the deeper parts of the channel between the shoals, where maximum rates of about 7 m per tide were attained. Only in a small area in the lee of the 4th shoal is a slight dominance of the flood current exhibited.

Example 6: The western part of the crescent bars.

The 6th shoal of the ebb delta during the time of the survey period consisted of a double crescent bar sand bank (fig. 122). The shoal was surveyed only during the first two low tides. It is situated at the turning point of the ebb delta where the strike direction of the shoals changes from N-S to NE-SW. The aircraft also had to change direction at this point which made it difficult to obtain usable photographs.

Comparison of the two surveys available demonstrates that on this shoal ripple migration was also mainly directed into the tidal inlet. This fact was not reflected by the ripple orientation at low tide (fig. 122). Maximum migration rates of about 10 m were noted. Only a small area between the two crescent bars demonstrated ripple migration with the ebb current.

Example 7: The eastern part of the crescent bars.

Only relatively small megaripples with lengths of 1–2 m and widths of 2–10 m were observed here during the survey period. These were unsuitable for comparison over successive tidal cycles (fig.

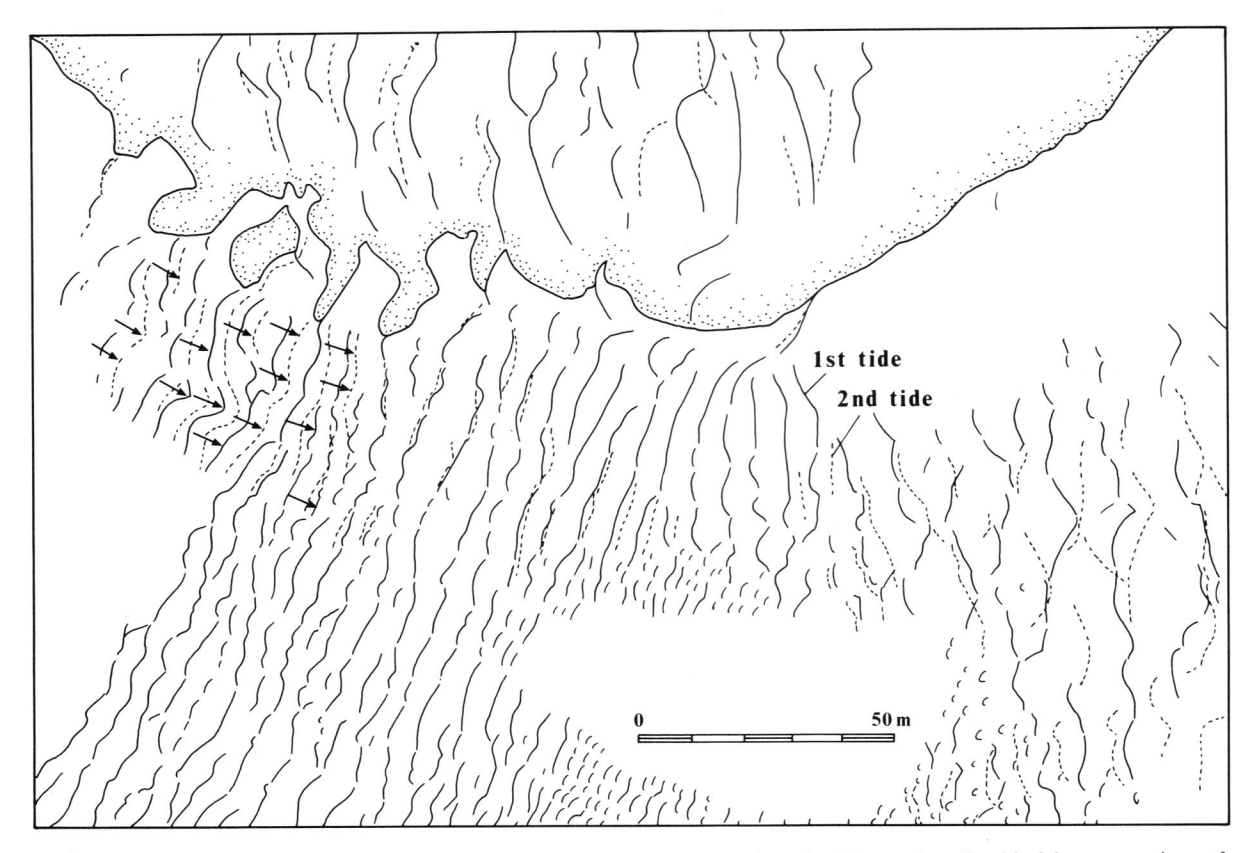

Figure 120: Ripple migration in the channel between the second and the third shoal of the Wichter Ee ebb delta; comparison of aerial photographs of the 1st and 2nd surveys (evening of 14.6.84 and morning of 15.6.84). The direction of ripple migration is indicated by the arrows.

123). The clearly visible ebb orientation gives no suggestion of the actual direction of movement. The influence of the ebb current appears to have been stronger on the shoals than in the channels. Ripples on the bar margins outrun channel ripples in ebb direction.

Major changes, however, were observed at the eastern end of the crescent bars during the survey. A new elongated sand bank formed north of the easternmost crescent bar, the surface of which was completely levelled by the surf. The formation of this shoal seems to have been a result of strong sediment movement in the surf zone at the seaward margin of the ebb delta. Sediment transport was certainly intensified by the strong northwesterly winds.

The influence of the surf on the shaping of the other shoals seems to have been of relatively minor importance. During fair weather, the shoals are completely covered by megaripples from the inlet to the seaward margin. Strong winds (for example during the 1984 surveys) result in the flattening of the seaward parts of the shoals (cf. section 4.3). This, however, is only a secondary process. The final conclusion must therefore be that when the shoals are inundated, ripple-forming processes prevail. The sequence of events seems to be similar to that described by Reineck (1963), concerning the formation of longshore bars.

Conclusions

The observations and measurements presented above permit, for the first time, the detailed reconstruction of the sand transport mechanism in the reaches of a tidal inlet (fig. 124). This reconstruction supplements Hanisch's observations (1981) on the Harle ebb delta. It incorporates a number of parallel processes.

The aerial photograph mosaics of all four sur-

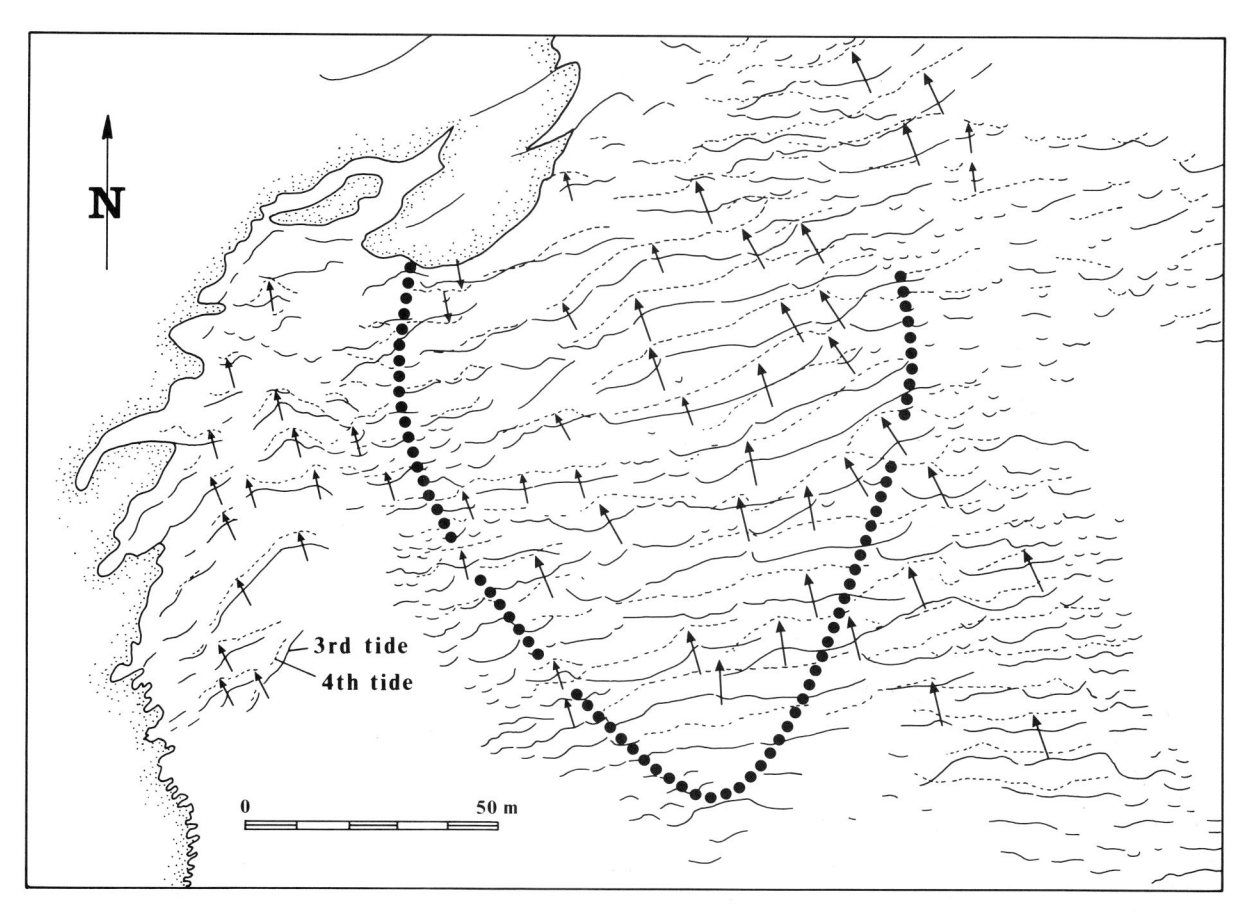

Figure 121: Ripple migration in the channel between the third and the fourth shoal of the Wichter Ee ebb delta; comparison of aerial photographs of the 3rd and 4th surveys (evening of 15.6.84 and morning of 16.6.84). The direction of ripple migration is indicated by the arrows.

veys demonstrate that the extent and form of the major ripple fields in this area remained almost static throughout the survey period. This indicates that the form assemblages do not owe their development to rapidly changing conditions, but that they are in a state of equilibrium with long term processes. The fact that a number of the smaller form elements also remain almost unchanged, through a succession of tides, supports this assumption.

Even very detailed investigation of a single aerial photograph fails to reveal the dominant ripple migration directions. Ebb current modification

Figure 122: Ripple migration on the crescent bars of the Wichter Ee ebb delta; comparison of aerial photographs of the 1st and 2nd surveys (evening of 14.6.84 and morning of 15.6.84). The direction of ripple migration is indicated by the arrows. The shaded area was formed mostly by the ebb current.

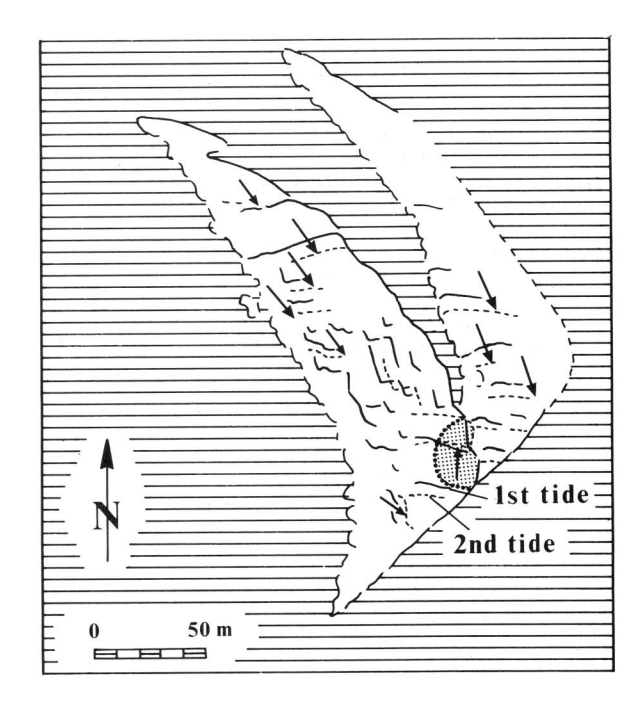

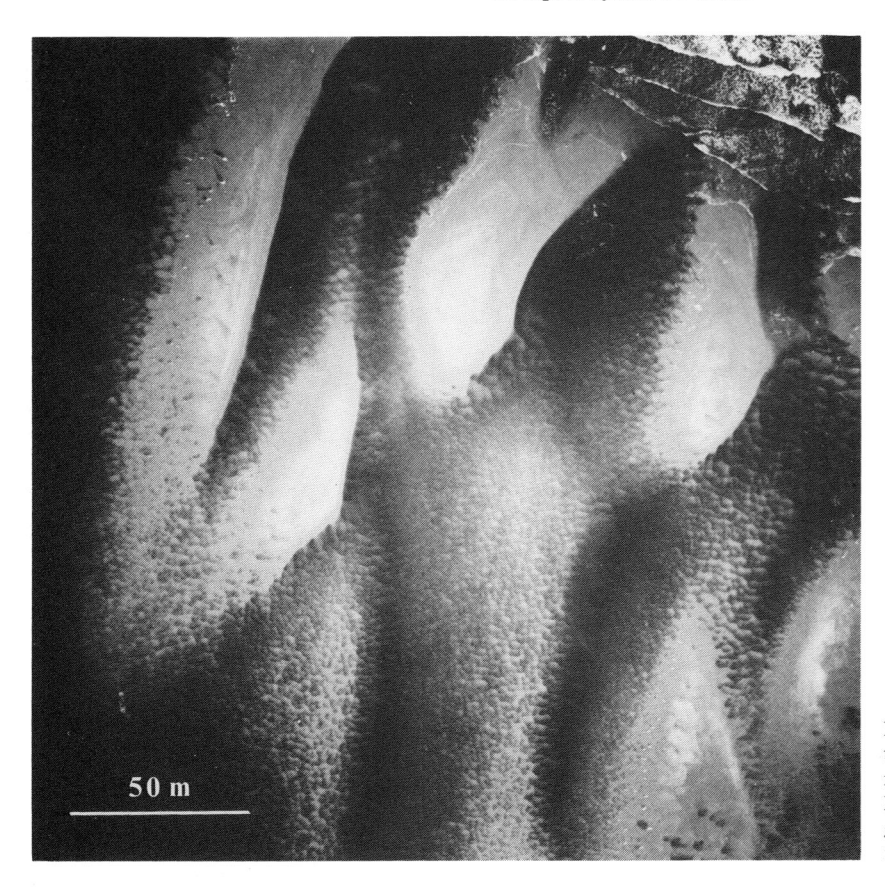

Figure 123: Aerial photograph of the crescent bar area, where small megaripples predominate (ripple length: about 1–2 m). Aerial photograph of F.Böker of 15.6.1984, freigegeben Brg. Nr. 5765/601.

was so strong in each case that all traces of the preceding flood phase were either eliminated or changed beyond recognition.

Hanisch (1981), while investigating the bedding structures of the shoals of the Harle ebb delta, came to the conclusion that the sand on the shoals was mainly transported by the ebb current and that the flood current was more active in the channels. This is in contradiction to the general appearance of the ebb delta morphology. The position and shape of the channels, which resemble the ebb and flood channels of the tidal flats, strongly suggest that sediment transport in some channels is directed seaward and in others into the tidal inlet. Any interpretation of the internal bedding of the shoals is likely to be misleading because in many cases migration rates are larger than ripple lengths which leads to a strong dominance of the ebb-orientated bedding in 40 cm long box cores (fig. 125).

It should also be remembered that box cores can only give information about the uppermost decimetres of the shoals; they cannot penetrate the deeper parts of the sediment bodies. Judging from the channel depths between the shoals, the transported bedforms are at least 6 m thick in the ebb tidal deltas.

The shoals in effect are large ramps. Sediment is transported over their surface mostly through ripple migration. The form and orientation of the ramps provide evidence of the most recent dominant transport directions. The migration of the major shoals seems to occur by a delta-like progradation of the ramps.

The shoal ramps migrate in a given direction until the opposite-running tidal current removes sand from its face. Scarplets form where erosion is dominant. These form along both the inner and the outer margin of the ebb delta (fig. 126) indicating that sand transport on the shoal is not dominantly directed seawards. The ebb and flood current transport rates are approximately of the same magnitude.

Only a minor volume of the sediment transport-

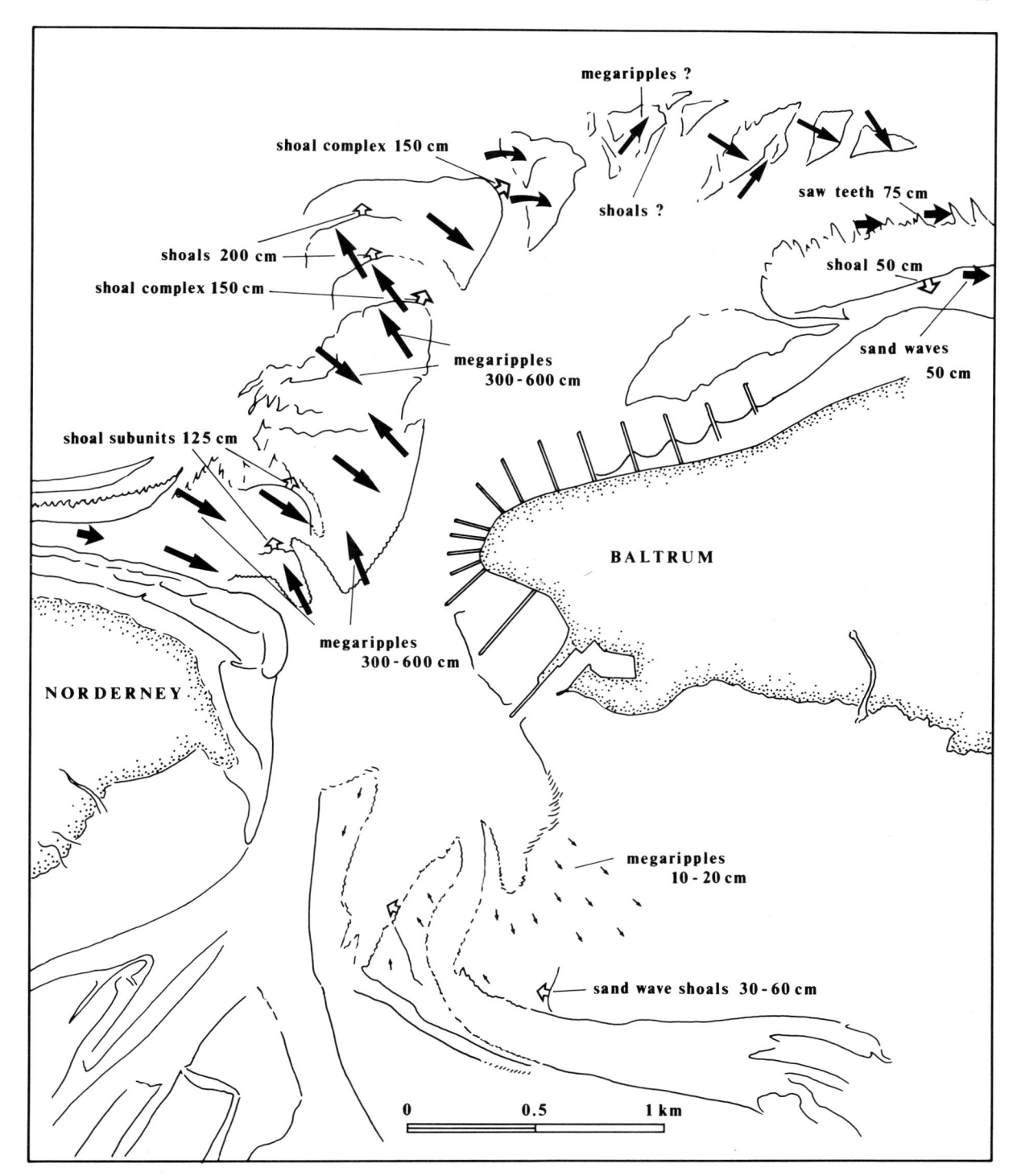

Figure 124: Movement directions of bars and sand waves around the Wichter Ee tidal inlet for a single tide; the size of the arrows represents the speed of the process.

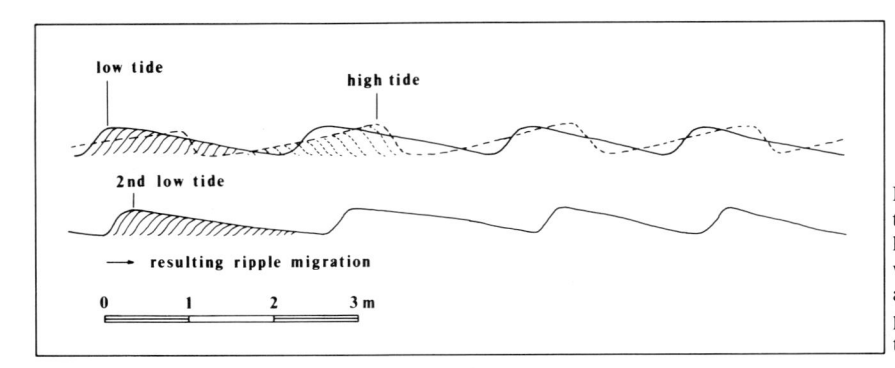

Figure 125: If the migration rate per tide is greater than the ripple length, only ebb-orientated forms with ebb-orientated bedding occur at low tide, even if the resulting ripple migration is in the direction of the flood current.

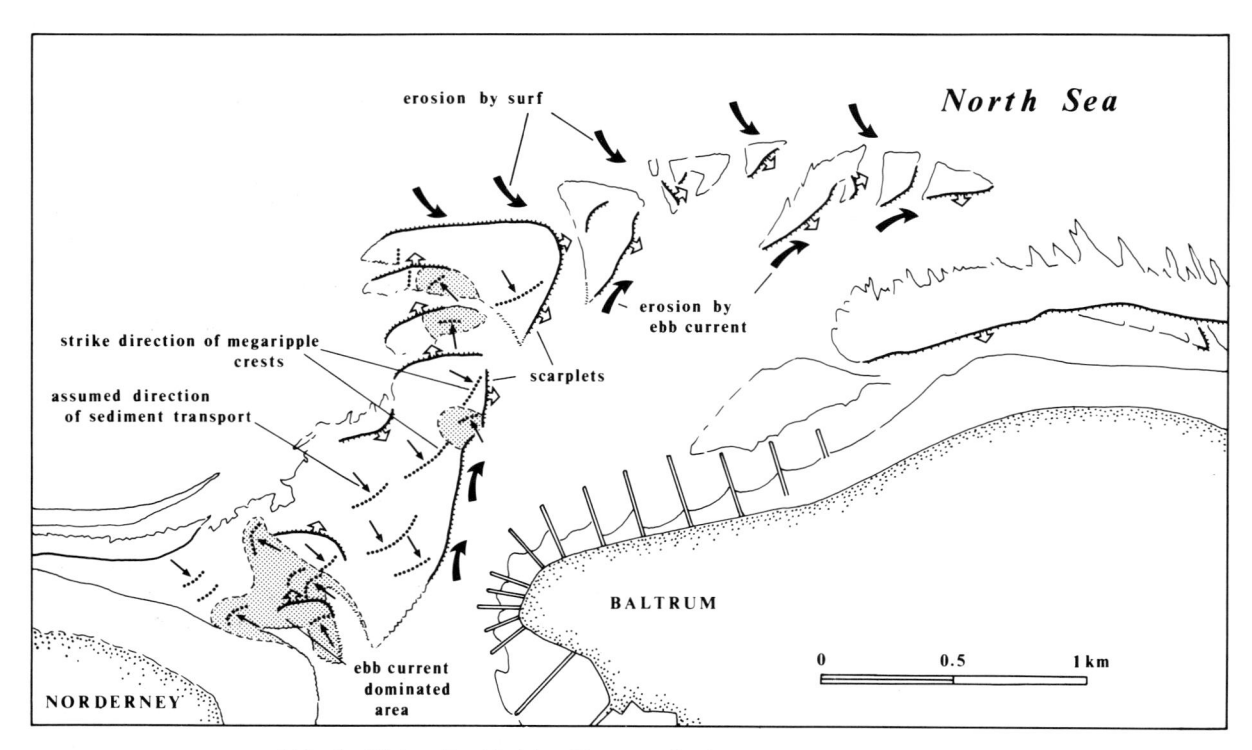

Figure 126: Escarpments within the Wichter Ee ebb delta. The generalized curvature of ripple crests indicates the dominant directions of sediment transport (drawn from aerial photographs taken on 10.9.1982).

ed by these processes appears to actually move through the tidal inlet. Oscillating megaripples are the most important agents of sand transport in the ebb delta regions. In the case of the Wichter Ee inlet, material transport in the flood deltas constitutes only about 10 % of the transport effected in the ebb delta.

5.2.5 Expansion of the tidal deltas

The Holocene sea level rise has continued up to the present day. Gauge measurements taken during the last 100 years suggested that a rise of 20–30 cm per century were to be expected (e.g. Kramer, 1983). Recent investigations, however, have rev-

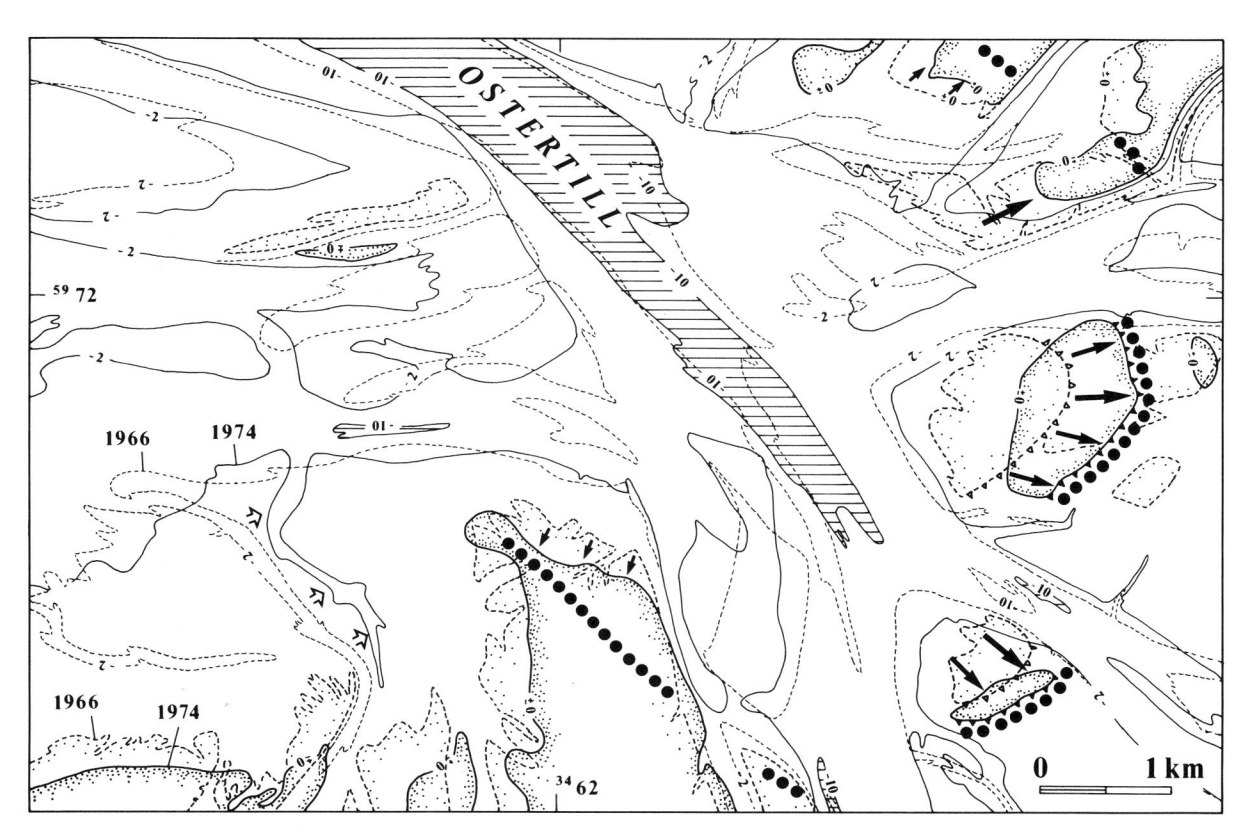

Figure 127: Morphological changes in the flood delta area of the Ostertill, older stage shown with broken lines. Sources: Topographische Wattkarte 1:25,000 Niedersächsische Küste, Sheet 14 (1967) and Küstenkarte 1:25,000, Sheet 2116 K Knechtsand (1974).

ealed that this assumption may be much too low. The average tidal high water levels of twelve gauges from the German North Sea coast have undergone a rapid rise during the last 25 years (1959-1983). This would amount to 64 ±15 cm per century (Jensen, 1984: 509).

After evaluating the long-term records from 19 gauges in the German Bight, Liese (1979) noted that the tidal range in this area had increased between 1946/55 and 1966/75 by an average of 10.5 cm. This implies that the discharge of the inlets had also risen accordingly. This should in turn result in morphological changes, although modifications of the ebb deltas have not yet been observed. One reason for this may be that the influence of the surf is strong enough here to prevent a marked seaward extension of the deltas.

Quite the opposite trend, however, was observed after the damming of the Lauwerszee in 1969. The drainage area of the Eilanderbalg was reduced from an original 350 million m^3 to a mere 200 million m^3. As a result, the shape of the ebb delta changed. The seaward slope reduced its gradient from 1:300 in 1971 to 1:350 in 1981. Whilst the outer margin of the shoal arc remained in the same position, the ebb delta crest migrated about 400-500 m landward and now lies slightly higher than before (Postma, Rijkswaterstaat, Sect. Groningen, Delfzijl, personal communication).

In the flood deltas morphological changes should be more significant as a result of the lower water depths. In fact, a number of flood deltas in the area investigated show a clear tendency towards extension, as the three examples cited below demonstrate:

Example 1: Ostertill
The Ostertill is one of the tidal streams in the unprotected tidal flats between the Elbe and Weser estuaries. It is situated between the Großer Knechtsand in the south and Scharhörn in the north. The Topographische Wattkarte 1:25,000 Niedersächsische Küste, Sheet 14 (1967) and the Küstenkarte 1:25,000, Sheet 2116 K Knechtsand

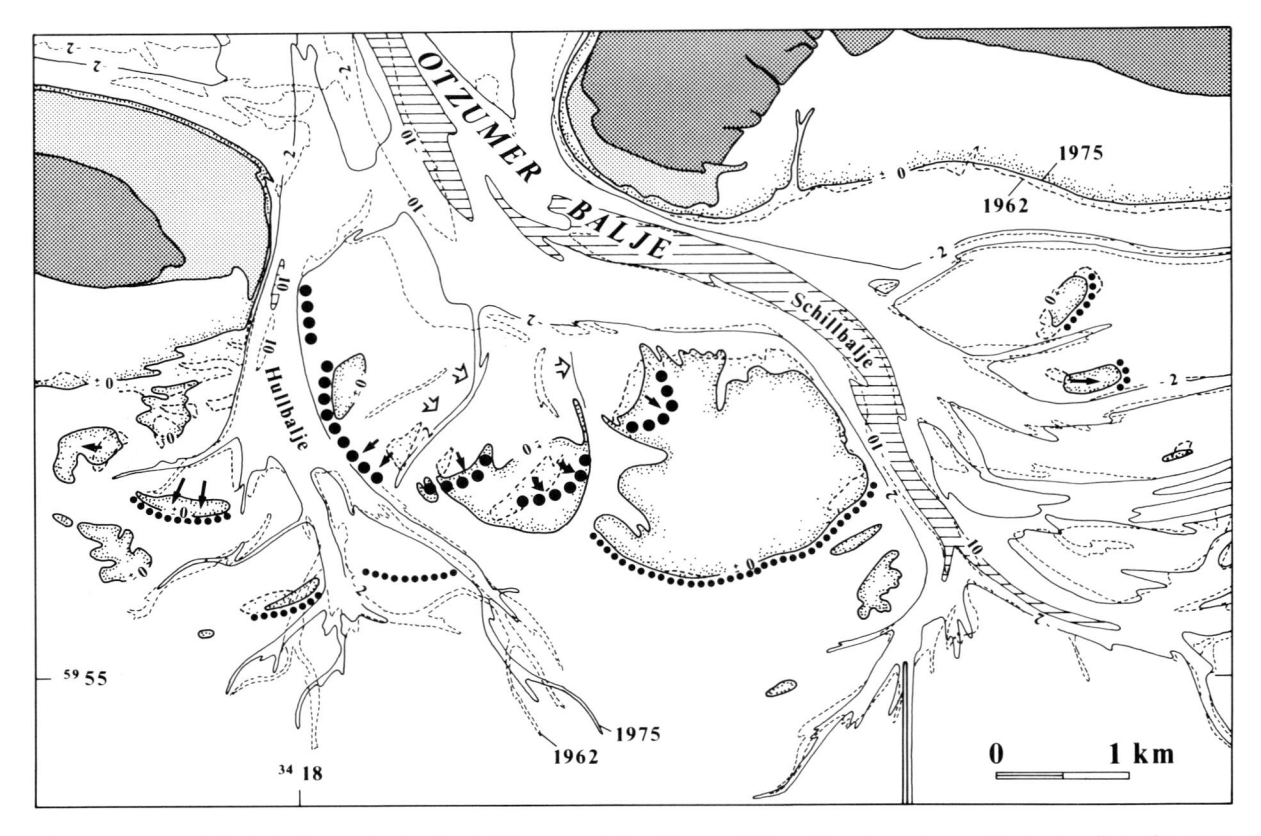

Figure 128: Morphological changes in the flood delta area of the Otzumer Balje, older stage shown with broken lines. Sources: Topographische Wattkarte 1:25,000 Niedersächsische Küste, Sheet 7 (1964) and Küstenkarte 1:25,000, Sheets 2210 K Langeoog and 2212 K Wangerooge (1975).

(1974) were used to evaluate the morphological changes.

In fig. 127 the changes of the 0 m, -2 m and -10 m contour are demonstrated. The dotted line shows the position of the flood delta margin in 1974. The southern end of the deepest part of the channel was found 1 km further north in 1974 than 1966. The deepest part has widened and at the same time rotated anticlockwise towards the northeast.

The strongest changes can be observed in the shape of the 0 m contour. The proximal slopes of the strongly developed flood ramps have migrated landward at a rate of 25–60 m/year. The -2 m contour has changed in the same way as the 0 m line, but to a much lesser extent. Here the migration rates vary between 6–12 m/year. Only a single shoal shows markedly higher migration rates. The strong morphological changes in the Hohenhörn Sande area is reflected in the irregular pattern of the isobaths.

Example 2: Otzumer Balje

The Otzumer Balje, the inlet between Spiekeroog and Langeoog, consists of two major branches, the Hullbalje in the west, and the Schillbalje in the east. During the 13 year observation period, the drainage areas of both channels have only slightly altered (fig. 128), as a comparison of the Topographische Wattkarte 1:25,000 (1964) and the Küstenkarte 1:25,000, Sheets 2210 K Langeoog and 2212 K Wangerooge (1975) reveals.

The flood delta of the Otzumer Balje can be subdivided into an inner delta (large dots) and an outer delta (small dots). A similar subdivision can also be made in other tidal inlets, e.g. Harle, Blaue Balje (cf. map in Ehlers & Mensching, 1982). While the flood ramps of both deltas are clearly developed and easy to identify on aerial photographs, the forms of the outer delta are far less distinct. They are possibly only activated during storm surges.

Changes in the flood delta areas here are considerably smaller than in the Ostertill case. They

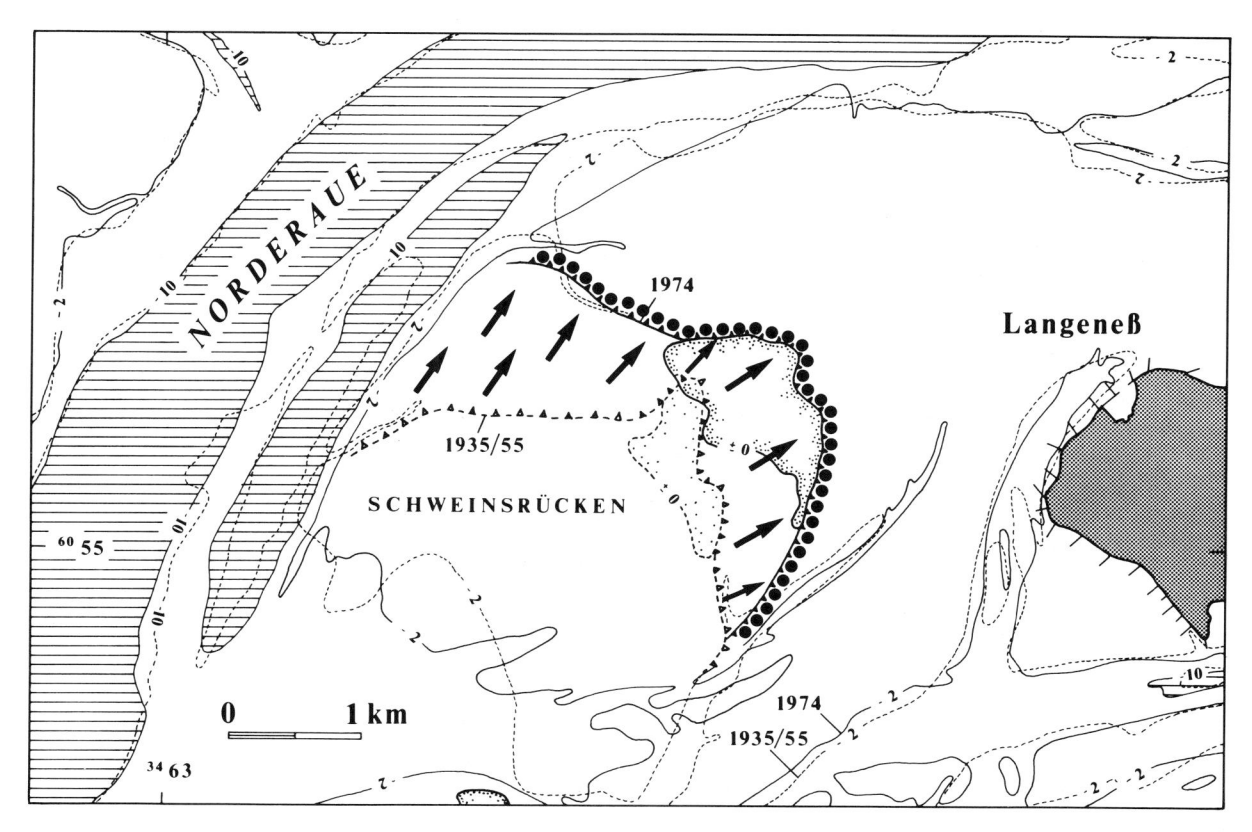

Figure 129: Morphological changes in the flood delta area of the Norderaue, older stage shown with broken lines. Sources: Müller & Fischer (1955), Sheets 13 and 14, 1:25,000 and Küstenkarte 1:25,000, Sheet 1316 K Amrum (1974).

amount to an expansion of 0 - 200 m during the observation period, i.e. 0–15 m/year. These changes have not affected all parts of the flood delta, but in all cases where changes have been observed (black arrows), it has led to expansion of the flood delta area.

Changes are more clearly limited to the upper parts of the flood ramps than in the case of Ostertill. The -2 m contour shows no clear tendency, and at depth the most striking change is the westward shifting of the Otzumer Balje main channel. This is also reflected in the changes of the -10 m contour. The main channel has migrated over a distance of 300 m in 13 years.

Example 3: Norderaue

In the Norderaue area between Amrum, Föhr, Langeneß and Hooge, a very large flood delta has formed. Map comparison of the sheets published by Müller & Fischer (1955), Sheets 13 and 14, Scale 1:25,000 as well as the Küstenkarte 1:25,000, Sheet 1316 K Amrum illustrates this change. The mapping of the Küstenkarte was undertaken in 1974, but as Müller's and Fischer's maps were produced between 1935 and 1955, only very rough quantifications of morphological changes result. The tendencies, however, are clearly visible.

Fig. 129 shows changes in the area of the large flood ramp to the west of Langeneß. The map is the same scale as figs. 127 and 128, demonstrating the huge size of this particular flood ramp, which is protected from the ebb current by a well-developed ebb shield. On both ends of the flood ramp long levée-like ebb spurs are found.

The dimensions of this flood ramp are equal to the entire flood delta systems of smaller inlets. It has migrated northeastwards during the observation period, a distance of 1–1.5 km. This means that expansion rates of 25–40 m/year to 50–80 m/year can be envisaged. As in the other two cases, the morphological changes mostly occurred in the highest parts of the form; the seaward -2 m contour, however, has not followed the highest parts of the flood ramp, but remained stable.

The comparatively long observation period provides a better demonstration of the recent mor-

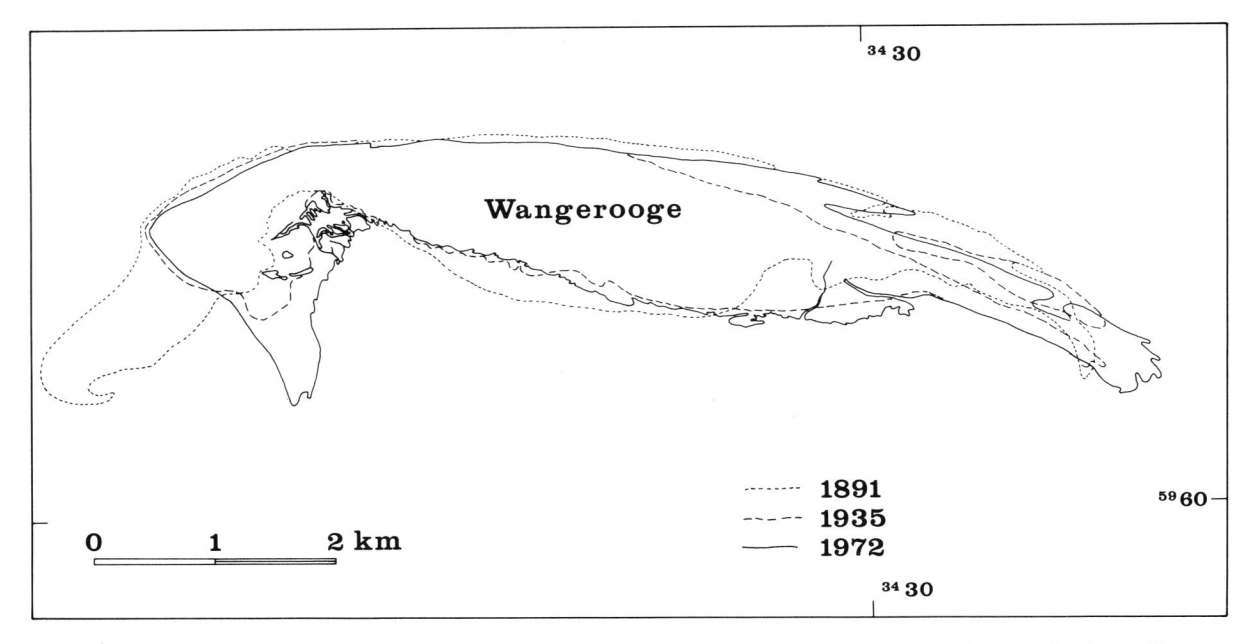

Figure 130: Development of the Isle of Wangerooge, demonstrating the mobility of the spits compared to the island core. Sources: Topographische Karte 1:25,000, sheet 2213 Wangerooge, issues of 1891, 1935 and 1972 (from Ehlers & Mensching, 1982).

phological changes. The other form elements of the Norderaue flood delta, not shown on fig. 129, have either taken part in the great expansion, or remained *in situ*. No shrinkage of a flood delta element was observed in any of the areas.

5.3 BARRIER ISLANDS

5.3.1 SUBDIVISION OF THE ISLANDS

The islands of the Wadden Sea can be subdivided into two major groups; marsh islands and barrier islands.

The Pleistocene cores of the Isles of Sylt, Amrum, Föhr and Texel represent highs in the pre-Holocene subsurface which remained unsubmerged by the Holocene transgression. Subsequently these Pleistocene cores have been supplemented by recent salt marsh areas.

The islands of Pellworm and Nordstrand, the Halligen (Norderoog, Süderoog, Südfall, Hooge, Langeneß, Oland, Gröde, Habel and Nordstrandischmoor), Jordsand (NE of Sylt) and Griend (between Terschelling and Harlingen) are remnants of old marsh areas. These marsh islands have been repeatedly flooded and covered by several metres of marine sediment, but at no time have they ceased to be islands.

The barrier islands form the main group and, in contrast to the marsh islands, owe their sole existence to landward directed sand transport, and the accumulation of shoals. After these emerged above high water level they were stabilized by dunes.

Landward and coast-parallel sand transport remain active today. Spits have developed at both ends of these islands, such as those on Rømø, Sylt, Amrum, Trischen and on the Eiderstedt peninsula. Where the marine influence has acted perpendicular to the coast, the spits are equally well developed at both ends. In the area of the West and East Frisian Islands, east of Terschelling, this pattern is modified by the easterly coast-parallel component of sand transport. The western spits are therefore mostly underdeveloped and in some cases have migrated around the western ends of the island cores so that today they are found on the landward sides of the islands. This process has only occurred very recently on Wangerooge, as indicated by comparison of maps of different ages (fig. 130).

Man has caused modifications of the natural system since the Middle Ages. Whilst coast-parallel sand transport remains almost unhindered, coast-normal sand transport is today largely restricted to the tidal inlets. Therefore, the natural progression, which led to a slow landward migration of the islands as an adjustment to rising sea

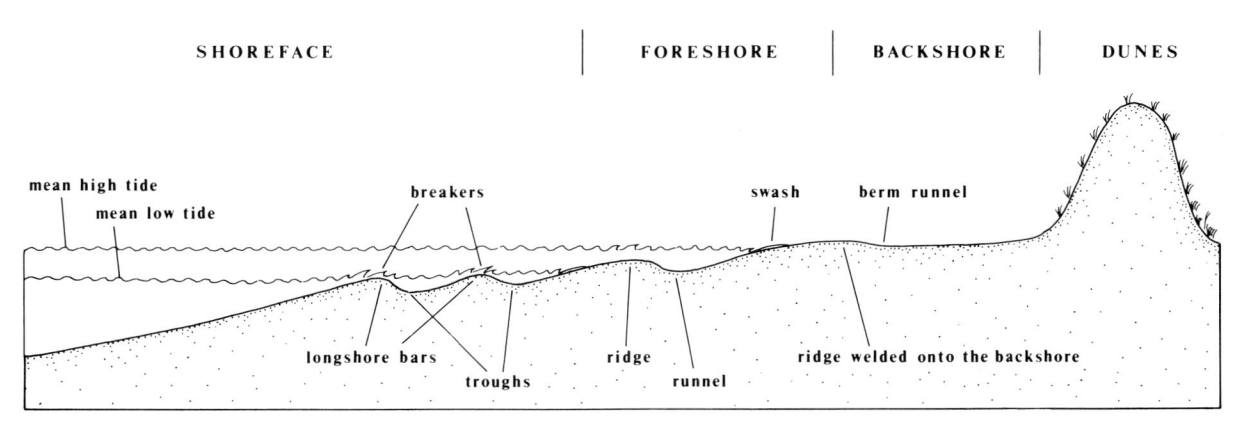

Figure 131: General profile of the beach and nearshore zones, illustrating the terminology used in this section.

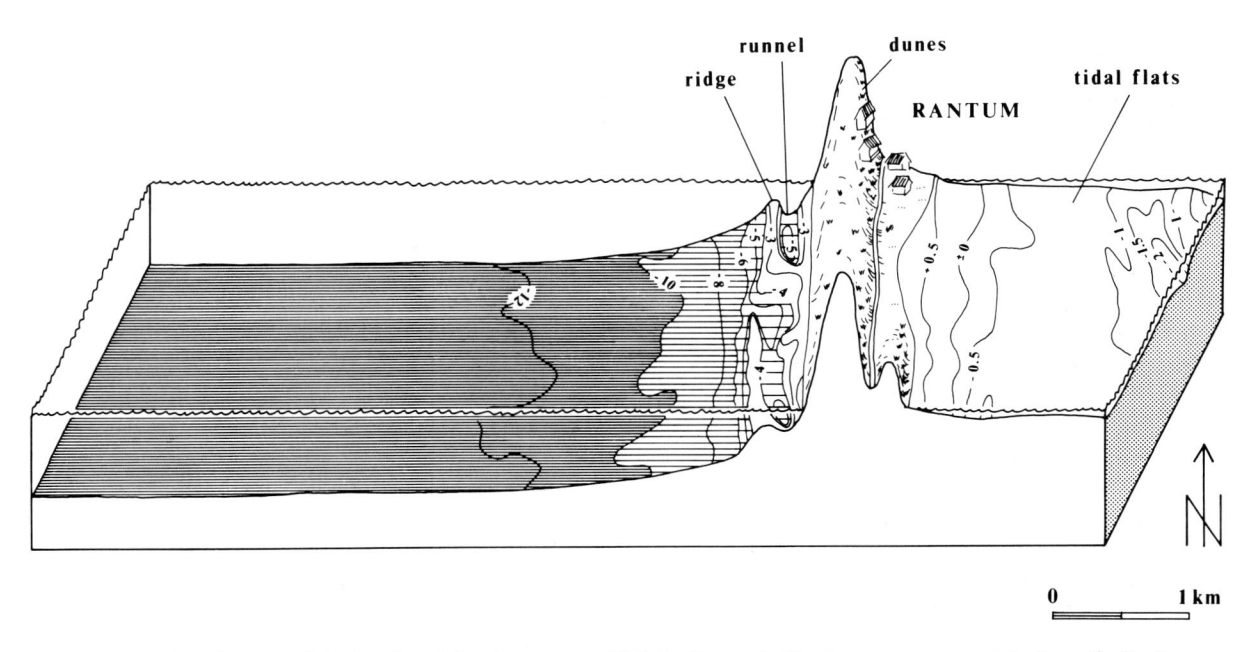

Figure 132 a: Block diagram of the beach and foreshore areas of Sylt (at Rantum). The dunes are represented schematically. Source: Küstenkarte 1:25,000, Sheet 1114 K Rantum.

level, has been interrupted almost completely. The dunes have been fixed and washover processes are now limited to a few marginal areas. Stabilization of the islands has been achieved as a result, but natural accretion has not been promoted. The opposite is in fact the case; large parts of both the seaward and landward coasts are currently being slowly eroded. The dominant processes responsible, and the resulting human responses, will be considered in the following chapters.

5.3.2 Beach processes

The barrier island beaches (fig. 131) vary considerably in width. The strongest contrasts are found on some of the East Frisian Islands (e.g. Borkum, Baltrum, Wangerooge). Some parts of these islands have no beach at all, whilst others have beaches several hundreds of metres wide.

The width of a beach is dependant on the dip angle of the shoreface, the sediment supply and

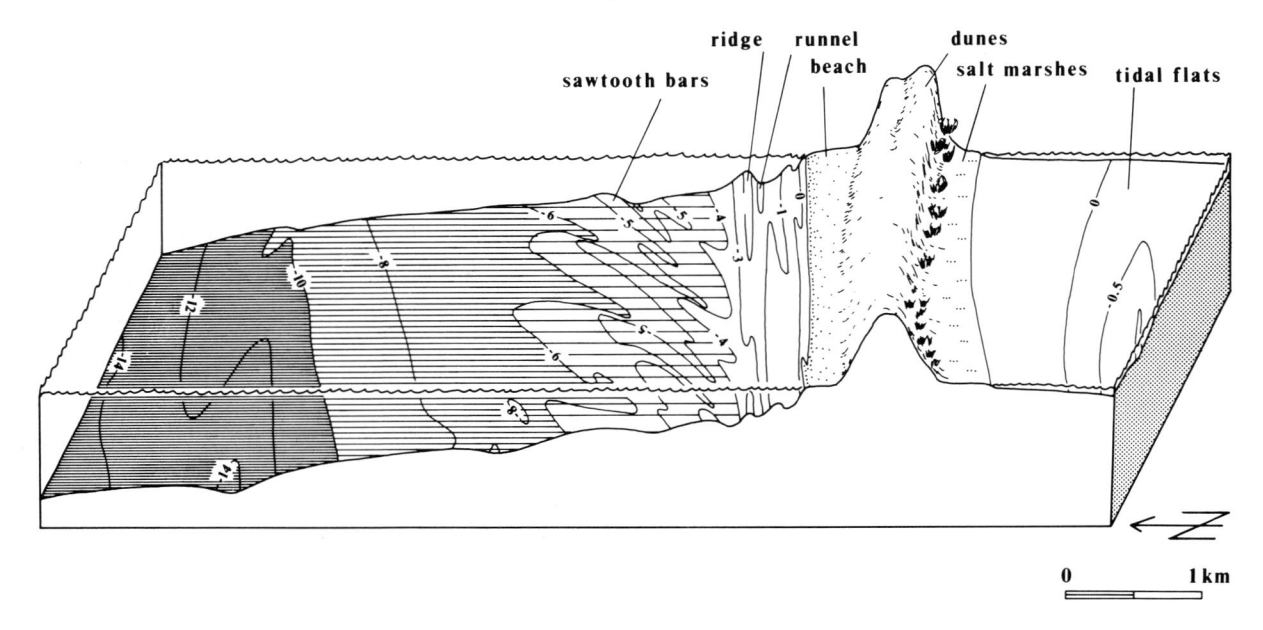

Figure 132 b: Block diagram of the beach and foreshore areas of Langeoog. The dunes are represented schematically. Source: Küstenkarte 1:25,000, Sheet 2210 K Langeoog.

the tidal range (Davis, 1978: 239). The Isle of Sylt, for example, has a steep shoreface (fig. 132 a) and, at this most exposed point of the North Frisian coast, sediment is supplied only by cliff erosion. As a consequence the beach is very narrow.

In contrast, on Langeoog, the shoreface slopes gently towards the sea (fig. 132 b). The -10 m contour is situated about 3.5 km from the coast (at Rantum, Sylt: 1 km), and the sediment budget is well-balanced. Until recently it had not been found necessary to erect engineering structures to protect the island. The beach is about 200 m wide (at Rantum, Sylt: 50 m).

The beach area can be subdivided into shoreface, foreshore and backshore, each of which is characterized by a number of typical landforms.

Shoreface

Most of the coast-parallel sand transport occurs in the shoreface area. The mean spring tidal low water line (SKN) defines the boundary between the shoreface and the foreshore. The upper shoreface can be subdivided into a number of coast-subparallel striking ridges, the crests of which approximate sea level at low water (fig. 133). The western ends of the ridges are normally in contact with the beach or foreshore (Reineck, 1963; 1979). The ridges comprise sediment deposited by breaking waves (cf. section 4.3).

The lower part of the shoreface is characterized by other types of ridges, which are shorter, broader and trend obliquely towards the beach. The so-called 'sawtooth bars', found at a depth of about 4 m, are particularly well-developed. They strike NE-SW along the coast of the East Frisian Islands. These forms are large enough to be accurately mapped on the 'Topographische Wattkarte' (fig. 132 b) since they are approximately 1.5 km long. It can be assumed that these features migrate eastwards, because there is an easterly current seawards of the East Frisian Islands.

Radar images of parts of the Dutch coast were taken on three occasions during the short active period of the SEASAT satellite in the autumn of 1978. One of this series reveals the underwater topography (fig. 134). Conventional aerial photographs demonstrate that the wave pattern reflects the underwater topography down to a depth of about 4 m. This is particularly true during periods of strong swell with a large effective wave depth (fig. 135).

Other bar systems running obliquely towards the coast are found at greater depths off the East Frisian Islands. These strike almost at right angles to the sawtooth bars mentioned above, i.e. about NW-SE. These forms are much larger, up to 15 km long, and can be traced down to water depths of about 25 m (Reineck, 1979 a: 294). Reineck (1979 a) points out that none of these forms are station-

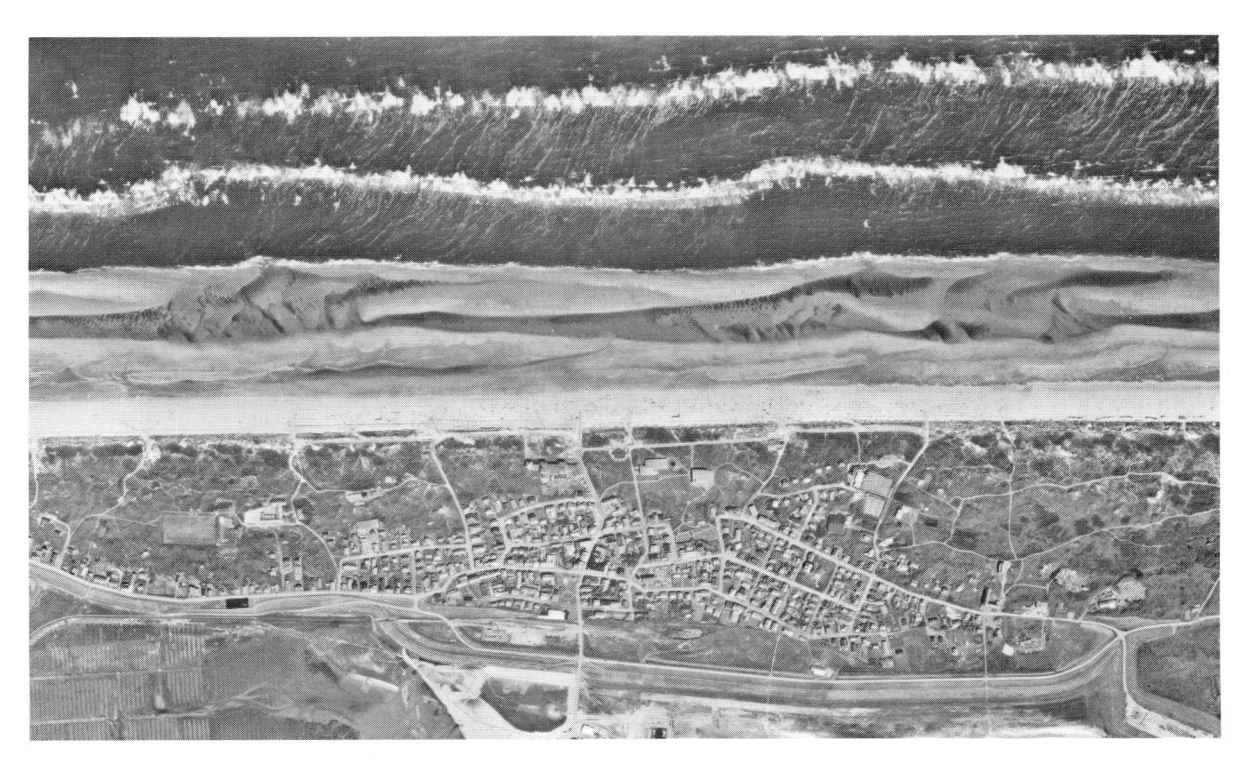

Figure 133: Longshore bars and ridge and runnel on the beach and foreshore area of Juist. Two of the longshore ridges have emerged at low tide. The aerial photograph demonstrates that they are formed of a number of sandwaves, the western (left) ends of which are always closer to the beach than the eastern ends. The two longshore bars are delineated by the lines of breakers. The distance of the outermost bar from the dune foot is about 700 m (Aerial photograph of the Niedersächsisches Landesverwaltungsamt, Landesvermessung, of 18.6.1983, Bildflug Juist (1966), Str.1/0460, freigegeben durch NLVA–Landesvermessung –, Hannover, unter Nr. 33/83/1966).

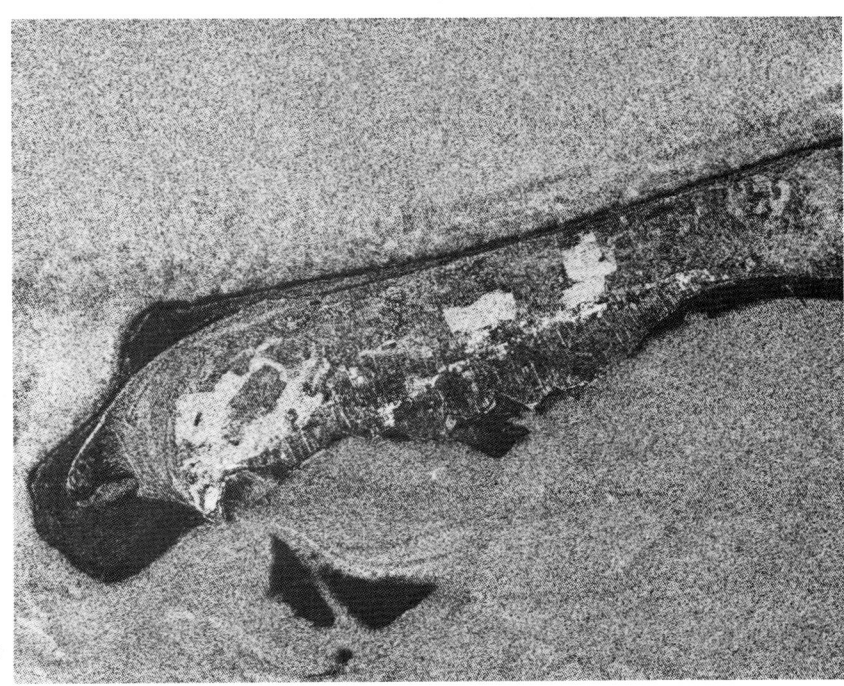

Figure 134: SEASAT image of the Isle of Terschelling at high tide. The sawtooth bars on the foreshore (at about -4 m) are clearly visible (SEASAT SAR image S 0891 N 05323 E 00531 of the DFVLR, date: 28.8.1978).

Figure 135: Conventional aerial photograph of the Norderney coast during a strong swell. The sawtooth bar trend is reflected in the wave pattern. On the right a turbidity cloud is being transported seaward by a rip current (Aerial photograph of the Niedersächsisches Landesverwaltungsamt, Landesvermessung, of 29.5.1966, Bildflug Ostfriesische Küste (419), Str.2/275, freigegeben durch Regierungspräsident Münster/W. unter Nr. 2316/66).

ary. Aerial photographs that provide an opportunity for the measurement of migration rates of these features are unfortunately not available.

By comparison of precise charts from several successive years, Krüger (1911; 1937: 249) was able to measure the migration rate of the bars off Wangerooge at about 300 m/year. However, he was only able to measure the migration of the major form elements (fig. 136) and not the individual sawteeth, which would have involved much more effort. According to Krüger (1937: 247), Gaye measured the migration rate of the sawtooth bars to be about eight times higher than that of the larger features.

The sawtooth bars found on the beach of Baltrum are an exception. They occur at a much shallower water depth (about -1.5 m), within the system of beach-parallel ridges, and may partially emerge at low tide (fig. 137). This may be because the ebb deltas of the Wichter Ee (in the west) and the Accumer Ee (in the east) are almost in contact, because of the small size of Baltrum. The intervening beach area is thus protected against strong waves, and the ridges are forced further seaward than on the other islands. The migration rates of the sawtooth bars off Baltrum have been measured by aerial photograph evaluation (cf. section 5.2.2) to be about 1.5 m per day during the investigation period.

Foreshore

The foreshore consists of subparallel ridges and runnels (fig. 133). There are one or two ridges in most cases and these, rather than forming continuous sediment bodies several kilometres in length, are subdivided into numerous sand waves with minor intervening gaps (fig. 138). These gaps provide channels for the rip currents to carry the beachward directed water of breaking waves back towards the sea. The sand waves of the ridge systems migrate landwards onto the beach, and also downdrift along the beach simultaneously.

The beachward migration of ridges on Scharhörn was investigated by Göhren (1970). The migration rates vary over a wide range (fig. 139). Measurements taken on Norderney and Baltrum during this study show migration rates of 50–500 m/year to be most frequently found.

Small run-off channels are formed when the ridges are exposed and dry at low tide (fig. 140). The terracettes on their slopes reflect the decreasing amount of water passing through the channels as the water table and base level fall.

Minor forms on the foreshore

The ridges and runnels are covered by a series of characteristic small forms (Brenninkmeyer, 1982), the most significant of which are summarized below:

Swash marks and current lineations

Incoming waves create swash marks on the seaward slopes of the ridges (fig. 141) which are preserved only during the falling tide. Primary cur-

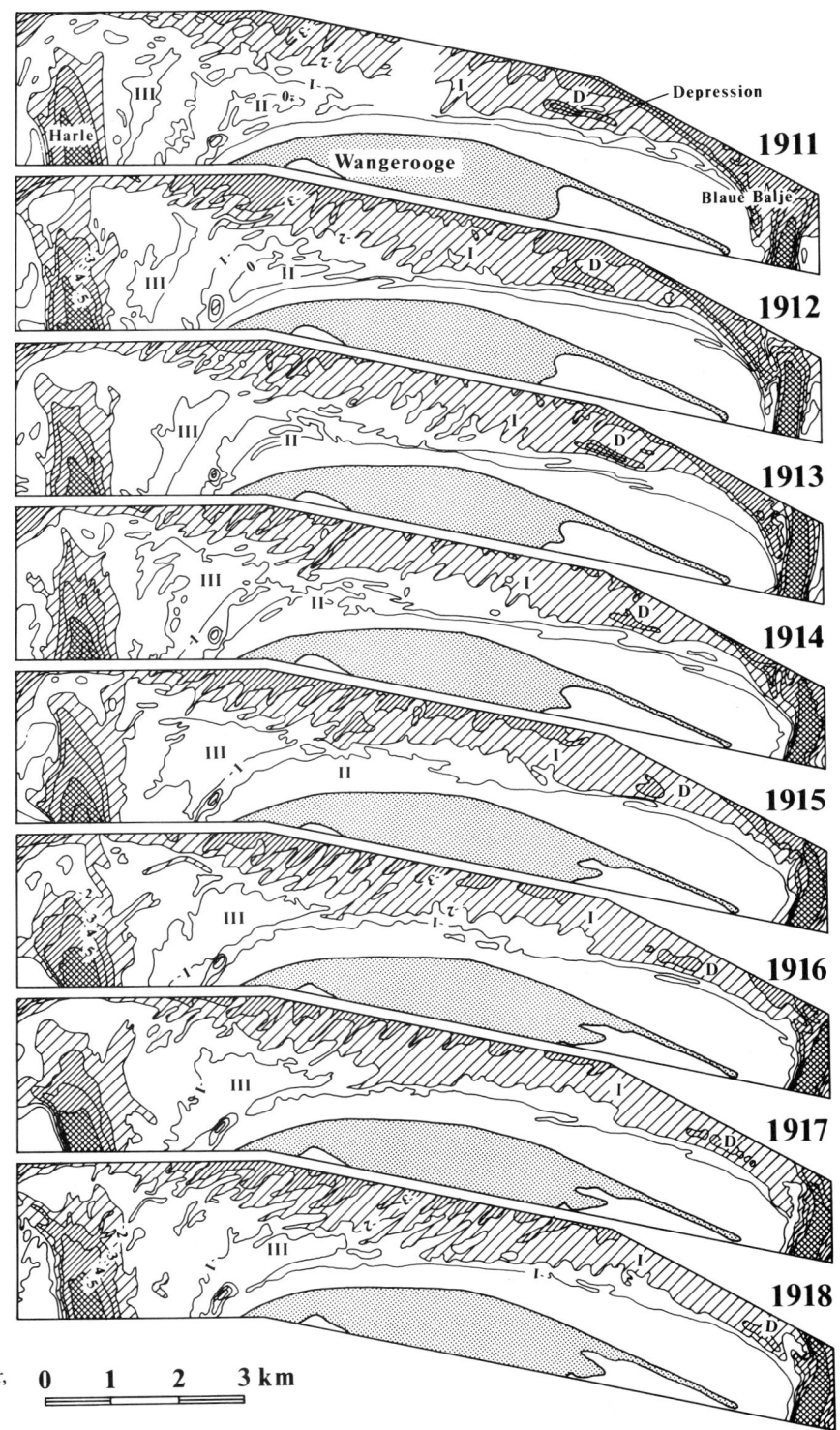

Figure 136: Bar migration on the Wangerooge shoreface (after Krüger, 1937).

Figure 137: An emergent sawtooth bar on Baltrum beach at low tide (1984).

Figure 138: A 'sand wave' in the Baltrum ridge system (1980).

rent lineations are frequently found where ridges are washed by fast currents (fig. 142) (Reineck & Singh, 1978: 55; Allen, 1984 b: 525). These consist of parallel-trending ridges and furrows, with a height difference of less than 1 mm in general.

Small ripples

The most widespread micro-features on the beach are small ripples (Johnson, 1919: 489). They cover the lower parts of the runnels as well as small depressions and troughs on the foreshore. The small ripples strike mainly parallel to the long axes of the runnels. They have long continuous crests (fig. 143) and are mostly oscillation ripples. The megaripples which may occur in combination with these bedforms (fig. 143) are transport ripples. Their crests strike at right angles to the runnel axes (cf. Moore, Fritz & Futch, 1984).

Reineck & Singh (1978: 31) distinguish between 'undulatory small ripples' and pure 'wave ripples'. 'Undulatory small ripples' are thought to be formed by weak currents in contrast to oscillation ripples. Without investigation of their internal structure, however, a field distinction is not possible (Reineck, 1961 b: 58; Reineck & Singh, 1978: 26). The formation of oscillation ripples has been

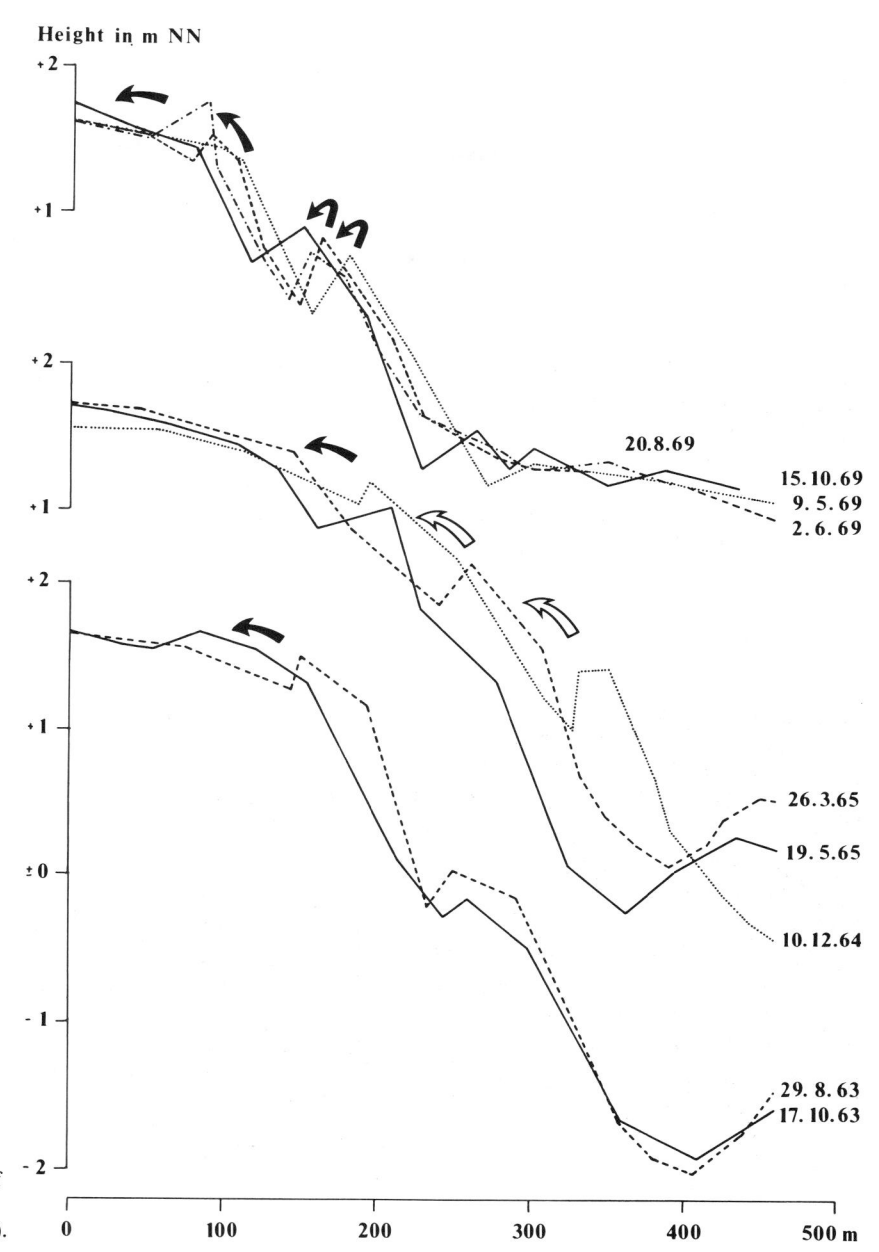

Figure 139: Beachward migration of ridges in the Scharhörn foreshore area (Source: Göhren, 1970: fig. 43).

studied by Miller & Komar (1980 a, b) and Dingler & Clifton (1984). The interpretation of oscillation ripples is dealt with by von Grafenstein (1984). Various types of oscillation ripples and their differentiation are discussed by Allen (1984 a: 419).

The small ripples found on the ridges are transport ripples. Observations of emerging ridges on Spiekeroog have demonstrated that they are formed by the swash of a single wave. The pre-existing ripple field is erased by each new wave. When the current velocity decreases a new ripple field is formed. It may be of different orientation, depending on direction and strength of the incoming waves.

Antidunes

The seaward slopes of the ridges are often covered with backwash ripples. These long parallel-striking ripples migrate upslope against the flow direction of the water, and are therefore antidunes (fig. 144) (cf. Broome & Komar, 1979). Antidunes can also be formed in small fast flowing channels

Figure 140: A drainage channel in the gap between two 'sand waves' of the Wangerooge ridge system. Falling water level has initiated small terraces in the channel (1979).

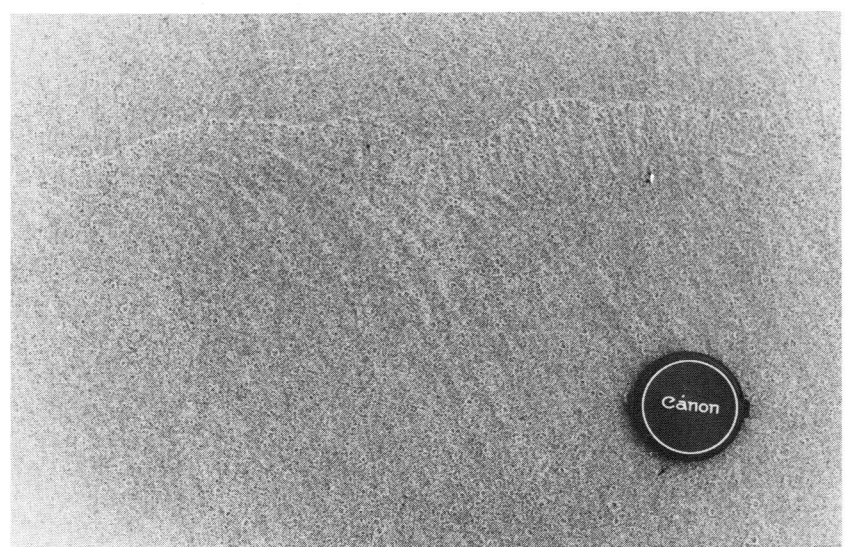

Figure 141: Swash marks on the Norderney beach (1983).

Figure 142: Primary current-lineation on the Norderney beach (1982).

Figure 143: Oscillating small ripples on the landward margin of the runnel at Wangerooge. The weakly developed megaripples striking at right angles to the small ripples are forms related to the coast parallel sediment transport (1980).

Figure 144: Backwash ripples on the beach at Ameland (1981).

Figure 145: Antidunes forming in a small fast-flowing creek on the beach at Borkum (1984).

Figure 146: A shallow creek draining a depression on the beach at Borkum has eroded part of a small ripple field, and formed a series of elongated lingoid ripples. The flow direction was from lower right to upper left (1984).

Figure 147: In the drainage channel of the creek in fig. 146, islands of the former small ripple field have been preserved. Streamlined flow lines have developed around these islands (1984).

on the beach (fig. 145). These are rarely fossilized because they are transformed into normal current ripples with decreasing flow velocity.

Lingoid small ripples
Lingoid ripples are formed in fast-flowing shallow water, instead of well-developed small ripples (fig. 146). In extremely elongated forms the troughs are reduced to hollows only a few centimetres deep and the crests of the ripple tongues form subparallel-striking steps only a little over 1 mm in height.

Figs. 146 and 147 show how water draining a puddle on the beach has eroded parts of a small ripple field, and has transformed it into strongly elongated lingoid ripples. Streamlined flow lines have evolved around small islands in the creek (fig. 147).

Rhomboid ripples
Rhomboid ripples are frequently found on the landward slope of the foreshore ridges as well as on the steeply sloping parts of the beach (fig. 148). These are formed where shallow water flows at high velocities. This occurs when the swash of incoming waves flows across the ridge, or when

Figure 148: Rhomboid ripples on the recurved flood spit at the eastern end of Norderney (1982).

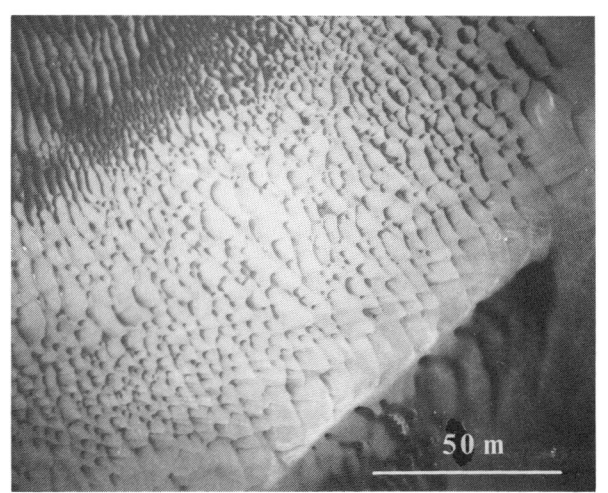

Figure 149: Rhomboid megaripples on the ebb delta of the Wichter Ee inlet. These forms develop under the influence of strong currents from normal megaripples. Upper picture: Othello Shoal Shield; aerial photograph of F.Böker of 15.6.1984, freigegeben Brg. Nr. 5765/603; lower picture: Norderney, margin of the Othello Shoal Shield (1982).

Figure 150: Rhomboid rill marks on the beach at Wenningstedt (Sylt) (1984).

Figure 151: Sand tongues on the lee side of the ridge at Rømø beach (1982).

backwash runs down the beach slope. Johnson (1919: 517) therefore called these bedforms 'backwash marks'.

Rhomboid ripples are commonly found in fine to medium sand, and the sorting of the sand according to its specific weight creates the characteristic light and dark pattern (fig. 148); cf. Komar (1976: 366).

Most rhomboid ripples are a few millimetres high and several centimetres to decimetres in length. Where the bedforms are several decimetres to more than 1 m in size Reineck & Singh (1975: 37) refer to 'rhomboid megaripples'. On the shoals of the ebb deltas, however, they can occur as forms over 10 m long with crests several decimetres in height (fig. 149).

Rhomboidal features may also occur in coarser sediments (fig. 150) but they are comparatively poorly developed and mostly restricted to shallow grooves. Otvos (1964, 1965) described them as rhomboid-deltoid shaped patterns, formed by a close network of intersecting shallow miniature grooves. They differ from rhomboid ripples by being erosional rather than accumulative in origin. Otvos therefore refers to them as 'rhomboid rill marks'.

Sand tongues

The small-scale morphology of the beachward slopes of ridges and berms is a product of current

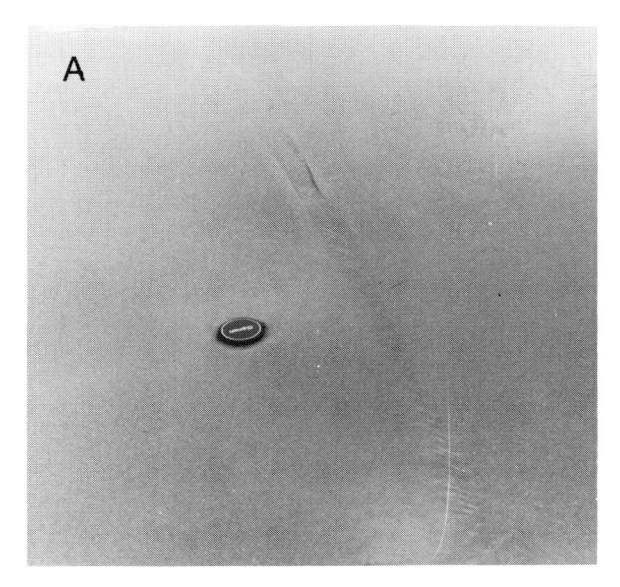

Figures 152 a–c: Formation of rill marks on the beach at Hörnum (Sylt) (1984).

conditions and sediment supply. The erosional rill marks may be largely replaced by sand tongues in areas of positive sand budget; these may extend for several decimetres from the ridge into the adjoining runnel (fig. 151).

Rill marks
On tidal coasts rill marks are formed in the transitional area between the emergent and flooded parts of the beach. Irregularities in the beach slope may lead to local concentrations of runoff which would otherwise occur as sheet flow. Erosion occurs leewards of irregularities such as small stones or shell fragments.

An elongated straight rill is usually formed at first (fig. 152 a). With increased erosion, a small delta will develop at the lower end of this rill. This sediment accumulation causes the linear drainage course to divide into several branches. This bifurcation may develop into multifurcation if enough sediment is supplied.

The rill is further deepened and extended by backward erosion (fig. 152 b). Small dendritic tributary rills form at the upstream end of the rill, and increase in size as rill formation proceeds (fig. 152 c). Rill formation continues until either water supply ceases or they are destroyed by the rising tide.

Rill marks are formed where runoff occurs as a water film of less than 2 cm (cf. Reineck & Singh,

1978). At the time of rill inception, runoff is concentrated within the rill and small adjoining areas in its upper course fall dry. This is accelerated as tributary side rills develop.

Cepek & Reineck (1970) distinguish between a number of different rill marks (cf. Reineck & Singh, 1978: 58-62). Most of these, however, can be easily classified into major bedform groups. 'Branching rill marks', 'meandering rill marks', 'bifurcating rill marks' and 'rill marks with accumulation tongues' all of which are special examples of rill marks formed on gently sloping beaches, as described above. 'Tooth-shaped rill marks', 'comb-shaped rill marks', 'fringy rill marks' and 'conical rill marks' are formed where small steps are dissected by erosion.

Such small steps, up to 2 decimetres in height, are found on the beachward edge of ridges. Fig. 153 shows comb-shaped rill marks at the edge of the foreshore ridge of Ameland. In contrast to the rill marks on the beach, these are purely erosional in character, developing small accumulation cones at the base of the step rather than deltas.

Adhesion ripples
Where sand is trapped by a wet surface when strong winds cause sand to drift into water-soaked areas, adhesion warts or adhesion ripples form (fig. 154). Well-developed features are found on

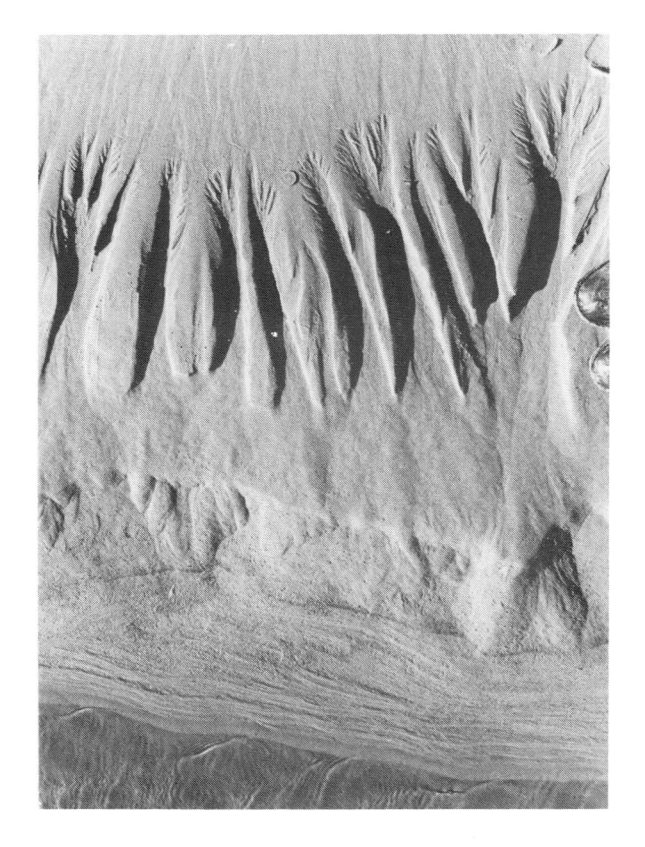

Figure 153: Comb-like rill marks on the beachward slope of a ridge at Ameland (1981).

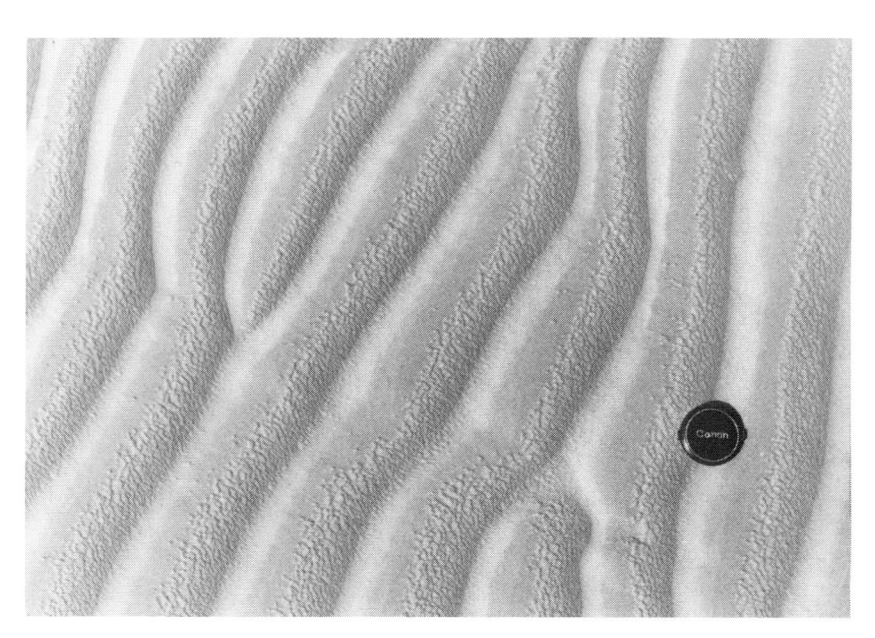

Figure 154: Adhesion ripples on the windward slopes of a small ripple field on the beach at Borkum (1984).

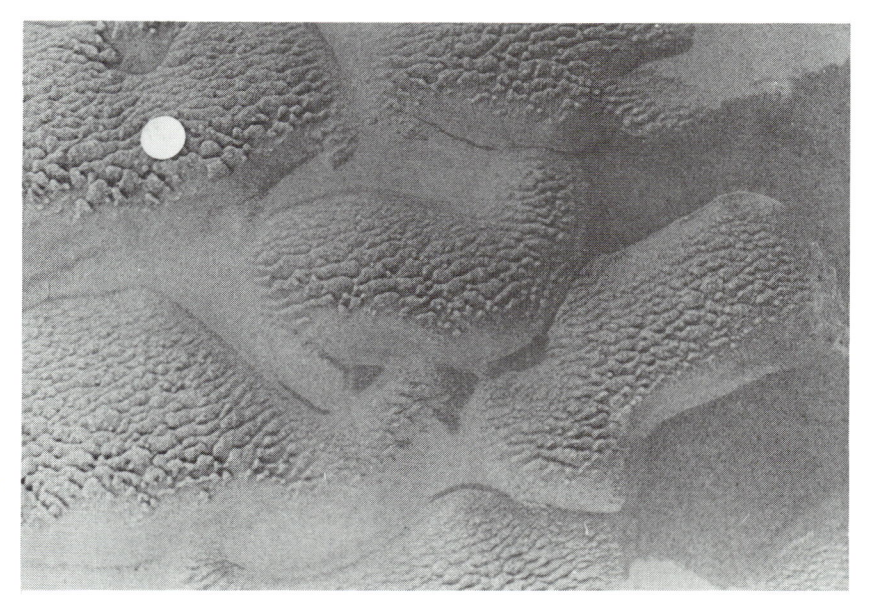

Figure 155: Adhesion ripples on the beach at Vlieland. In the ripple troughs a sand-water mixture with no characteristic form is deposited (1981).

Figure 156: Cavernous beach sand on the beach at Mellum (1981).

the ripple crests, whilst in the troughs the accumulated sand tends to diverge and no distinct surface forms develop (fig. 155).

Cavernous beach sand

Cavernous beach sand is formed where air becomes trapped within rapidly deposited sand (Reineck & Singh, 1978: 57; Allen, 1984 b: 535). The cavities have a diameter of few millimetres, giving the sediment a sponge-like texture (fig. 156) and a very low degree of compaction (fig. 157).

Backshore

The sediment supply of the backshore is maintained by the ridge system of the foreshore. The foreshore ridges migrate slowly upwards onto the shore, until they are finally welded onto the backshore and cease to exist (Abele, 1977).

Spring tides and storm surges are the only marine events which actively influence the backshore. For most of the time it constitutes a dry plain dominated by aeolian processes (cf. section 5.3.3).

Figure 157: Cavernous beach sand on Norderney. As a result of its sponge-like texture the cavernous sand is very poorly compacted (1982).

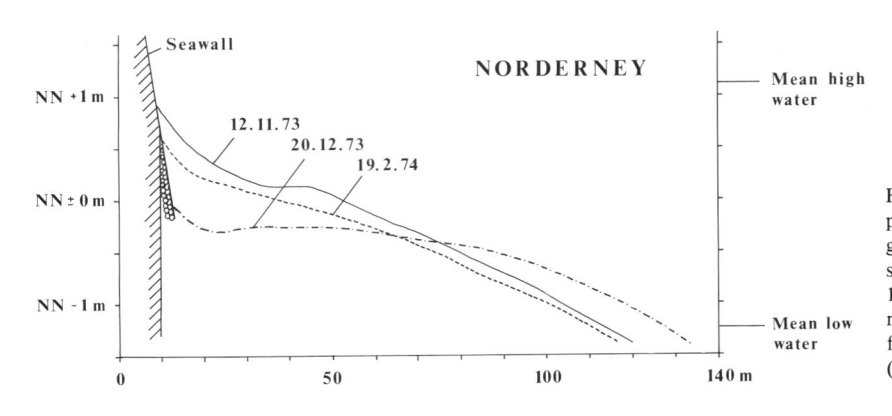

Figure 158: Reaction of a beach profile on Norderney (between groynes G_1 and H_1), to a series of storm surges between 13.11.–14.12.1973 and its subsequent restoration. The profile is taken from an area of net sand loss (Source: Homeier, 1976: Anl. 21).

Beach response to storm surges

Storm surges lead to a lowering of the backshore, the eroded sand being accumulated on the foreshore or even on the shoreface. In subsequent periods of calm weather the ridges of the foreshore migrate upwards and replenish the beach, until the initial situation is reached again. This process has been measured in a number of profiles on Juist and Norderney (fig. 158).

The pre-storm conditions are not always restored. In areas with strong coast-parallel littoral drift, a storm surge may lead to a net loss of sediment at the updrift end of the island, and a net gain at the downdrift end. This is the process observed by Hempel (1980), who described the apparently anomalous situation where storm surges subsequently led to the formation of new dune ridges on the eastern ends of Spiekeroog and Wangerooge.

5.3.3 DUNE FORMATION

A belt of dunes several kilometres wide forms the coastal barrier along the southern North Sea coast. It originally began to develop during the Middle Ages. The huge parabolic dunes on Sylt, Amrum and Fanø represent the most fully developed features. On most of the islands at present, however, only irregularly shaped dune ridges are found, which are subdivided into a number of crests and blowouts.

Reinke (1909) outlined the fundamental conditions required for dune formation on the Wadden Sea barrier islands. He recognized that dune growth can only occur with the active participa-

Figure 159: Migrating dunes in the Listland area on Sylt (1982).

Figure 160: Coarse sand and fine gravel sample from the crest of a migrating dune in the Listland area (Sylt) (1982).

tion of vegetation. He defined the following stages of dune formation:

1. Primary dunes: recent sand accumulations around *Triticum junceum* several centimetres to decimetres high.
2. Secondary dunes: deep accumulations of sand around *Ammophila arenaria* (formerly: *Psamma arenaria*).
3. Tertiary dunes: heath dunes with *Calluna vulgaris* and *Empetrum nigrum* which displace the original grass vegetation.

Dune development from the primary to the tertiary stages has been described on a number of occasions (e.g. Jessen, 1914: 66). This classification is still in use today, but for both vegetational and morphodynamic studies a more sophisticated classification is desirable.

Reinke's original scheme does not include sand accumulations on the barren beach, the morphodynamics of which resemble desert dunes (Solger et al., 1910: 30). The special group of migrating or parabolic dunes are also excluded. The term 'parabolic dunes' was introduced by Steenstrup (1894; *op. cit.* in Solger, 1908). Migrating dunes form mostly from incompletely vegetated secondary dunes. They can also be formed, however, by the destruction of the vegetation cover on tertiary dunes (e.g. by overgrazing).

In this text the following subdivision is proposed:

Figure 161: Sand being driven in long streaks over the beach of Spiekeroog during a storm. No significant landforms are created, and the sand accumulates locally in irregularly shaped patches (1984).

Wind ripples
Beach dunes (without vegetation)
Primary dunes
Coast-parallel dune ridges
Migrating dunes and
Old dunes.

Deflation

Huge quantities of sand are required to form dune complexes. This sediment is supplied by barren sand flats on the beach. Beach width therefore delimits the spatial potential for dune genesis, but a broad beach does not necessarily guarantee dune formation. For instance, active dune formation has occurred on Amrum and Rømø (both with beach widths of about 1 km) but on the Außensände (Outer Sands), which are sand banks 1.5–3 km in width, no dunes have developed. Directly east of the village on Wangerooge, a new dune ridge has begun forming. Further to the east, however, where the beach width increases, the dunes are stable. Dune inception seems to be limited to areas with a strongly positive sand budget, located at the downdrift ends of islands and adjacent to the ebb tidal delta.

The grain size of the dune sands normally ranges between fine to medium sand, the same size range incidentally, which occurs on most of the beaches in the study area. Coarser-grained material can be deflated where fine to medium sand is unavailable. The huge migrating dunes of Sylt consist partially of coarse sand to fine gravel (figs. 159 and 160). The median value of the grain size distribution of these dunes is about 0.4 mm according to Veenstra (1984: 49).

Heavy storm tides appear to have a largely destructive impact on coastal dunes. However, this is only the case in areas with a negative sand budget. After extensive investigations on Spiekeroog and Wangerooge, Hempel (1980) was able to demonstrate that new dune ridges formed in front of the old backdunes as a result of severe storm surges.

The dunes are replenished both from the backshore areas, and also, to a lesser extent, from the foreshore. At wind velocities of 6–7 Beaufort the sand dries rapidly and begins to drift. The whole upper sand layer is in movement during storms and drifts in long streaks over the beach (fig. 161). Sand drift occurs almost exclusively in a 10-cm zone above ground level, even when wind velocities exceed 9 Beaufort.

Plate 8: Row of barchan-like sandwaves on the Fanø beach off Fanø Bad (1980).
Plate 9: Deflation landscape with seasonally high groundwater table in the North Fanø summerhouse area (1980).
Plate 10: Migrating dune with blowout in southern Fanø looking towards the northeast (1980).

PLATES 8-10

PLATES 11-12

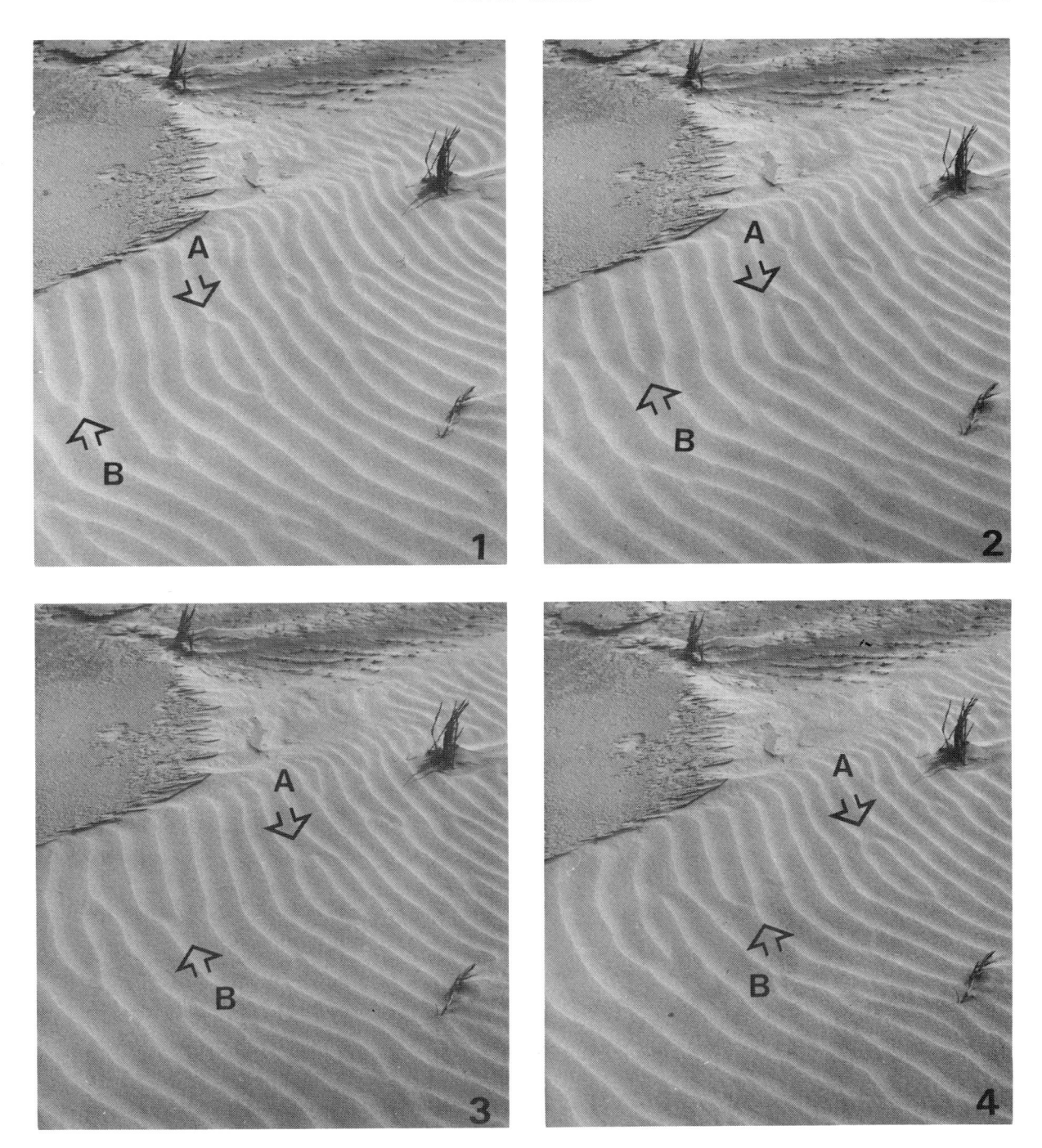

Figure 162: Wind ripple migration on Borkum. Photographs taken at 5 minute intervals. The development of the junctions at A and B reveals faster and slower migration strips within the ripple field (1984).

Plate 11: View across a sandy plain on eastern Norderney with a central washover channel and a dune core in the background (1982).
Plate 12: Spiekeroog's Ostplate sand flat flooded during spring tide (1984).

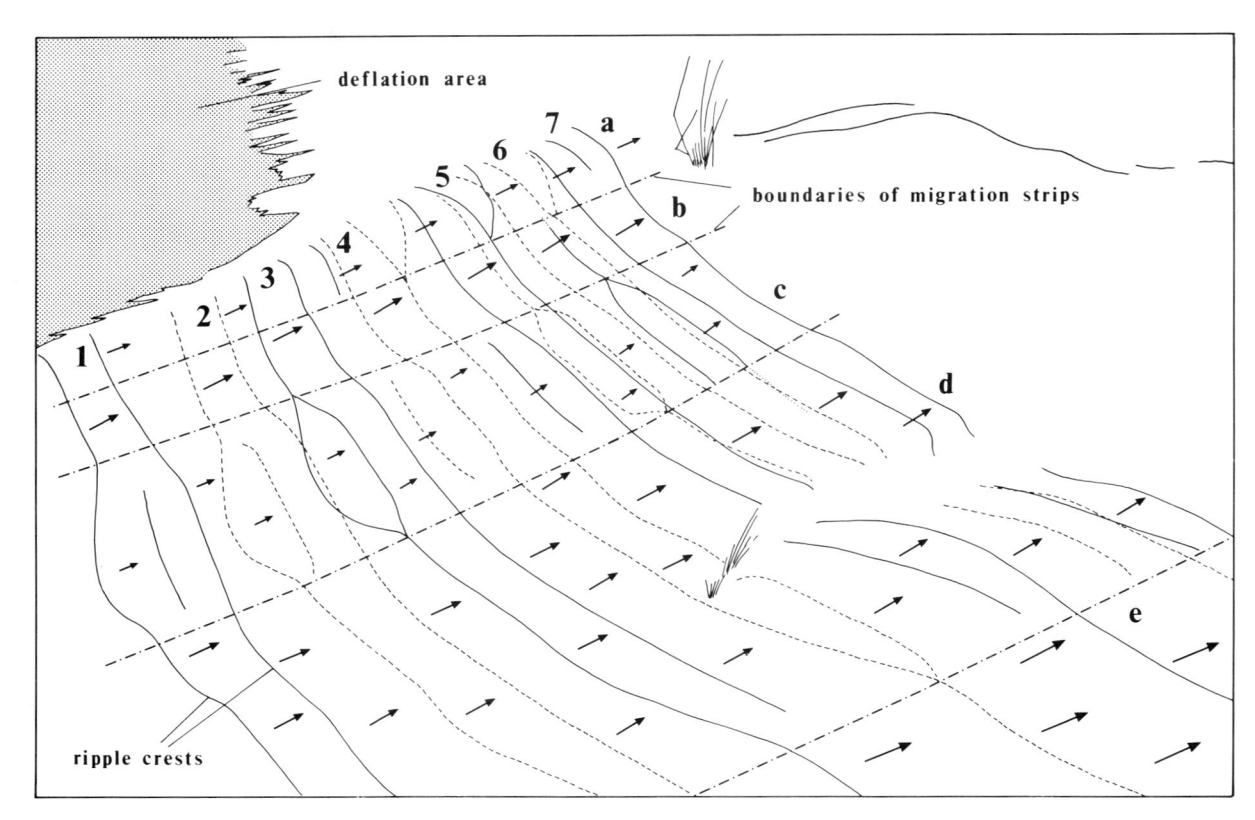

Figure 163 a: Migration of a group of wind-ripple crests within the ripple field shown in fig. 162 over a period of 30 minutes. Numbers 1 to 7 show ripple crest positions at 5 minute intervals. The size of the arrows indicates the relative migration velocity. Strip a has migrated at a slower rate than b; c was the slowest and e the fastest strip.

Wind ripples

Wind ripple fields, like other ripple fields, consist of a large number of generally parallel striking ripple crests which are occasionally truncated or connected with neighbouring crests. Field observations have demonstrated that the truncations and junctions are not randomly distributed, but follow certain alignments. Although the ripple junctions have been described in a number of publications (e.g. Allen, 1984 a: 308; Reineck & Singh, 1975: 42) no explanations have yet been proposed for their origin.

Wind velocity immediately above the ground is strongly dependant on surface micro-relief and causes the ripples in a ripple field to move at different rates. During our own investigations on Borkum beach it was possible to distinguish between zones of faster and slower ripple migration (figs. 162, 163 a and b). Ripple length remained constant in all zones, although truncation of the ripple crest occurred when part of the ripple lagged behind. It was found that as soon as the deformation exceeds about one third of the ripple length, the connection between the ripple and the lagging section is severed. Its crest becomes connected to the following ripple after a short transition period. Continuous ripple crests are maintained for most of the time. Junctions always result at boundaries between areas within the ripple field which move at different velocities.

Dunes on beach backshore

Beach barchans, which may be aligned in rows (plate 8), normally develop during low wind velocities. The barchans tend to be approximately 10 cm high, and up to several metres broad. Our own measurements on Borkum and Spiekeroog have demonstrated that at a wind velocity of 6 Beaufort small barchans migrate at about 2–3 cm per minute. This corresponds closely with a rate of 123-210 mm/h for barchan migration measured by Baschin (1903), during wind velocities of 6-7 Beaufort on the beach at Fanø.

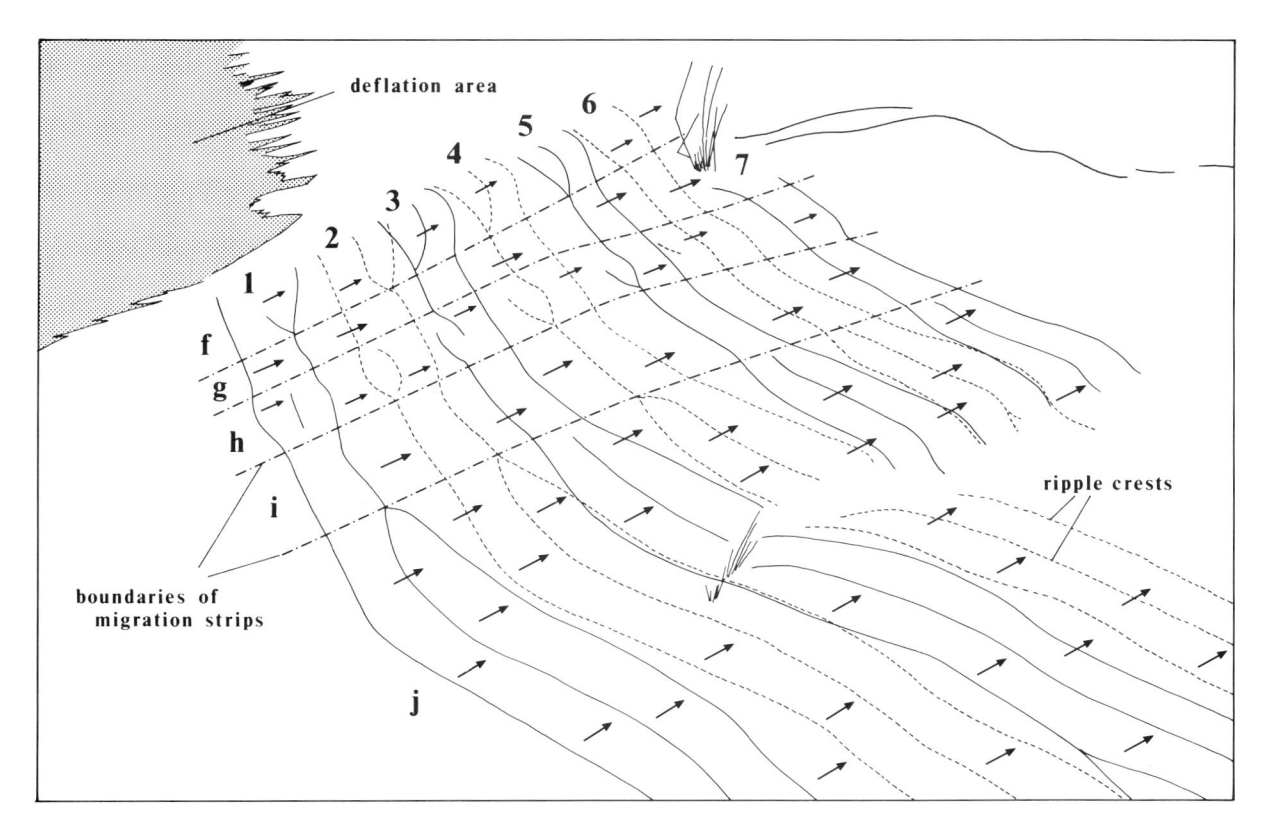

Figure 163 b: Migration of a group of wind-ripple crests within the ripple field shown in fig. 162 over a period of 30 minutes. Numbers 1 to 7 show ripple crest positions at 5 minute intervals. The size of the arrows indicates the relative migration velocity. Strip f has migrated at a slower rate than g, whilst h has been slower than g and i.

Figure 164: Dune stripes at the eastern end of Norderney during a period of strong northwesterly wind (6–7 Beaufort). Aerial photograph of F.Böker of 15.6.1984, freigegeben Brg. Nr. 5765/602.

Figures 165 a and b: Migration of beach barchans on Spiekeroog beach during a period with wind velocities of 8–9 Beaufort (Photographs: 1984).

The stripe-like alignment can occasionally become dominant, and the barchan-like form is suppressed. Such beach dune stripes are shown on the aerial photograph of a section of the eastern end of Norderney (fig. 164). Sand transport is limited to the higher, i.e. drier, parts of the beach ridge in this case. The distance between the two white marks is 100 m, and the strike direction of the line between the marks is 105°. Sand movement is fastest on the highest parts of the ridge causing zig-zag truncations of the dune crests. The distance between the dunes is greater on the ridge tops than on the flanks. On the SW flank there are 9 stripes, whereas over the same distance on the ridge there are only 7. This demonstrates that the size of the beach barchans and dune stripes depends on wind velocity. During storm periods the barchans and stripes increase in size and height (figs. 165 a and b).

The barchan surfaces are not smooth, but are covered with fields of small wind ripples (fig. 166). An accumulation rim forms at the windward end

Figure 166: Small beach barchans on Borkum, covered with wind ripples (1984).

Figure 167: The upstream end of a small beach barchan on Borkum, showing the accumulation rim and initial ripples (1984).

of the barchan behind which initial ripples develop. After a few decimetres these grade into small ripples (fig. 167). These small ripples migrate at faster rates than the barchan itself. For instance, at wind velocities of 6 Beaufort the small ripples move at more than 5 cm per minute. The migration rate of the saltating sand grains is, however, much faster.

Smooth barchan surfaces replace the small ripples at higher wind velocities. A centimetre-thick surface layer of drift sand is permanently mobile under these conditions, and the accumulation rims are in most cases missing. Some small ripples can be found on the barchan flanks. Their shape and instability closely resembles the initial ripples mentioned above (cf. figs. 165 a and b).

The beach does not consist entirely of easily deflated loose sand. Some resistance to the wind can occur when the surface is cemented by a weak crust of salt crystals (Solger et al., 1910: 18). This

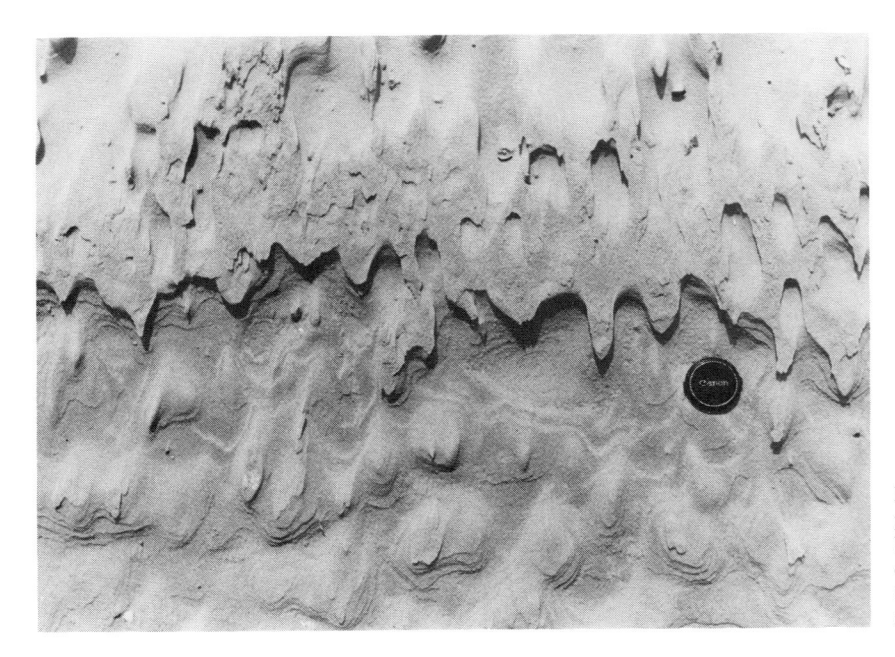

Figure 168: Deflation exhuming a pre-existing small ripple field on the beach of Borkum. The adhesive crust on the surface leads to the formation of an escarpment-like micro-relief (1984).

Figure 169: Micro 'mushroom' forms on the beach at Amrum: deflated footprints in a former beach barchan (1981).

causes the formation of miniature escarpment-like features (fig. 168). Where the barchans become reactivated after a period of stagnation the more consolidated parts, for example footprints of tourists, are modelled by scouring into mushroom-like shapes (fig. 169). Deflation can lead to overhangs where the salt crust is developed most strongly, and a type of aeolian breccia is formed where parts of the crust break down (fig. 170).

Solger *et al.* (1910) and van Dieren (1934) both recognized that backshore beach barchans do not represent early stages in the development of coastal dunes. The barchans mainly migrate laterally along the shore, without transporting significant amounts of sand inland. Only where the deflated sand contacts vegetation can primary dunes develop. These may be generated by barchans, but in most cases they will result from sand drifting freely across the beach.

Figure 170: Fragmentation of a salt-encrusted dune surface by deflation; Schiermonnikoog (1985).

Figure 171: Primary dunes on the beach at Mellum, which have been damaged by deflation (1981).

Primary dunes

In protected localities with a positive sand budget, a pioneer vegetation of salt-tolerant plants, for example *Triticum junceum*, stabilize the sand and encourage the development of higher dunes (plate 27). Parts of these summer sand accumulations are removed by subsequent winter storm tides, but in areas with a dominantly positive sand budget they are preserved and further developed in the following years.

The construction of primary dunes occurs predominantly during periods of strong wind. Deflation is dominant during storms with wind speeds greater than 8 Beaufort. Prolonged storms often drive the sand out of the vegetation hummocks, but a considerable proportion is immediately re-deposited as a dune-shadow in the lee of the obstacle (fig. 171). The deflated area can be replenished soon after the end of the storm, so that such an event may eventually lead to a further growth of the primary dunes.

Shadow dunes

Shadow dunes represent a special form of the primary dunes (fig. 173). Sand drifting across the beach may be fixed by vegetation or also deposited in the lee of other obstacles. The formation of shadow dunes was described by Sokolow (1894: 62) who refered to them as 'tongue hummocks'. Solger et al. (1910: 23) also cited several examples, and refered to them as 'sand tails' or 'wind shadows'.

The formation of this dune type has been described in detail by Hesp (1981). Dune development of this kind is used to construct artificial dune ridges in the lee of sand fences. The dune dykes (Dutch: stuifdijken) of the West Frisian Islands, and the dune ridge on the eastern end of Wangerooge, have been constructed principally by this technique.

The development of one of the youngest stuifdijks, built after 1931 on the Boschplaat (Terschelling), has been documented by Visser (1947). Fig. 174 shows how the dune ridge developed in the lee of protecting brush fences and pine brushwood. Only after the planting of *Ammophila arenaria* did sand also begin to accumulate on the windward side. The sand dyke reached a height of about 4 m above ground level within 6 years.

Obstacles do not always result in the accumulation of sand; in dominant deflation situations they may either lead to the formation of major blow-out areas (fig. 175) or negative shadow dunes, as often seen in the lee of shell fragments (fig. 176). Allen (1984 b: 200) calls these features scour-remnant ridges.

Coast-parallel dune ridges

The formation of coast-parallel dune ridges occurs wherever sand drifting along the beach is trapped by obstacles. Under natural conditions this accumulation is usually related to vegetation, but modern human interference is an important additional factor. The dune ridges created by sand fences (stuifdijken) represent a special case of coast-parallel dune ridge.

Recent dunes accumulating close to the beach beyond the influence of the tides, but which are still not completely vegetated, are called 'white dunes' (plate 28). These are easy to identify on aerial photographs. They do not form a continuous barrier, being subdivided into a series of individual crests and blow-outs.

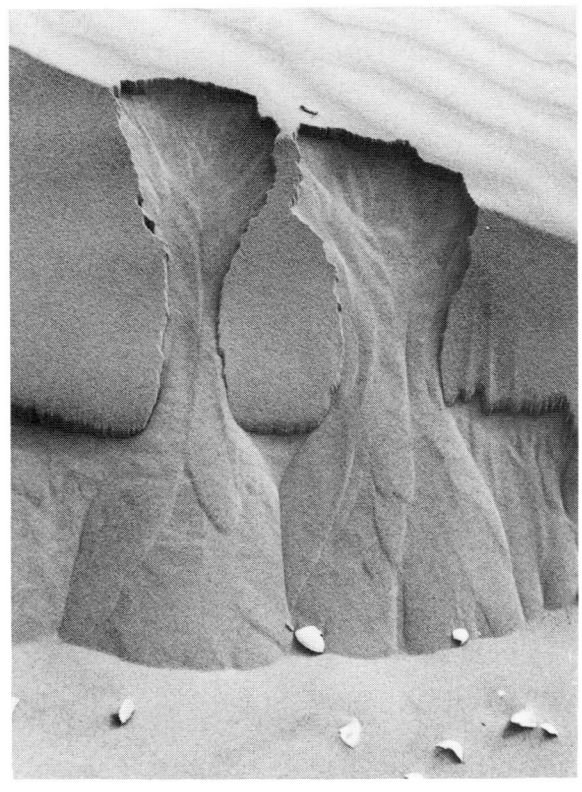

Figures 172 a and b: Tongue-like avalanches on the leeside slope of a beach barchan at Borkum (1984).

Where only isolated grass hummocks occur and the dunes are not fixed by a continuous vegetation cover, barchan-like dune forms can grow to a height of over half a metre. The steep windward faces may be subdivided by avalanche tongues, as frequently found on the slopes of full-developed 'white dunes' (see below) (figs. 172 a and b).

Figure 173: Formation of shadow dunes in the lee of a sand fence on Vlieland (1981).

Figure 175: Deflation mould in the lee of a remnant sand block on the beach at Borkum; wind direction was towards the camera (1984).

Terschelling

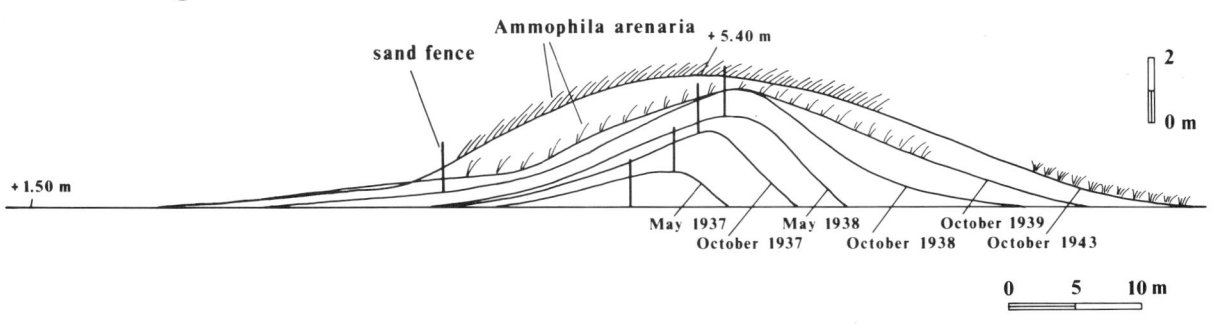

Figure 174: Development of the stuifdijk on the Boschplaat, Terschelling, at Paal 26, between 1937 and 1943 (Source: Visser, 1947: fig. 4).

As sand is preferentially deposited in the lee of obstacles, sand shadows over 10 m in length can be formed behind isolated dunes, which may reach heights of over one metre (fig. 177). These shadows may ultimately connect with neighbouring dunes and are therefore important in the formation of continuous dune ridges.

Beach-parallel lines of coastal dunes form in areas with a positive sand budget. The coastal dunes on eastern Langeoog (fig. 178) have formed during the second half of the 19th century.

Wherever the coastal dune ridge is interrupted

Figure 176: Negative shadow dunes forming in a deflation area on Spiekeroog beach. Long erosional sand keels are formed behind obstacles, mostly shell fragments (1984).

by blow-outs, sand from the beach is deflated into the white dune areas. The sand migrates via small ripple fields into the deflation pits and is then blown up the windward slopes and across the crests, to be deposited in tongue-like sand accumulations in the lee of the dunes. Dune lee slopes always maintain an equilibrium profile because they are developed by sand grains rolling downslope, windward slopes may become oversteepened and are therefore unstable.

Instability is greatest in situations where vegetation on the dune crests prevents sand transport across the ridge. Sand accumulation at the base of the vegetation leads to an oversteepening of the slope, and sediment tends to glide downslope in avalanche tongues only to be blown back up by the wind. Observations on Borkum have demonstrated that even during wind velocities as low as 6–7 Beaufort several such avalanches may occur per minute (fig. 179). Avalanche tongues of this type have been described by Kaiser (1971) and Jäkel (1980) from barchans in the Sahara.

Migrating dunes

Most of the backdune areas are inactive, but migrating dunes may develop where a dense vegetation cover is not encouraged. Well developed active migrating dunes are only found today on the

Figure 177: A sand keel (shadow dune) has formed in the lee of a recent white dune on the Boschplaat (Terschelling). Such keels can connect isolated beach dunes given a sufficient sand supply (1981).

LANGEOOG

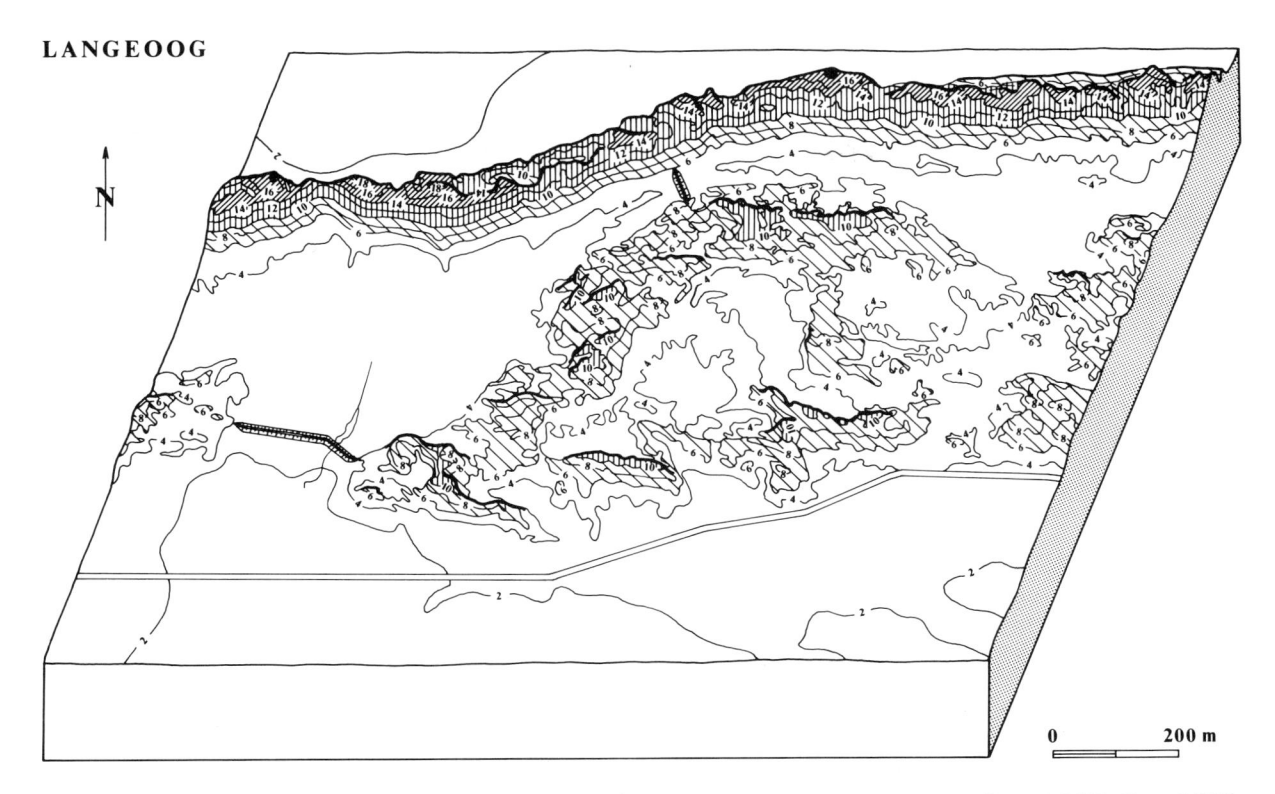

Figure 178: A coast-parallel dune ridge and an old dune core on Langeoog (Source: Deutsche Grundkarte 1:5,000, Sheet 0457,8 Langeoog-Dreebargen).

Listland of Sylt. Passive forms are present on Rømø, Fanø and Amrum, and in many cases the dune arcs of the West and East Frisian Islands consist of old migrating dunes.

The migrating dunes of the Listland on Sylt (fig. 180) are actively moving towards the east. Only the central parts of the dune ridges are being displaced, however, the marginal parts being retained by the vegetation. Grass causes these areas to grow higher than the active central parts, finally leading to the formation of elongated dune ridges flanking the migrating dunes. The crests of these ridges are irregularly sculptured depending on local variations in the degree of vegetation cover.

The depressions between the large migrating dunes are characterized by a number of low swells striking at right angles to the prevailing wind direction (Kolumbe, 1943). These swells represent the former windward slope bases of the migrating dunes fixed by vegetation during periods of calm weather and are normally only a few decimetres high (fig. 180). The migrating dunes of the Listland on Sylt reach heights of about 30 m, the active parts rarely being higher than 24 m above NN.

The base of the dunes is between 2–4 m above NN, depending on the groundwater level.

The migration rate of the dunes in the Listland can be established by the comparison of old maps (fig. 181). Comparison of the 1880 first edition of the List Meßtischblatt 1:25,000 with the 1932 edition shows that in this period the dunes migrated up to 200 m. The map of 1880 was surveyed in 1878. The date of surveying is not given on the 1932 map, but comparison with the Grundkarte 1:5,000 indicates that the contour lines were adapted from this large-scale map. The contours were photogrammetrically surveyed by the Reichsamt für Landesaufnahme, Berlin, Photo-Abteilung in 1929, the aerial photographs being taken in 1925 (Priesmeier, 1970: 14). The maximum migration rate of the dunes was therefore about 4.3 m per year.

It seems unlikely that this migration rate has continued up to the present because parts of the dunes have subsequently been artificially stabilized and the sand supply for the active dunes restricted. It is unfortunate that both the latest issues of the topographical map 1:25,000 (1984) and of

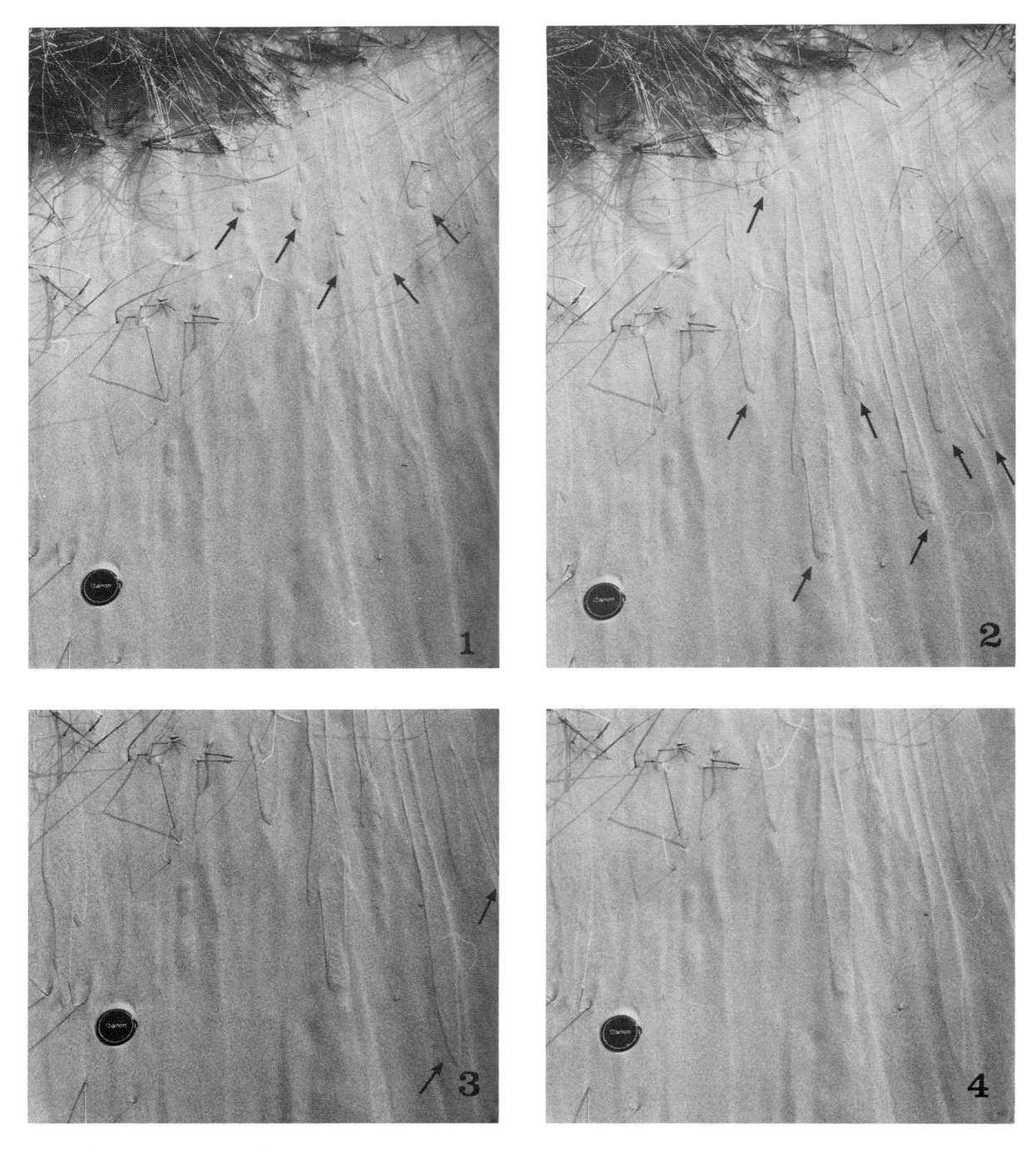

Figure 179: Sand-flow tongues (arrows) on the oversteepened slope of a white dune on Borkum (wind velocity: 6–7 Beaufort). The vegetation prevents the sand from being blown over the dune crest. Sand accumulation at the foot of the vegetation leads to oversteepening of the slope and causes the tongue-like avalanches. Photographs taken with an interval of 10 seconds (1984).

the Deutsche Grundkarte 1:5,000 (1984) still show the contour lines of 1925; in the case of the Grundkarte this results in the bizarre situation that some of the dunes, portrayed in great detail, are no longer represented on the respective map sheet.

The old maps were compared with a Landsat-3-RBV image taken in 1981 since no aerial photographs were available. This image lacks contrast, as with most RBV frames, but the barren areas of migrating dunes are still clearly visible.

SYLT

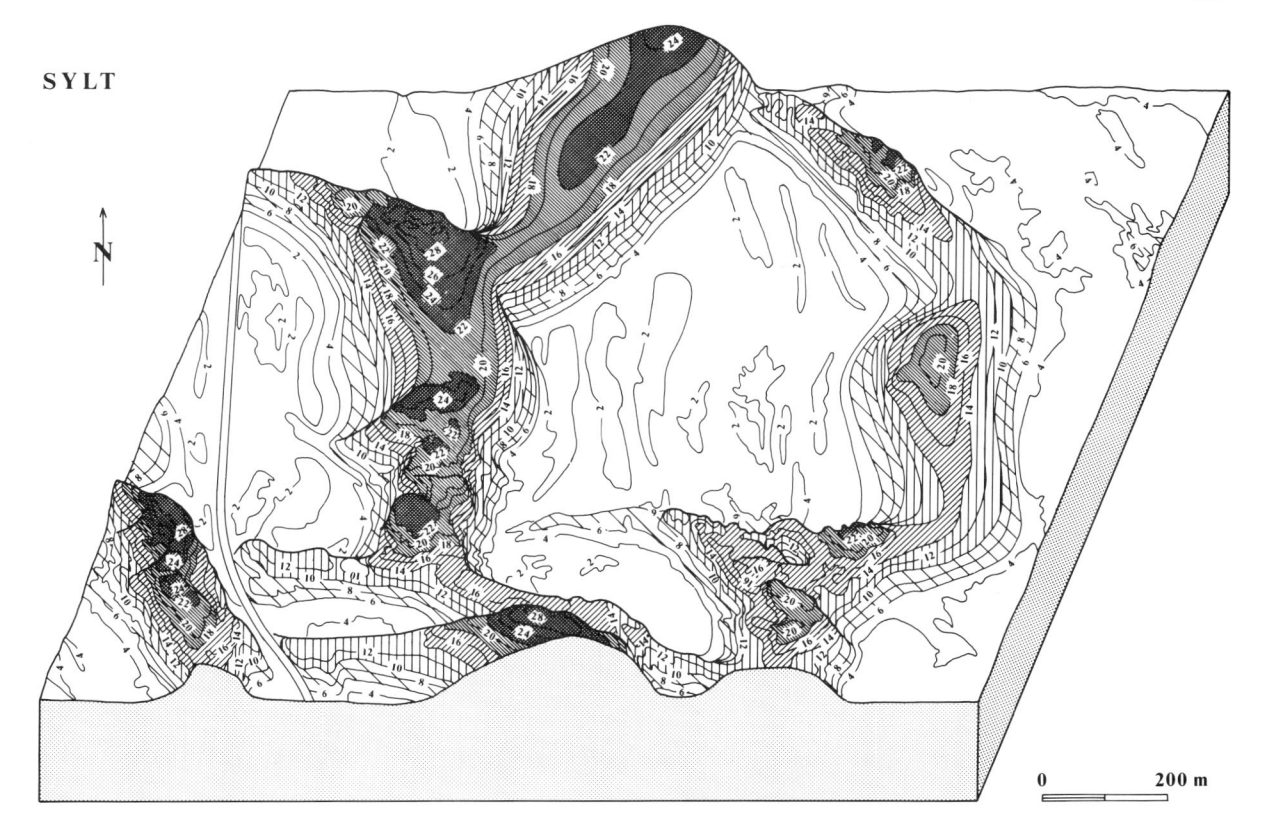

Figure 180: Block diagram of migrating dunes in the Listland on Sylt. Source: Deutsche Grundkarte 1:5,000, Sheet 6098 Sylt-Mannemorsumstal.

The scale of the RBV print (about 1:125,000) was sufficient to allow a comparison with the 1:25,000 maps. Using these data it appears that between 1929 - 1981 the dunes migrated at an average rate of 4 m/year. This coincides well with the migration rate of 3.75 m measured by Priesmeier (1970: 49) for the period between 1925–1965.

This figure corresponds with older records representing the period before the first accurate surveys of this area, quoted in the literature. Mager (1927: 93) states that 'Baudissin (1865) mentioned a migration rate of 14 feet (4.07 m) per year for the west-easterly progression of the Sylt dunes'.

Fig. 182 shows part of the dune core of Norderney. These dunes, like many others, were formed in a similar way to the migrating dunes of the Listland. The extent of the Norderney dune arc resembles that of the great migrating dunes on the Listland, and their maximum heights are also comparable (22 m on Norderney compared to 30 m on Sylt). The more intensive dissection of the Norderney dune ridges is a result of their greater antiquity.

Not all migrating dunes have necessarily moved at the same rate. Van Dieren (1934: 182) reports that the 10–20 m high barchans on Terschelling, east of Lies, have migrated at a rate of 15–20 m per year during the l9th century. Mager (1927: 97), however, states that the dunes at Rantum on Sylt, in the l8th and l9th centuries, migrated at a rate of 2.91 and 2.33 m per year respectively.

Hempel (1980, 1983) recognized the importance of storm floods in providing sand for dune formation. There seems to be a connection between the formation of the great migrating dunes of the barrier islands and the great storm floods of the Middle Ages. Assuming that the positions of the migrating dunes of the North Sea barrier islands were largely fixed by the middle of the l9th century, they have had about 500 years to reach their present positions since the great storm floods of the l3th-l5th centuries. At a migration rate of

166 *Morphodynamic units*

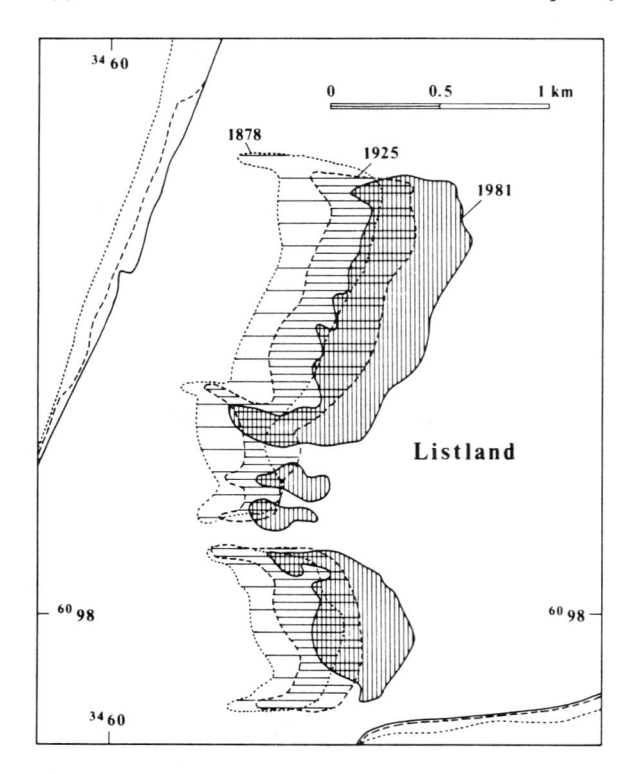

Figure 181: Displacement of migrating dunes in the Listland on Sylt from 1878 to 1981. Sources: Meßtischblatt 1:25,000 List (Insel Sylt), issues 1880 (mapped 1878) and 1932 (topography of 1925); Landsat-3-RBV image 212-22 1 of 6.8.1981.

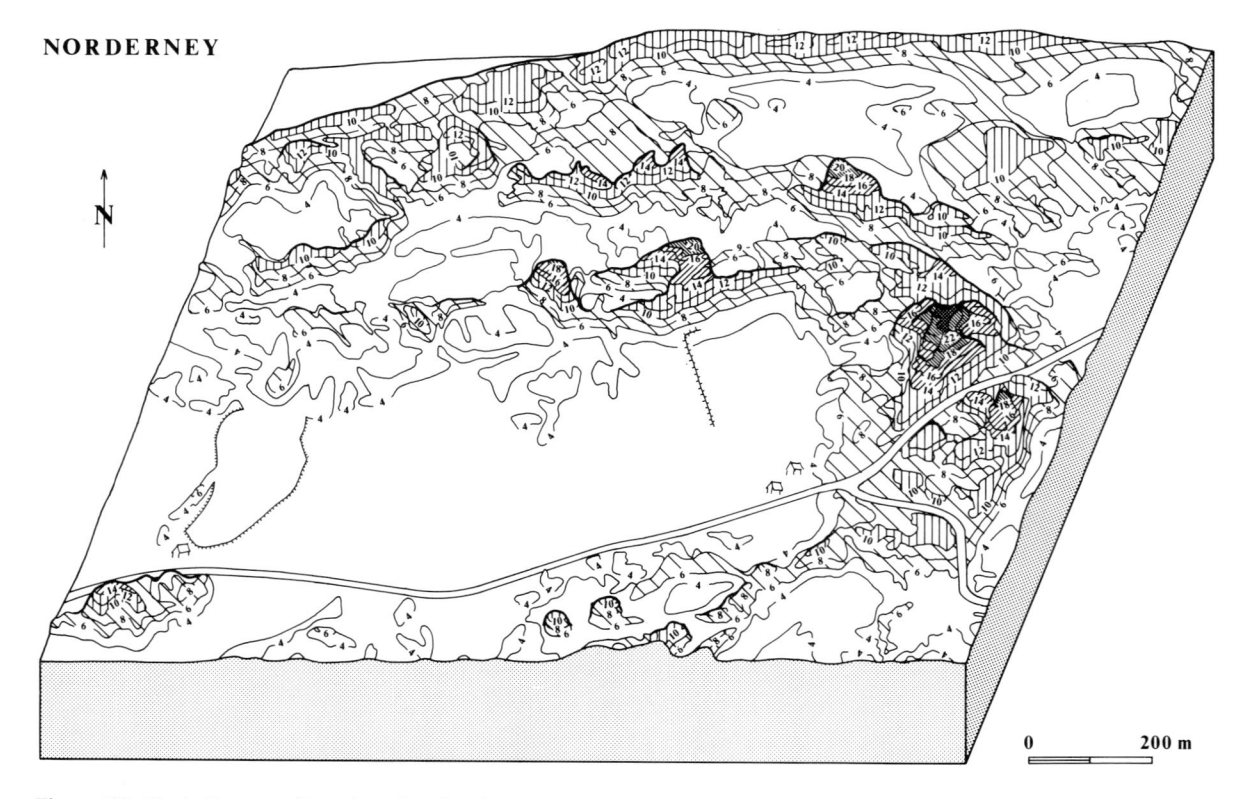

Figure 182: Block diagram of inactive migrating dunes on Norderney. Source: Deutsche Grundkarte 1:5,000, Sheet 7754 Norderney-Meierei and Sheet 7954 Norderney-Weiße Düne.

Figures 183 a and b: Sand accumulation on the old dunes of Wangerooge. Sand has been blown across the main dune ridge during periods of strong wind; (a) dunes at the eastern end of Wangerooge; (b) sand accumulation in front of the 'Saline' (1979).

2–20 m per year they potentially travelled over a distance of 1–10 km which would correspond well with their present positions.

The following average speeds can be calculated for the aeolian displacement of sediment accumulations along the Wadden Sea coast:

Dune form	Distance between crests	Velocity
Wind ripples	5 cm	5 cm/min
Beach barchans	5 m	150 cm/h
Migrating dunes	1000 m	4 m/a

Old dunes

The backdunes in the interior parts of the islands, the so-called 'grey dunes', are covered with dense vegetation. This does not mean, however, that they are completely inactive. During periods of strong wind, especially in winter, sand is easily deflated across the main dune ridge (figs. 183 a and b). This sand accumulates in the lee of obstacles and contributes to the slow growth of the dunes.

The process is, however, more than balanced by the erosion in the backdune areas, especially where

Figure 184: A blowout in a grey dune on Baltrum. Deflation was initiated by the destruction of vegetation by pedestrians (1980).

vegetation has been damaged by pedestrian trampling (fig. 184) and burrowing animals, mainly rabbits.

Dune destruction by deflation

Active aeolian accumulation on the old dunes tends to be restricted, on most islands, to a modification of the relict migrating dune relief. The dune arcs are carved by notches and blowouts and segmented into individual summits. Fig. 185 shows part of the frontal crest of the former Norderney migrating dune (fig. 182) intersected by three strong blowouts. Such erosion can ultimately lead to complete destruction of the original dune relief. Fig. 186 shows arcs of migrating dunes on Fanø which were completely destroyed by blowouts.

Most blowouts tend to be about 50 m in length and 20 m in width. They are not, therefore, large enough to be shown in any detail on the 1:25,000 topographical maps or the 1:12,500 aerial photographs. As the contour lines on the 1:5,000 Grundkarte have not generally been revised in more recent editions, comparison of these maps cannot be used to trace recent changes in the grey dune areas. The major remodelling of the topography can, however, be easily detected in the field.

The morphology of the blowouts generally resembles that of migrating dunes (Goldsmith, 1985: 340). The proportions, however, differ markedly; whilst migrating dunes are between 500 to over 1000 m long, the blowouts are usually less than 100 m in length.

Figs. 187 a and b show the same area in the Baltrum dunes facing east. Photograph (a) was taken in 1980, photograph (b) in 1984. The blowout in the foreground has enlarged by several metres in a south-easterly direction. In contrast, the small ridges on the northeastern margin of the blowout seem to be less affected by deflation. The vegetation cover is also much more extensive in 1984. The coast-parallel dune ridge in the background was only breached at one site, where a sediment tongue began to accumulate in the backdune area.

Figs. 188 a and b show another section of the Baltrum grey dune area, a little further to the east. The freshly accumulated sand in the foreground of photograph (a) was eroded by 1984. The sand accumulation in the neighbouring dune valley had been vegetated. The photographs record the rapid vegetational changes which often occur in dune areas. These changes are clearly exemplified by the different plant distributions on the dune slope on the left sides of the photographs. The footpath in the centre of figure (a) is half vegetated by 1984 in figure (b), which results from more rigorous dune protection.

By 1984 a new wind breach had formed in the group of dunes in the left background, and a single hillock had been separated from the rest of the group. In the central part of the dune group a sand

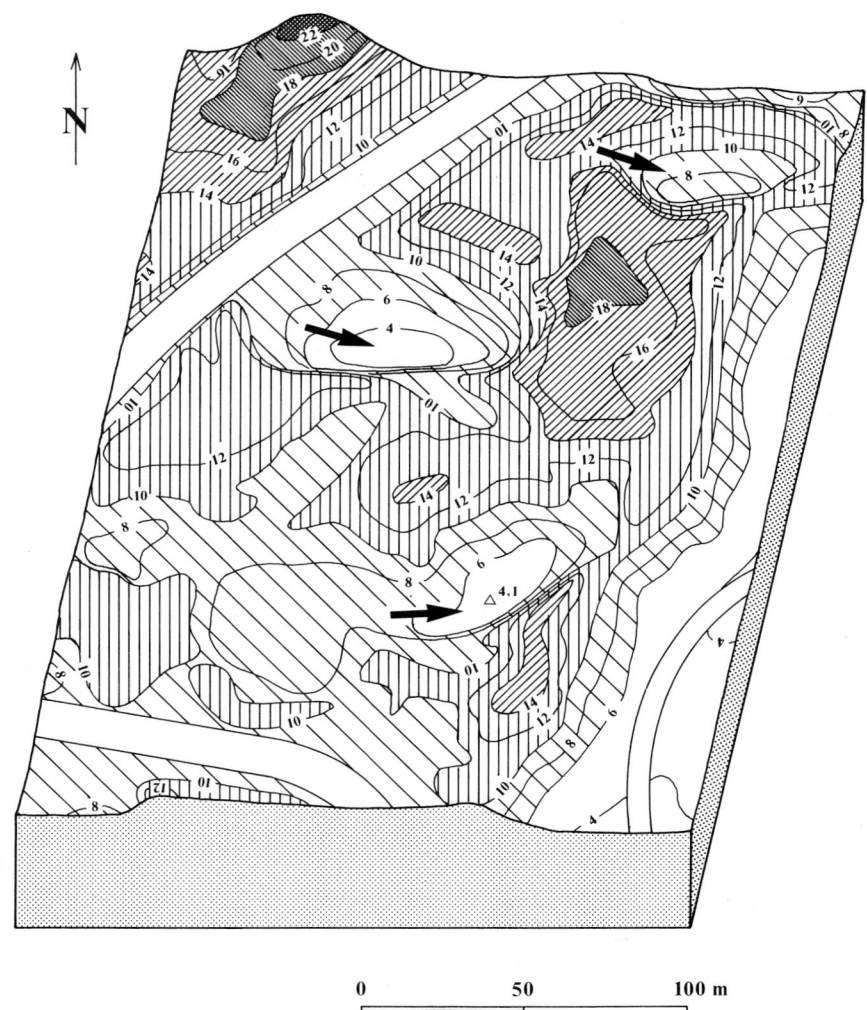

Figure 185: The proximal side of a former migrating dune on Norderney, dissected by a number of blowouts (arrows). Compare with fig. 182. Source: Deutsche Grundkarte 1:5,000, Sheet 7954 Norderney-Weiße Düne.

tongue is seen to have prograded several metres across the leeside slope. The southeastern part of the dune group, which was still active in 1980, is completely fixed and vegetated in figure (b); the wind breach seen in figure (a) has given rise to the separation of the dune on the right in figure (b).

No rates can be given for the degradation of the dune crests by wind breaching and blowout. These forms tend to extend by about 0 - 5 m per year. It must be remembered, however, that these processes are not continuous and may be interrupted by long periods of relative stability.

Dune crest displacement by blowouts occurs on most of the islands. These blowouts can be deflated down to the groundwater table in summer and autumn. In spring, when the groundwater table rises to its maximum elevation, major areas of the blowouts become water-covered. The summer house settlement on Fanø is located in an inactive deflation landscape of this type (plate 9).

Dune destruction by pedestrian trampling

The influence of trampling on the vegetation of white and grey dunes was investigated at selected test sites on the Skallingen peninsula (Hylgaard, 1980; Hylgaard & Liddle, 1981). It was discovered that the passages of only 200 people can initiate a reduction in plant cover by up to 50 %. Only *Empetrum nigrum* and *Hypnum cupressiforme* were able to survive the passage of over 2,500 people, whilst *Festuca rubra*, *Ammophila arenaria* and

Figure 186: Aerial photograph showing horseshoe-shaped migrating dune ridges in southern Fanø (Vindgab-Bjerge) dissected by numerous blowouts (Aerial photograph: Geodætisk Institut, Denmark. 1.9.1964, D 339-F Nr. 03. Reproduced by permission (A. 532/85) of the Institute.

Veronica officinalis were destroyed.

A number of dune stability problems also arise as the age of the dunes increases. The loose sand deteriorates rapidly as nutrients are leached, and the only remaining vegetation is often *Empetrum* or *Calluna* on the windward side of the dunes and *Corynephorus canescens* on the leeside (plates 29 and 30). *Corynephorus* in particular is extremely susceptible to erosion. Trampled and eroded areas cannot be revegetated without artificial assistance. Unnecessary pedestrian traffic must therefore be limited beyond the marked footpaths.

In the Netherlands, in 1977, the 'Werkgroep Rekreatie van de Landelijke Vereniging tot Behoud van de Waddenzee' published the results of investigations on the degree of trampling in the dune areas of the West Frisian Islands. The number of overnight stays per year on the West Frisian Islands increased from 3 million in 1963 to 8 million in 1976, and the environmental stress increased accordingly. Aerial photograph interpretation and field investigations revealed that footpath length in the dunes increased considerably in the period between 1969/70 and 1975/77:

Increase in path length in the dune areas of the West Frisian Islands (Werkgroep Rekreatie, 1977: 8):

Island	Path length in m/ha 1969/70	1975/77	Increase
Texel	345	(375)*	(10%)
Vlieland	305	415	35%
Terschelling	345	480	40%
Ameland	395	630	60%
Schiermonnikoog	525	605	15%

*The figures for Texel based only on field studies (1977).

Figures 187 a and b: Dune landscape in the eastern part of Baltrum; (a) 1980, (b) 1984.

Comparison of these results with aerial photographs taken in 1983 clearly illustrates that the deterioration of the dune landscape has largely ceased. The same is also true of the East Frisian Islands. At the beginning of the investigation period the frequency of tourist-induced dune destruction was high, but most of the vegetation damage has now been repaired and the number of footpaths in most areas has decreased rather than increased.

The dune area between Wenningstedt and Kampen on Sylt (fig. 189) is particularly prone to pedestrian trampling because of the large numbers of visitors. The aerial photograph shows the high path density. As the dune crests constitute preferred paths, deflation begins here and leads to the formation of a series of characteristic secondary blowouts on the crests, a number of which can be seen in this picture.

According to Hylgaard (1980: 17) a destroyed vegetation cover can be partially restored in about one year. However, the new vegetation will differ markedly from the original plant cover. It may take several years before the original state is achieved.

Figures 188 a and b: Dune landscape in the eastern part of Baltrum, further to the east of fig. 187; (a) 1980, (b) 1984.

Dune destruction by surface runoff

Natural erosion by running water plays only a minor role in the dune areas. Precipitation usually percolates immediately through the loose sand of the dunes and causes no subaerial run-off. During heavy showers, however, this is not the case (cf. Klijn, 1981: 109). In summer thunderstorms, water flows in small rills down the grass-covered slopes and does not percolate until it reaches areas of open sand, such as in a blowout. During a heavy shower on 20.5.1984 (according to the List weather station 4.8 mm of rain fell in 30 minutes), traces of erosion and deposition by running water could be found in a grey dune area of about 2 km^2 west of List (Sylt) in more than 10 localities.

Small footpaths represent preferred zones for the initiation of erosion. One reason for this may be the greater consolidation of the sand on the paths, which leads to reduced percolation when compared with other sand areas.

The rain-water run-off flushes out sand from the grass hummocks and transports it downslope. Where grass cover extends into the adjoining dune valley sand can be transported down as far as the valley bottom, but where the water meets areas of open sand it percolates rapidly. The sand accumulates in rims which ultimately form small deltas if the process lasts long enough (figs. 190 and 191).

Such runoff processes are also active in winter

Figure 189: Secondary blowouts on the dune crests between Wenningstedt and Kampen on Sylt. Aerial photograph of Fa. Rüpke, Hamburg, of 2.11.1984, Bildflug Westküste Sylt Nr. 20 3564 3417, Bild Nr. 129, freigegeben durch Luftamt Hamburg unter Nr. 1674/84.

Figure 191: Close-up view of a delta accumulated by rain runoff on a vegetated dune slope on Sylt. As the runoff decreases the delta becomes inactive and pitted by raindrop impact marks (1984).

Figure 190: Dune-slope erosion by heavy rain in the backdune area of Listland, Sylt. A small delta has accumulated on the unvegetated sand in the foreground (1984).

Figure 192: A small delta on the beach at Wangerooge which developed during a period of snow melt. The meltwater was unable to percolate because the ground was frozen. Sag features have developed from the melting of underlying snow (1980).

Figure 193: Oblique aerial photograph of a washover area of the 'wash-in' type on eastern Norderney. Aerial photograph of F.Böker of 16.6.1984, freigegeben Brg. Nr. 5765/595.

when the dune surface is sealed by frost. Fig. 192 shows a small delta which has accumulated at the foot of a dune slope during winter snow-melt on frozen ground on Wangerooge. The sag features indicate that part of the delta accumulated on snow, which later melted out.

5.3.4 Washover areas

Along the Dutch coast south of Den Helder, and the Danish coast north of Blåvandshuk, the coastal dunes form a continuous barrier which is rarely interrupted, for example at major river mouths. On the barrier islands the situation is different, however. Only dune cores of several hundred metres in diameter and, in exceptional cases, several kilometres, were formed, These were separated from each other by dune-free lowlands, the washover areas. This landform assemblage of dune cores and washover areas results from the occasional breaching of the dune ridges during storm tides, and leads to considerable sand transport

Figure 194: Washover channel at the eastern end of Ameland looking towards the north. The replenishment of the adjoining recent dunes occurs both directly from deflation of sand from the washover area, and also by sand blown through the washover channel from the beach into the interior parts of the island (1981).

Figure 195: Seaward end of the washover channel featured in fig. 194 looking towards the south. Transition from the overdeepened channel trunk to the alluvial fan (in the background) on which the recent dunes have formed (1981).

176 *Morphodynamic units*

Figure 196: Washover channel in the eastern part of Norderney during spring tide. The rising water flows through the channel towards the tidal flats and divides Norderney in half for several hours (1982).

NORDERNEY

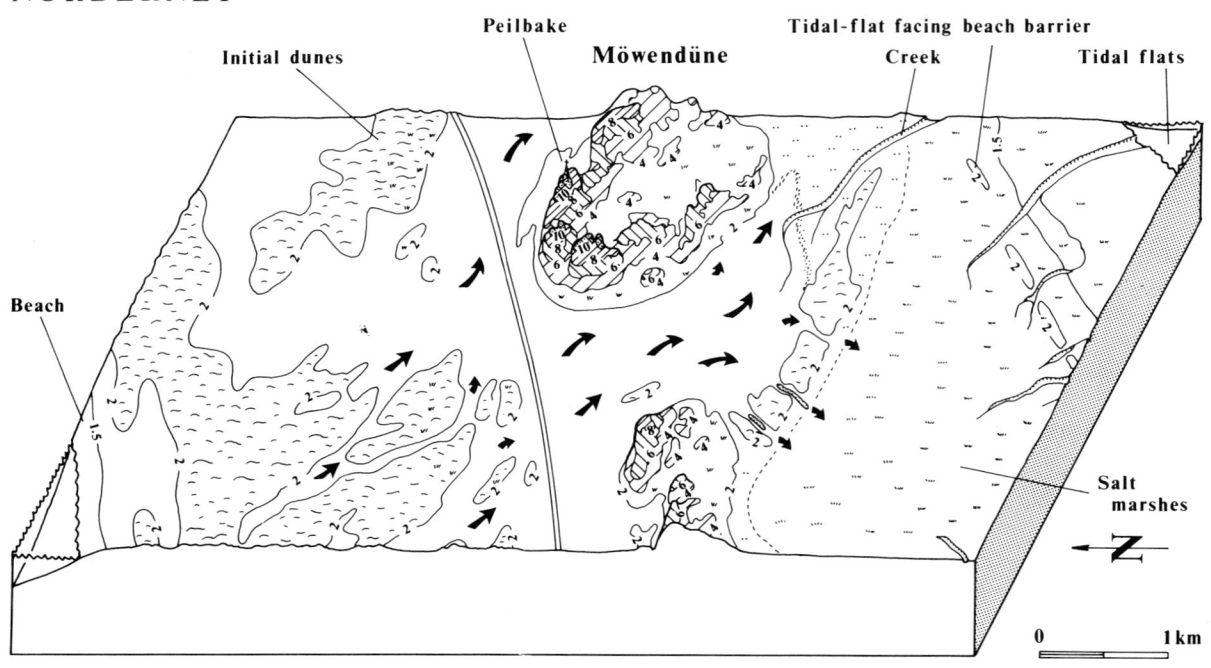

Figure 197: Block diagram showing dune cores and washover areas on eastern Norderney (cf. plate 11). Source: Deutsche Grundkarte 1:5,000, Sheet 8554 Norderney Nordstrand-Ost.

cross-island from the beach to the tidal flats (Hayes, 1979: 10).

There are barrier islands elsewhere, the sediment budget of which is strongly influenced by washover processes. Armon (1979: 77) for instance, estimates that storm overwash accounts for 48 % of the landward directed sediment transport on the Malpeque Islands (Gulf of St.Lawrence), compared with only 52 % for transport through the tidal inlets.

However, washover processes are much less pronounced along the North Sea coast. In many cases

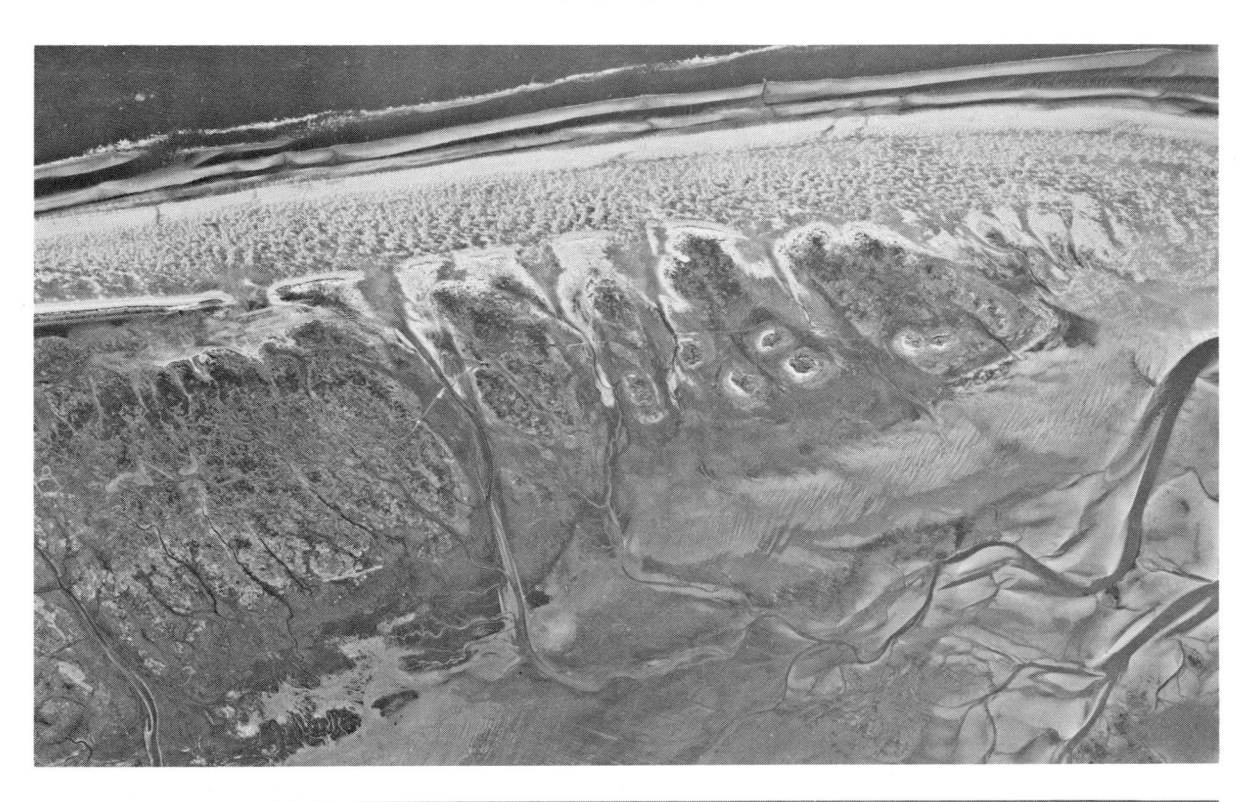

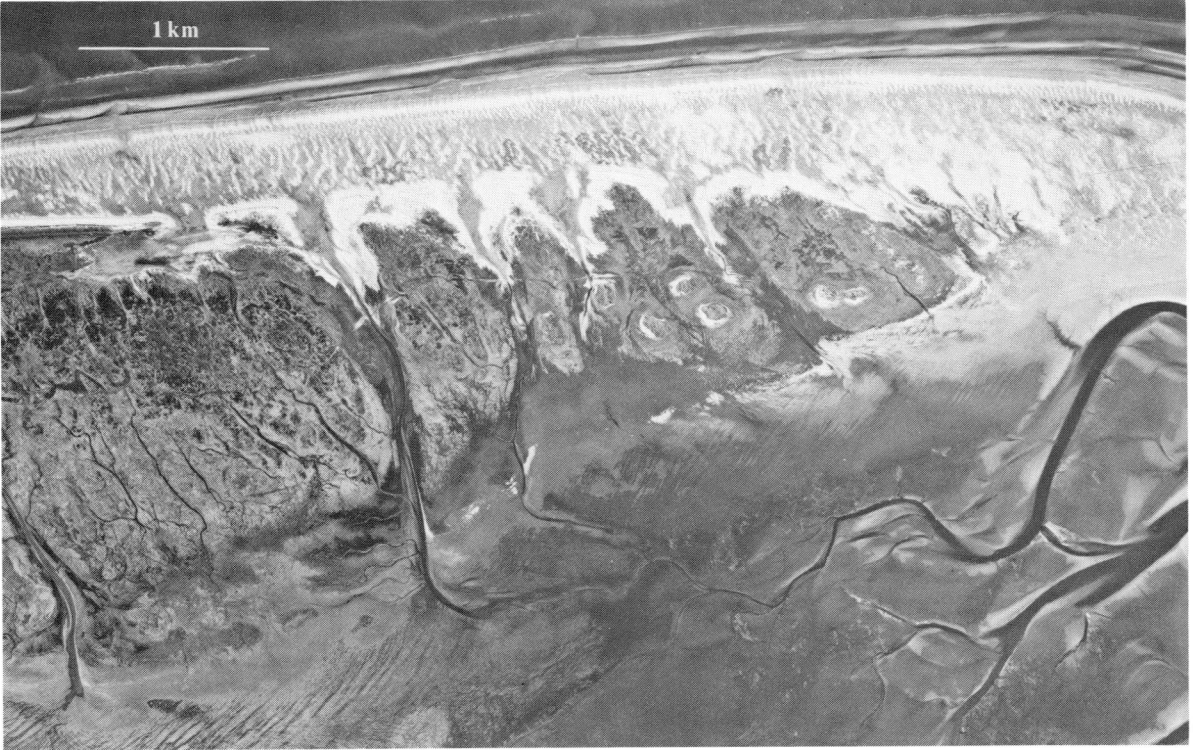

Figures 198 a and b: Two aerial photographs of the eastern part of Schiermonnikoog, taken in the autumn (a) and the following spring (b). The sand deposited by the winter storms on the margins of the washover areas is clearly visible in the lower photograph (Scale: ca. 1:50,000). Aerial photographs taken by KLM Aerocarto Holland on 2.10.1979 and 13.4.1980; flights No. 5-15-29 and 5-15-32 for Rijkswaterstaat.

only a 'wash-in' takes place instead of a 'washover' in which sediment is washed from the beach through the dune belt towards the tidal flats; most of it is deposited in the interior parts of the islands (fig. 193).

Originally, washover areas existed on all the East and West Frisian Islands. Juist, for example, consisted of three separate dune cores until the 17th century. All three are now connected by sand dykes or artificial dune ridges.

The active washover areas consist of
1. broad, funnel-shaped mouths,
2. a narrow, overdeepened trunk,
3. a poorly developed sand fan,
4. one or more run-off channels towards the tidal flats

Small pools remain in the overdeepened central parts of the channels even during low tide (figs. 194 and 195).

Washover areas along the Wadden Sea coast are nearly all restricted to the West and East Frisian Islands. They are absent in the North Frisian and Danish Wadden Sea, with the exception of a small area on the Skallingen peninsula which is no longer active (Jacobsen, 1980: 54).

Sand transport through the washover channels

The washover channels of the West and East Frisian Islands are not only inundated during storm surges, but also during each exceptionally high tide (fig. 196). However, the morphological consequences of such floods are minor. Major sediment movement is limited to the winter months when sand is transported into the interior parts of the islands. At the same time the adjoining dune cores are eroded, often extensively. This leads to the formation of elliptical dune cores within wide dune-free sand plains.

The eastern end of Norderney provides a good example of such washover areas (figs. 193, 197 and plate 11). Between the beach and the old dune cores a broad initial dune field has developed, which is only flooded during spring tides and winter storm surges. During the floods the sand is transported beyond the old dune cores and deposited in large alluvial fans to the south of the dunes. Recent dunes have formed at the margin of the sand fans on the salt marsh. They are delineated in fig. 197 by the 2 m contour. The sediment for these recent dunes is provided from the washover area, especially the sand fan.

The importance of sediment supply from the washover channels for the dunes in the inner parts of the islands is demonstrated by the comparison of two aerial photographs of Schiermonnikoog (fig. 198). One photograph (a) was taken in the autumn, the other (b) during the following spring. The differential distribution of loose sand, which appears as light areas in the photographs, is clearly visible. Large amounts of sand have been transported into the inner parts of the island along the washover channels during the winter storms.

Damming of washover areas

The repeated flooding of the islands via the washover channels seriously affected the agricultural use of the marshes. The fields and meadows became inundated by sand, and the soil was damaged by salt. The gaps in the dune ridges also heightened the flood risk for the settlements. Early attempts were therefore made to close the washover channels between the dune cores. The following are examples:

1630: the two dune cores of Texel (Texel and Eijerland) were connected via a dune dyke formed with the help of sand fences (van Straaten, 1979: 7).

1860: the two dune cores of Borkum were connected by damming the Tüskendör area with an artificial dune belt (Wegmann, 1982: 9).

1874: the gap between the two dune cores of Wangerooge was closed by the 'Reichsdeich'.

1904: the building of an artificial dune ridge on the eastern part of Wangerooge was begun to eliminate any washover processes in that area.

1906: the 'Große Slop' washover area on Langeoog was finally closed by a dam (fig. 199).

1930: the Flinthörn dunes in southern Langeoog were connected by a dam to the northern main dune core (Backhaus, 1943: 81). Active washover areas occur only in very few places on the West and East Frisian Islands today:
1. at the easternmost end of Ameland,
2. at the eastern end of Schiermonnikoog,
3. at the easternmost end of Borkum,
4. at the eastern end of Norderney and
5. on the long eastern sand flat of Spiekeroog.

One of the best developed washover areas, the Boschplaat on Terschelling (figs. 200 and 373), became inactive after the erection of a sand dyke in 1936 (van der Molen, 1978 b: 17).

Consequences

Although the damming of the washover areas al-

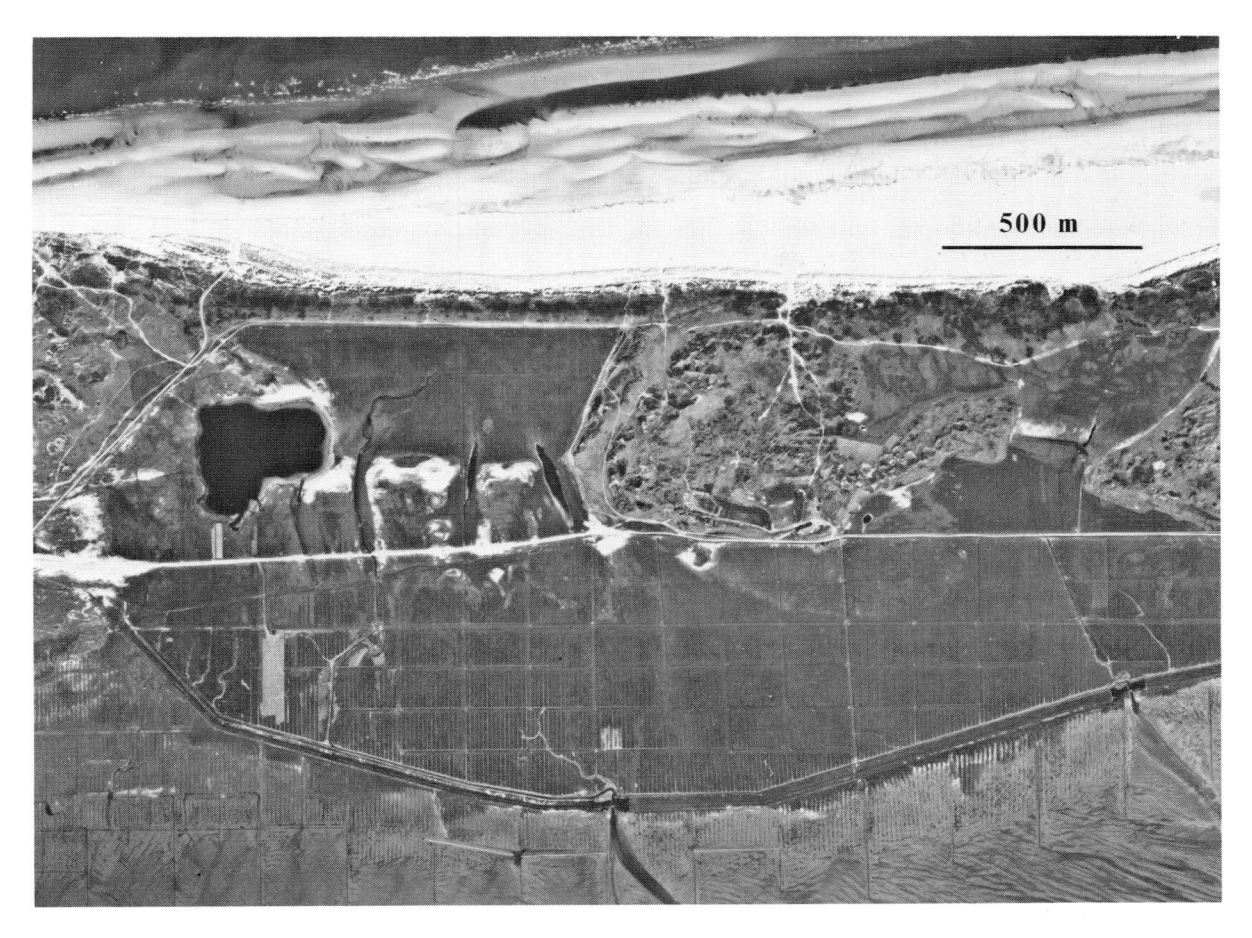

Figure 199: Aerial photograph of the 'Große Slop' washover area on Langeoog. The Slop has been blocked by a sand dyke since 1936, but the channels and sand fans of the washover area are still visible. The lake to the left is a dredging pond. Aerial photograph of Niedersächsisches Landesverwaltungsamt, Landesvermessung, of 27.5.1978, Bildflug Langeoog (1497), Str.1/058, freigegeben durch NLVA – Landesvermessung –, unter Nr. 17/78/1497.

leviated the flood hazard for the islanders, it also created a serious problem. The damming stopped the sand supply to the Wadden Sea facing shores, and thus contributed to high erosion rates along the salt marsh shoreline (cf. section 5.3.5).

The importance of the washover areas for the continued existence of the islands should not be underestimated. A landward sand transport across the islands is necessary during a transgressive period to maintain the island barrier by landward migration. The islands may ultimately be inundated by the sea if this sand transport is eliminated (Leatherman, 1982).

5.3.5 Natural accretion

Given the positive sediment budget of the Wadden Sea (cf. chapter 3) one might expect strong and extensive accretion. The opposite is the case, however. Salt marshes are actively developing in only a very few areas of the Wadden Sea. It is conspicious that even in comparatively protected areas erosion is often very effective (cf. section 5.3.6).

In 1934 Mungenas (p. 60) observed the signs of erosion, particularly in the form of 'flood cliffs', predominantly occurring along the coast of the East Frisian mainland. Indications of coastal erosion include small cliffs at the margin of the salt marsh, margins decaying into discrete grass hummocks and the existence of beach ridges along the salt marsh margins. These signs are particularly noticeable in places where, as indicated by the presence of salt marshes, accretion prevailed relatively recently.

Areas of accretion are exceptional in the Wad-

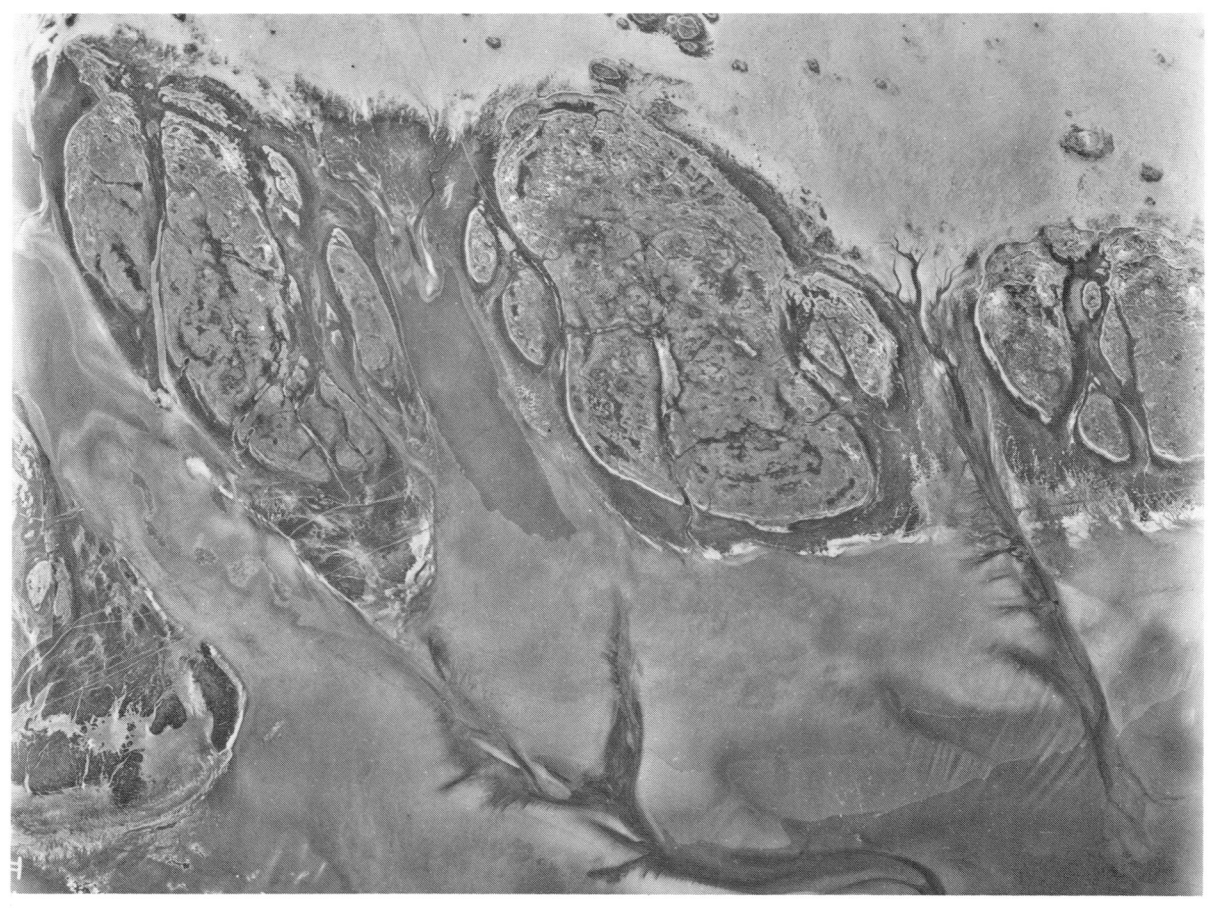

Figure 200: Dune cores and washover channels on the Boschplaat, Terschelling. Aerial photograph of the Fototheek Topografische Dienst, Emmen, of 1949, Bl Run II 30.

den Sea today. They occur only under particularly favourable conditions. These include:

1. Areas with calm water
   a. in the coastal lowlands behind large bays breached by the medieval storm surges, e.g. Dollart, Jade Bay, Ley Bight,
   b. in the protected areas along embankments connecting the islands and Halligen with the mainland, e.g. Hindenburg Dam, Rømø Dam, Nordstrand Dam,
   c. in protected areas in the lee of the barrier, e.g. parts of Baltrum, Spiekeroog, the Danish mainland coast and the Skallingen peninsula.
2. Areas with a positive sand budget
   a. in areas protected by beach barriers, e.g. Eiderstedt and Rømø,
   b. on the eastern ends of some of the West and East Frisian Islands, e.g. Spiekeroog, Terschelling, Schiermonnikoog.

Accretion in areas of calm water

Where currents are weak in protected areas in the lee of the islands, or along the mainland coast, transported sediment is deposited and accretion begins. This process is actively promoted by the spread of vegetation.

The seaward margin of the vegetation occurs at about 40 cm below mean high water (Dijkema, 1983 a: 189). Single tussocks of *Spartina anglica* represent the furthest seaward vegetation (plate 5). This grass is not indigenous along the coasts of the German Bight but originated from the coast of southern England. It was first introduced into the Netherlands for land reclamation purposes, and later also in North Germany and Denmark (Meyer-Deepen & Meijering, 1979: 111; Dijkema, 1983 a: 188). The sediment-trapping potential of this grass is limited, however. Except in well-protected areas *Spartina* does not form closed meadows but grows in single hummocks. Erosion-

Figure 201: Hummocks of *Spartina anglica* with neighbouring erosion hollows on the tidal flats at Neßmersiel (1982).

al crescents are frequently found around these hummocks (fig. 201). In general, however, the grass has some effect in reducing current velocities and therefore in promoting sedimentation.

In most areas of the Wadden Sea *Spartina anglica* grows in association with *Salicornia europaea* (fig. 202). This halophyte inhabits large areas of the Wadden Sea sides of the islands, and along the mainland coasts. *Spartina* prefers wetter sites, whilst *Salicornia* prefers better drained soils. Both plants are most successful on clay-rich soils in sheltered areas (Dijkema, 1983 a: 189).

Suaeda maritima is often found in association with *Salicornia europaea* (plate 6). Both plants are annuals. *Salicornia* and *Suaeda* both turn orange to red in autumn (September/October) before withering (plate 11). With the exception of these withered remains the *Salicornia* meadow is completely bare in winter and its sediment-trapping potential is therefore restricted to the summer months.

As soon as enough sediment has accumulated to ensure that the soil is no longer regularly flooded with salt water, *Salicornia* and *Spartina* are replaced by *Puccinella maritima*. This grass is halophytic, but is not dependant on salt water like *Salicornia*. *Puccinella* is an outstanding mud gatherer. Unlike the *Salicornia* meadow, a *Puccinella* meadow provides a perennial, almost totally

Figure 202: *Salicornia europaea* at the Sehestedter Außendeichsmoor, Jade Bay (1984).

closed, vegetation cover which is only occasionally interrupted by minor creeks and bare hollows.

A large number of other halophytic plants occur in association with *Puccinella*. These plants all have a high nitrogen demand which is satisfied by

Figure 203: Aerial photograph of the Pohnshallig Koog (Nordstrand). Former levées on Pohnshallig visible in centre. Aerial photograph of 5.4.1980, freigegeben durch das Landesvermessungsamt Schleswig-Holstein unter Nr. SH 510/80, Bildflug Husum/Brunsbüttel, Str.4/909, vervielfältigt mit Genehmigung des Landesvermessungsamtes Schleswig-Holstein vom 3.12.1985.

the permanent mud supply. These plants include *Aster tripolium* and *Halimione portulacoides* (plate 7). Extensive flooding, after spring tides for instance, results in the entire meadow being covered with a thin grey layer of mud.

Higher on the salt marshes the *Puccinella* meadow is replaced by a *Festuca* community with *Festuca rubra arenaria* and *Limonium vulgare*. *Artemisia maritima* is also found, notably on the more elevated levées of small salt-marsh creeks and on small barrier beaches (plate 7). All these plants fix mud and contribute considerably to the salt marsh accumulation. In general the salt marsh grows up to a height of about 50−80 cm above mean sea level (Linke, 1947: 10).

Salt marshes are characterized by a strong local relief of creeks and hollows. Overdeepened channels more than 1 m in depth are common, which remain water-covered at low tide. These overdeepened features are only formed in the salt marsh environment. Grass tussocks and hummocks cause turbulence, and therefore erosion during floods, which may result in the formation of these features.

The formation and modification of the creeks and channels is restricted to periods of exceptionally high water when large areas of the salt marshes are flooded. Though water subsequently drains through the salt marsh creeks causing some erosion, accumulation generally prevails. Comparison of aerial photographs taken over a period of time suggests that the drainage network gradually decays; continuous creeks become intersected, and closed depressions fill with sediment and become covered by vegetation.

Example 1: Ley Bight

A number of the deep bights created by storm floods have been completely reclaimed (Middelzee, Harle Bight), whilst others have been dammed (Zuider Zee, Lauwerszee) or at least partially reclaimed (Dollart, Jade Bay). Complete dyking of the Ley Bight was not considered because this is an area of ecological importance.

In the future, however, the adjoining Hauener Hooge area will be dyked in order to improve the drainage of the adjoining marshlands and to permit tide-independent boat traffic. This will undoubtedly lead to some accretion in the innermost parts of the bight, but any further accretion will be limited. Since 1950, when large parts of the Ley Bight have been dyked by the Störtebeker Dyke, accretion has proceeded at a very slow rate. In contrast, erosion has even occurred in some areas (Michaelis, 1984).

Example 2: Nordstrand

The Isle of Nordstrand was connected by a causeway to the mainland in 1906/07 (Müller & Fischer, 1936 b: 201). With the blocking of the then 3.7 m deep Pohnsley tidal creek and the resultant cessation of large-scale overflow across the water divide during storm tides, strong accretion was initiated on both sides of the dam. Between 1920/25 the Pohnshalligkoog, at the eastern margin of Nordstrand, was dyked (Müller & Fischer, 1936 b: 226), and between 1933/35 the causeway to the mainland was reinforced and raised above high water level, leading to a further increase in sedimentation.

An aerial photograph of the former island of Pohnshallig shows ground striping similar to the features found in the beach barrier-created salt marshes off Eiderstedt (fig. 203). This particular case, however, is on the landward rather than the seaward side of the island. Beach barriers could only form before the damming of the Pohnsley when large quantities of water could flow over the water divide.

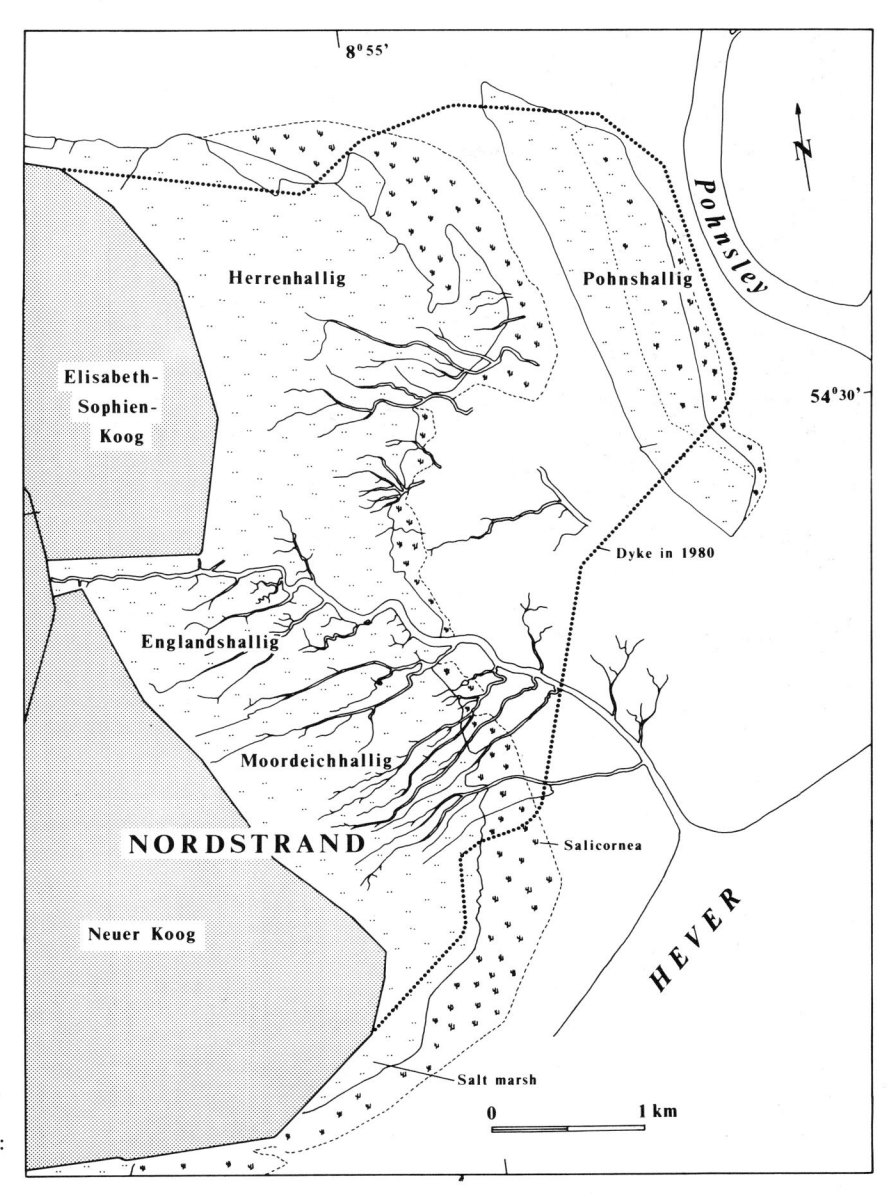

Figure 204: The drainage system and accretion area between Nordstrand and Pohnshallig in 1857 (Source: Müller & Fischer, 1936 b: Tafel VI).

The 1857 map in Müller & Fischer (1936 b: Tafel VI; fig. 204) shows that the accumulation of the Pohnshallig was replenished by sediment transport from the Holmer Fähre in the north. The northern margins of the former Herrenhallig and Pohnshallig both constituted flat flood spits, which afforded the protection required for further sedimentation in the south. In the latter areas, the so-called Englandshallig and Moordeichshallig, normal salt-marsh drainage systems developed, whereas the Pohnshallig shows no traces of any such drainage creeks. Following the damming of the Pohnsley, sediment transport continued from the north. This is reflected in the different accretion rates north and south of the causeway. The young foreland in the south is only about half as wide as that in the north.

Accretion in areas with a positive sand budget

Example 1: Eiderstedt

The accretion off Eiderstedt is controlled by the formation of large beach barriers. In 1938 only one such barrier lay off St.Peter-Ording, but by 1958 another barrier of the same size had formed seawards of the first one, separated from the latter by only a minor runnel (König, 1972). By 1980 both barriers had coalesced into a single, broad

Figures 205 a and b: The rising tide inundates the barrier beach off St.Peter-Ording. The surface of the 1 km broad barrier beach is levelled by the frequent washovers (1981).

beach barrier in the north.

The beach barriers are flooded during every normal high tide (fig. 205). During such transgressions a gradual landward directed sand transport occurs across the barrier. This transport becomes especially important during periods of exceptionally high tides and strong westerly winds. Sand transport causes the beach barrier to advance eastwards in broad lobes.

The aeolian component is of considerable importance in the migration of the beach barriers. The broad sand plains are major source areas for aeolian transport. Sand is predominantly blown landwards by the prevailing westerly winds, leading to additional sedimentation in the sheltered lagoons behind the barriers. Salt marsh growth in these lagoons originates in the north where the barriers impinge on the mainland coast, and proceeds southwards. The lagoons are well protected against strong currents because the barriers have

Plate 13: The severely damaged groyne J on Wangerooge (1979).
Plate 14: Partially eroded old dyke of the Juvre Koog directly north of the breach (1984).
Plate 15: Revetment at the western end of Wangerooge (1977).

PLATES 13-15

PLATES 16-17

developed recurved spits at their southern ends.

Accretion occurs in a number of ways:

a. Small creeks descend into the lagoon from the margin of the beach barrier forming alluvial fans at their mouths (fig. 206).

b. Washover and aeolian processes transport sand across the barriers and into the lagoons (fig. 207).

c. From the higher vegetated parts of the lagoons in the north, sediment is transported southwards through the tidal creeks and deposited in the form of deltas and levées in the lagoon (fig. 206).

d. From the mainland coast *Spartina* and *Salicornia* vegetation colonizes the lagoon and enhances mud deposition (fig. 206).

The beach barriers do not form a closed barrier seawards of the Eiderstedt coast, but are interrupted by minor inlets in two places (at the Tümlauer Bucht and south of St.Peter). Recurved spits are being formed at both ends of the barriers, demonstrating that sand approaching the coast is being transported both towards the Eider mouth as well as, though to a lesser extent, to the north towards Husum.

The beach barriers finally make contact with the mainland and are vegetated in the same way as the sediment-filled lagoons. The vegetational develop-

Figure 206: Aerial photograph of the barrier beaches off St.Peter-Ording. Small channels with alluvial fans at the foot of the inner barrier (Aerial photograph of 23.4.1980, freigegeben durch das Landesvermessungsamt Schleswig-Holstein unter Nr. SH 510/80, Bildflug Husum/Brunsbüttel, Str.13/092, vervielfältigt mit Genehmigung des Landesvermessungsamtes Schleswig-Holstein vom 3.12.1985).

Figure 207: Sand blowing into the depression between the barrier beach and St.Peter-Ording, blocking the old salt marsh drainage system (1981).

16	Plate 16: Seawall at Westerland. The wall was built in 1912, has had to be reinforced several times and has only been preserved during recent decades by repeated beach replenishments (1982).
17	Plate 17: Seawall at Baltrum during a summer storm (1984).

Figure 208: Aerial photograph of the Terschelling Polder. The old salt marsh drainage system is visible as a result of the sediment moisture content differences (Aerial photograph of the Fototheek Topografische Dienst, Emmen, 11.3.1983, 05 F 06).

ment of the beach barriers at St.Peter-Böhl has been studied by Menke (1969).

The beach barriers of Eiderstedt are not exceptional, but represent one component of the general landward directed sand transport. They can be formed without being attached to any Pleistocene core area. Generally, however, all barrier islands consist of similar barriers, although in most cases the beach barriers are directly attached to the islands from the very beginning of their formation. Lagoons, usually short-lived features, form only rarely. Typical examples of the latter were formerly existent on Rømø and Amrum (cf. sections 7.2.4 and 7.3.2).

Example 2: Schiermonnikoog
In 1952 some small dune cores, separated from each other by broad washover areas, formed on the highest parts of the eastern end of the island, which had previously always been a barren sand flat. In order to promote accretion the Rijkswaterstaat built a 'stuifdijk' (dune dyke) from the island core to the easternmost dune complex, the then

over 9 m high Willemsduin (Abrahamse & Luitwieler, 1985: 9).

The 7 km long dyke, however, did not survive for long and was rapidly breached. As a result, the eastern flat of Schiermonnikoog has remained a washover area. This is dissected by a number of channels through sediment is transported towards the tidal flats during higher tides. In place of a row of coast-parallel striking dune ridges along the linear 'stuifdijk', a number of broad dune cores have been formed, protected against the sea by horseshoe-shaped dune ridges. Salt marsh growth has increased dramatically on the Wadden Sea side of the flat. Almost the whole of the eastern flat now forms part of the recent accretional area of Schiermonnikoog.

The drainage systems of the accretion areas

Natural salt marsh formation is barely active at present. Brushwood groynes and ditches (Grüppen) have been constructed in any area, showing a tendency towards accretion, thus creating a com-

Figure 209: Aerial photograph of the northern part of Texel (Eijerland). This polder was a separate island until the late 17th century. The fossil drainage system displayed through differences in soil moisture content (Aerial photograph of the Fototheek Topografische Dienst, Emmen, 11.3.1983, 09 B 14).

pletely new drainage system, which has nothing in common with the original.

The natural drainage systems of the salt marshes are rather contorted and irregular (Wadsworth, 1980). Since the entire North Sea coast has been settled for about 2,000 years, these natural drainage systems have, in most cases, disappeared. The Halligen of North Frisia provide rare examples where the old drainage systems can still be seen.

The present day Halligen are relatively young islands which have formed on old marshland. Large areas of the North Frisian Wadden Sea were joined to the mainland in the Middle Ages and were used for agriculture. The straight ditches intersecting the old creeks were first excavated at this time. Remnants of the old agricultural drainage systems can still be found today at low tide far out on the tidal flats (cf. section 7.3.4). Since the Middle Ages the salt marsh surface on the Halligen has risen by more than one metre, for example on Hooge it has risen by about 1.20 m. The old ditches were nevertheless used continuously (cf. Bantelmann, 1966: 76).

The old marshes which formed behind protecting dunes possessed a similar drainage system. Under favourable conditions this may be revealed in aerial photographs. Fig. 208 shows a fossilized drainage system of the 'old marsh' type on Terschelling. The Terschelling Polder was dyked in the Late Middle Ages (Smit, 1971). The similarity in form and size of the creek patterns on Hooge and Terschelling is striking.

This form of drainage system is only rarely found on the barrier islands, and usually only occurs in totally calm sedimentary environments. These environments occurred in North Frisia because of the existence of the seaward barrier. In Terschelling they occurred because of the great length of the island and its continuous protective dune belt uninterrupted by washover areas.

Where washover processes dominate, a different drainage pattern is found. This is almost exclusively the case in the recent marsh regions of the islands, exemplified by the accretion area at the eastern end of Spiekeroog. Instead of irregular creeks, rather straight, linear drainage channels are being formed which have only a few side branches.

The situation is similar in the marshes of northern Texel. Until 1629 this area was an independent island separated from Texel by a small tidal inlet, the so-called 'Eijerlandse Gat'. The small dune core of the Isle of Eijerland was not large enough to protect the young accretional area in its lee against washover, so a mildly dendritic elongated drainage network formed. The principal pattern of this was represented on a map drawn in 1854 (cf. van der Molen, 1978a: 96). Although the area was dyked in 1835, the old drainage system can still be detected today on aerial photographs taken under favourable conditions (fig. 209).

Consequences

Natural accretion in the investigation area occurs in very few locations at present. Apart from in large bays (e.g. Dollart, Ley Bight, Jade Bay, Tümlauer Bucht), accretion is concentrated where beach barriers and spits are developing as a result of a strongly positive sand budget, under the protection of which salt marsh growth can be initiated (e.g. Eiderstedt, Rømø).

The building of brushwood groyne systems along almost the entire Wadden Sea mainland coast (cf. Dijkema, 1983 b: 305) has caused accretion as far to seaward as is technically possible. As a result of high costs, reclamation work today is only undertaken where it is necessary to create a sufficiently broad foreland for dyke protection.

Since almost all the potentially reclaimable areas of the Wadden Sea coast have been dyked, there are no longer any broad forelands which rep-

188 Morphodynamic units

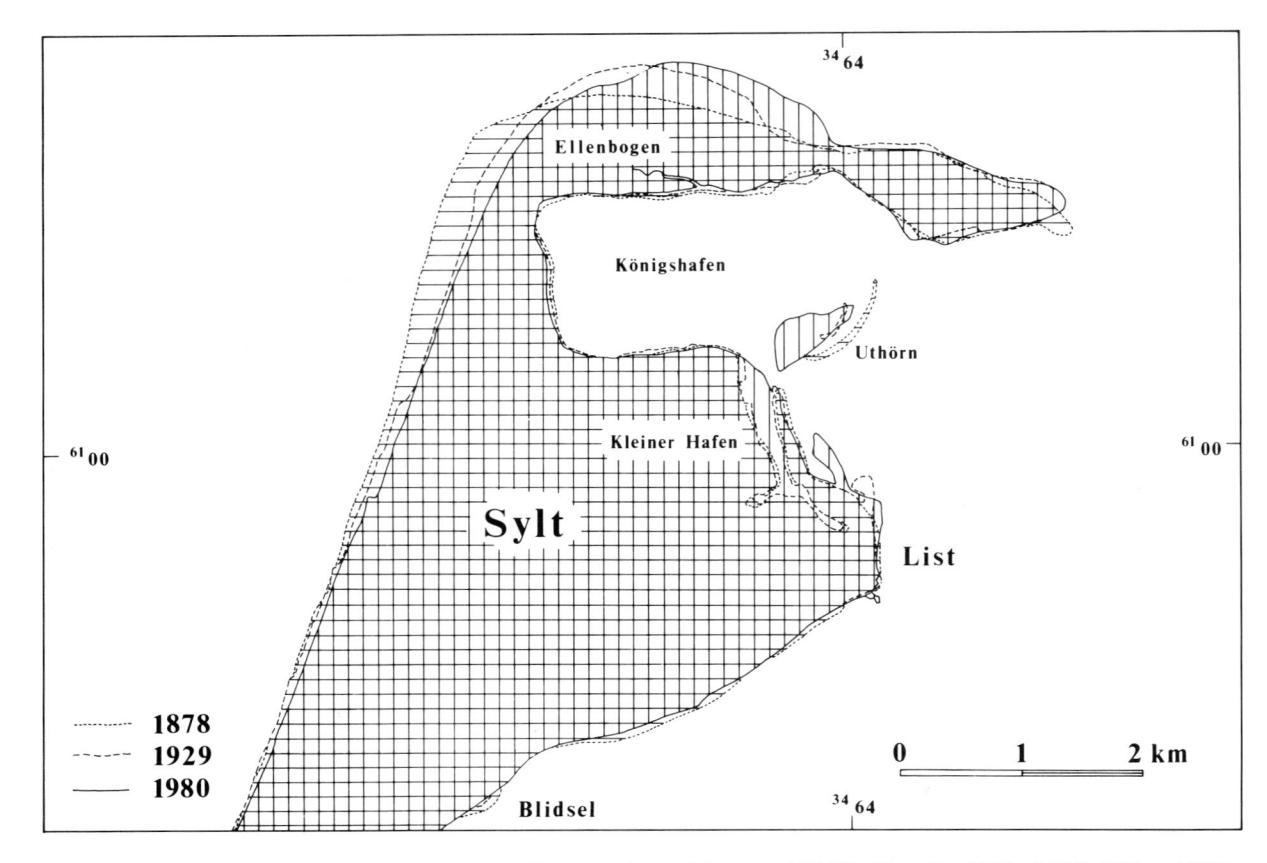

Figure 210: Coastal erosion on the northern part of Sylt; comparison of the maps 1:25,000, Sheet List (Sylt) of 1878, 1929 and 1980.

resent additional storage capacity for rising water during storm surges. This causes storm tides to attain higher levels than would otherwise be the case. Additional coastal protection measures have therefore been necessary, including projects to dyke additional, non-reclaimable areas with the aim of shortening the main dyke lines.

The increased dyking of salt marshes and tidal flats leads to a strong reduction of these ecologically important areas of the Wadden Sea (cf. Hamel, 1981; Heydemann, 1981 a; Hartung, 1983). Heydemann (1981 b) has pointed out that undyked marsh with an overall area of about 20,000 ha constitutes only about 2.7 % of the Wadden Sea area. One third of this (6,800 ha) was threatened by dyking projects, two of which (Rodenæs–Emmerlev and Nordstrander Bucht) are now either completed or under construction.

The widespread lack of actual accretion is not simply a result of the exhaustion of potentially reclaimable land. Field observations demonstrate that the equilibrium between accretion and erosion has been passed. Large areas which were dominated by accretion only a few decades ago are now being actively eroded.

5.3.6 Coastal erosion

There are signs of coastal erosion along almost the entire seaward coast of the Wadden Sea barrier islands. Even on Juist, where a seawall and 7 groynes were constructed between 1913 and 1920, later to be sanded up and completely engulfed in sediment, recent erosion is a cause of serious concern (Luck & Stephan, 1983).

The Wadden Sea shore is retreating by about the same order of magnitude as the island barrier off the eastern coast of the USA (e.g. Virginia: 2–6 m per year; Leatherman, Rice & Goldsmith, 1982; Rice & Leatherman, 1983). The highest rates are found on the Outer Sands (Taubert, 1982) and the young barrier islands in the inner parts of the German Bight (cf. chapter 4). Sylt is the most exposed point within the chain of the barrier islands and experiences the strong coastal retreat. Erosion is most obvious at the Rotes Kliff between Westerland and Kampen but is even stronger on the extremities of the elongated island – along the western coasts of the Ellenbogen (Elbow) Peninsula (fig. 210) and the Hörnum Peninsula (fig. 211).

The Ellenbogen Peninsula is a typical recurved spit formed by the flood current. Whilst steady erosion occurs on the seaward side, sediment is contemporaneously deposited on the northern flank of the spit. The spit consists of dune-capped subparallel beach ridges, occasionally supplemented by new ridges in the north. Coastal erosion along the western side is so strong at present that there is a very real danger that the peninsula might be severed in the Königshafen Bay area in the foreseeable future. The island width has almost halved between 1878–1980 at this point (fig. 210), amounting to no more than 400 m today.

The situation on the Hörnum Peninsula is equally unfavourable (fig. 211). No recurved spit has formed here because the peninsula is flanked on the Wadden side by the deep channel of the Hörnum-Tief, which runs directly past the coast. Until 1930 seaward erosion of the peninsula resulted in its southward growth (Gripp, 1968), but this has since ceased. The highest recent rates of erosion have been measured south of the tetrapod wall which protects part of the Hörnum settlement (cf. section 5.3.7). In the period between 1848 and 1980 coastal erosion amounted to about 250 m, i.e. approximately 2.5 m per year.

Coastal erosion rates differ regionally depending on the degree of exposure and the properties of the substratum. The coastline changes of the Wadden Sea area have been mapped by Mrozcek (1980) and briefly discussed by Bird (1985: 51). Fig. 212 illustrates the present centres of erosion and salt marsh formation in the Wadden Sea area. More details are given in the regional descriptions (chapter 7).

Cliff retreat

Ideas concerning the mechanisms which cause the erosion of coastal cliffs have been subject to repeated changes during the last 150 years. Cliff erosion was originally explained exclusively in terms of wave action (e.g. Cuvier, 1827: 9; Penck, 1894 (2): 555). Hansen (1859: 45), however, gave a more complete account of the factors affecting cliff retreat. In his description of the Rotes Kliff on Sylt, he pointed out that strong rain showers contributed to cliff destruction, and he also noted the importance of landslides in producing debris aprons at the foot of the cliff, providing sediment for subsequent wave erosion.

After investigations of a Baltic Sea cliff, Mortensen (1921) stressed the importance of atmospheric factors in cliff erosion. He concluded that cliff retreat was mainly the result of rain water run-off (Mortensen, 1921: 40). Erosional influences from the landward side are also considered dominant in Martens' paper (1927), which provides a critical review of Mortensen's ideas.

Both authors admit they did not have the opportunity to observe their cliffs under storm conditions (Martens, 1927: 8; Mortensen, 1921: 14), which Kannenberg believes explains their misinterpretation (1951: 21).

Wasmund (1936) went even further than Mortensen in postulating that internal material proper-

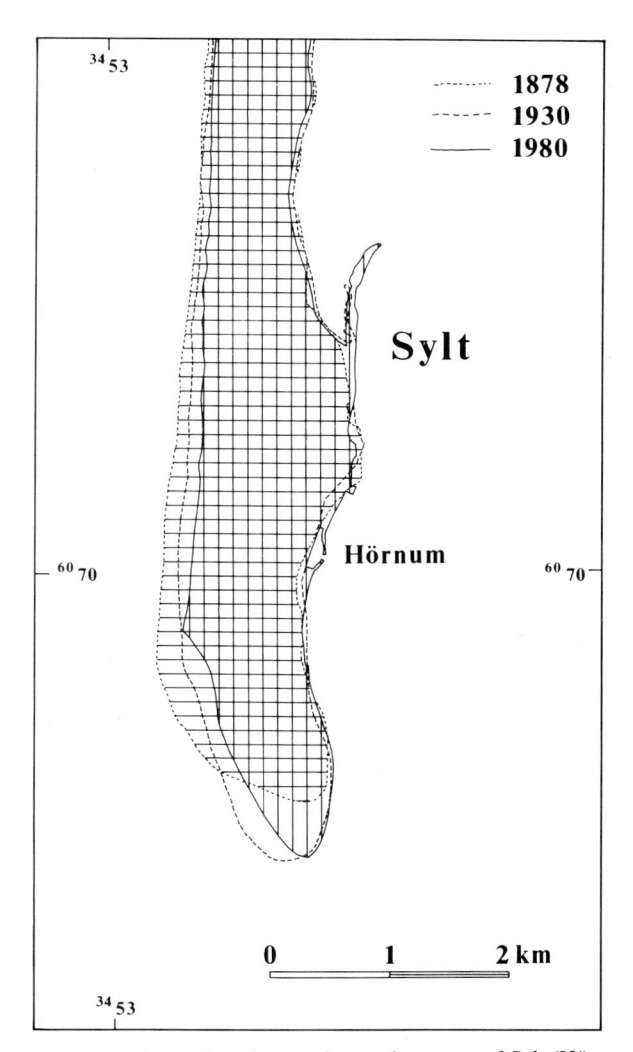

Figure 211: Coastal erosion on the southern part of Sylt (Hörnum Peninsula); comparison of the maps 1:25,000, Sheet Hörnum (Sylt) of 1878, 1930 and 1980.

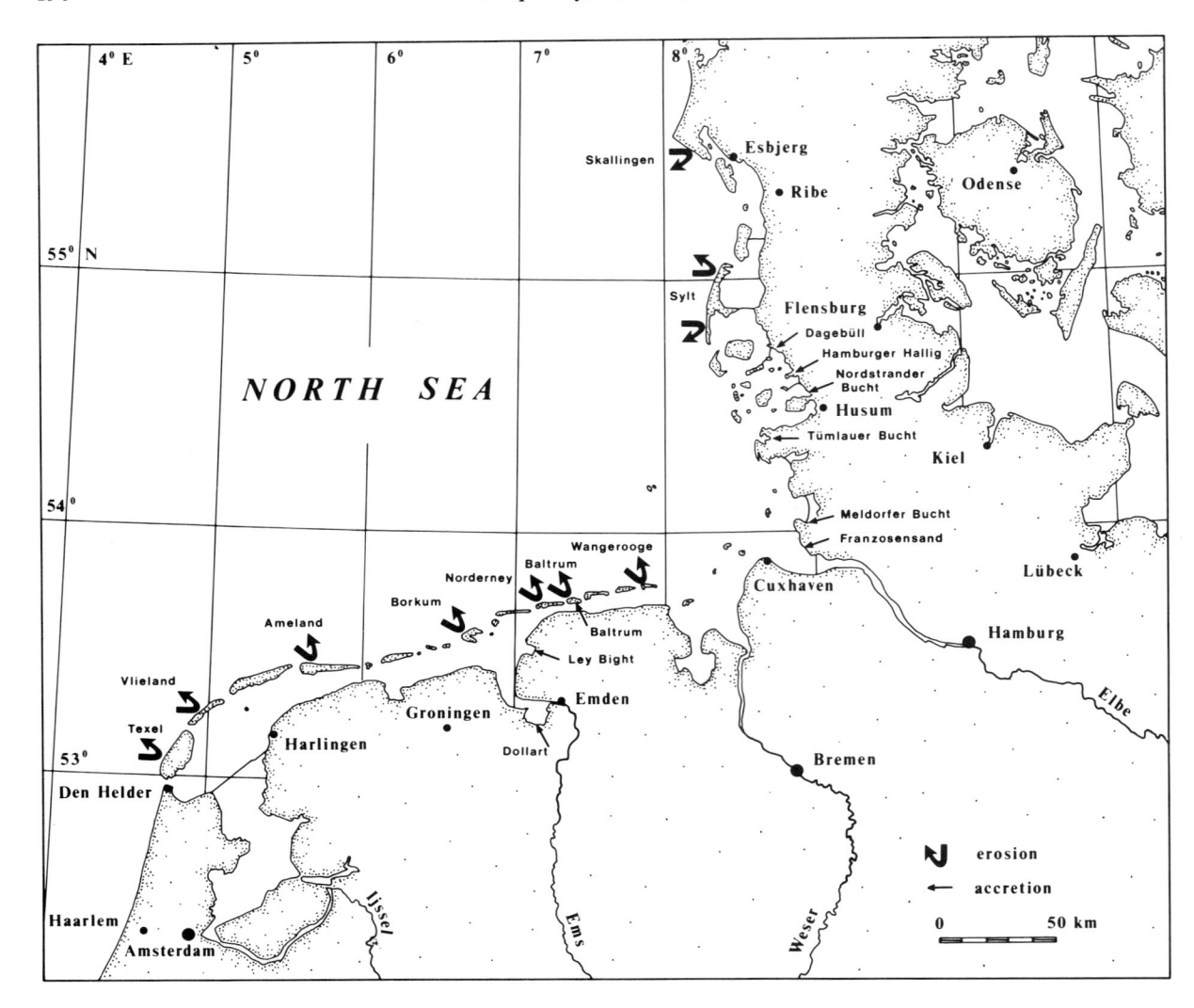

Figure 212: Coastal erosion and salt marsh formation in the Wadden Sea area.

ties were responsible for this erosion. He thought cliff erosion was a result of weathering, for which he used the term 'internal coastal decay', and that the role of waves was restricted to subsequent sediment removal.

Kannenberg (1951: 22), who investigated the Baltic Sea cliffs of West Germany, re-emphasized the importance of marine abrasion as the major factor in cliff retreat. In active cliffs the role of subaerial factors is restricted to furthering erosion, and in 'fossil' cliffs (more correctly: passive cliffs) to protection from further erosion.

Cliffs composed of Pleistocene sediments are only found in a few places along the Wadden Sea coast:

1. On the islands:

a. Sylt: Rotes Kliff; in the northern part it consists of thrust till material, in the southern part of undisturbed till, overlying Pliocene kaolin sands.

b. Sylt: Morsum Kliff on the Nösse Peninsula; this consists of glacially thrust strata of Pliocene sands, Limonite Sandstone and Mica Clay.

c. Amrum: 'Ual anj' cliff north of Steenodde, which is composed of periglacially disturbed Pleistocene sandy deposits.

d. Föhr: Goting Kliff southwest of Nieblum, which is composed of thrust Pleistocene sands and till, together with rafts of Mica Clay.

2. On the mainland:

a. The cliff north of Esbjerg, which is composed of Pleistocene sediments.

b. Emmerlev Klev north of the German/Danish

Figure 213: Rotes Kliff on Sylt at Wenningstedt; winter erosion has formed a broad debris apron at the foot of the cliff (1983).

border; the cliff is cut in Saalian till.

c. The cliff at Arensch, south of Cuxhaven, which is composed of Pleistocene sands; today it is almost inactive.

The retreat of these coastal cliffs is among the most spectacular erosional processes operating along the Wadden Sea coastline. The Rotes Kliff between Westerland and Kampen, retreated 5–7 m during the storm surges of January 1976 (Ministerium für Ernäherung, Landwirtschaft und Forsten, 1977: 14). On average, however, the erosion rate of the Rotes Kliff is about 0.7 m per year (cf. Müller & Fischer, 1938: 246).

Coastal cliffs can remain passive for months and even years. Most of the time, therefore the cliff toe is covered with a thick debris apron (fig. 213) which has to be removed before wave erosion can attack the actual cliff base. The cliff base is rarely attacked directly, and in most cases only part of the debris apron is removed by wave action. This causes an oversteepening of the cliff profile which leads to intensified slumping and sliding of sediment down the slope. Most of the landslides occur as slab slides, several metres to tens of metres in length; these glide, partly *en échelon*, down the cliff face.

Of the atmospheric factors, frost action and rainfall run-off have some influence on cliff erosion in till material. Rainfall run-off leads to the formation of long rills (fig. 214) which mostly fol-

Figure 214: Rotes Kliff on Sylt; rainfall run-off has dissected the slope with numerous small rills (1983).

Figure 215: Goting Kliff on Föhr; protrusions are created in areas with more resistant clay and till rafts; sandy layers are more easily eroded (1981).

Figure 216: Dune erosion south of Hörnum (Sylt) (1983).

low pre-existing faults or joints in the till. Where the run-off transcends a lithological contact, such as the boundary between till and underlying softer sediments along the southern part of the Rotes Kliff, overhangs of several decimetres sometimes form. Bottom seepage from the cliff face, from the top of clay layers, for instance, promotes cliff retreat. The natural angle of slope of till cliffs is little more than 30°, but this is never attained in active cliffs because renewed erosion always leads to an oversteepening of the profile.

Areas of varied lithology lead to a more differentiated shoreline, because cohesive sediments are more resistant to erosion than loose sand, e.g. the Morsum Kliff on Sylt and the Goting Kliff on Föhr. In both cases comparatively hard clays form nose-like protrusions of the cliff, whereas easily eroded sands are found in the coves and bays (fig. 215).

Dune retreat

Higher rates of erosion than those recorded along the cliffed coasts are reached where dune coasts with narrow beaches are directly exposed to wave attack. For instance, the dunes along the west coast of Sylt retreated by up to 30 m during the storm surges of January 1976 (Ministerium für Ernährung, Landwirtschaft und Forsten, 1977: 14).

The maximum slope angle of the dunes is normally limited by the natural angle of the internal friction of dry sand, which is about 30°. In vegetated dunes, roots fix the sand so that steeper slopes may be formed. Where unprotected by artificial structures, storm surges attack the dune foot, causing slope slab failures (fig. 216).

Where the sand budget is positive, the dune margin can once again advance seawards. This process, of course, takes much more time than the regeneration of the beach (cf. section 5.3.2). According to the results of investigations by the 'Forschungsstelle Norderney', it is considered that dune recovery normally takes place over a number of years (Homeier, 1976: 116).

The dunes are further eroded by aeolian processes. Unvegetated steep dune slopes suffer severe deflation during periods of strong winds, thus contributing to the erosion of the cliff face. Strong winds create deep breaches between individual dune summits in the coastal 'white dune' zone, through which sand can be transported into the interior of the islands. This process may lead to the formation of migrating dunes in extreme cases.

The resultant natural landward migration of the dune barrier is usually blocked today by the 'fixing' of the dunes (cf. Dolan, 1972). These are eroded by storm surges but the released sediment cannot be transported into the inner parts of the islands (van Dieren, 1934: 221). Sand is therefore lost from the island sediment budget, or redeposited in undesirable or unnecessary places, such as at the eastern ends of the East and West Frisian Islands.

Salt marsh erosion

Along the salt marsh coasts, for instance the landward sides of the islands and the mainland coast, soil erosion is hindered by the vegetation cover. Coastal erosion is generally much slower here than along the seaward, steep cliffed coastline of the barrier islands. Salt marsh erosion is indicated by either the formation of small cliffs up to about a metre in height (fig. 217), or the disintegration of the salt marsh margin into individual ridges and furrows (fig. 218; cf. Allen, 1984 b: 43). The salt marsh debris is able to form clay balls at the foot of the cliffs as a result of its strong cohesion (fig.

Figure 217: Cliff on salt marsh shore at Kampen (Sylt) with clay balls at the scarp toe (1984).

217). Similar features have been described by Lüders (1937: 16) from the Oberahnesche Felder in Jade Bay. The clay balls may cause the development of potholes and contribute to cliff erosion in this way.

Cliff retreat can be caused by the stripping of vegetation cover, as well as by undercutting and slope failure (fig. 219). This is most likely to occur during the early stages of flooding of the salt marshes, when the surf is still active along the cliff margin and when the topsoil, with its air-filled pores, becomes buoyant. Both types of erosion normally occur together. Wave backwash also contributes to the erosion and dissection of the salt marsh margin.

Unlike the high cliffs of the Pleistocene island cores, which are protected against wave action by a backshore of variable breadth during normal tides, the salt marsh margins are exposed to erosion during every ordinary tide. Significant debris toes cannot therefore develop, and equilibrium profiles are inhibited.

In areas which have only recently changed from

Figure 218: Disintegration of the salt-marsh margin into isolated elongated grass hummocks, separated from each other by decimeter-deep furrows; Elisabeth-Außengroden, east of Harlesiel (1979).

Figure 219: Effects of erosion on a salt-marsh cliff on Sylt; Königshafen Bay, west of List. The grass cover is being stripped by the waves and the cliff is retreating by overhang collapse (1982).

accretion to prevailing erosion, the salt marsh is often juxtaposed against the tidal flats with very little height difference, and so well-defined cliffs are missing. However, features which Sindowski (1973) called 'tidal-flats-facing beach ridges' can be found in these areas (cf. Hanisch, 1981). These slightly elevated, largely sandy, ridges are often colonised by *Artemisia maritima* which, being a relatively tall plant, contributes to the small height differences.

There is very little reliable data concerning salt marsh erosion. Despite its relatively protected position, the cliff of the Großes Oberahnesches Feld, a submerged marsh island in the Jade Bay, retreated at a rate of 2–2.5 m per year between 1872 and 1934 (cf. Lüders, 1937: fig. 6).

Retreat of the salt marsh margins is not limited to the islands, but also affects the mainland coast. In order to obtain reliable information about this process, aerial photographs of the Danish mainland coast south of Esbjerg were evaluated, taken in 1946, 1964 and 1984. This period of about 40 years was characterized by a marked retreat of the salt marsh margin, protrusions being preferably eroded (fig. 220). The erosion rate was about 25 cm/year.

Until a few years ago, very strong coastal erosion was also observed along the shores of the unprotected Halligen. These are all covered today by revetments, and some, such as Hooge, even have summer dykes. Fig. 221 shows the coastal erosion of the Hallig Norderoog between 1927 and 1980. Approximately 130 m were lost from the most exposed western end during this period, i.e. about 2.5 m per year.

Such rates of between 2–2.5 m per year are apparently extremes, and only valid for areas with an obviously negative sediment budget. These rates may appear insignificant when compared with the maximum rates attained along dune coasts, but the salt marshes are not directly exposed to wave attack from the open sea, and their relatively greater stability is a result of their semi-protected position.

5.3.7 COASTAL PROTECTION

Archaeological investigations indicate that human settlement along the Wadden Sea coast began relatively late. Whilst traces of settlements of the Neolithic Vlaardingen Culture have been found in the Dutch marsh areas, the oldest proven settlement in the German North Sea marshes is from the Bronze Age (west of Rodenkirchen at the River Weser). The vast majority of settlements date from the Iron Age (after 700 B.C.). The earliest settlements were all on the flat marshes. The only protection against inundations here was to settle on slight elevations. Three of the Iron Age settlements are situated on the levée of the Lower Ems at heights between −1.0 and −0.3 m NN. Botanical evidence suggests that this area was then a freshwater marsh; no indications of tidal influence have been found (Behre, 1970).

These early settlers had no recourse to coastal protection, and the Early Iron Age settlements soon had to be abandoned because of the rising sea level. These sites were covered with a layer of marine clay between 300–100 B.C. during the course of the Dunkerque-I-Transgression. The flat marsh was resettled during the subsequent regression, even extending seawards beyond the present coastline (Haarnagel, 1961). The only available explanation for this seems to be that at this time storm surges were absent, or at least infrequent. This renewed colonization began earlier in Lower Saxony than in Schleswig-Holstein where the oldest settlements in the marshes date from the 1st and 2nd centuries A.D.

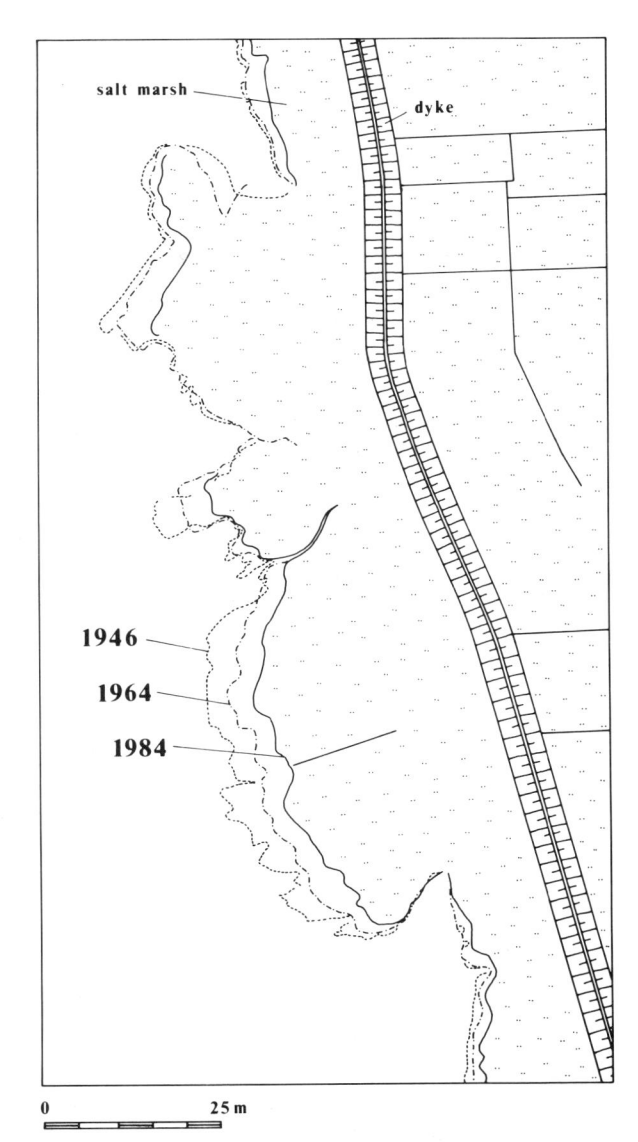

Figure 220: Salt marsh erosion at the Danish mainland coast, about 1 km north of the Kongeå mouth. Sources: Aerial photographs of the Geodætisk Institut, København, of 1946, 1964 and 1984.

Dwelling mounds

Some of the settlements had to be abandoned as the influence of the storm surges once again increased. Others were protected by artificially constructed dwelling mounds. These mounds had to be repeatedly raised, finally reaching heights of up to + 7 m NN. In the Netherlands these dwelling

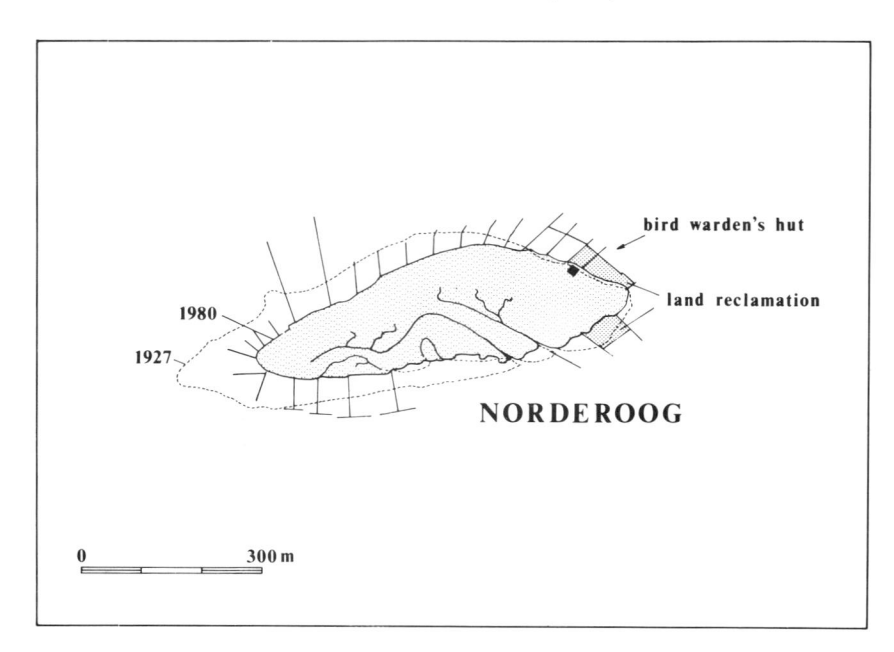

Figure 221: Coastal erosion at the Hallig Norderoog between 1927 and 1980, mapped after comparison of the aerial photograph published by König (1972), with more recent aerial photographs of the Landesvermessungsamt Schleswig-Holstein (1980).

Figure 222: Traces of old salt peat excavation pits in the tidal flat area south of Hallig Langeneß (1984).

mounds are called Terpen, in Lower Saxony Wurten and in Schleswig-Holstein Werften or Warften. The construction of the dwelling mounds and the abandonment of the flat settlements began in the 1st century A.D. (Behre et al., 1979: 104). Many dwelling mounds, however, are much younger. The farm mounds on Pellworm, for example, were built in the 12th century, those on Nordstrand during the 14th century (Müller-Wille, 1982: 259), whilst on the undyked Halligen islands the construction of Warften continued into the 19th century. The last Warft was built in 1890 on Langeneß (Neu-Peterswarft), after its predecessor had been destroyed by coastal erosion (Müller-Wille, 1982).

For almost 1000 years the dwelling mounds provided the only means of protection against

Figure 223: Former farmland in the North Frisian Wadden Sea (south of Hallig Hooge), seen at ground level (1981).

storm surges but today this is only true on the undyked Halligen. Of the 41 Warfts on the Halligen of North Frisia, 38 were still inhabited in 1979 (Riecken, 1982: 147).

Dyke construction

Large-scale dyking of the coastal marshes began at about 1000 A.D. Small creeks were closed and the neighbouring Halligen joined to form larger islands. It is likely that a closed dyke system has existed along the North Sea coast since the 12th/13th centuries (Kramer, 1983), thus protecting vast marsh areas from regular inundations. Dyking, however, also had adverse consequences. Natural settlement of the soil could no longer be compensated for by additional accretion, and the reduced storage area resulted in higher floods which repeatedly breached the dykes.

Arable as well as pastural agriculture was active in the freshly dyked areas; this required drainage of the land. The drainage ditches increased the natural settlement of the marshes and the polders finally fell below mean sea level. This development was exacerbated by extensive peat digging, the peat being required for both heating and salt production (fig. 222). Salt, as a precious commodity, could be extracted comparatively easily from peat saturated during previous inundations.

Severe storm surges, in particular those of 1362 and 1634, resulted in numerous dyke breaches and

Figure 224: Aerial photograph of the Juvre Koog, Rømø. After strong coastal erosion the dyke had to be rebuilt about 200 m further inland. Aerial photograph of the Geodætisk Institut, Denmark, Flight 7609 G 027, 17.5.1976, reproduced by permission (A. 532/85) of the Institute.

Figure 225: The destroyed old Juvre Koog Dyke on Rømø (1984).

the loss of large areas of land which had been so lowered by cultivation that they were not reclaimable (fig. 223) (Petersen & Rohde, 1977).

Dyke construction technology is much more advanced today, and it is possible to rapidly repair damage caused by the most severe storm surges, such as that of 1952 in the Netherlands (cf. Dendermonde & Dibbits, 1956). Experience has demonstrated, however, that dykes can only fulfil their protective function in areas where the coast is not subject to strong retreat.

An example of this is the recent development on the Isle of Rømø. The causeway construction between Rømø and the mainland (1939-48) altered the drainage areas of the tidal creeks so that north of the dam the Juvre Priel strengthened and approached the dyke of the Juvre Koog. After the failure of a number of defence strategies, which included the blasting of a new Priel bed, building of brush groynes and a revetment, and land reclamation measures in order to reduce the drainage area, a second dyke was finally constructed in 1965, about 200 m inland of the old dyke (fig. 224). The fields between the dykes had to be abandoned on 24.11.1981 when the Juvre Priel finally breached the old dyke (fig. 225 and plate 14) (Jespersen & Rasmussen, 1984).

Dune protection

The first attempts at dune protection concentrated on strengthening the existing dune belt. Intensive livestock grazing, however, made it difficult to maintain a closed vegetation cover in the dune areas and aeolian sand drift continued until the 19th century.

The planting and fixing of the dune areas on Sylt was initiated in 1788. On many islands rabbits were exterminated in order to prevent them from tunnelling into the dunes. With the exception of a few migrating dunes in the Listland area of Sylt, all the major dune areas were fixed by the early 20th century. In addition, dune dykes were erected in many places, following the Dutch experience, in order to connect the dune cores and so prevent the islands from inundations during storm surges. As early as 1633 the islands of Eijerland and Texel in the Netherlands were connected by an artificially raised dune ridge, a so-called 'stuifdijk'. These measures alone, however, were insufficient to protect the islands effectively against erosion.

Groynes

The first engineering measures to interfere with the natural processes of the Wadden Sea were constructed along the German North Sea coast during the 19th century. Between 1810 and 1834 an attempt was made for the first time to stop the erosion of the western head of Wangerooge by constructing 10 brush groynes. Two brush groynes were constructed on Norderney in 1846. These structures, however, were too light to resist the force of breaking waves, and since 1860 stone groynes were constructed. The first one was built on Norderney, and others of the same type soon

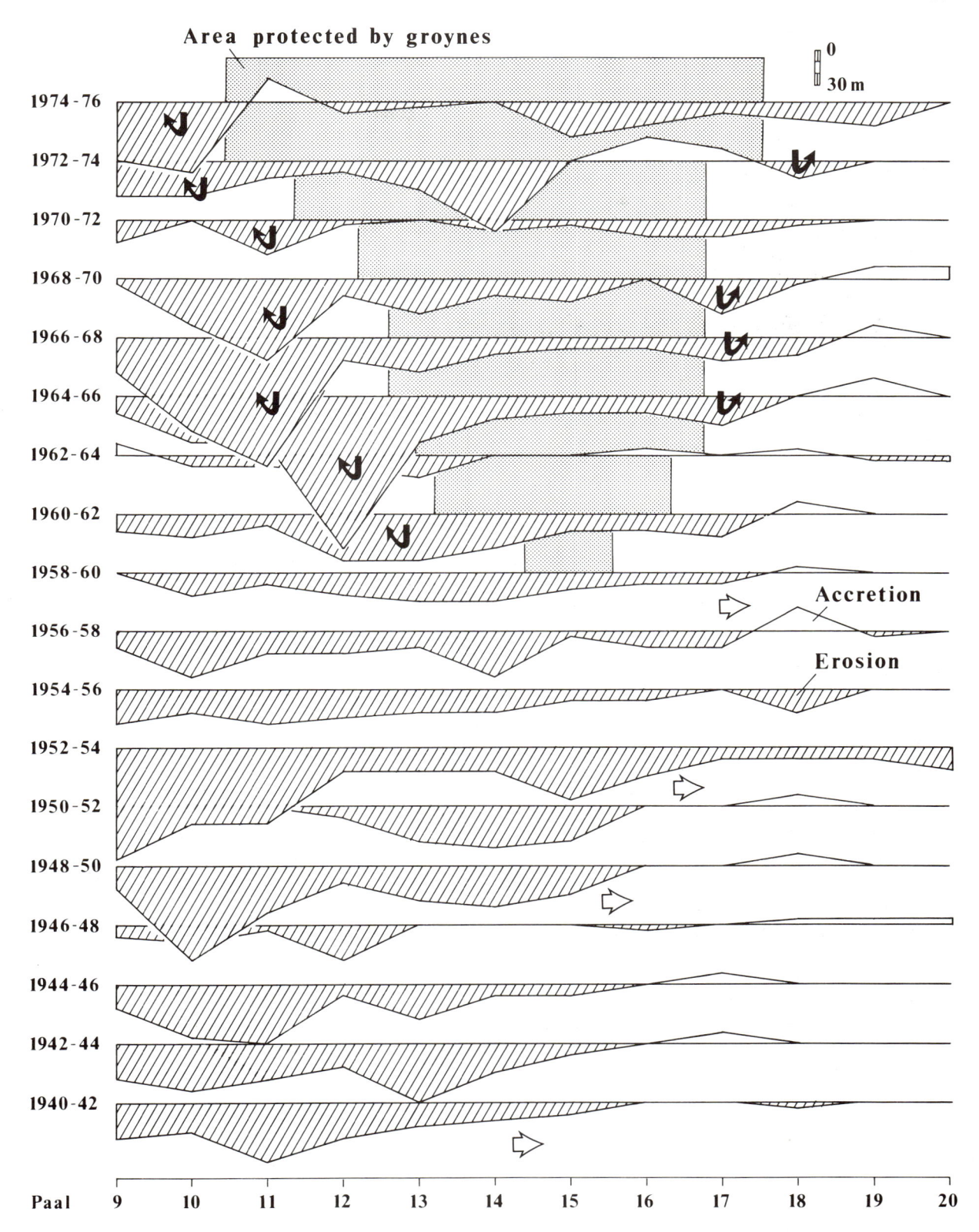

Figure 226: The effect of groynes on erosion and accretion on the southwestern shore of Texel (for location see fig. 386). White arrows: northeastward extension of the erosion area before groyne construction. Black arrows: leeside erosion (Source: Elorche, 1982).

followed on most of the other German islands (Küstenausschuß Nord- und Ostsee, 1981).

Generally three different types of groynes can be distinguished:

1. Beach groynes: groynes on the beach, which mainly prevent beach erosion by waves or surf currents.

2. Current groynes: groynes, often in combination with a revetment, which protect the beach or foreland against erosion by longshore currents.

3. Underwater groynes: underwater extensions of groynes which protect the foreshore slope against erosion.

Experience has demonstrated that the positive effects of groynes on beaches are extremely limited. As a rule single groynes do not have any positive effect, and in most cases groups of groynes are necessary. Groynes were often built to protect the foot of seawalls and revetments against erosion. Where installed as a group, however, groynes act as a revetment. Lee-side erosion is initiated downcurrent of the group, which normally necessitates the construction of additional groynes in the adjoining beach areas (Ausschuß 'Küstenschutzwerke', 1981: 265).

As a result, the effects of groynes on stabilizing the islands have to be judged in the light of the different functions and degree of exposure of the various structures. The large current groynes at the western heads of Borkum, Norderney, Baltrum, Spiekeroog and Wangerooge have in every case halted the threatening approach of the tidal inlets. In some cases, however, this has caused extreme overdeepening of the inlets. For example, the Norderneyer Seegat has increased its depth by about 3 m between 1859 and 1973 (Luck, 1976: Anl. 25). The depth of the Harle Inlet (between Spiekeroog and Wangerooge) before the construction of the dam-like Groyne H in 1940 was SKN−12 m, in 1945 SKN−25 m and in 1973 SKN−32 m (Homeier, 1973: 23).

The construction of beach groynes has been much less effective. On Sylt in particular, but also on other islands such as Texel and Vlieland, the groynes have not significantly reduced the rate of beach erosion. Where groynes are in disrepair they soon become uprooted and lose contact with the backing cliff. In particular the steel bulkhead groynes built before the Second World War, which have been corroded by salt water into bizarre rusty sharp-edged fences, form hazardous obstacles for swimmers.

The effect of groynes on coastal retreat can be exemplified by Texel (fig. 226). During the last few decades the southwestern shore of Texel has suffered a negative sand budget. The Rijkswaterstaat attempted to counter this development with the construction of beach groynes. The first groynes were built in 1959. Leeside erosion began at both ends of the groyne field. This was countered by repeated extensions of the groyne system to the south and north, but while erosion of the protected beach sector was slightly reduced, leeside erosion continued at both ends. The last groynes were built in 1978 in the south, and in 1979 in the north (Elorche, 1982). At present beach replenishment is used to stabilize the shore (cf. section 7.5.8).

Cost-intensive repair works are necessary to maintain the groynes on most islands. Where neglected the groynes soon become ineffective as the stones sink into the beach and collapse due to undercutting by runnels. This is exemplified by the recent development of Groyne J on Wangerooge between 1978 and 1984 (figs. 227 a–d and plate 13).

Seawalls

Walls were constructed simultaneously with groynes to protect the barrier coasts endangered by erosion. In 1857/58 the first 'dune revetment', in effect a 975 m long seawall with an S-shaped profile (the so-called 'Norderneyer Profil'), was constructed on Norderney (fig. 228). The structure had to be protected by additional groynes, and during subsequent years it had to be repeatedly extended in order to counter the consequences of the resulting lee-side erosion.

Similar problems were encountered with the Westerland seawall of 1907, which had to be extended as early as 1912 (plate 16). Between 1936 and 1938 the wall was extended by a basalt revetment, and further extension was undertaken in 1946 at the same time as a damaged section of the seawall was replaced by a revetment of concrete slabs. The Westerland structures required further extension in 1954, a so-called rough revetment being constructed. In 1960, after repeated alterations and extensions, the foot of the wall had to be reinforced. Large granite blocks were placed in front of the structure in 1969, additionally protected by massed tetrapods (Küstenausschuß Nord- und Ostsee, 1981: 339).

Revetments

The steep profile of seawalls increases the power of

Figure 227 a: Groyne J on Wangerooge (built 1895-97) has been severely damaged by undercutting and sagging. In the breach in the foreground, the basal brush layer was exposed under rock slabs. The flank protection of the groyne is missing (1978).

Figure 227 b: Two more slabs are beginning to slide down (in left foreground) (1979).

waves. Therefore, since about 1936, more smoothly sloping revetments have been built, the earliest of basalt, and later of asphalt mounted with stones in order to increase the roughness of the surface and decelerate waves. These structures, however, also suffer foot erosion, albeit at a slower rate than the seawalls. The revetment constructed on the Ellenbogen Peninsula on Sylt is typical of such damage (constructed in 1938-40; Lüpkes & Siemens, 1940).

Tetrapods

Repeated attempts have been made to secure endangered coastal areas using tetrapod walls. The irregular form of tetrapods makes them more successful than smooth seawalls, but they are susceptible to settlement because of their weight. They may also cause lee-side erosion in the same way as revetments, well illustrated by the tetrapod walls on Sylt (figs. 229 a and b).

Figure 227 c: The slab on the far right, opposite the breach end, has rolled to the right (westward) (1980).

Figure 227 d: The beach surface has been lowered by about 30 cm. In the fore- and middleground the ripraps are also sagging. The landward end of the groyne has been protected by rocks. The seawall, partially sand-covered until 1980, lies entirely exposed (1984).

Case study Wangerooge: Coastal protection and its consequences

The examples above demonstrate that shore protection measures rarely constitute any final solution. In most cases the measures taken lead to adverse consequences in adjoining coastal areas, thus necessitating additional countermeasures. These problems are typified by the Isle of Wangerooge (fig. 230):

1874: After severe dune losses along the western end of the island the so-called 'Reichsdeich' was built. The first groynes were built at the same time.

1879: The stone pavement of the 'Reichsdeich' in the west had to be supplemented by a seawall.

1894–1897: In order to prevent a potential breach of the dunes further to the east, in the lee of the 'Reichsdeich', the dyke was extended by the 'Reichsmauer' seawall.

1899: The seawall had to be extended towards

Figure 228: Seawall of the Norderney Profile at the western head of Norderney, built in 1857/58 (1982)

Figure 229 a: Tetrapod groyne off Hörnum (Sylt) during storm; view towards the south. The tetrapods are undercut by the currents and sink into the sand (1983).

the east; additional new groynes were also necessary.

1905: The seawall was further extended, and additional groynes constructed.

1928: Further extensions of 500 m had to be made on the seawall towards the east.

1938–1941: Groyne H at the western end of the island had to be extended to a length of 1.4 km in order to defend the approach to the Harle tidal inlet.

1962: The storm surge in February destroyed the western part of the seawall, which had to be replaced by a revetment. This revetment soon required extension towards the east because of strong lee-side erosion.

1974: The revetment had to be reinforced and extended towards the south in order to counter lee-side erosion.

1979: In front of the new 'Kurhaus', which was

Figure 229 b: Tetrapod groyne off Hörnum in the background; view towards the north. Strong lee-side erosion has breached the dune belt south of the structure and endangers the settlement (1983).

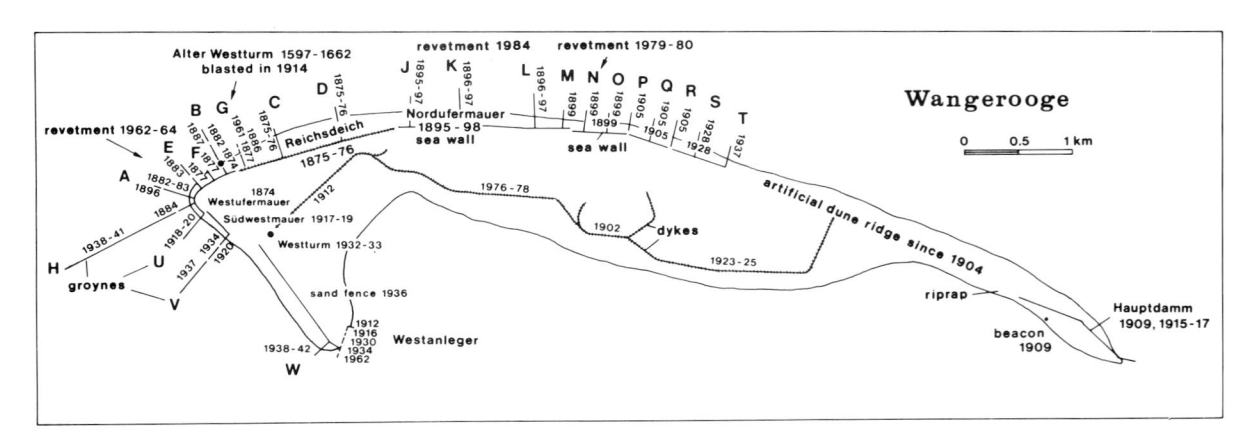

Figure 230: Map of the coastal engineering structures on Wangerooge (from Ehlers & Mensching, 1982).

built directly behind the seawall, another revetment had to be built.

1982: Beach replenishment in front of Wangerooge village became neccessary to protect the revetment and seawall.

1984: Because of continued beach losses in front of the remaining part of the 'Reichsmauer', a connecting revetment between the existing ones is being constructed.

In addition, the dune ridge in the eastern part of the island has been protected from wave attack from the tidal flat side by a riprap (fig. 231). There is no doubt that additional structures will soon be required.

Beach replenishment

There has been a growing tendency in recent years to introduce active shore protection measures, rather than to erect massive structures (passive coastal protection) (cf. Stephen, 1981: 179). Beach replenishment provides such a method without damaging the recreational landscape by the presence of massive structures (figs. 232 a and b).

The chief disadvantage of beach replenishment is that, like all other protection measures, it cannot actually prevent natural coastal changes. While a beach may be maintained at a stable level by replenishment, erosion may continue at greater depth offshore.

Figure 231: Stone riprap protecting the tidal flat facing dunes on the eastern spit of Wangerooge (1984).

Measurements taken during the first sand replenishment on Sylt have shown that the positive effect of the exercise was limited to the area above NN -5 m, whilst erosion continued at greater depths (fig. 233). This led to an oversteepening of the foreshore profile which was balanced by seaward sediment transport after a couple of years (Wenzel, 1979). At worst such a steepening of the foreshore slope may lead to instant erosion of the freshly replenished beach.

Even without the adverse results mentioned above, the sand demand for beach replenishment remains very high:

Beach replenishments on Norderney:

Year	Sand in m^3
1951/52	1,800,000
1967	240,000
1976	400,000
1978	400,000
1982	475,000
1984	400,000

As the table indicates, on average more than 200,000 m^3 of sand has been fed to the Norderney beach each year since 1951. Nevertheless, it has been possible to avoid additional dune erosion east of the shore protection structures.

Fig. 234 shows the typical development of a beach replenishment. In 1976 the spaces between the seven westernmost groynes on Norderney were replenished with 400,000 m^3 of sand. This sand started to migrate eastwards and southwards during the following months. About 50 % of the supplied sand was lost to adjoining groyne fields within 4 years (Pätzold, 1982).

A major problem is to find an appropriate source area for the sand. The sand used for the first beach replenishments on Norderney (1951/52, 1967), Sylt (1972) and Wangerooge (1976) was taken from the neighbouring tidal flats. This created large holes in the tidal flats which refilled very slowly.

The sand for the recent replenishments on Norderney was dredged from the Robbenplate shoal in the middle of the Norderneyer Seegat inlet, i.e. directly from the shoal arc leading from Juist to Norderney. Some of the sand which would otherwise migrate to the middle and eastern part of Norderney has therefore been diverted to the western beach of the island. Adverse consequences for the eastern parts are not expected, however, because the sand will be deposited there sooner or later anyway. The sand for the latest replenishment on Sylt was also taken from the adjoining ebb-tidal delta south of the island.

Consequences

Coastal erosion on the barrier islands is a consequence of changing current conditions and of the rise in sea level. The stabilizing of the western ends of the islands and of the exposed beaches while the sea level continues to rise must lead to a permanent lack of sand along these most exposed parts

Figure 232 a: Beach replenishment on Norderney; the freshly supplied sand is smoothed by bulldozer (1984).

Figure 232 b: Beach replenishment on Norderney; a freshly filled groyne field north of the beach promenade (1984).

of the coast. As the causes of coastal erosion cannot be eliminated, the available countermeasures do little more than treat the symptoms rather than the cause. Coastal engineering structures and sand replenishment can slow down the process, but cannot stop it completely.

Under natural conditions the system of barrier islands would adjust to the rising sea level by simply migrating landwards, but this is prevented by the fixing of the coastline. At the same time the migration of sediment from the sea towards the tidal flats by washover processes across the islands and by dune migration has been halted by human countermeasures. In addition there is therefore a lack of sand on the tidal-flat facing shores of the islands as well. Attempts have been made to counter this by extensive protective structures (figs. 235 and 236).

Luck (1976: Anl. 9) has attempted to estimate the total value of coastal engineering structures on

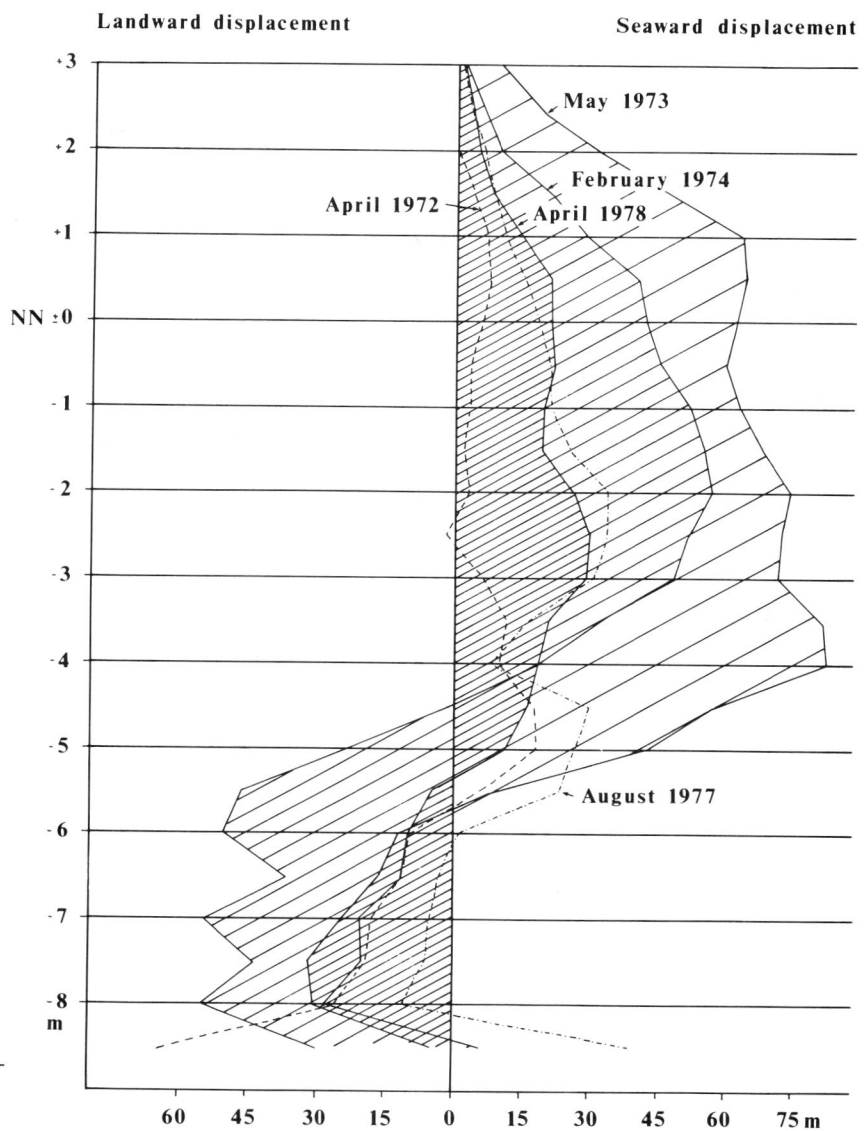

Figure 233: Shifting of the beach and foreshore contour lines, compared with the 1970/71 situation before the sand replenishment on Sylt (Source: Wenzel, 1979: fig. 10).

the East Frisian Islands, compared with the total value of the buildings on these islands (prices of 1974). Whilst on Langeoog the ratio of coastal engineering structures to buildings is 5,000,000 : 237,000,000 DM (1:47), on Baltrum the ratio is only 21,000,000 : 84,000,000 DM (1:4).

Given the high costs of coastal engineering and the subsequent countermeasures necessitated by this vigorous interaction with natural sediment transport, a more flexible approach to coastal protection problems should possibly be considered. However, this flexibility will only be possible, if in future erection of buildings is prohibited within a specified safety zone adjacent to the beach, in contrast to building a few metres behind the seawall as on Sylt and Wangerooge.

Reynolds (1982: 326) has summarized some of the results of his investigations about 'the societal response to beach erosion at Marco Island (Florida)', as follows: 'New construction shoreward of a properly calculated setback line must be avoided at all costs. Protective structures such as bulkheads, seawalls, and revetments, when warranted to protect property, will result in the loss of a recreational beach. These structures also have a tendency to grow in size, transforming sandy beaches into engineered shorelines.' This statement applies equally to the Wadden Sea barrier coast.

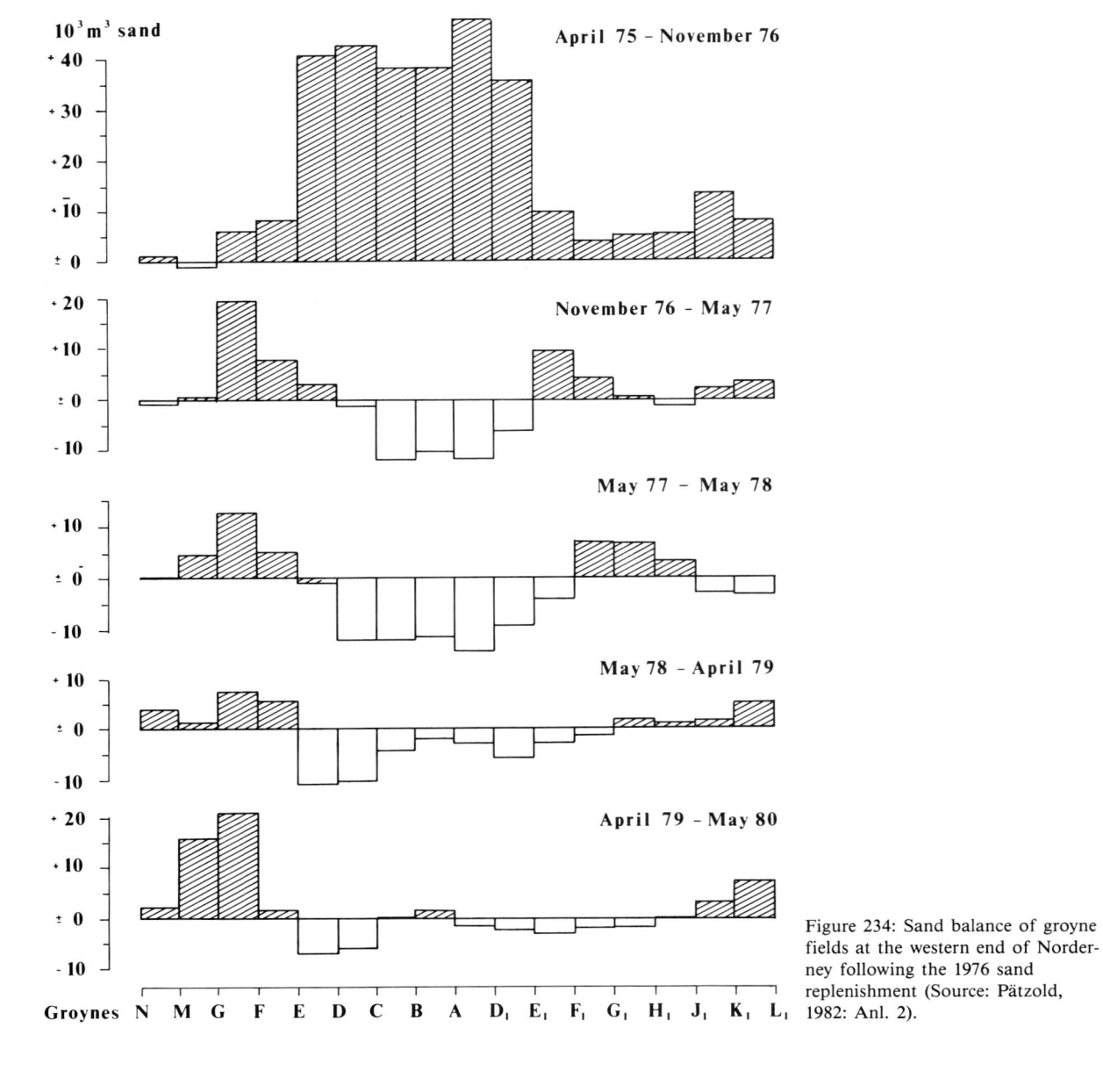

Figure 234: Sand balance of groyne fields at the western end of Norderney following the 1976 sand replenishment (Source: Pätzold, 1982: Anl. 2).

Figure 235: Southward extension of the tidal-flat facing revetment at List (Sylt) (1984).

Figure 236: Revetment at Blidsel (Sylt); lee-side erosion has caused an increased retreat of the adjoining dune area (1984).

CHAPTER 6

HISTORICAL DEVELOPMENT

A knowledge of the general historical background is essential for an understanding of the Wadden Sea landscape. Whilst a detailed description would be beyond the scope of this study, an attempt is made here to summarize the main phases of historical development.

Early trade

Trade began to play a major role in the Wadden Sea area during the early Middle Ages. Shipbuilding had developed to such an extent that it was possible to transport goods both rapidly and safely by sea, offering a favourable alternative to the poor road conditions of the coastal lowlands.

After a few centuries the Frisian trade, which covered the entire Wadden Sea area, lost its importance to the growing Hanseatic network. The Frisians' main trade was agricultural products, such as butter, honey, hides and, above all, fish. The Hamburg duty lists (Zollisten) of 1399/1400 report considerable quantities of fish imported into Hamburg by merchants from Wangerooge (Scheurlen, 1974: 80).

Salt production

One of the most important Frisian trade products was salt, extracted in North Frisia from peat saturated with salt water during earlier floods (Bantelmann, 1966; Marschalleck, 1973). North Frisian salt production began around 500 A.D. and reached its zenith in the 11–14th centuries. Peat extraction led to a widespread lowering of the marshlands. This, in turn, increased the impact of the 'Mandränke' flood of 1362, which finally put an end to large scale salt industry. Prange (1982: 30) reports that in the area SW of Niebüll alone, 800,000 m^3 of peat had been dug within a period of 200 years, yielding about 20,000 tons of salt. In spite of the known detrimental effects to the landscape, small-scale salt production continued after the 'Mandränke' for several centuries.

Salt production in North Frisia ended in the 18th century, when the higher quality salt from the Lüneburg salt works supplanted the rather bitter, greyish Frisian salt. Hansen (1865: 50) reports that around 1727, eighty-four people were still engaged in salt production in Galmsbüll, and that 16 ships were being used for salt transport. By 1762 peat digging had declined to only 160 tons, and in 1782 salt production at Galmsbüll finally ceased. The poor, however, were still digging peat in the tidal flats in 1865; they used the so-called 'Tuul' as fuel (Hansen, 1865: 49).

Tribal feuds and piracy

During the Middle Ages a number of local feuds and tribal wars in West and East Frisia hindered trade and economic development. After 1390, during the course of these conflicts, piracy flourished along the East Frisian coast. The so-called 'Vitalienbrüder', expelled from the Baltic Sea by the combined actions of the Hanseatic League and Teutonic Knights, turned their attentions to the North Sea. It was not until 1433 that the Hanseatic League was finally able to defeat the pirates, who had operated under the protection of various Frisian chieftains (Scheurlen, 1974).

Fishery

A new source of income for the islanders emerged in about 1425. The herring schools, instead of migrating annually southward along the Swedish coast into the Baltic Sea, began to move into the North Sea instead (Hansen, 1877: 72). Large scale fishing of these started around Helgoland. Pontoppidan (1785; *op. cit.* in Quedens, 1968: 8) indicates that about 2000 fishermen were employed in the seas around Helgoland around 1530. By the middle of the 17th century, however, the resources became exhausted and the herring fishery around Helgoland was abandoned.

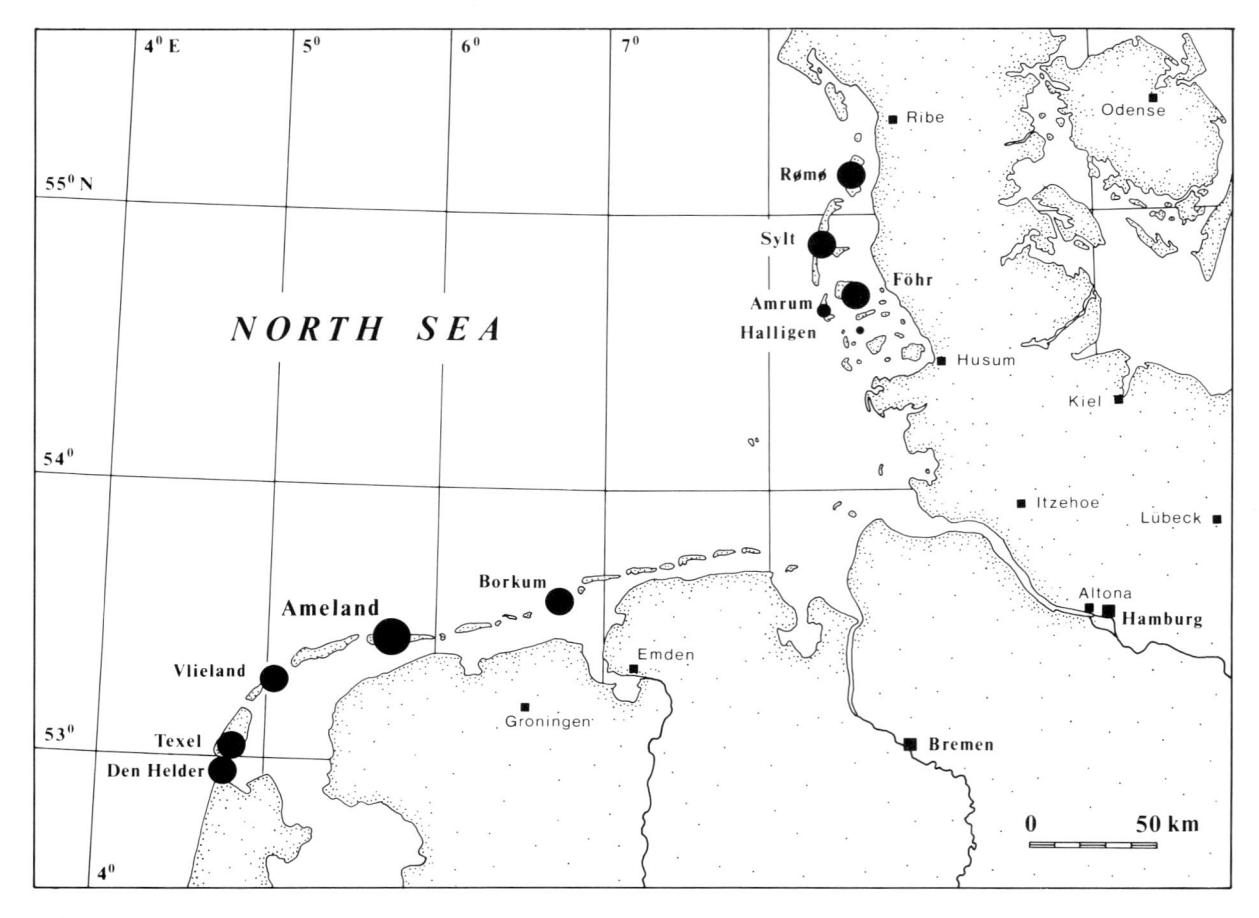

Figure 237: Centres of whaling on the islands of the Wadden Sea in the 17th/18th century.

Whaling

There were other examples of the over-exploitation of natural resources by the coastal dwellers causing final collapse of an industry and loss of income. Around 1610/11 the British began whaling in the seas around Spitsbergen and Greenland, followed by the Dutch in 1612 who soon established their supremacy. The French also joined them in in 1613 (Falk, 1983: 13). Hamburg was the first German town to equip ships for whaling, in 1643 (Oesau, 1955: 65). Danish towns followed, Copenhagen in 1763, the then Danish towns of Altona, Glückstadt, Tønder, Ribe, Kiel, Elmshorn, Kollmar, Itzehoe and Brunsbüttel also participating (Falk, 1983: 45).

The seas around Spitsbergen and Greenland were rich in whales during the early years of whaling, and this rapidly became an extremely profitable business. For the Wadden Sea islands this initiated a period of widespread prosperity, often referred to as the 'Golden Era' (fig. 237). The islanders provided a large number of sailors, often in positions of responsibility such as helmsmen or skippers ('commandeurs'). The degree to which various islands were involved in whaling varied. On the Isle of Föhr, for instance, about 1,500 men went to sea in 1760 of a total population of about 4,500. About 70 % of the male population were therefore occupied in sea-faring (Quedens, 1985: 31; Falk, 1983: 37; 1984).

As the majority of the men were at sea for most of the year, farming was left to the women and the few remaining men. One of the reasons why dyking on Föhr began so late is that farming was regarded as a secondary activity of minor importance. In addition to this, there were also few workers available on the island to undertake such projects (Quedens, 1985: 34).

Whaling did not only bring prosperity, however, it was dangerous and the resulting human losses were considerable. Krohn (1949: 23) has shown

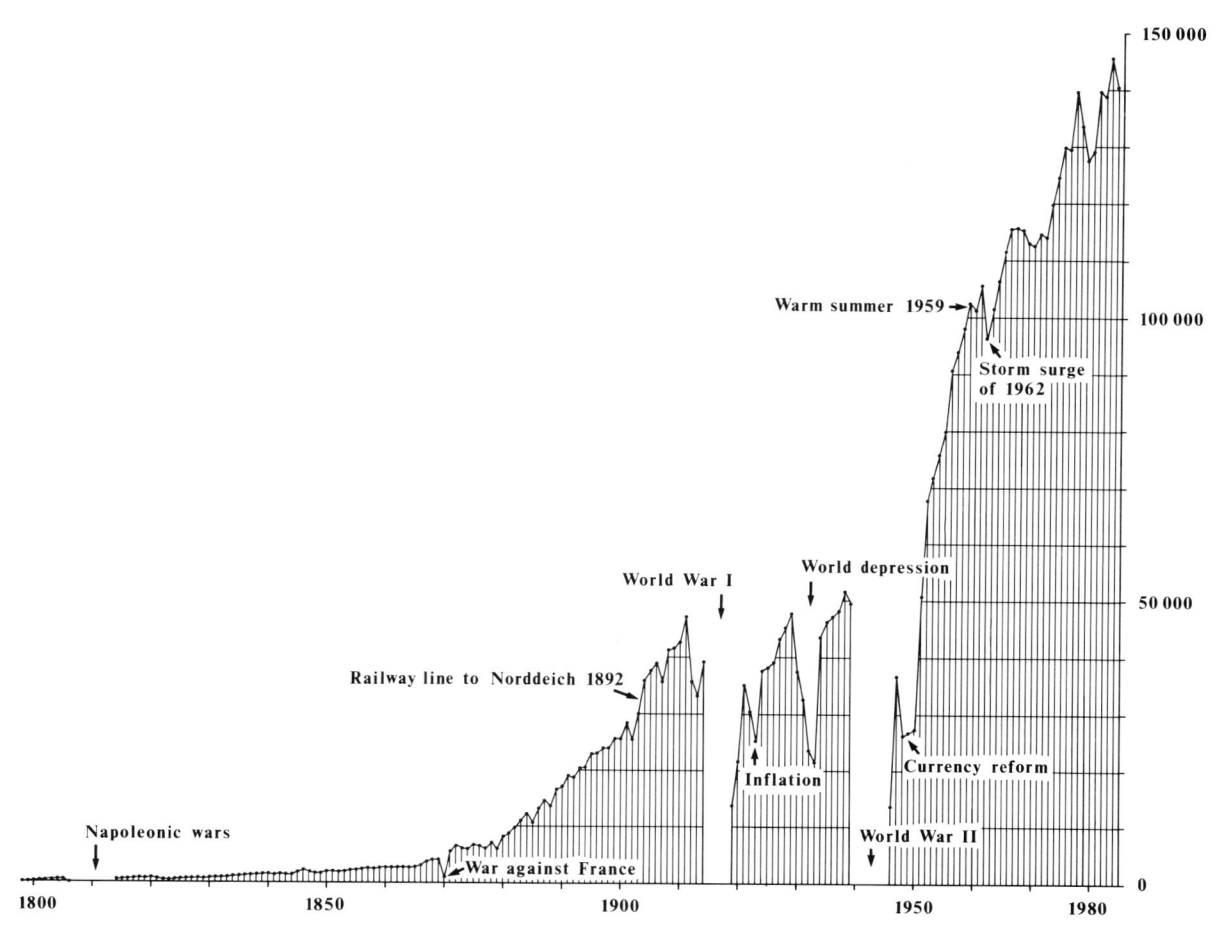

Figure 238: Development of tourism on Norderney; numbers of visitors per year (Sources: Kulinat, 1969; personal information by Kurverwaltung Norderney).

that the death rate of the male population on Sylt was 4 per thousand at this time, indicating that about 50 % of the male population died at sea.

Commercial shipping

A number of the islands were little concerned with whaling. Nevertheless they played a major role in the sea trade, especially Vlieland and Terschelling, situated on the deep water of the Vlie inlet. Their dominance was superseded by Amsterdam, the centre of commercial power in the Netherlands, at the end of the 17th century.

The Amsterdam trade, which in previous years had entered via the Vlie, later became increasingly concentrated in the Marsdiep. The destruction of West-Terschelling by the English in 1666 confirmed Amsterdam as the dominant port in the region (de Boer, 1980: 226). The sea route continued to cross the Zuiderzee. After 1825 the North Holland Canal to Den Helder was used. In 1875 the Noordzeekanaal was opened, connecting Amsterdam directly with the North Sea at Ijmuiden.

Whaling ceased being profitable during the second half of the 18th century and most islands turned to commercial shipping, despite the fact that the era of commercial shipping had itself passed its climax. Sønderho on Fanø was one of the places which succeeded in commercial shipping, acting as a port for the trading town of Ribe, whose port was only accessible via the meandering course of the Nipsau River. In 1800, at the climax of its prosperity, Sønderho had about 1,000 inhabitants, reaching 1,057 in 1890. Around 1850, however, Sønderho stagnated and Nordby in northern Fanø took over its function. The trade here reached its acme in 1870, shortly afterwards to be replaced by the newly founded port of Esbjerg (Tougaard & Meesenburg, 1974; Meesenburg et al., 1977).

During the 19th century new ports were founded on the mainland, rapidly attracting all the trade because of their more favourable locations. Bremerhaven was founded in 1826; after 1851 Den Helder flourished briefly as a commercial port, until in 1875 the Noordzeekanaal was opened. Wilhelmshaven was founded in 1853 as Prussia's naval port and Esbjerg was founded in 1868, after Denmark had lost her North Sea ports in Schleswig-Holstein to Prussia during the war of 1864. Lastly, during 1890 the fishing port at Cuxhaven was opened.

Poverty and emigration

For most island communities in the Wadden Sea area the 19th century was a time of poverty and decay. With the Napoleonic wars and the continental blockade, commercial shipping foundered, never fully to revive. The islanders turned increasingly to farming, but since the islands were too small to feed the population which had increased considerably during the 'Golden Era', many people were forced to emigrate. The population of Rømø, for instance, decreased from 1,800 inhabitants in 1737 to a mere 925 in 1895 (Sørensen, 1982: 9).

Coastal shipping continued to play a minor role, particularly on those islands which were too small for farming. For example, the occupation list of Spiekeroog for 1885 names 22 sailors among the 53 male adults. As sailing ships were rapidly replaced by steamers after 1870, work for the islanders progressively decreased. They turned to fishing at first, but by the end of the 19th century this had been superseded by modern trawlers operating from the mainland ports (Meyer-Deepen & Meijering, 1983: 99).

The first sea-side resorts

Tourism started on the islands at this time. The first seaside resort on the Wadden Sea coast was Norderney, founded in 1797 (fig. 238). Sea bathing had become fashionable in England during the 18th century, following the example of the Royal Family, and in 1770 the physician Russel in his book *'De usu aquae marinae in morbis glandularum'* had recommended bathing in the sea as a remedy. The first German seaside resort was founded in 1794 at Doberan on the Baltic Sea coast (Mecklenburg) (Kulinat, 1969: 46).

There were obvious reasons why Norderney was the first resort on the North Sea coast. The island was situated close to the mainland and was therefore accessible at low tide to horse-drawn coaches across the tidal flats (Kobbe & Cornelius, 1841: 51). Several resorts were founded subsequently:

Wangerooge	1804
Spiekeroog	1809 ?
Wyk auf Föhr	1819
Langeoog	1830
Juist	1840
Borkum	1850
Nes (Ameland)	1854
Westerland (Sylt)	1857
Schiermonnikoog	1866
Wittdün (Amrum)	1890
De Koog (Texel)	1896
Lakolk (Rømø)	1898

At first, sea bathing was largely restricted to the upper classes. The rich, government-run, resorts therefore flourished, whilst other places developed rather slowly. The resort at Norderney was taken over by the Royal Hannover Government in 1814, and the resort at Wangerooge by the Grand Duchy of Oldenburg Government in 1829 (Kulinat, 1969: 49). In 1834 King Ernst-August of Hannover paid a visit to Norderney; in 1844 Bismarck spent his holidays there. Georg V, when he became King of Hannover in 1851, moved his summer residence to the island (Nordseebad Norderney, 1980). Wyk auf Föhr only began to flourish after King Christian VII of Denmark started to spend his annual 4–5 week holidays there (1842 to 1847; Quedens, 1985: 75).

The other seaside resorts experienced considerable initial problems. After only five years, the resort in Nes (Ameland) had to be abandoned through lack of visitors, and it was not until 1902 that another attempt was made (Bakker, 1973: 87). Similar factors caused the closure of the resort at Juist in 1858, but it re-opened in 1866 (Kulinat, 1969: 55). The 'Aktiengesellschaft Wittdün-Amrum', which ran the resort at Wittdün, went bankrupt in 1906 (Quedens, 1979: 67); as did the resort on Lakolk in 1903. Tourism continued on a small scale until the First World War, however, it only began to flourish after the Second World War when accessibility was considerably improved by the newly built dam to the mainland (Sørensen, 1982: 213).

Poor traffic infrastructure was a severe handicap to the development of many of the resorts. In Germany, Norderney and Borkum occupied favourable positions because of their tide-independent shipping routes. Emden, Borkum's mainland port, was the first to be connected by railway in 1856, encouraging the development of the resort. Wangerooge (via Carolinensiel) followed in 1890, Norderney and Juist (via Norddeich) in 1892 (Kulinat, 1969: 55).

Onset of coastal protection

The development of the coastal resorts had an important impact on the morphological development of the islands. The resort in Norderney was soon threatened by coastal retreat, forcing the rapid construction of coastal protection works. Despite the Norderney experience, Borkum, originally well-protected behind dune ridges, expanded into the open beach areas (fig. 239). Wangerooge, which had been completely demolished by the storm floods of 1854/55, was originally rebuilt in a protected position in the centre of the island. However, shortly afterwards hotels and boarding houses were built by the beach, forcing the construction of new coastal defences. The complex coastal protection schemes undertaken in these places, and especially in Westerland (Sylt), will be examined in more detail in the regional section.

World War I and its consequences

Outbreak of Word War I brought an immediate end to tourism. Fortifications were built on Borkum, Wangerooge and Sylt because an enemy landing from the sea seemed possible. However, no fighting took place on the coast, although an advance of strong British naval forces into the German Bight on August 28th, 1914 had revealed a weakness in the German defences. The German fleet, which was anchored in the Jade, could not sail because the Jade bar was less than 10 m below Mean Low Water and could only be crossed at high tide (Costello & Hughes, 1978: 66).

Tourism continued to develop after World War I, but at a slower rate than before and interrupted by a number of severe crises. Changes in German society following the 1918 revolution now made it possible for more people to spend their holidays at the coast, benefitting the smaller, simpler resorts (Kulinat, 1969: 57).

After the 1920 referendum, Nordschleswig was annexed by Denmark. Emotions ran high; strict passport and visa regulations were introduced by Denmark, virtually severing Rømø from its predominantly German tourist base. This also hindered traffic to Sylt which ran via a ferry service from now Danish Hoyerschleuse to Munkmarsch on Sylt. The situation only improved on Sylt following the completion of the railway embankment (Hindenburg-Damm) in 1927 (Voigt, 1977: 38).

After the National Socialists came to power in 1933 coastal protection was intensified and land reclamation increased. Economic stringency was no longer an important consideration (Schmidt, 1937: 87). In addition to work creation schemes, a dominant factor was the 'Erweiterung des Lebensraumes' (expansion of the living space) programme. Between 1933 and 1935 more money was spent on coastal protection and land reclamation than in the entire period from 1900 to 1932 (32,000,000 Reichsmark vs. 30,000,000 Reichsmark) (Stadermann, 1937: 65).

These projects were obviously much less cost-efficient than the earlier schemes. The Adolf-Hitler-Koog (now: Dieksander Koog), 1,350 ha in size, was planned in 1933 and completed by 1935, costing 4,000,000 Reichsmark (Stadermann, 1937: 62). The Hermann-Göring-Koog (now: Tümlauer Koog), 585 ha in size, also planned in 1933 and complete by 1935, cost 2,400,000 Reichsmark (Fischer, 1956: 311). The basic costs for land reclamation, but without subsequent measures such as drainage, were about 3,700 - 5,200 Reichsmark per ha. Land reclamation was therefore clearly unprofitable (Stadermann, 1937: 62).

The arms race before the Second World War also led to changes in the coastal areas. New barracks and settlements were built on the previously almost uninhabited ends of Sylt. In 1935 the construction of the Wehrmacht Hörnum sea-plane base began, including a repair yard for aeroplanes, docks, hangars, the largest plane-lifting crane in Europe, a power plant and the Hörnum officers' families settlement (Pahl & Carstensen, 1973). The Listland in northern Sylt, a nature reserve since 1923, was partly expropriated. The town of List became a naval and air force base, and a network of concrete roads and bunkers, firing positions and depots was constructed throughout the dune area (Oestreich, 1976: 303).

The Second World War

At the beginning of the Second World War the military defences along the German North Sea coast were strengthened. However, after 1940 there was little activity following the realization that the threat of coastal landings was unrealistic. Therefore the guns were increasingly withdrawn from the coastal batteries and moved to the more threatened sectors of the 'Atlantikwall' (Rolf, 1983:

1891

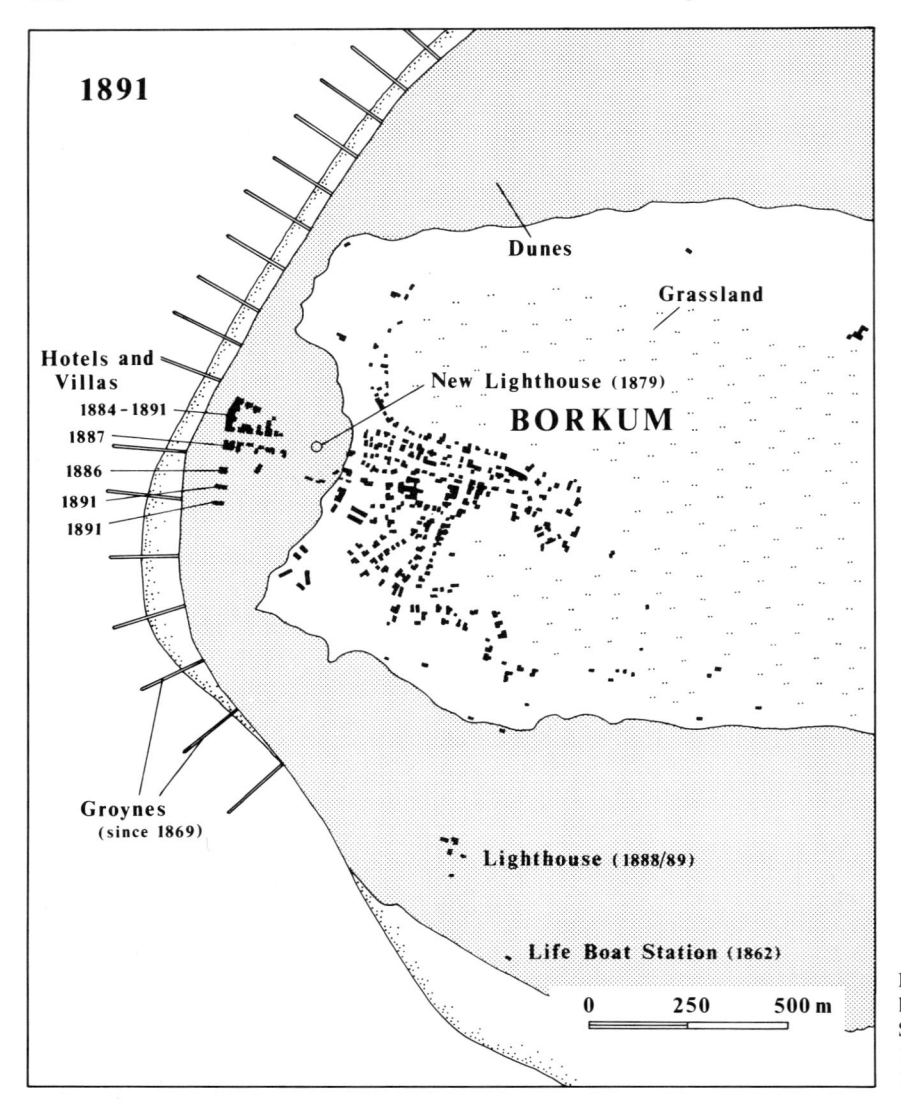

Dunes
Grassland
Hotels and Villas
1884-1891
1887
1886
1891
1891
New Lighthouse (1879)
BORKUM
Groynes (since 1869)
Lighthouse (1888/89)
Life Boat Station (1862)

0   250   500 m

Figure 239: Built-up areas on Borkum 1891 (left) and 1976 (right). Sources: Topographische Karte 1:25,000 Sheet Borkum, Feldmann, 1977; 1980; Witte, 1970).

149). The islands, however, continued to play an important role in the air defence system.

The strengthening of the 'Atlantikwall' was only intensified after the spring of 1944, when the landing of American and British forces became imminent, leading to a continuous fortress system from southern France to northern Norway. In the Wadden Sea area special attention was paid to the defence of the port entrances, of vital importance in case of an invasion. The existing bunkers and coastal batteries on Texel (protecting Den Helder), Borkum (Emden) and Wangerooge (Wilhelmshaven) were extended, and new defences were built on Fanø and the Skallingen peninsula (Esbjerg).

These military activities led to extensive interference with the natural landscape, leaving deep scars, many permanent. An aerial photograph of the 'Robbe' radar station on Rømø, as it appeared immediately after the war, gives an impression of the size of these installations (fig. 240).

With the exception of the Georgian Uprising on Texel in 1945, there was no fighting in the Wadden Sea area during the Second World War. The islands were only occupied by Allied troops after the end of the war, although some islands had been the targets of air raids. The attack on Hörnum during the night of 18./19.3.1940 inflicted little damage (Pahl & Carstensen, 1973) and the same was true of an attack on Norderney, aimed at the military air field in the Südstrandpolder. Wangerooge was attacked by 480 bombers as late as

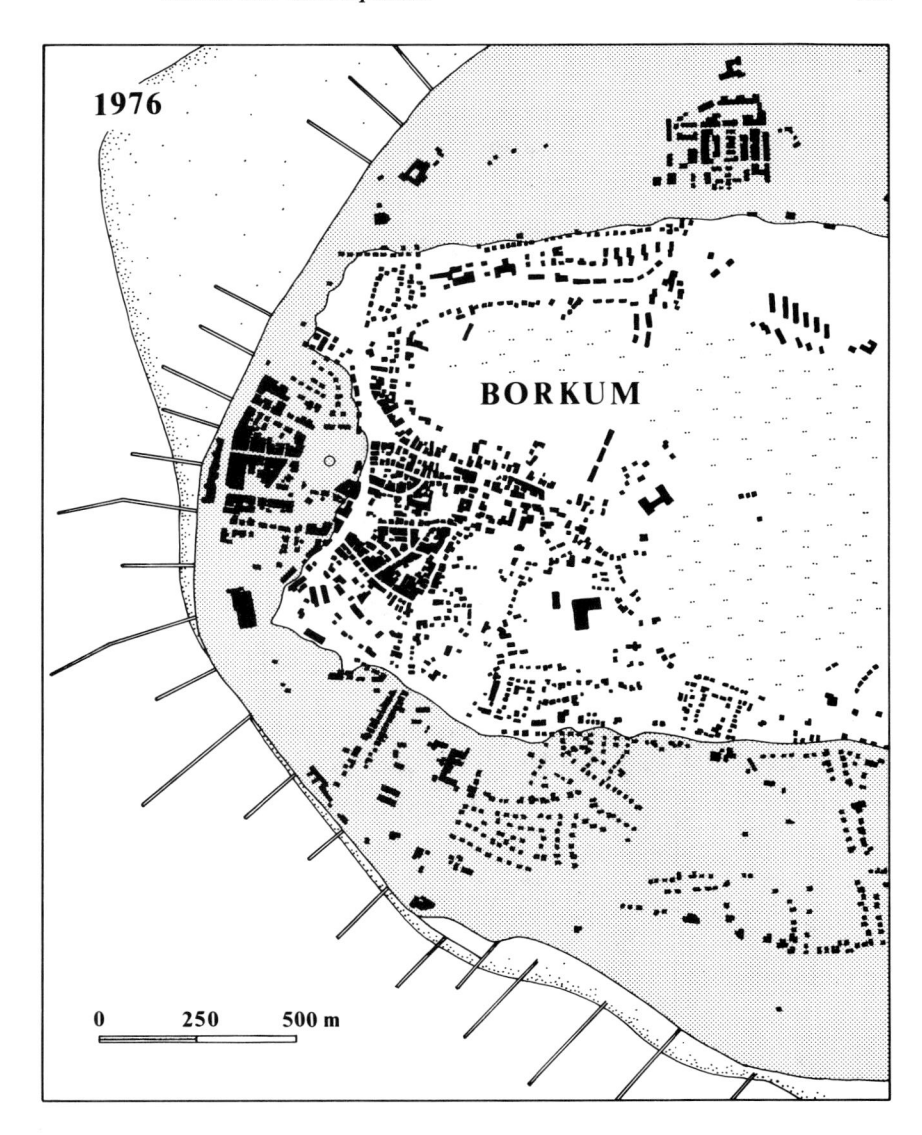

April 25th, 1945; the 5,000 bombs dropped turned the island into a landscape of craters, still largely visible today (Jürgens, 1970; 1985, Nicolaisen, 1985). Presumably it is the only such existing crater field in Germany.

Age of large-scale tourism

An era of large-scale tourism started after the Second World War in the whole Wadden Sea area, on the islands as well as on the mainland coast. The number of people who stayed overnight on the Lower Saxonian coast increased fourfold on the pre-war figures as early as 1960. The climax of development on the Dutch Wadden Sea islands was reached after a very sharp rise in 1975 (Waddenzeecommissie and figures provided by Central Bureau voor Statistiek). On the German islands, however, the numbers of day visitors and those staying overnight are still increasing (fig. 238). Nevertheless, experts agree that the limited infrastructure and natural conditions will restrict future development (Buchwald, Rincke & Rudolph, 1985: E 3). There are practical problems with water supply, sewerage and waste disposal. The decisive factor, however, is the undesirability of unbridled expansion.

The increase of tourism and the resultant improvement of the islanders' economy has led to an enormous expansion of the settlements. The built-up area on Langeoog increased from 14.6 ha (1891) to 34.4 ha (1940) and to 105.1 ha (1984), reflecting

Figure 240: 'Robbe' radar station on the Isle of Rømø with surrounding mine fields. Aerial photograph of the Geodætisk Institut, Denmark, Flight E 54 5020, 1945, reproduced by permission (A. 532/85) of the Institute.

a sevenfold increase within less than a century (fig. 241). Demand for building land is especially high on those islands where there are already extensive summer house developments, notably on the Danish islands of Fanø and Rømø. Planning restrictions, however, increasingly limit further expansion.

The shortage has artificially raised the price of land and houses on the islands to a level beyond the means of people of average income, affecting in particular young islanders who want to start their own businesses. A considerable proportion of the available houses are already used as holiday homes by rich people from the mainland. As early as 1974 there were already 3,213 summer houses and holiday apartments in Westerland alone (Oestreich, 1976: 208). The low occupation rate of these apartments causes the settlements to be deserted out of season. 74.5 % of the holiday apartments in Kampen (Sylt), for example, are occupied for less than 3 months each year, 59.5 % for less than 2 months each year (Oestreich, 1976: 524).

The considerable proportion of holiday apartments on the East Frisian Islands is emphasized by the figures published in the 'Umweltprobleme der Ostfriesischen Inseln' report (environmental problems of the East Frisian Islands):

Island	Inhabitants	Holiday apartments
Borkum	8,272	1,100
Juist	2,626	300
Norderney	8,122	1,200
Baltrum	777	70
Langeoog	3,011	550
Spiekeroog	1,140	95
Wangerooge	1,938	650

Source: Buchwald, Rincke & Rudolph (1985: B 57).

Together with the expansion of housing there has been an increasing demand for new transport facilities. All the islands which permit car traffic (all except Schiermonnikoog, Juist, Baltrum, Langeoog, Spiekeroog and Wangerooge) have had to expand their road networks considerably. The area used for parking spaces alone amounts to about 9 ha on Norderney (Sources: Deutsche Grundkarte 1:5,000 of 1984). This was the size of the entire settlement on Norderney in 1820 (Source: von Halem, 1822). In the case of the car-free islands, these problems are merely transferred to the adjoining mainland.

The demand for private airfields on the islands has been even more intense (on Norderney: 15 ha). Airfields now exist on Texel, Ameland, Borkum, Juist, Norderney, Baltrum, Langeoog, Wangerooge, Eiderstedt, Föhr and Sylt.

The yacht harbours, at present relatively small, occupy additional space on the Wadden Sea sides of the islands.

The space available for recreational activities is further limited by military training grounds (on Vlieland, Terschelling and Rømø) and by nature reserves which are at least seasonally closed to the public.

Nature conservancy

The rapid growth of tourism on the islands and in the Wadden Sea in general has interfered considerably with the natural landscape. Intensive pedestrian trampling of the dune landscape on Sylt, for instance, has destroyed the role of the nature reserves

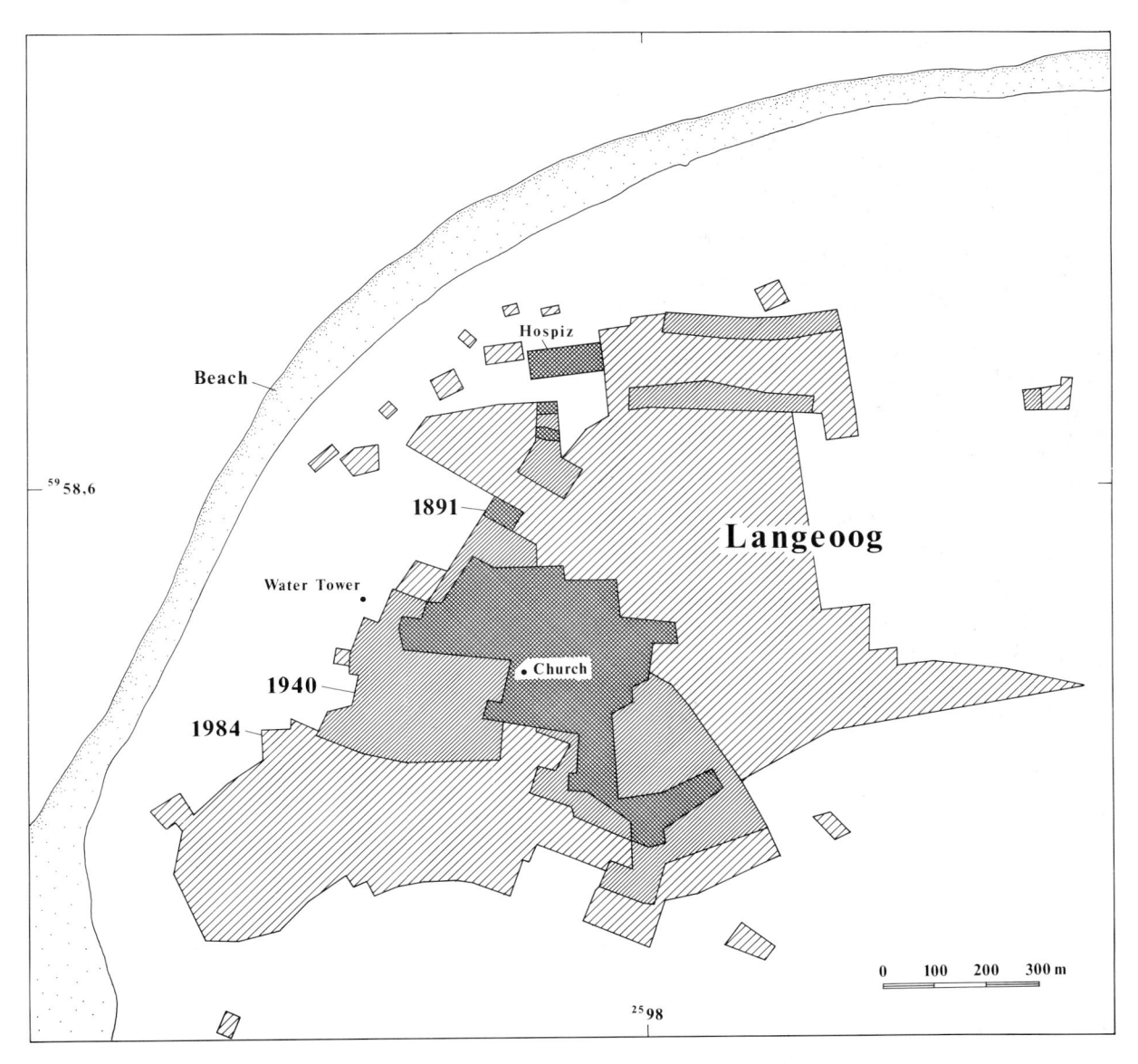

Figure 241: Development of the built-up area on Langeoog from 1891 to the present. Sources: Topographische Karte 1:25,000, Sheet Langeoog, of 1891 and 1940, Deutsche Grundkarte 1:5,000, Sheets 9657,5 Langeoog and 9957,5 Langeoog-Ost of 1984.

there as breeding grounds for birds (Kruse, 1981: 91). Where tourist pressure becomes excessive, unregulated pedestrian traffic destroys dune vegetation, creates footpaths and promotes erosion especially along dune crests, most visited by tourists. The extensive use of barbed wire to protect the dunes greatly detracts from the natural beauty of the islands.

There is a continuing tendency to equate nature conservation with the protection of plants and animals, but not with the landscape as a whole. Landscape conservation, however, is only fruitful if it relates to an area as a whole and not to limited sanctuaries within that area. The wisdom of transforming those islands threatened by erosion into artificially defended 'fortresses' seems questionable. Any installation of massive coastal engineering structures should be preceded by discussion to determine whether or not the desired protection of an island by sea-walls, groynes and tetrapods may actually constitute its ultimate destruction as a part of the natural landscape.

Figure 242: Foam created by the alga *Phaeocystis pouchetii* in May on the beach at Amrum (1980).

Pollution

The pollution of the North Sea endangers the natural landscape of the Wadden Sea as well as its use as a recreational area. The most significant damage is caused by oil pollution. Tanker accidents have so far been insignificant in this context. The present oil pollution results entirely from the regrettably common practice of cleaning the tanks at sea, which even when detected and punished, is cheaper for the ship-owners than the proper removal of the oil in port. Some of the oil is degraded biologically in sea water but largely insoluble tar lumps still remain to be deposited on the beaches. The increasing numbers of dead sea birds proved to be casualties of this oil pollution reflect the increasing severity of this problem (Buchwald, Rincke & Rudolph, 1985: C 16)

Environmental problems also result from the increasing eutrophication of the North Sea. A conspicuous result of this is the formation of large quantities of foam in spring times since 1978 (fig. 242). The foam is produced by the mass appearance of the alga *Phaecocystis pouchetii*. Extensive foam formation has also been observed occasionally in earlier years, such as in 1938, but these were exceptions. The great increase during recent years is regarded as a result of the increasing supply of nutrients, especially phosphate and nitrate (Buchwald, Rincke & Rudolph, 1985: C 9).

Effects of heavy metal pollution are not so apparent, but equally damaging. Rather than being broken down, these are actually enriched in the sediments. Samples from the Wadden Sea around the East Frisian Islands showed that the natural heavy metal content of the sediments was anthropogenically concentrated by the following factors:

Copper	280 %
Zinc	820 %
Mercury	950 %
Lead	1,030 %
Cadmium	1,100 %

(Source: Schwedhelm, 1984: 47). The pollution of the North Sea by chlorinated hydrocarbons also leads to detrimental consequences for the fauna and flora of the Wadden Sea in certain cases. Investigation of fish has demonstrated beyond doubt that their intake of pollutants is greatest in the estuaries of the large rivers Elbe and Weser (Buchwald, Rincke & Rudolph, 1985).

CHAPTER 7

REGIONAL DESCRIPTIONS

7.1 DENMARK

7.1.1 SKALLINGEN

The Skallingen peninsula forms the northernmost end of the Wadden Sea barrier coast (fig. 243). A cape (Blåvandshuk) has formed along the boundary between the smoothed coastline of Jylland and the Wadden Sea. Old cliffs further inland demonstrate the former existence of a straight shoreline (fig. 244) whilst the protrusion of Blåvandshuk is the product of later sedimentation.

The formation of the Blåvandshuk cape

A double dune ridge extends from the north southwards towards Blåvandshuk, consisting of the Havsande dunes in the west and a second dune ridge almost 1 km further inland; the Blåvands Fyr lighthouse is situated at the southern end (Hornsbjerge) of the latter ridge. The coastline has been displaced seawards in the area north of Blåvandshuk during the last few thousand years, a process which continues today. Comparison of the Videnskabernes Selskab map of 1804, the first accurate map of the area, with recent topographical maps demonstrates that during the last 170 years the coastline has shifted about 250 m seawards, i.e. an average of 1.5 m per year (fig. 245).

The former cliff line of Varde Bakkeø is still visible at Grærup, 3 km inland from the North Sea shoreline (Bartholdy & Pejrup, 1980: 63). This cliff has been stable for about the past 6,000 years (Jacobsen, 1969: 15). The cape of Blåvandshuk was formed by sand transport parallel with the coast and directed towards the southwest. Aeolian redeposition formed a broad dune belt. Parabolic dunes have migrated as far inland as Bordrup, about 8 km from the coast. With the exception of the outermost coastal dune ridge, the dunes became stabilized and are widely forested today. Part of the area is being used as a military training ground.

Following the formation of the Blåvandshuk cape the coastline first extended to the southeast via Blåvand and Langli. Originally, Langli was a peninsula, that has subsequently been severed from the mainland (Bartholdy & Pejrup, 1980).

The formation of the Skallingen peninsula

The Skallingen peninsula is the most recent example of coastal accretion in the area. On a map of 1605 it appears only as a high-lying shoal called the 'Schalling-Sandt', probably resembling the existing outer sands of Peter Mejers Sand and Koresand. By 1804 dune formation had started on the shoal, which was only connected to the mainland via a small land bridge, the Hobo Dyb bay still extending far to the northwest (fig. 245) (Jepsen, 1978; Tougaard & Meesenburg, 1978). During the course of the 19th and 20th centuries this bay gradually filled with sediment, and today Skallingen is connected with the Blåvandshuk cape throughout its full length.

In 1870 the Skallingen peninsula consisted of barren sand flats. A few small dunes became established on the seaward side, protected by vegetation which by the beginning of the 20th century had started to spread extensively over the peninsula. As a result, accretion increased on the tidal flat side of the peninsula (Jakobsen, 1953). The vegetational succession through the recent salt marsh area has been thoroughly studied (Gabrielsen & Iversen, 1933; Iversen, 1936, 1953; Mentz, 1952; Larsen, 1953 a, b; Meesenburg, 1972).

The recent development

Before construction of a sand embankment in 1932-33 the peninsula has been frequently flooded during higher tides. The old washover topography can still be seen on recent aerial photographs.

In contrast with this area of recent accretion north of the Blåvandshuk cape, the coastline south of Blåvand is experiencing steady erosion. This

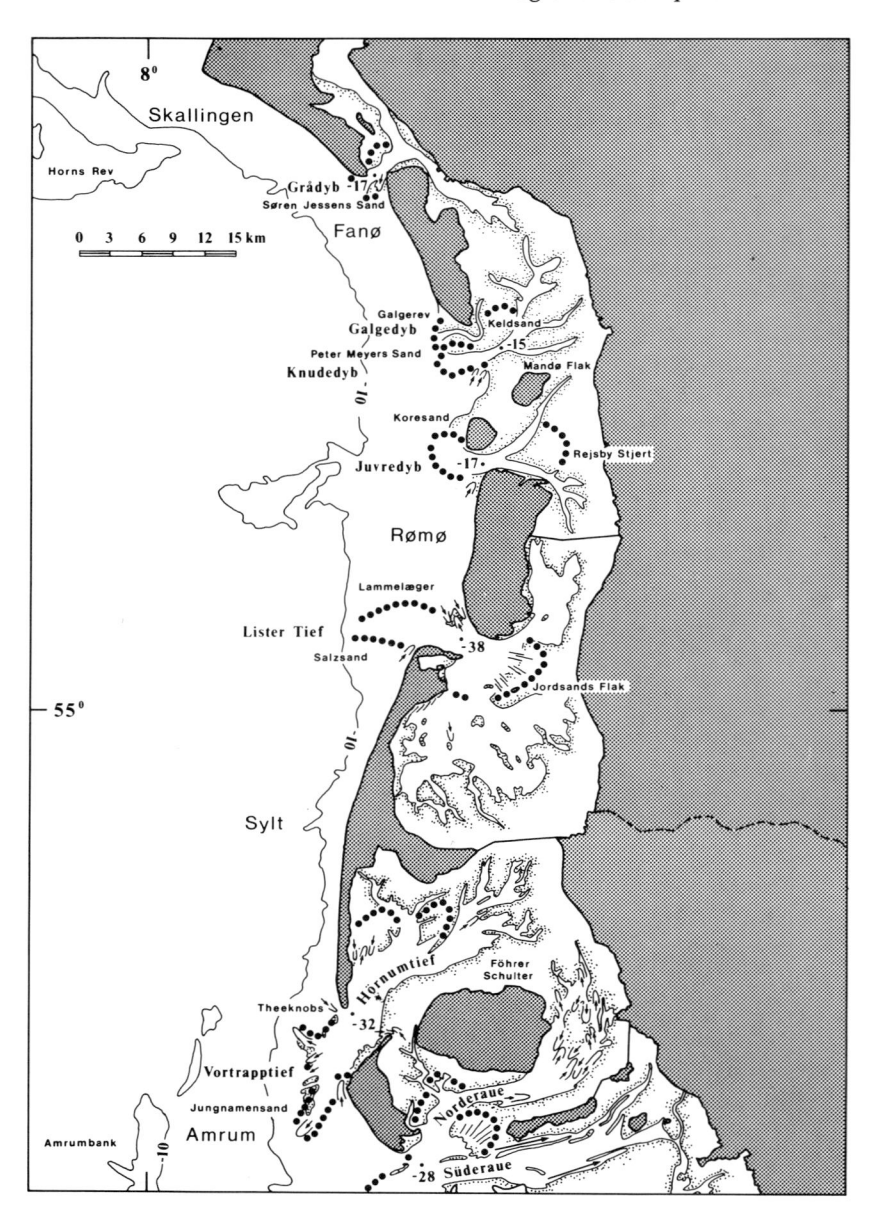

Figure 243: Map of the Wadden Sea area between Skallingen and Amrum. The maximum depths of the tidal inlets are given in metres. Dotted lines: boundaries of the ebb and flood tidal deltas. Arrows: distinct ebb and flood channels.

can be demonstrated by the present position of the bunkers which were erected in the dunes towards the end of the Second World War (fig. 246). Whilst the remains of the 'Flugabwehr-Leitstelle' (Radar station) at Blåvands Fyr are still visible on the dune surface, the bunkers 1.5 km further to the southeast lie on the beach now, indicating retreat of the dune belt by about 30–45 m.

To remedy this retreat 11 wooden pile groynes have been constructed south of the Blåvand settlement, 300–400 m apart. The major groynes are supplemented in the southwest by additional smaller, lighter brushwood groynes. The effect of the groynes is limited; although they do retain some sand from the coast-parallel drift and are able to accumulate aeolian sand, at least temporarily, at the landward ends. The ends of the groynes are undercut on their downdrift, southeastern side. As in most places along the Wadden Sea coast the groynes can retard the rate of coastal erosion, but they are unable to prevent it.

The marshland southeast of Oksby is today only separated from the sea by a small dune ridge. In order to prevent breaching, this dune ridge has

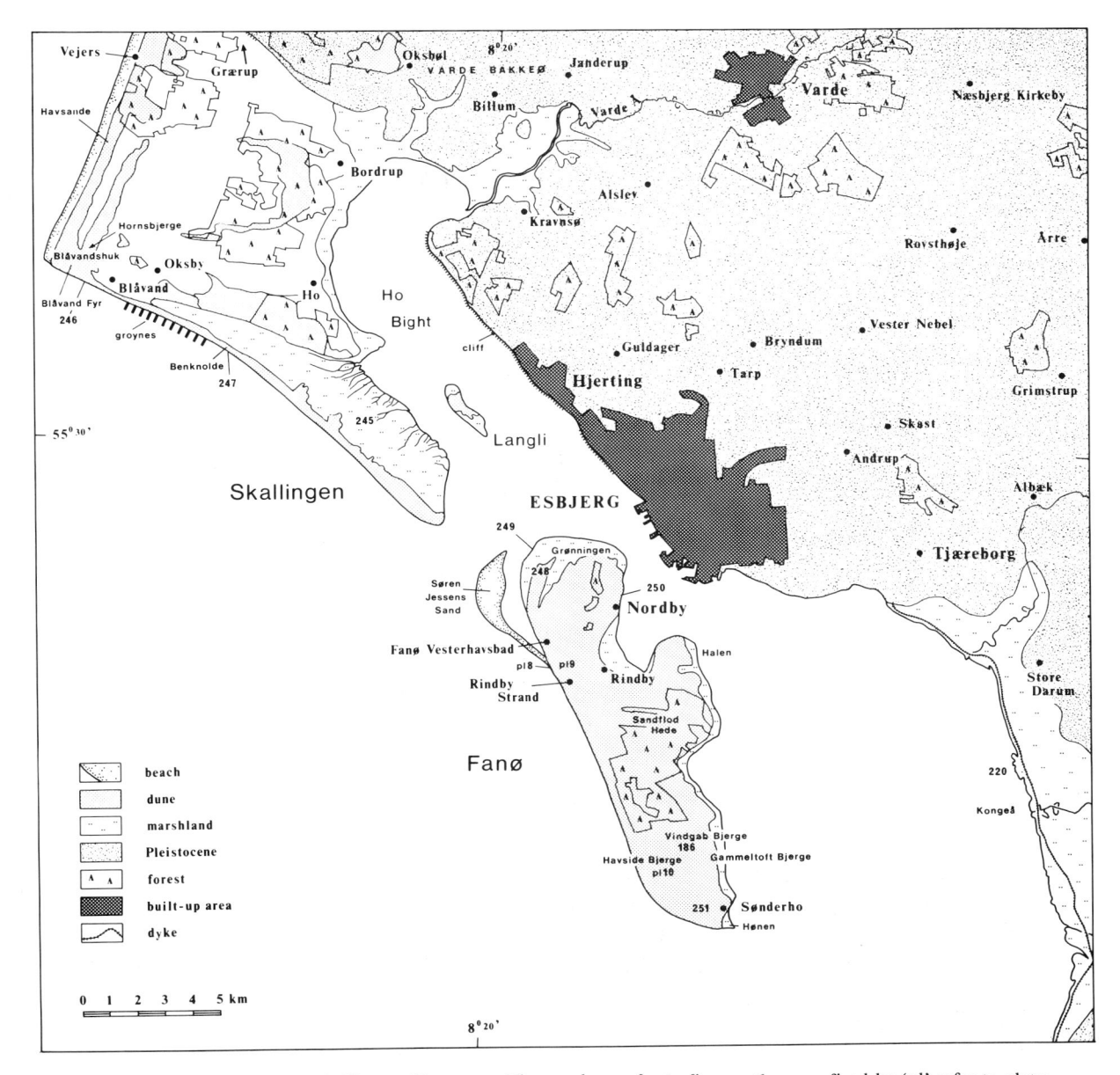

Figure 244: Location map of the Skallingen–Fanø area. The numbers refer to figures, those prefixed by 'pl' refer to plates.

been replenished with artificial sand supply. Further to the southeast the dunes are eroded (fig. 247).

The southern end of Skallingen is undergoing extensive changes. Developments here between 1870–1973 were discussed by Tougaard & Meesenburg (1978). Between 1974 and 1984 this area was subject to heavy erosion. The tip of the peninsula has been cut back by about 1.5 km. In 1983 two new groynes were constructed in order to control erosion, so far with only limited success (Tougaard & Frikke, 1984: 132).

Because of the accessibility of its 18 km long North Sea beach, the Skallingen peninsula is a favourite place for tourist day visitors. The fragile vegetation cover of the young dunes is suffering considerably from human trampling, especially around major car parks (cf. Frederiksen, 1977; Thiessen, 1977).

224   *Regional descriptions*

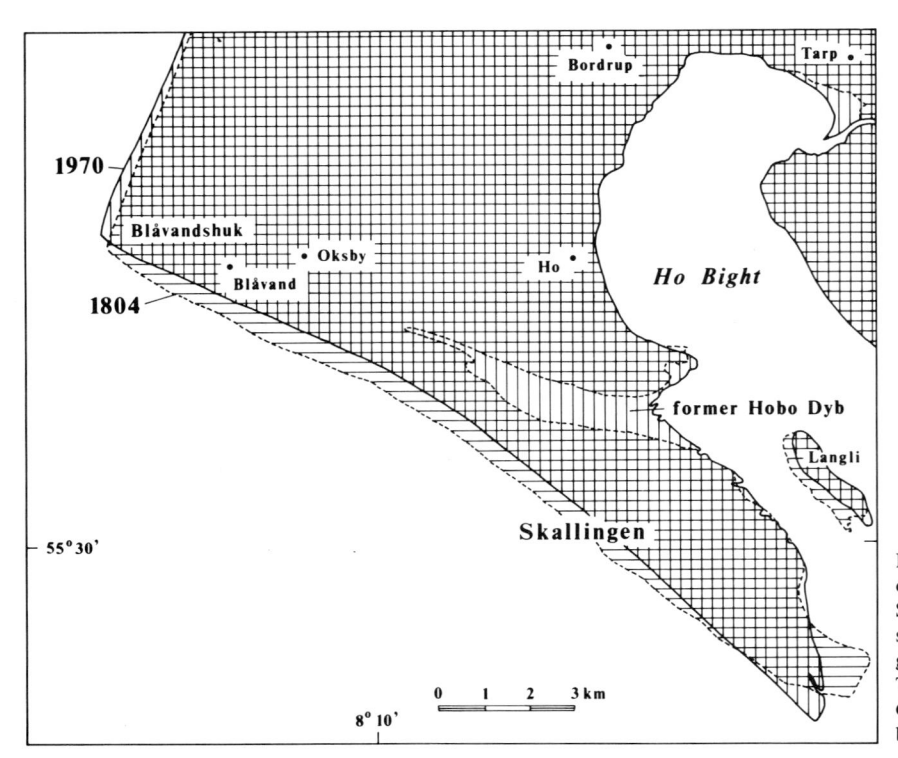

Figure 245: Erosion and deposition on the Skallingen peninsula. Sources: Map of the Videnskabernes Selskab of 1804 (in: Tougaard & Meesenburg, 1978) and Topographical Map 1:100,000 of the Geodætisk Institut, Sheet 1113 Esbjerg of 1970.

Figure 246: Bunker ruins from the Second World War document the extent of coastal erosion, about 1.5 km south of Blåvandshuk (1980).

7.1.2 Fanø

A positive sand budget dominates the Isle of Fanø. Accretion is revealed by recent fan-like dune ridges in the north, and has led to the formation of a number of low dune ridges parallel with the beach in the central part of the island. Minor traces of recent erosion can be found only at the southern end of the island, where the Galgedyb inlet has approached the village of Sønderho.

Northern Fanø

Large scale sedimentation in northern Fanø started around 1700 (Meesenburg et al., 1977: 23). The development of the Skallingen peninsula during

Figure 247: Dune erosion at Benknolde, Skallingen peninsula (1980).

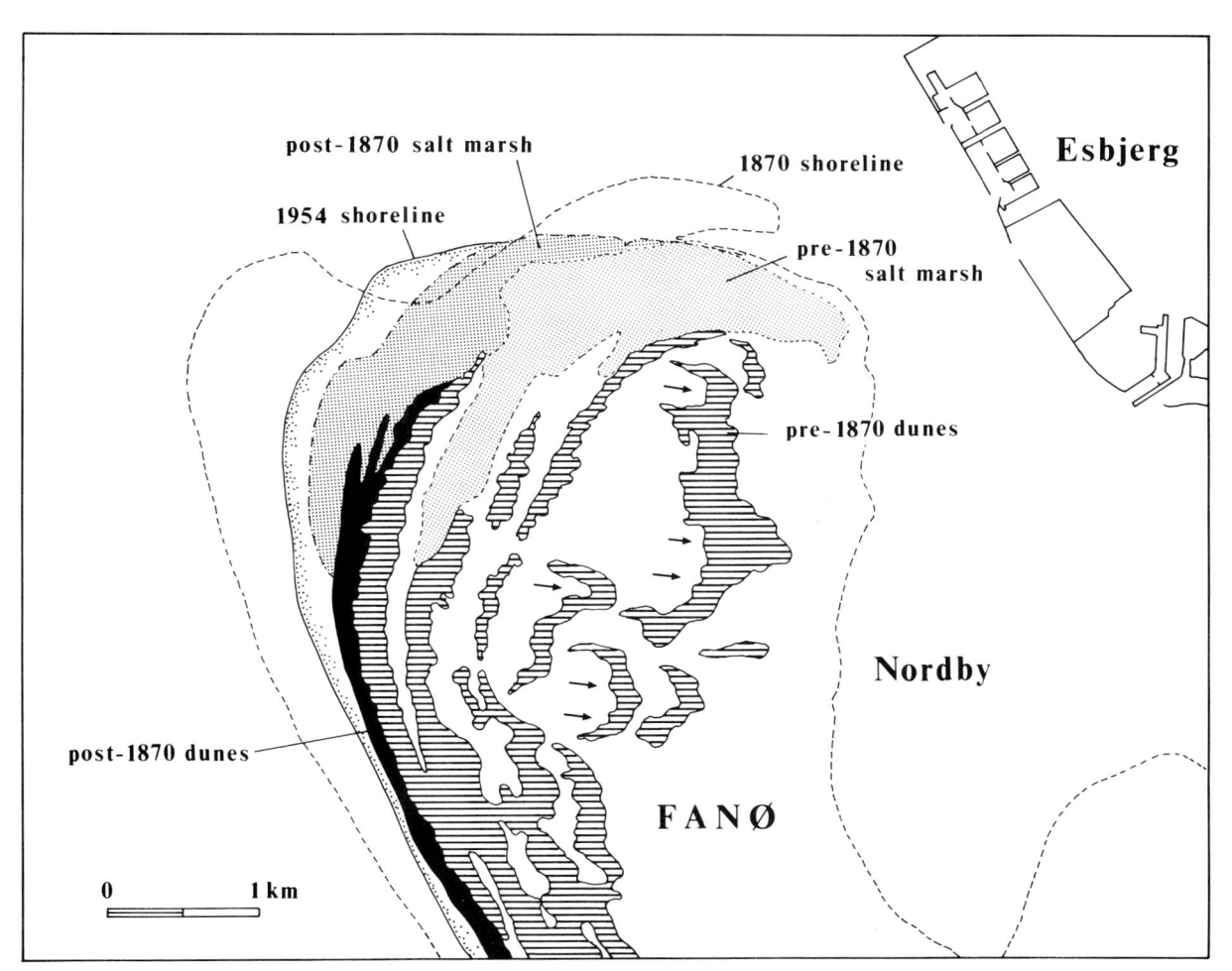

Figure 248: Recent accretion at the northern end of Fanø. Sources: Map 1:40,000 of the Geodætisk Institut, Sheet Esbjerg (1873) (in: Meesenburg, et al., 1977: 34) and Map 1:25,000 of the Geodætisk Institut, Sheets 1113 III SV Grådyb (1969) and 1113 III SØ Esbjerg (1967).

Figure 249: Active erosion of beach ridge on northern Fanø; the grass cover is stripped and the sand washed further inland (1980).

Figure 250: Tidal-flat facing beach ridge on Fanø, looking towards Nordby (1980).

the last 300−400 years effectively reduced the catchment area of Grådyb Inlet, which shifted northwards as a result; cf. Jepsen (1978: 15). Hobo Dyb Bay started to silt up, and the delta of the Varde Å to extend further seawards, leading to a further decrease in the size of the Ho Bight tidal basin.

The coast-parallel component of the sand transport is directed southwards off Fanø (Bruun, 1978: 399). At the northern, updrift end of the island this leads to the formation of new beach ridges seawards of the old ones and, protected by these,

salt marshes develop. The first part of this area, called the 'Grønning', was farmed as early as 1745 (Meesenburg, *et al.*, 1977: 23), the adjoining dunes being fixed. Seawards of the old dune core two new broad and high dune ridges had formed by 1875. Additional land of more than 1 km in width, subdivided by two broad beach ridges, was in existence here until recently (fig. 248).

The outer beach ridge is experiencing severe erosion on the northwestern flank (fig. 249). Sand eroded from the ridge is redeposited as small washover fans on the landward side. Less active forms of beach ridge migration are also found on the tidal-flat facing coast of the island (fig. 250).

South of this area accretion replaces erosion. Sand progradation occurs seaward of the beach ridge. Reduction in wave energy by shoals of the Grådyb ebb tidal delta permits new salt marsh growth along northwestern island shore. Søren Jessens Sand has approached the island in recent years; in 1960 it was still separated from Fanø Bad by the 4 m deep channel of the 'Hamborg Dyb' (Meesenburg et al., 1977: 21) but in 1980 a connection developed leading to an increase in the beach width from 100 to the present 300 m (plate 8).

The Søren Jessens Sand is only flooded by the highest tides. As it migrates further landwards, the sand moves onto the sea-facing salt marsh off Fanø Bad, smothering the young vegetation.

The summer house colonies

The coast-parallel dune ridges south of Fanø Bad at Rindby are occupied by summer houses for a distance of three kilometres. The building of summer houses on a large scale started at its earliest in 1953. In the period between 1933-35 only a few basic weekend cottages were constructed but the building boom of 1957-61 resulted in a considerable increase in the number of buildings. Building continues at a somewhat slower rate up to the present day (plate 9). In 1977 there were 2,344 summer houses on Fanø with an additional 800–1,000 in the planning stage (Meesenburg et al., 1977: 47).

The summer house area between the dune ridges consists of moist deflation depressions, which are submerged at least seasonally as a result of the high groundwater table. The drainage of these areas was already a major problem when agriculture was important on the island; for example drainage ditches over 2 km in length had to be excavated, from Sandflod Hede to the Wadden Sea (cf. topographical map).

The southern end

The southern end of Fanø abuts directly against the Galgedyb, a small tidal inlet north of the major Knudedyb inlet between Fanø and Mandø. Galgedyb and Knudedyb are separated by the Keldsand and Peter Meyers Sand shoals. The main channel of the Galgedyb tends east–west, but east of the southern spit of Fanø it turns abruptly north. This northern inlet terminus threatened to undercut the easternmost houses of Sønderho at the end of the 19th century. By about 1900 a new sand spit started to form at the southern end of Fanø, covered by sand dunes. Spit growth shifted the channel of Galgedyb Inlet further away from the Sønderho settlement.

Aerial photographs indicate that recent developments have reversed this trend (fig. 251). In 1945 (fig. 251 a) parts of the Galgedyb flood delta were still situated west of the tidal creek, appearing as light, dry sand flats in the photograph. About twenty years later (fig. 251 b) this western part of the flood delta was again reduced. By 1976 (fig. 251 c) the flood delta was restricted to the area east of the Galgedyb, whilst parts of the Hønen peninsula and the young salt marsh in its lee have been eroded. Erosion still continues (fig. 251 d), and the inlet is threatening the easternmost houses in Sønderho again. However, the width of Galgedyb Inlet steadily decreased since 1945. The northern end has been considerably curtailed, and the morphological changes southeast of the Hønen peninsula indicate a progressive reduction in its discharge. Galgedyb Inlet will completely disappear in a short time, and Keldsand shoal will soon weld itself onto southern Fanø.

Under the protection of the ebb delta of the Galgedyb–Knudedyb inlets the positive sand balance has led to the formation of a new, several metres high coast-parallel dune ridge, separating a 200 m wetland area from the beach. Map comparison indicates that this dune and marsh must have formed after 1870.

Young dunes are actively forming seaward of the dune ridge in a 5 - 10 m broad strip on the beach, but there are also indications of major erosion during recent years. The Second World War bunkers now lie about 10 m seawards of the dunes, demonstrating a retreat of the dune foot of about 15 m since 1944.

The large migrating dunes

The main coast-parallel dune ridge has a height of about 13 m. Another, much larger, dune ridge is found about 1 km further inland (height: up to 23 m). These, the so-called Havside Bjerge, are former migrating dunes dissected by numerous blowouts (plate 10).

The Havside Bjerge dune field has not changed position significantly since the first large-scale topographical survey in 1870. Before the formation of the Havside Bjerge, two older generations of migrating dunes had already started to cross the deflation lowlands of southern Fanø: the Vindgab Bjerge preceded by the Gammeltoft Bjerge dunes.

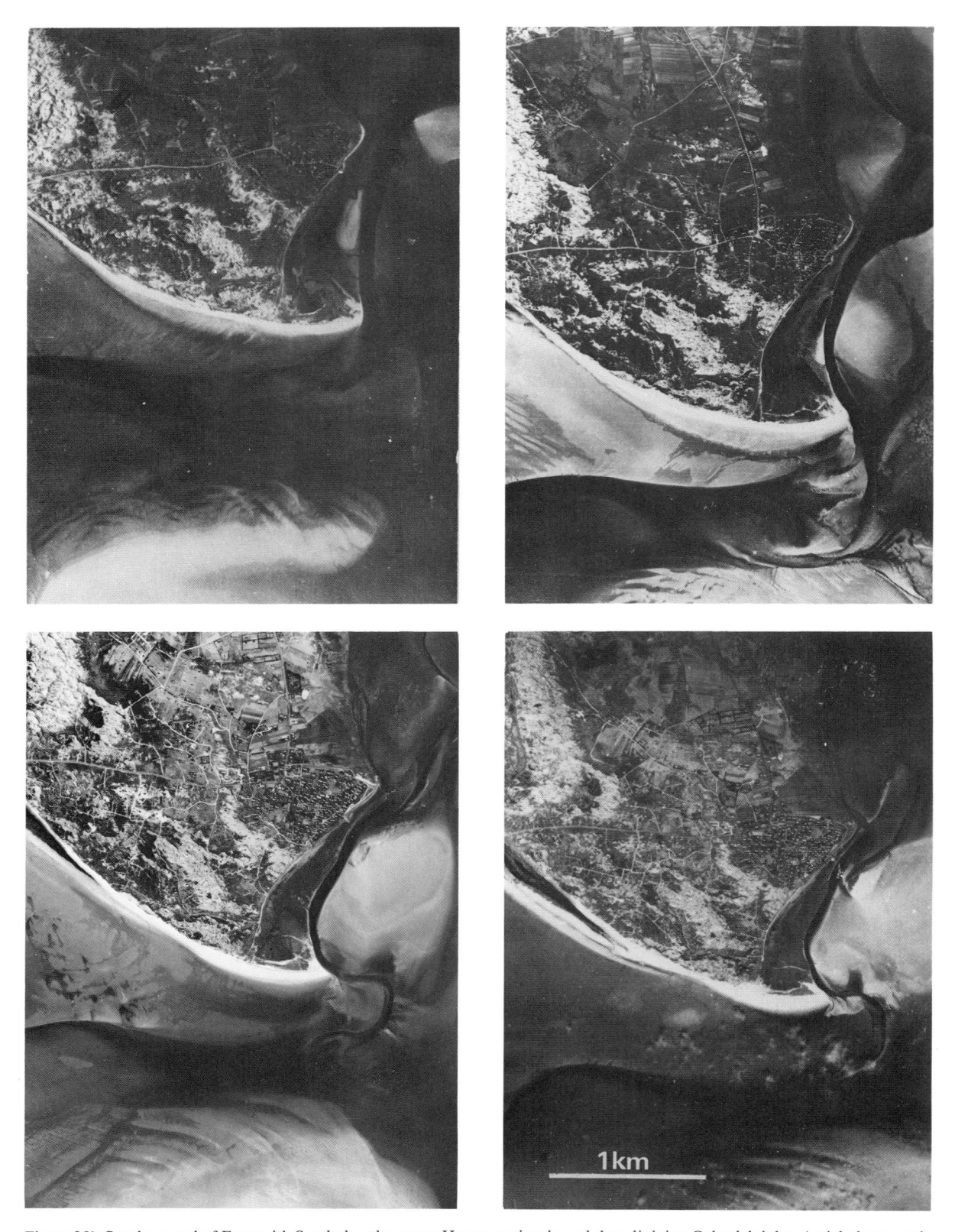

Figure 251: Southern end of Fanø with Sønderho, the recent Hønen peninsula and the adjoining Galgedyb inlet. Aerial photographs of the Geodætisk Institut, Denmark; a: 16.7.1945, E 44 Nr. 5002; b: 1.9.1964, D 339-E Nr. 300; c: 19.4.1976, D 7609-B Nr. 358; d: 24.4.1984, D 8403-B Nr. 184, reproduced by permission (A. 532/85) of the Institute.

Today they are situated on both sides of the Sønderho–Nordby road, about 2 km north of Sønderho. The greater age of these dune ridges is indicated by the degree of dissection by blowouts. Former migrating dunes are also found in the forests of central Fanø.

Dune migration on a large scale started two hundred years ago, because of land use practices. The sheep were left to graze on the dunes during the summer whilst the hay for winter was grown on the salt marshes; even the dune grass was itself harvested occasionally. The natural consequence of this land-use was that the sand started to shift. Names such as Sandflodhede (Sand flood heath) south of Rindby stem from this period.

Coastal erosion control

The overall positive sand budget of Fanø has obviated the need for protective structures such as groynes or revetments. Only Sonderhø town itself is protected, by a 3 m high dyke on its east and north sides. This relatively low dyke withstood even the strong storm surge of 24.11.1981.

Nordby is not protected by dykes. The old centre of the settlement was sited safely on a former migrating dune about 5 m high. The more recent parts of the town, however, have extended onto lower-lying land. During the 1981 storm surge water rose to an extremely high level and inundated the settlement including an area on the landward side of the old dune ridge. About a third of the houses in Nordby were flooded (Meesenburg & Tougaard, 1982: 26).

7.1.3 MANDØ

The Isle of Mandø (formerly Manø) is located over 5 km east of the seaward margin of the tidal flats. Although not actually a part of the island chain, Mandø has a dune belt on its western side. In the southern part of the double island, Ny Mandø, the dune belt is about 200 m broad and 12 m high, protecting the village of Mandø By (figs. 252 and 253). The dune belt on the northern part, Gammel Mandø, is less than 100 m broad and only 6 m high. The rest of the island consists of marshes. Mandø therefore closely resembles the Halligen and marsh islands between Föhr and Eiderstedt.

The tidal flat area between Mandø and the open sea, over 2 km in width, gives the island good protection against storm surges and coastal erosion. It is not surprising therefore that since the publica-

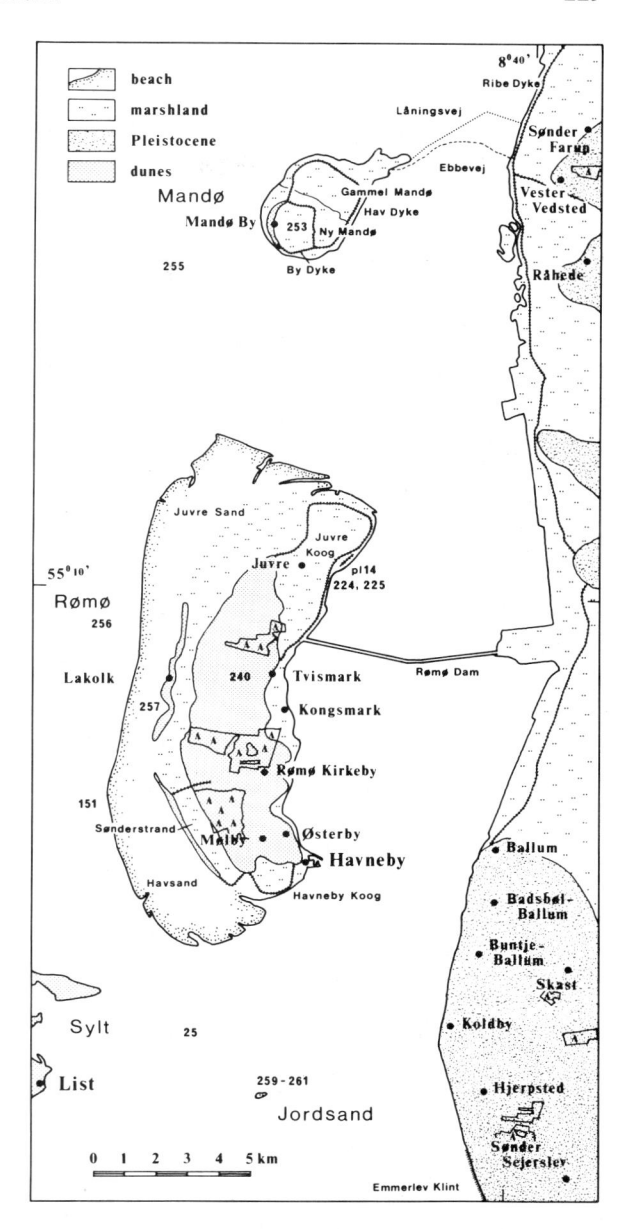

Figure 252: Location map of Mandø, Rømø and Jordsand. The numbers refer to figures, those prefixed by 'pl' refer to plates.

tion of the first accurate maps in 1794 there have been no records of any significant land losses. Today there is some minor erosion at the southwestern and southeastern ends of the island, which, since the storm flood of 1981, has been countered by the construction of brushwood groynes (fig. 253). In the northeast, in contrast, accretion is rapid.

The displacement of the island through the centuries resembles that of the Halligen. Jacobsen (1953: 143) proved the existence of a 200–500 m

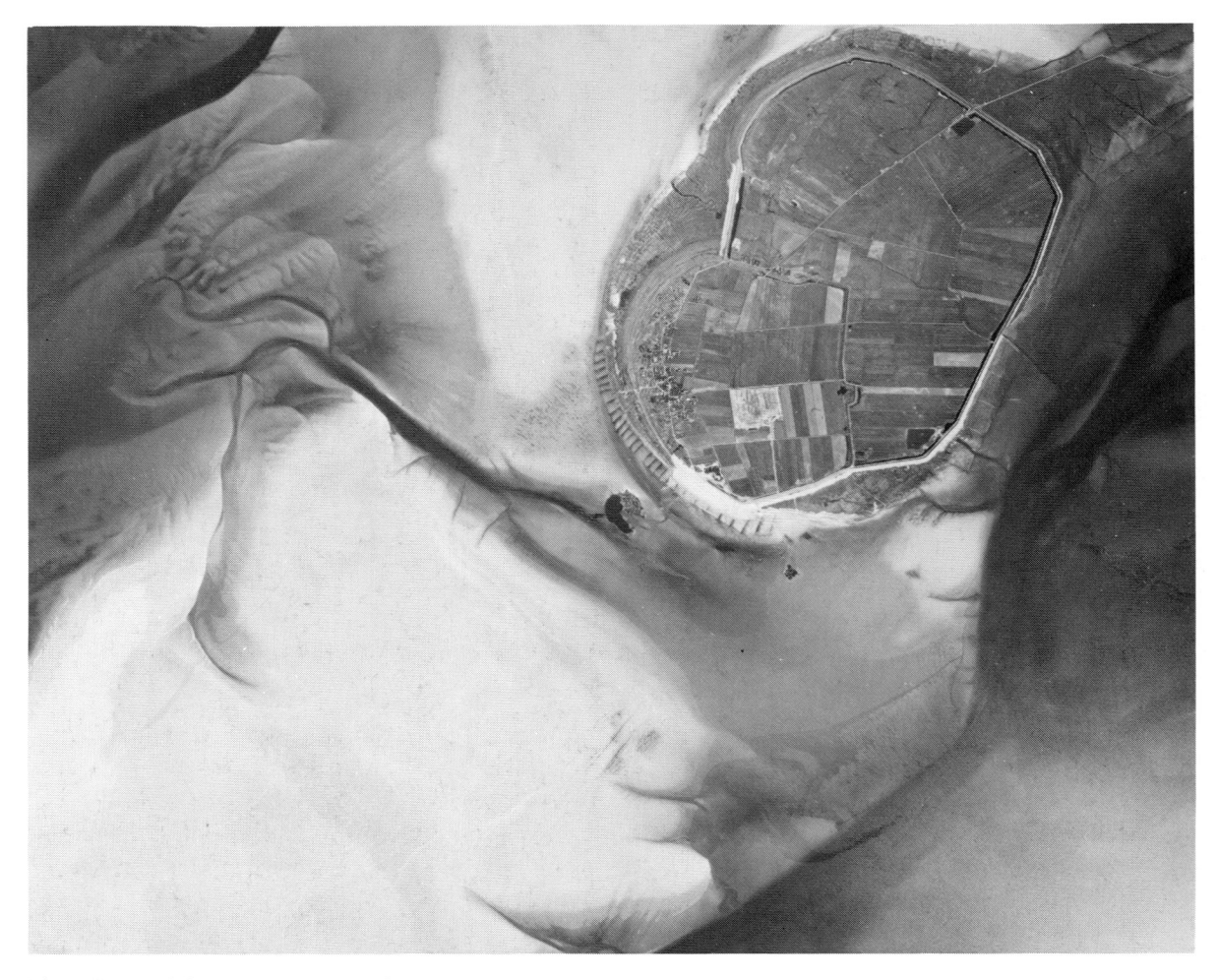

Figure 253: Aerial photograph of Mandø. Photograph of the Geodætisk Institut, Denmark, of 25.4.1984; Flight 8403 C Nr. 033, reproduced by permission (A.532/85) of the Institute.

broad belt of old marine clay, a part of the old island core, by drilling beneath Gammel Mandø in the northwest. This clay layer adjoins the tidal flat sands in the east which, like the old island core, have later been covered by more recent marine clay.

The barrier of the outer sands

The available topographical maps give the impression that there is a broad gap of over 14 km in the barrier island chain between the southern end of Fanø and the northern end of Rømø. Charts and satellite images (fig. 254) both demonstrate, however, that this is not the case. The outer Peter Meyers Sand (south of Fanø) and Koresand (off Mandø) represent the missing links in the barrier chain.

Peter Meyers Sand is separated from Fanø by the Galgedyb. This sand body and the Koresand both closely resemble the outer sands off the North Frisian Halligen islands. In both areas the sands are not inundated during ordinary high tides, but none of them have dunes. On the satellite image the dry sand flats appear in a light tone, contrasting well with the adjoining tidal flats.

No published data are available concerning the displacement of the Danish outer sands, therefore the map published by Moritz (1903) has been compared with the 1978 1:100,000 map of the Geodætisk Institut (fig. 255). These maps show great differences between the tidal flat areas but only a limited evaluation is possible because their land/sea boundaries differ. An interpretation of the steep sea-facing slopes of the outer sands, however, appears to be straightforward.

Fig. 255 demonstrates that the seaward margin of the Koresand was displaced over 1.5 km land-

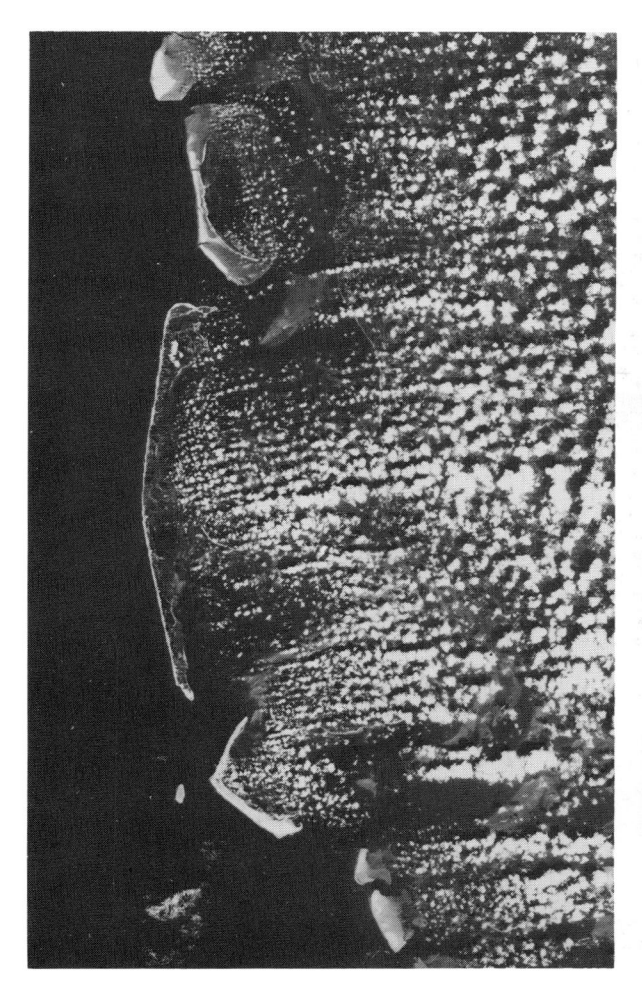

Figure 254: Satellite image of the Danish and North Frisian Wadden Sea coast between Mandø and Süderoogsand. The position of the Koresand in the north is comparable to that of the North Frisian Außensände (Outer Sands) south of Amrum. Landsat 5 TM image of 16.4.1984, Track 197, Frame 022, Q 2, processed by ESA at Fucino. Left: Band 1; right: Band 5.

wards during the period in question. Peter Meyers Sand also shows a retreat of about 1 km. This suggests migration rates of about 15–20 m per year implying that the outer sands are much more mobile than the islands of the Danish Wadden Sea. These migration rates are of the same order of magnitude as those of the North Frisian outer sands (Taubert, 1982).

Historical development

Mandø was presumably once considerably larger. After a severe storm surge in the 16th century, probably in 1558, the original settlement on the northern island (Gammel Mandø) had to be abandoned, leading to the foundation of Mandø By on a more protected site on Ny Mandø. The new village already appears on a map produced by Johannes Meyer in 1643 (Zenius, 1983). The size of Mandø is given as 4 km long and 2.5 km wide in Danckwerth's topography of 1652 (Jacobsen, 1953).

The main source of income was fishing until about 1600 when it was replaced by seafaring. Agriculture only became important in the 19th century, but has continued to be so (Jacobsen, 1953: 144). A weak summer dyke, the so-called Tofte Dyke, was built before 1797, but was breached during the storm flood of 1881. In 1887 arable land was increased to 118 ha by the construction of the new By Dyke. By 1870 the tidal creek which had orginally separated Gammel Mandø from Ny Mandø had largely silted up. In 1928 the two islands were connected by a road dam, and in 1937 they were finally dyked together by the construction of the Hav Dyke to form one

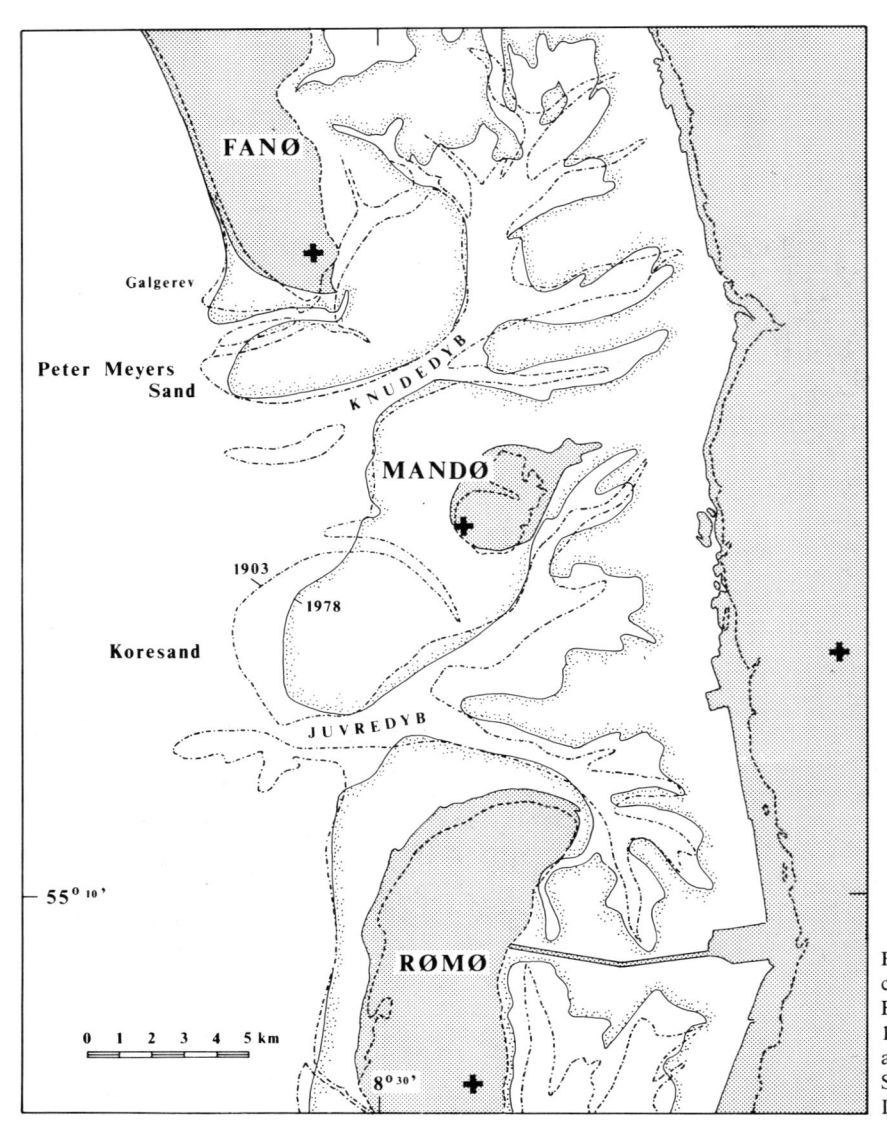

Figure 255: Map showing the changes of the outer sands between Fanø and Rømø between 1903 and 1978. Sources: Moritz (1903, map 2) and Topographical Map 1:100,000 Sheet 1112 Ribe of the Geodætisk Institut, Denmark (1978).

large polder (Zenius, 1983).

Tourism is still only of limited significance on Mandø. There are no hotels and only 36 summer houses on the island (Jacobsen, 1977: 56). In the long run, however, tourism promises most for the future. One of the main disadvantages is the long connection with the mainland (shortest distance: about 4.5 km). Following the construction of the Ribe Dyke on the mainland in 1914, access was maintained via the so-called 'Ebbevej' (Ebb Road) across the sandy tidal flats. In 1978 this road was supplemented by another road, 70 cm higher, the so-called 'Låningsvej', which makes the connection with the mainland largely independent of the tides (Zenius, 1983). The construction of this dam was accompanied by some controversial land reclamation measures (Jacobsen, 1978: 405) which began in the lee of the island in 1962. Since 1938 a salt marsh over 2 km wide has developed there.

Whilst the generally positive sand balance of the Danish North Sea islands has led to the formation of broad beaches and high dunes on the western coasts, the eastern sides of the islands are actually more vulnerable to storm surges. Mandø took a se-

Plate 18: Satellite image of the Schleswig-Holstein North Sea coast; Landsat-5 TM image of 25.4.1984, Track 196, Frame 022 (DFVLR).

PLATE 18

PLATES 19-21

vere battering during the storm surge of 24.11.1981 when both the Hav Dyke and the By Dyke were breached in several places. The largest breach in the Hav Dyke was 250 m wide and 15 m deep. The water rose to a level of + 3.4 m in the By Koog and in the outer polder to as much as + 4.4 m (Meesenburg & Tougaard, 1982: 16).

7.1.4 RØMØ

Rømø, like Fanø, has no Pleistocene core and is composed solely of sand. In contrast with the neighbouring island of Sylt, no prehistoric finds have been reported from Rømø. It must therefore be concluded that the island in its present form is of recent origin. The island can be subdivided into three major morphodynamic zones (fig. 256); in the east an 'old' dune core with a narrow fringe of marshes, in the centre a recent marsh 0.5 – 1.5 km wide and subdivided by beach ridges, and to the west the over 1 km wide sand beach of Havsand and Juvre Sand.

The positive sand budget

The present form of Rømø is determined by a strongly positive sediment budget. It resembles Amrum with its abundant supply of sand. In both cases this results at least partially from the erosion of the Isle of Sylt. The actual transport pattern, however, is far more complex and not yet sufficiently investigated. Aerial photographs reveal that in the southern part of the beach on Rømø sand transport is directed southwards, into the Lister Tief, whereas in the northern adjoining parts of the beach the sand migrates northwards, towards the Juvre Dyb. This migration pattern causes both northeastward and southeastward directed accretion, whilst the west coast experiences slow retreat (fig. 256).

In the past the sediment budget of Rømø has not always been so positive. On a barrier island accretion and erosion are always acting simultaneously, with either the first or the latter prevailing. The relict cliff at the western end of the dune core reveals that prior to the positive sand supply the seaward margin of the island was originally situated much further to the east. Havsand and Juvre Sand were initially only local shoals at the updrift and downdrift ends of the ebb deltas of the Lister Tief and Juvre Dyb inlets. Until the early 19th century the Havsand was separated from Rømø by Havsands Lo channel; it became attached to the island by about 1850 at the earliest (Sørensen, 1982: 52).

Despite the fact that the Havsand is only inundated during especially high tides (e.g. spring tides), there is almost no dune formation on the sand plain, which reaches a maximum width of about 4 km. Only small localized areas of primary dunes have formed (fig. 257).

Where the processes are undisturbed by human interference, the sand blown across Havsand and Juvre Sand finally accumulates in low ridges on the margins of the vegetated interior part of the island. In favourable conditions these ridges may even develop into dunes. The entire recent marsh area in the west consists of a system of sub-parallel sand ridges, separated from each other by moist depressions. The patterns of deposition are similar to those found on the Eiderstedt peninsula at St.Peter-Ording (cf. section 5.3.5).

The dykes

During the 19th century an over 1.5 km wide area of salt marsh had developed on the bank of the Havsands Lo. This area, protected by the high Havsand in the southwest, was used as pasture. However, when the Havsands Lo started to infill around 1850-65, the marsh was threatened by drifting aeolian sand from the Havsand. A brushwood fence was therefore erected in 1867 in order to halt the sand. Within a few decades several dune ridges 6 – 7 m high had developed along this fence, forming a natural dyke. The construction of additional dykes on the flanks enabled the Sønderstrand marsh to be completely poldered in 1926.

The other dykes on Rømø are also relatively recent. The small Havneby Koog was dyked in 1912, only to be lost during a storm flood in 1916. It was re-dyked in 1926, when a sluice for the combined drainage of Sønderstrand and Havneby Koog was also installed (Meesenburg, 1978: 11). The Juvre Koog was dyked in 1928.

Plate 19: Rotes Kliff, Sylt; Westerland in the background. The cliff retreats at an average rate of about 0.7 m/year (1982).
Plate 20: Beach steps at Wenningstedt (cf. fig. 213). The widening gap between the steps and the cliff face illustrates continuing coastal retreat (1984).
Plate 21: House south of the beach steps in Wenningstedt, now only 5 m from the cliff (1984).

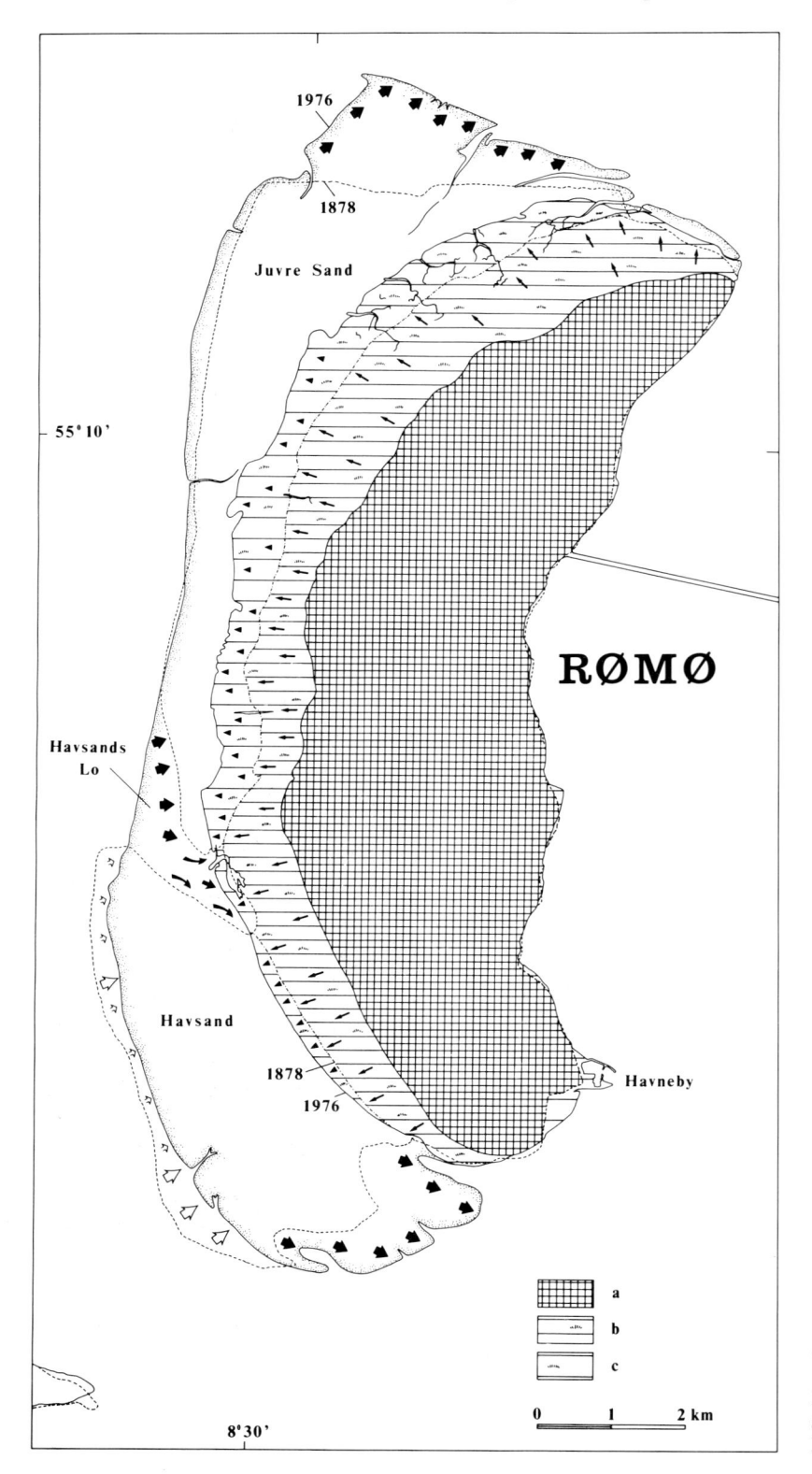

Figure 256: Morphodynamic subdivision of Rømø; a = old island core; b = recent beach ridge and dune areas, prior to 1878; c = recent beach ridge and dune areas, after 1878. Black arrows: accretion; white arrows: erosion. Sources: First edition of the Topographical Map 1:25,000, Sheets Haff-Sand (1880), Kirkeby (1880), List (1880), Jerpstedt (1880) and Specialkort over Rømø 1:25,000 (1978).

Figure 257: Primary dunes on the Havsand at Lakolk (1984).

Figure 258: Coastal erosion on the east coast of Rømø, east of Østerby (1982).

The dunes

The island is well protected against floods from the west by a broad dune belt. The highest dunes on Rømø are 17 m in height, but in contrast to Fanø and Sylt, only a very few well-developed parabolic dunes are present (cf. Jacobsen, 1977 b: map 1). This may be because the efforts to curb the shifting sand were more intense here than on the neighbouring islands. Sand movement was most active during the 'Golden Era' of whaling (ca. 1680–1740) when the population of Rømø reached its maximum. At that time 1800 people lived on the island, in contrast to the 786 inhabitants of today (1977). Intensive land use and grazing of the dunes caused serious damage to the vegetation cover and promoted blowouts (Sørensen, 1982).

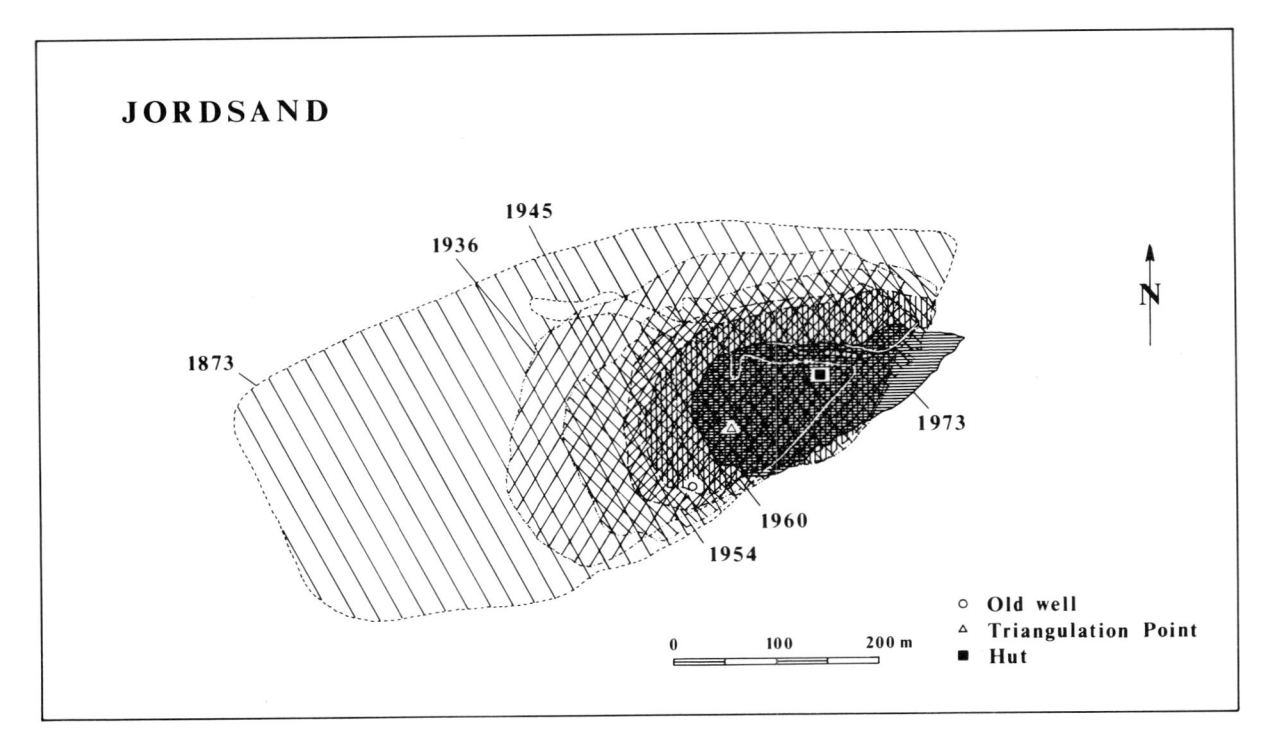

Figure 259: The displacement of Jordsand. Source: Jepsen, 1977.

Rømø's east coast

The east coast of the island experiences only minor change. Erosion is dominant at present (fig. 258), and the construction of the Rømø Dam only prompted deposition in a narrow area close to its termination. It has even led to increased erosion in northern Rømø, and the dyke of the Juvre Koog has had to be abandoned and rebuilt further inland (plate 14; cf. section 5.3.7).

Tourism

The foundation of the resort at Lakolk in 1898 marked the start of tourism on Rømø, but only on a limited scale, mainly because of poor infrastructure and accessibility. Tourism only became important after the opening of the Rømø Dam in 1949. Like Fanø, large parts of the island are now occupied by summer houses. Jacobsen (1977 a: 56) reports 880 summer houses, 90 % around Kongsmark and Kirkeby. There are three camping sites, the largest close to the beach at Lakolk. Vacant land on Rømø has become increasingly scarce, especially as northwestern Rømø is used as a training area for the Danish Air Force and is closed to tourists.

The percentage of day visitors is especially high on Rømø because of the causeway road. In addition, there is considerable traffic of visitors who stay on Rømø for only a few hours and then take the cheap ferry crossing from Havneby to List (Sylt). In contrast to the Netherlands and Germany, in Denmark it is possible to take cars onto the beach. Large areas of the Havsand therefore become giant car parks during weekends.

7.1.5 JORDSAND

The Isle of Jordsand, Denmark's only undyked marsh island in the Wadden Sea, is at present uninhabited and used as a bird sanctuary. It is situated on the Jordsands Flak tidal flats between the Danish mainland coast at Hjerpsted and the Isle of Sylt. After the implementation of coastal protection schemes on Griend in the 1950's and on the German Halligen, Jordsand became the only place where the development of an undyked marsh island in the Wadden Sea could be observed.

Historical development

Historically, Jordsand was considerably larger

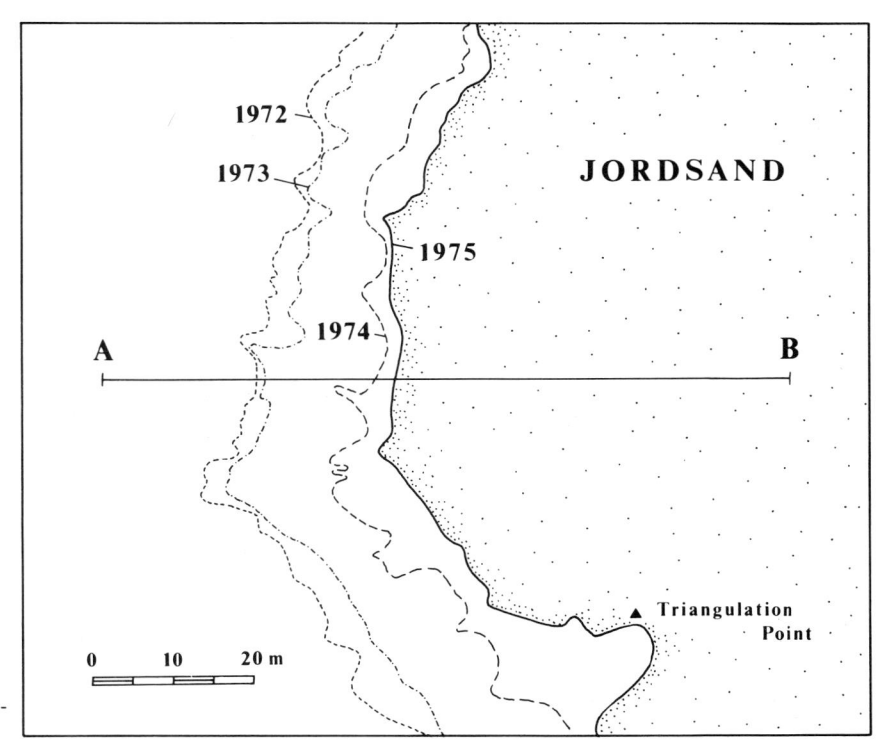

Figure 260: Erosion of the western shore of Jordsand between 1972 and 1975. Source: Jespersen & Rasmussen, 1976.

than at present. A map of 1695 shows two houses which were inhabited during the summer. The first accurate cartographic survey in 1805 represents Jordsand as an island of 43.2 ha, but by 1841 it had already diminished to 35.6 ha. The final house and its dwelling mound were destroyed by the storm surge of 7./8.12.1895. A bird protection scheme started in 1907, and the remaining pasture on the island was abandoned in 1923 after several sheep and horses were drowned by flooding during a summer storm (Jespersen & Rasmussen, 1976; Jepsen, 1977).

Recent morphodynamics

Like the Halligen, the Isle of Jordsand consists of a salt marsh area dissected by minor tidal creeks. The island is surrounded by a sandy beach ridge several decimetres in height indicating effective erosion. Erosion is most active in the west, some of the sediment being redeposited on the beach ridge, and some being washed in sand tongues around the most exposed western end of the island. This overall eastward movement of sand, supported by vegetational effects, leads to gradual accretion in the east, but this is unable to maintain an equilibrium with the erosion (fig. 259).

The basic morphological relationships of the island remain unchanged during this process, exemplified by the vegetational map drawn by Edelberg in 1944 (fig. 3 in Jespersen & Rasmussen, 1976: 14). Between 1972 and 1975, however, erosion in the west was so strong that the beach ridge was almost completely eroded. During these three years the coastline retreated 15 m on average (fig. 260; cf. Jespersen & Rasmussen, 1976: 15).

In 1973 drillings revealed that the western part of Jordsand is underlain by an area of old marine clay, over 1,000 m^2 in area, which represents a remnant of the original island (fig. 261). The recent accumulation in the east consists entirely of sandy beach ridge deposits rather than salt marsh as found on the larger Halligen. The original Hallig character of Jordsand disappeared many decades ago (Jespersen & Rasmussen, 1976).

A horseshoe-shaped depression can be found to the west of the island in the adjoining tidal flat. This feature is a characteristic of all islands which are situated on raised tidal flats (e.g. Scharhörn, Trischen, Griend). Hydrodynamically it resembles the current crescents formed by water flow around obstacles. Relatively deep water therefore occurs close to the island enhancing the deep erosion of the old marine clay. Until recently the erosion

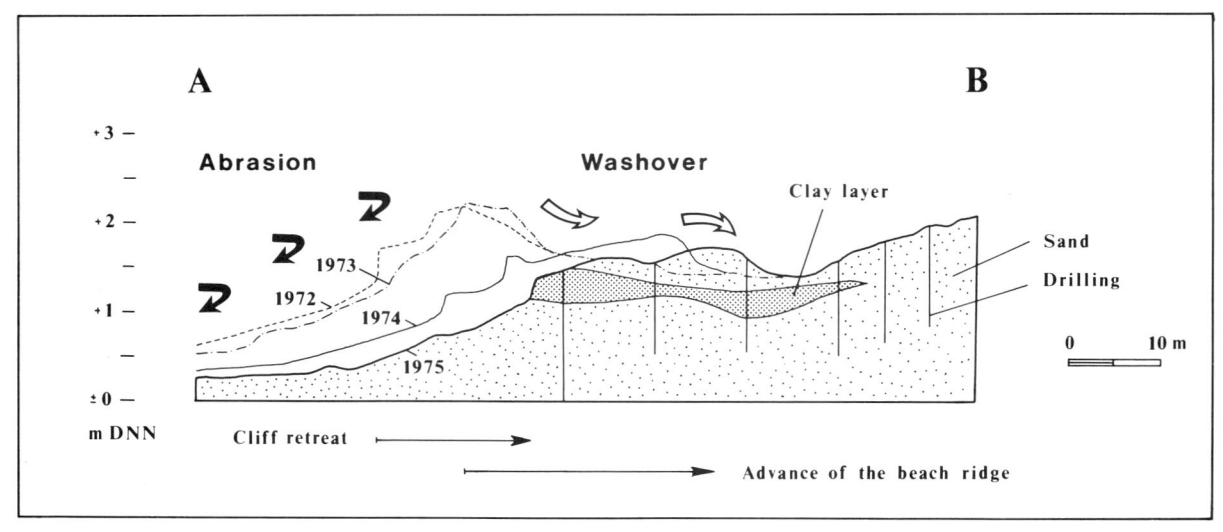

Figure 261: Cross section through the western part of Jordsand illustrating the stages of progressive destruction. DNN = Danish ordnance datum. Source: Jespersen & Rasmussen, 1976.

produced numerous scarp-like steps, several centimetres high, which were clearly visible on large-scale aerial photographs.

Coastal protection

As on all the other marsh islands, the natural development of the coastline on Jordsand has now been halted. The storm flood of 16./17.2.1962 caused 10 m of erosion on the west coast, and the severe storm flood of 3.1.1976 also resulted in marked erosion in the west. The latter left the whole island under a layer of sand, and reduced its size from 2.3 ha to about 1.9 ha. Following private proposals during the summer of 1976, a system of brushwood groynes was constructed based on the pattern of the Norderoog coastal protection scheme (Jepsen, 1977; Jespersen & Rasmussen, 1985). These measures can only be temporary at best; a more lasting solution would require much more fundamental, and expensive, engineering of the natural system.

7.2 SCHLESWIG-HOLSTEIN

7.2.1 SYLT

With a total length of over 40 km, Sylt is the longest of the Wadden Sea barrier islands. Like the neighbouring islands of Föhr and Amrum, Sylt has a Plio-Pleistocene core which forms the centre of the island (fig. 262). From this core spits over 10 km long have developed as a result of coast-parallel sand transport both north- and southwards. The west coast of the Plio-Pleistocene core and the spits are both continuously eroded, most of the sediment being carried away northwards towards Rømø or southwards towards Amrum.

The development of Sylt has always been characterized by progressive erosion in the west. Müller & Fischer (1938: 38) demonstrated that the retreat rate of about 0.86 m/year determined by Fülscher (1905), though measured over a relatively short period, is largely correct. The frequently quoted much higher rates are based partly on exaggerated reports and partly on interpretation of old unreliable maps, such as that of Meyer (1648). Mager (1927) was therefore incorrect in stating that the rate of coastal erosion on Sylt was many times higher than that quoted by Fülscher.

Geological setting

The core of Sylt consists of Pliocene kaolin sand, covered by till. In the Rotes Kliff (red cliff) section between Westerland and Kampen the Older Saalian till is exposed overlying the undisturbed kaolin sands for several kilometres. Thrust features can only be seen in the northern part of the cliff (cf. Felix-Henningsen, 1979); Elsterian sandy till is occasionally exposed between the Saalian till and Pliocene sands in the southern part.

Dunes cover the spits as well as the western part of the Plio-Pleistocene core and a dune belt up to 1 km wide also occurs on top of the Rotes Kliff.

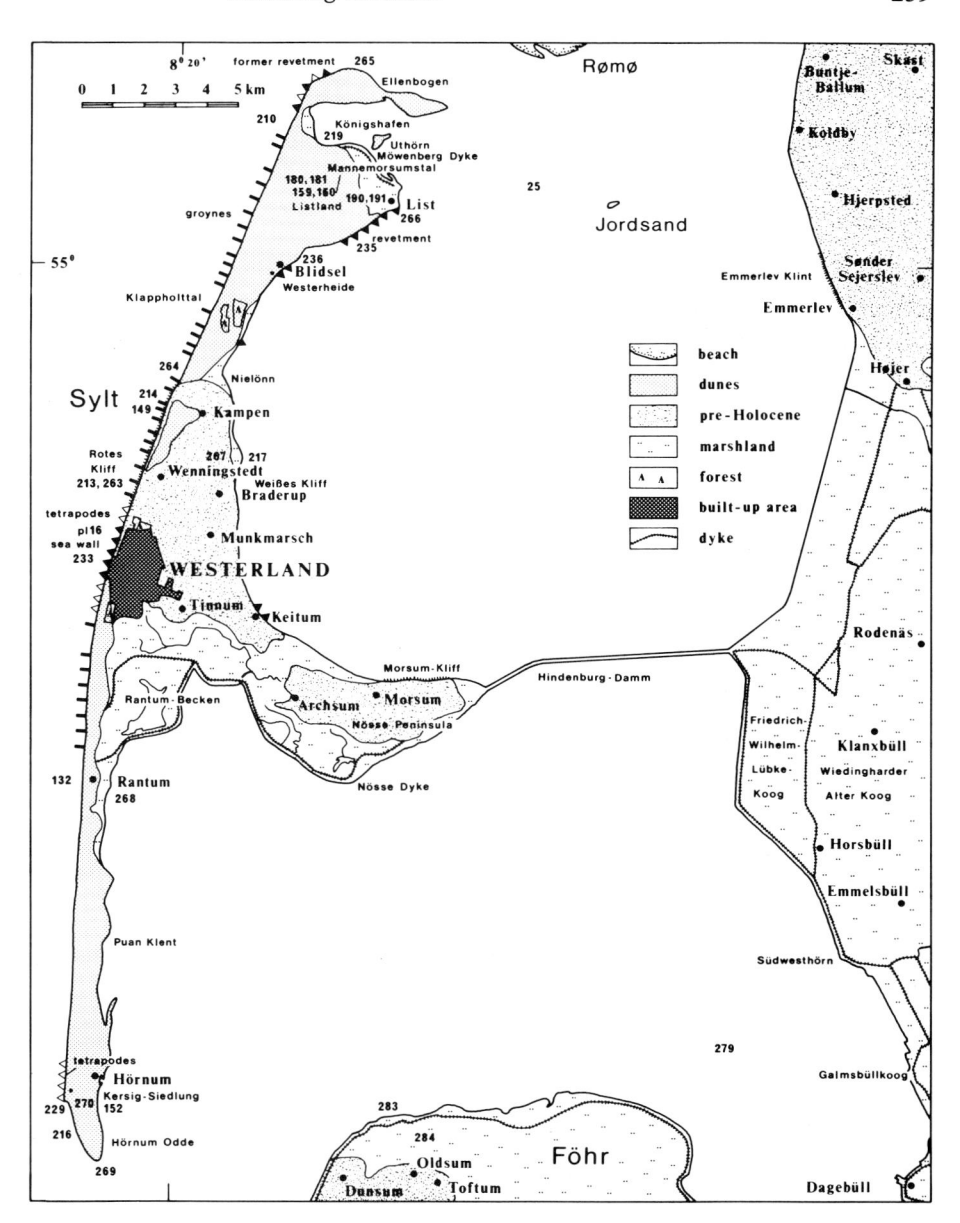

Figure 262: Location map of Sylt. The numbers refer to figures, those prefixed by 'pl' refer to plates.

There has been some discussion on how such dunes could develop on top of a high cliff. Gripp (1967) has demonstrated that no significant recent sand transport has occurred up the cliff face. The dunes therefore must have formed much earlier, that is prior to the formation of the cliff, when there was a gentle slope and broad beach in the west.

The fight against shifting sand

In contrast to all the other islands, a manuscript of an old chronicle of Sylt has survived. It was written during the first half of the 15th century and mentions topographical details. In these 'Silter Antiquitäten' Hans Kielholt reports contemporary extensive dune development. This marks the start of the formation of the 'Young Dunes'. Although no radiometric age determinations have been obtained so far, it seems plausible to correlate the beginning of the 'Little Ice Age' around 1450 with the beginning of Young Dune formation.

In the more fully documented time since the early 17th century, it is not so much coastal erosion which has threatened the human settlements on Sylt but the migration of the unfixed dunes. When

Hansen (1877: 122) described the storm surge of 1634, he also referred to the complete destruction of the existing dykes on the east coast. He did not even mention the considerable coastal erosion in the west, since it did not affect the settlements or cultivated farmland.

In 1634, however, the Eidum church, threatened by shifting sand, was demolished and subsequently rebuilt in the eastern part of Westerland. The church of Rantum suffered the same fate in 1757, and in 1801 it had to be replaced on a second occasion, again because it had become engulfed by dune sand (Müller & Fischer, 1938: 60)

Dune protection on Sylt started remarkably late. The first plantations were only started in 1788. Between the triangulations of 1778/82 and 1834 the arable land in the Westerland municipality had decreased from 103,059 Quadratruthen to 91,949 Quadratruthen, as a result mostly of the advancing migrating dunes. The school and three other houses in Rantum had to be abandoned in 1813 because of the shifting sand, and in 1821 the last house in Old Rantum, which was no longer inhabited, had to be demolished. In the meantime the village had lost a major part of its pasture and arable land to the advancing dunes, and had to be relocated southeast to the eastern end of the former farmland. It was not until 1837-45 that the dunes at Rantum were planted and stabilized (Müller & Fischer, 1938: 102).

Dune protection only became an official concern after Sylt was annexed by Prussia following the German-Danish War of 1864. Count Baudissin was appointed to administer the necessary measures. He introduced modern methods of dune protection in spite of strong resistance from the local population. He attempted to encourage the formation of foredunes in front of the dune cliffs instead of furthering dune migration with the construction of 'wind funnels' as had been done before (Müller & Fischer, 1938: 144). These measures mark the beginning of rigid coastal protection on the island.

Baudissin also introduced a tree nursery on Sylt, and in 1869-93 the Klappholttal dune valley was successfully planted with *Pinus montana*. Experimental plantations of other species such as oak and alder all failed (Müller & Fischer, 1938: 144).

Following the stabilization of the coastal dunes, some of the migrating dunes of the Listland were planted in 1930 in order to prevent them engulfing the settlement of List. The large migrating dune in Hörnum, which also endangered the settlement, was also stabilized in 1930 (Müller & Fischer, 1938: 181).

Coastal protection

The cessation of sand loss on the beach was a basic precondition for the successful construction of foredunes, and a low cost solution to this problem was required because of the considerable length of the shoreline. The first attempts to inhibit longshore sand transport were made in 1867 with light pole groynes. These structures were quickly destroyed by the strong surf, necessitating the construction of the first stone groynes off Westerland in 1872/73. In order to keep the distance between the expensive stone groynes as large as possible, lighter intervening groynes were built. The 22 km long groyne system along the west coast of Sylt was completed in 1899 (Müller & Fischer, 1938: 188). However, these structures were only able to decrease the rate of coastal retreat at most; they were unable to prevent it totally.

Before the foundation of the sea-side resorts on Sylt (Westerland: 1855), the villages of Westerland, Wenningstedt and Kampen were all situated at a safe distance from the coast, about 1 km inland. With the average coastal retreat of the Rotes Kliff being less than 1 m/year, the villages were secure for at least 1,000 years. There was not a single house south of Rantum on the Hörnum Peninsula at that time. To promote tourism, however, hotels and guest houses were built increasingly close to the coast, unnecessarily forcing the state to resort to rigid coastal defences for the protection of newly created buildings.

The Miramar Hotel in Westerland, built in the dunes in 1903, was only 12 m from the dune cliff following the storm surge of March 1906. The alarmed hotel owner ordered a 68 m long concrete revetment to be built in front of his property. In front of the nearby Baur-Breitenfeld House a wooden seawall had been constructed in the meantime, but this was destroyed during the storm surge of 5./6.11.1911 which also caused strong lee-side erosion at both ends of the Miramar seawall. This threatened the neighbouring houses, necessitating the rebuilding and extension of the wall, for 500 m to the north in 1912 at state's expense (Müller & Fischer, 1938: 228).

The subsequent build-up of coastal protection structures on Sylt is a continuous chronicle of reinforcements and piecemeal follow-up measures (cf. Krämer, 1984):

1. A storm surge on February 16th/17th 1916 de-

stroyed the extended seawall off the present site of the sea-water bath for a length of 50 m. The wall was subsequently rebuilt with additional footings.

2. The storm surge of November 15th/16th, 1920 lowered the beach considerably and destroyed the protection at the foot of the seawall.

3. This subsequently caused the collapse of more than 75 m of the wall in January 1921, and an additional 50 m during December of the same year.

4. The wall was rebuilt in 1922 with reinforced footings.

5. The seawall was given a hard-brick facing in 1923 (plate 16) and extended by 90 m to the south and 14.25 m to the north to counter lee-side erosion.

6. In 1935-37 the Nösse Dyke was built along the south coast of the Nösse Peninsula on Eastern Sylt. At the same time, the 2.5 km long Möwenberg Dyke was built northwest of List (height: 5.50 - 5.80 m).

7. The construction of the strong 2.24 km long basalt revetment along the western end of the Ellenbogen Peninsula was started in 1936.

8. The severe storm surges of October 1936 led to the construction of a 1:4 sloping basalt revetment north of the seawall at Westerland. The work was undertaken between 1936 and 1938.

9. The structures were neglected for years during the Second World War, and afterwards required reinforcement. Part of the seawall was destroyed in 1946-48 and subsequently replaced by a revetment of concrete slabs.

10. In 1954 the structures off Westerland were extended 200 m by the construction of a rough-surfaced basalt revetment.

11. The Kronprinz Hotel, built in 1894 behind the dunes, had to be demolished in 1955 because of continuous coastal erosion.

12. The first newly developed flounder groyne off Westerland was completed in 1958 (Zitscher, 1960). However, it subsequently proved to be no more successful than the pre-existing groynes.

13. In 1960/61 tetrapods had to be massed in front of the particularly endangered northern part of the old seawall off Westerland.

14. The storm surge of 17.2.1962 resulted in up to 10 m of dune erosion along the west coast of Sylt, and the coastal dunes were breached at the Kersig-Siedlung and south of Hörnum. Strong lee-side erosion occurred at the southern and northern ends of the Westerland seawall and revetment.

15. Additional tetrapods were installed at Westerland in 1967.

16. A tetrapod wall was built at Hörnum along the dune foot after the storm surge of 1967 which had again caused considerable dune retreat (cf. figs. 229 a and b).

17. In 1969 granite blocks weighing 4 – 6 tons each were installed to improve the protection at the foot of the Westerland seawall. Another 600 m long asphalt revetment was built on the Wadden Sea coast off Keitum.

18. The first beach replenishment was made at Westerland in 1972; the sand was supplied from a source area in the Wadden Sea off Rantum.

19. The revetment off Keitum had to be extended eastwards for another 250 m to the Tipkenhoog.

20. A second beach replenishment had to be undertaken in Westerland in 1978. 1,000,000 m³ of sand were again supplied from the Rantum Bight area.

21. In 1982 the dykes on the Nösse Peninsula had to be reinforced when the Nösse Dyke was almost breached during the storm surge of 24.11.1981. A 1.8 km long stretch of the dyke was raised to a height of 7.20 m, and the dyke foot was widened from 28 to 70 m.

22. Brushwood groynes were built in the Keitum Bight in 1982/83 in order to initiate mud accumulation.

23. For the first time a seaward source was chosen for a sand supply of 630,000 m³ at Hörnum in 1983 (the Theeknob Banks, south of Sylt).

24. Renewed beach replenishments became necessary in 1984/85 off Rantum, Westerland, Wenningstedt and Kampen. However, much of the sand was lost during the autumn storms of 1985.

Between 1962-83 alone about 64,000,000 DM have been invested in coastal protection on Sylt (Zitscher, 1985: 296). As already pointed out by Lamprecht (1957 a, b), massive coastal protection structures are not economic. They have been unable to halt coastal retreat along the west coast of Sylt; at best they have only slightly reduced the rate of retreat. The seawall off Westerland, which represents an unchanged coastline over the last 70 years, has only been preserved by repeated beach replenishments. The foreshore slope, however, has considerably steepened. The Ellenbogen revetment had to be abandoned after the war, and it has now been completely destroyed. The erosion of the Hörnum Peninsula only increased following the implementation of coastal engineering structures. Beach replenishment seems to be the most effective countermeasure against coastal retreat, but it has to be repeated every 6 – 8 years (Zitscher, 1985).

The 'Rotes Kliff'

Sediment movement in the surf zone off Sylt has been the subject of repeated major scientific investigations because of its immense importance for coastal protection (Köster, 1974, 1979; Anwar, 1974; Kirchner, 1974; Führböter *et al.*, 1976; Kachholz, 1982, 1984; Tiniakos & Kachholz, 1984). Wave and surf intensity investigations have been undertaken by Lamprecht (1957 a), Dette (1974) and Führböter (1974) and sand transport measurements conducted with tracers by Reinhard (1974 a, b).

The southern part of the Rotes Kliff is now included in the coastal protection of Westerland. Northward extension of the revetments by a tetrapod wall has stopped cliff retreat at this point, but lowering of the beach in front of the structures, and lee-side erosion causing considerable cliff retreat at the northern end of the tetrapod wall have also resulted.

The Rotes Kliff currently retreats at a rate of approximately 0.7 m/year (cf. also Müller & Fischer, 1938) (fig. 263, plates 19, 20 and 21). Erosion, however, is not a continuous process but occurs episodically. A series of years of inactivity can be followed by years of strong erosion. In 1981 off Wenningstedt, for instance, 9 m were lost in a single storm surge; another 5 m were similarly lost in 1984. Such high magnitude/low frequency events have led to widespread overestimations of the actual erosion rates.

Figure 263: Aerial photograph of the Rotes Kliff at Wenningstedt; the beach steps (fig. 213 and plate 20) and the endangered houses (plate 21) can be seen towards the centre. Aerial photograph of the Fa. Rüpke, Hamburg of 2.11.1984, Bildflug Westküste Sylt (20 3564 3417), Str.3/123, freigegeben durch Luftamt Hamburg unter Nr. 1674/84.

Figure 264: Dune retreat north of the Rotes Kliff as a result of lee-side erosion (1984).

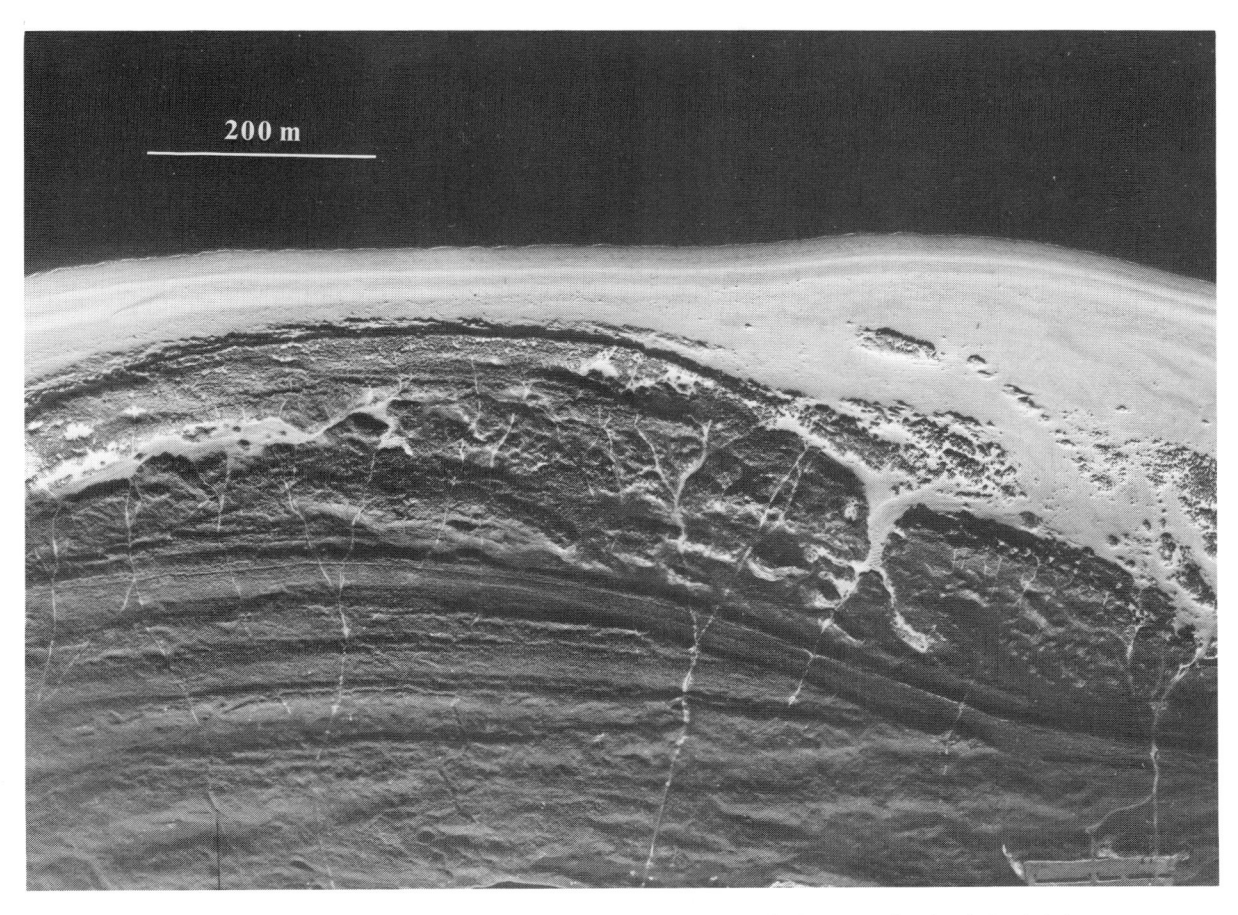

Figure 265: Aerial photograph of recent accretion along the north coast of the Ellenbogen Peninsula Listland, Sylt). Aerial photograph of the Fa. Rüpke, Hamburg of 2.11.1984, Bildflug Westküste Sylt (20 3564 3417), Str.4/195, freigegeben durch Luftamt Hamburg unter Nr. 1674/84.

Dune retreat has been considerable at the northern end of the Rotes Kliff, where the lee-side effect of the relatively resistant Plio-Pleistocene core is experienced (fig. 264). The position of the destroyed bunker north of the Rotes Kliff at Kampen allows a retreat rate of 2 m/year to be calculated for the last 40 years. An old house at the northern end of the cliff is only separated from the beach by a 20 m wide embankment of sand planted with *Ammophila arenaria*.

The northern spit

The spits of Sylt migrate eastwards across the core of the island. Wolff (1910: 64; 1922: 79) reported old salt marsh clay being exposed on the beach at the northern end of the Rotes Kliff after the storm surge of 1909. Jessen (1914: 331) and Bobzin (1926: 144) described similar phenomena at Rantum on the Hörnum Peninsula. Ordemann (1912: 138) was the first to identify this mechanism of eastward spit migration.

The old marsh, however, does not underlie the entire area covered by the spits. Hoffmann (1974) demonstrated that only the northern part of Hörnum Peninsula is underlain by marine clay and peat. The central part consists of dune sand overlying tidal flat sand, whilst the southernmost part comprises spit growth deposits. The greater age of the central parts of the spits relative to their outer ends has also been proven by archaeological investigations (Harck, 1974).

The Listland area on the northern spit of Sylt contains the only remaining active migrating dunes in the whole Wadden Sea area. The largest dune is about 1,300 m long and 400 m wide. The form and genesis of the migrating dunes have been investigated in detail by Priesmeier (1970). He dis-

Figure 266: Damage to bank protection in List, Sylt, on the Wadden Sea side of the island caused by the storm surge of 1981 (1982).

covered that the dunes migrated at an average rate of 150 m in 40 years (3.75 m/year). If one takes the coastal retreat of the west coast of the Listland as 46 m in 40 years (ca. 1 m/year), and assumes that the migration rate has remained constant, then the age of the landward dunes can be estimated to be 600-700 years, whilst the middle migrating dunes west of the Mannemorsumstal valley must have formed around 1650 A.D. (Priesmeier, 1970: 33).

The Listland was declared a nature reserve as early as 1923 but large parts of the area were expropriated for military purposes during the Third Reich. The state had to pay compensation to private land owners for this occupation after the war, and an agreement was reached whereby the landowners were given part of the nature reserve (three areas between the island railway, the road to List and the Blidsel Bight settlement). These areas, the present Blidsel–Westerheide settlement, were soon developed with single family houses. Attempts to restrict further the nature conservancy regulations failed, and in 1969 a new conservancy act became law. This prohibited any further building in the nature reserve. An exception was made as early as 1980, however, for an area of about 5 ha in order to create new car parks (Kruse, 1981: 86).

The northwestern end of the Ellenbogen Peninsula, which had been protected by a strong basalt revetment before the Second World War, is once more experiencing strong erosion. The revetment is now in disrepair. It is no longer maintained because built-up areas are not endangered. The peninsula continues to extend northwards (fig. 265) and may be severed from the rest of Sylt if countermeasures are not undertaken.

The large, shallow Königshafen Bay north of List consists of sandy tidal flats inhabited by *Arenicola*. Despite its relatively protected position in the lee of the Ellenbogen Peninsula and along the east coast of the island, no traces of active salt marsh formation can be found here. Indeed, the southern part of the bay is experiencing strong erosion.

Revetments on the Wadden Sea side

Despite early warnings, many buildings in recent years have been constructed much too close to the coast on the Wadden Sea side of Sylt. This has resulted in a demand for massive coastal protection structures on this side of the island as well.

List is now secured against erosion from the Wadden Sea side by a revetment. The storm surge of 1981 demonstrated that it occasionally has to weather strong attacks (fig. 266). In 1984 the revetment had to be extended to the south (fig. 235). Keitum was given a revetment in 1969. A similar revetment was built on the Wadden Sea coast of Blidsel, as well as along the road from Blidsel to Kampen.

The Blidsel revetment currently shows signs of lee-side erosion at its northern end (fig. 236), and the local population is now demanding a revetment along the whole Blidsel Bight (cf. Andersen, 1984: 8). At present the Blidsel Bight is the only point on Sylt where the former migrating dunes extend eastwards as far as the Wadden Sea side, where they form dune cliffs.

The cliffs of the Wadden Sea side

Between Kampen and Braderup on the Wadden Sea side of Sylt an inactive cliff about 12 m high is separated from the sea by a 300 m wide foreland of young marsh (fig. 267). This foreland, like the

KAMPEN

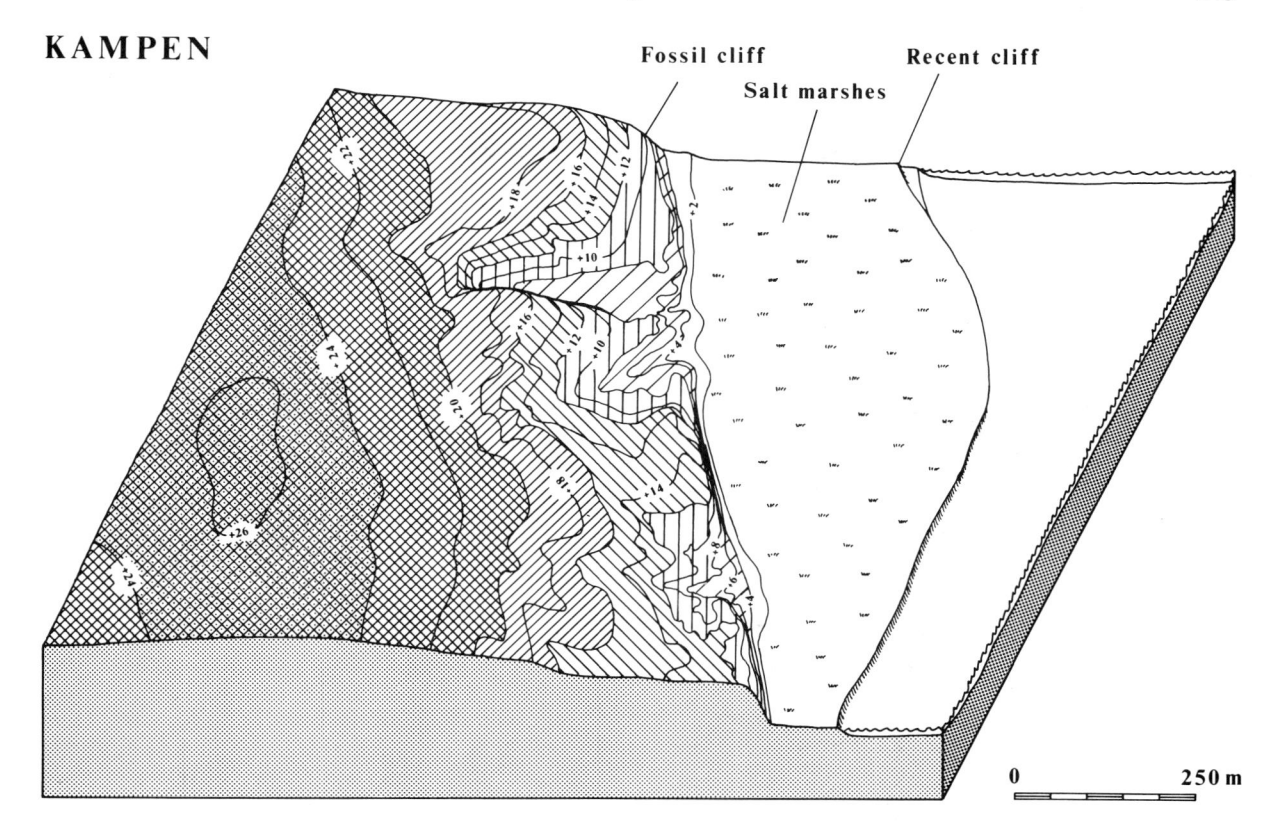

Fossil cliff **Recent cliff**
Salt marshes

Figure 267: Block diagram of the inactive cliff at Kampen. Source: Deutsche Grundkarte 1:5,000, Sheet 5890 Sylt-Kampen-Ost.

other major undyked salt marshes northeast of Kampen (Nielönn) is currently experiencing erosion rather than deposition (fig. 217). The geology of the marsh areas of Sylt has been investigated by Hoffmann (1969, 1975, 1981).

The Wadden Sea coast at Braderup consists of the so-called Weißes Kliff where Pliocene kaolin sand is exposed along the beach. At present the exposures are poor. The kaolin sand is extracted in several large pits near Braderup in the inner parts of the island (cf. von Hacht, 1979).

The spectacular Morsum Kliff at the north coast of the Nösse Peninsula is eroded only during major storm surges. It consists of several sequences of thrust Pliocene kaolin sand, limonite sandstone and mica clay, the thrust zone being overlain by young dunes.

The Hörnum Peninsula

Like the Listland in the north, the Hörnum Peninsula is covered by former migrating dunes, described by Gripp (1968). The west coast of the peninsula is experiencing strong erosion, which is not balanced by any accretion in the east. At Rantum, for example, the coastal dunes are heavily eroded, and the east coast also shows signs of erosion (fig. 268). The brushwood groynes installed earlier have failed to encourage any accretion.

The most vulnerable point is the southernmost end of the Hörnum Peninsula, where the Hörnum settlement is situated. The breach in the coastal dunes, south of Hörnum, during the storm surge of 17.2.1962 led to an initial attempt to close the gap using brush fences. At first this seemed successful (Czock & Wieland, 1965). However, in 1968 a tetrapod wall was constructed in an attempt to stop coastal retreat off Hörnum. The consequences for the area south of the wall have been disastrous.

Extremely strong lee-side erosion immediately caused rapid dune retreat (fig. 269 a). During the storm surges of January 1976 alone, 20 m of dune were lost north and south of the tetrapod wall (Zitscher, Scherenberg & Carow, 1979: 90). Attempts were made in 1983 to close the breaches by sand supplied from the Theeknob Banks, and brush fences were installed to encourage new dune

Figure 268: Coastal retreat on the Wadden Sea side of the Hörnum Peninsula at Rantum, Sylt (1984).

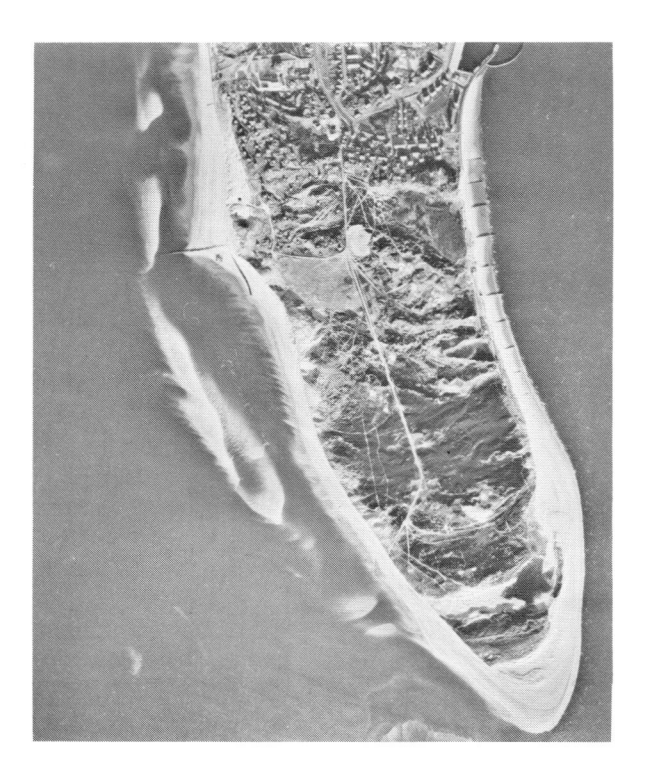

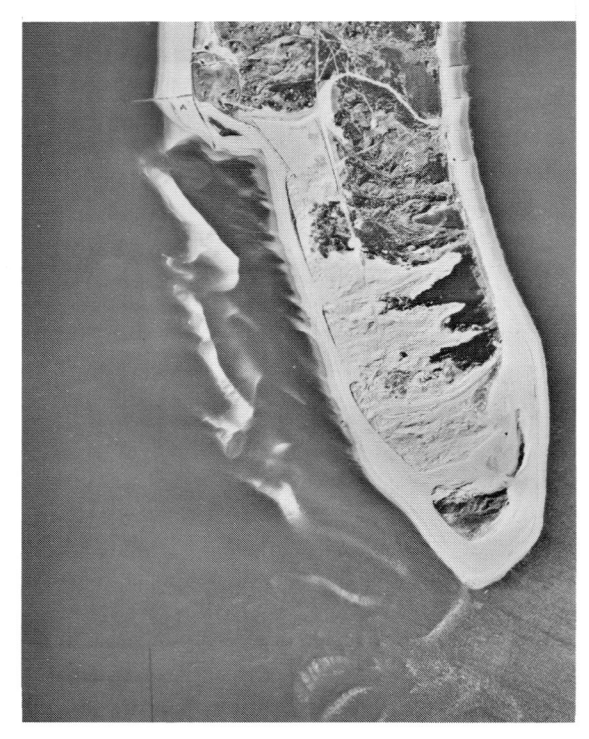

Figure 269: Coastal erosion at Hörnum Odde, Sylt, in the lee of the tetrapod wall; a = 1979, b = 1984. Aerial photographs of 7.12.1978 and 19.3.1984, freigegeben durch das Landesvermessungsamt Schleswig-Holstein unter den Nummern SH 510/79 und SH 510/84, Bildflug Schutzgebiete Westkürte, Sylt, Str. 1/410 und Bildflag Sylt, Str. 5/24, vervielfältigt mit Genehmigung des Landesvermessungsamtes Schleswig-Holstein vom 3.12.1985.

Figure 270: The dune breach south of Hörnum, looking towards the south (1984).

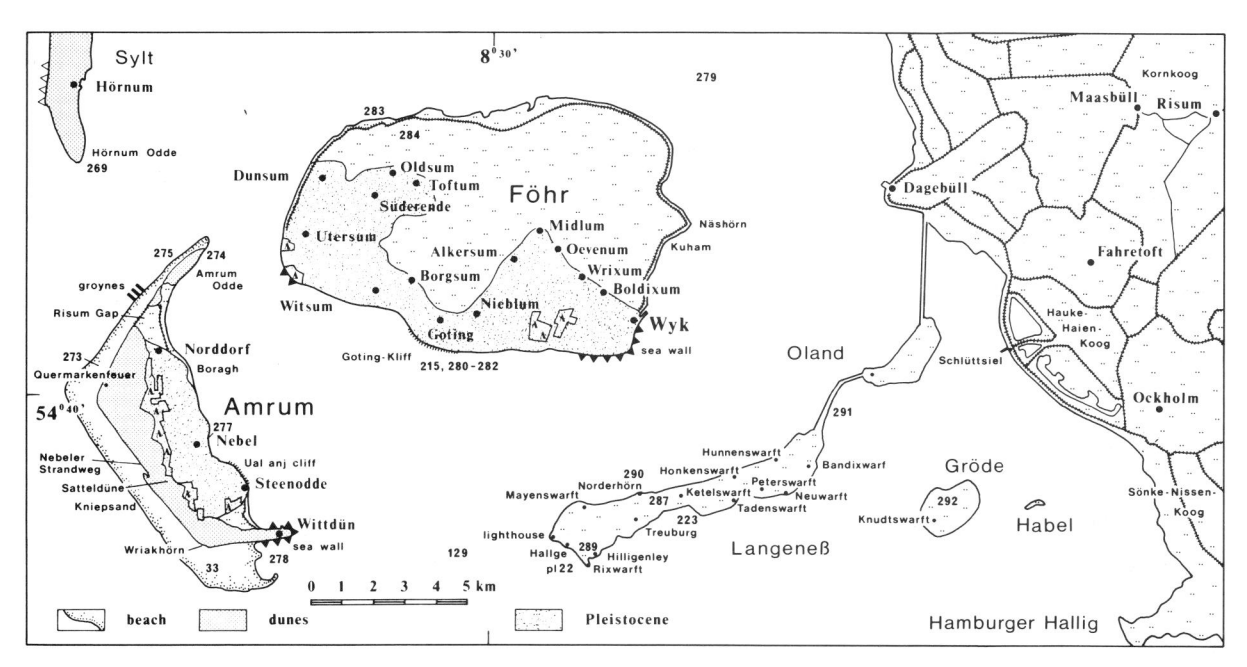

Figure 271: Location map of Amrum, Föhr and the Halligen Langeneß, Oland, Gröde and Habel. The numbers refer to figures, those prefixed by 'pl' refer to plates.

ridges. A sand wall was built in front of the most endangered houses at the southern end of the settlement, but erosion continued at a high rate (figs. 269 b and 270).

Between 1930 and 1968, before the construction of the tetrapod wall, about 150 m of dunes were lost off Hörnum, producing a coastal retreat rate of about 4 m/year. Whilst the tetrapod wall has completely stopped dune retreat off Hörnum, it has intensified it south of the structure. Between 1968 and 1984 about 200 m of dune were lost at this point, i.e. about 12.5 m/year. The peninsula is now only 600 m wide, and since the highest dunes in the west have been eroded, the peninsula itself may be completely lost in less than 20 years. The southeastern spit on Amrum, originally less than 300 m wide, disappeared in under 10 years in a comparable situation (Müller & Fischer, 1937 a: 137).

The countermeasures can only be regarded as

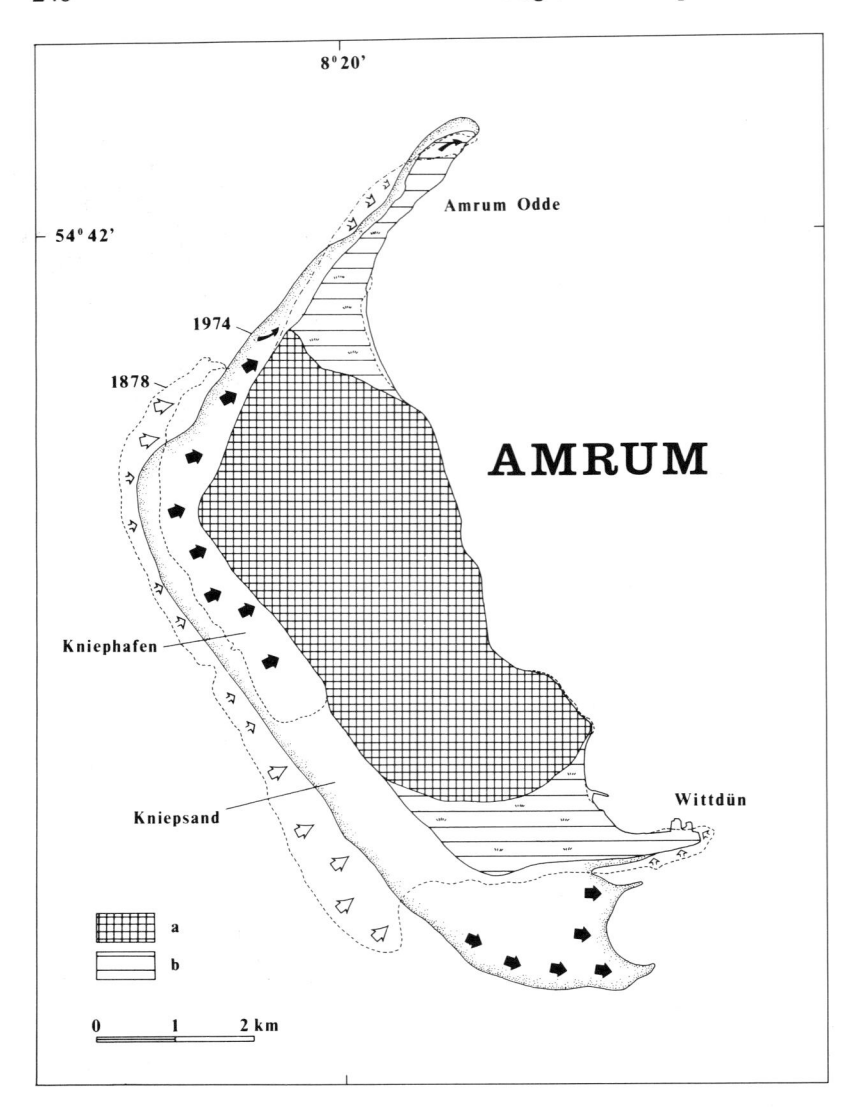

Figure 272: Morphodynamic subdivision of Amrum; a = old dune core; b = young spits and salt marshes. Black arrows: accretion; white arrows: erosion. Sources: First edition of the Topographical Map 1:25,000, Sheets Borgsum (1880), Kniep-Hafen (1880) and Nieblum (1880) and Küstenkarte 1:25,000, Sheets 1216 K Föhr (1978), 1314 K Amrumbank (1977) and 1316 K Amrum (1977).

provisional, and there is no solution that will be able to save the Hörnum Odde. The consequences of such a development for the northern end of Amrum, and perhaps for Föhr, which lie southeast of the Hörnum Peninsula, cannot be estimated.

7.2.2 AMRUM

Like the Isle of Sylt, Amrum has a core of older sediments, but unlike its neighbour to the north this core is not currently being eroded on the seaward side because it is protected by the over 1 km wide Kniepsand (fig. 271). Dune-bearing spits have developed north and south of the core and the western part of the Pleistocene core is itself covered by large dunes (fig. 272). Only the spits at both ends of Amrum show a clearly negative sand budget.

The 'Litorina Cliff'

The island's Pleistocene core is exposed forming a fossil cliff line along the west coast of Amrum to behind the first coast-parallel dune ridge. This feature is usually referred to in the literature as the 'Litorina Cliff'. Its exact form has been mapped by Jessen (1933: 15). The old land surface above the cliff is covered by a stone pavement.

The attachment of the Kniepsand to the island has halted further erosion of the cliff. This event started more than 200 years ago and since then re-

Figure 273: Densely vegetated primary dunes on the Kniepsand of Amrum (1981).

cent dunes have developed in front of the 3 m high cliff face, completely engulfing it in places.

The dunes

The recent coast-parallel dunes are not directly related to the old large dune arcs. The recent dunes were formed following the attachment to the island of the Kniepsand. This process was only completed during the 20th century. The coast-parallel dune ridge did not begin to form prior to 1870 (Voigt, 1969: 41).

It is difficult to assess the age of the main dune ridges, which extend inland as several large arcs. Morphologically they resemble the dunes on Fanø. Archaeological investigations have revealed that the dunes overlie Bronze Age tumuli as well as medieval fields. It is possible that overgrazing in the Middle Ages contributed to the destruction of the natural vegetation cover of the old coastal dunes and favoured their migration, although this cannot have been the original cause of dune formation. The simultaneous development of large dune ridges along the entire Wadden Sea coast from the 13th to the 15th centuries can only be explained by major climatic changes (cf. section 7.2.1).

The large migrating dunes of Amrum are inactive today, but the dune ridges are dissected by numerous blowouts down to groundwater level. These blowouts contribute to the continued modification of the dune landscape.

Young forest areas occur to the east of the dunes. Part of these extensive woodlands were planted after the Second World War. They are currently expanding because of the increase of wasteland; the original heath vegetation of the wasteland areas being quickly replaced by coniferous forest.

The Kniepsand

The Kniepsand is a large sand flat which is only flooded during exceptionally high tides. Indicators of aeolian redeposition such as sand stripes and beach barchans cover its surface. The margin of the white dunes against the Kniepsand is delineated by a distinct dune cliff. The coastal dunes between the Satteldüne and the Wriakhörn form a major dune ridge system, which in the lighthouse area consists of four coast-parallel ridges which reach a maximum height of 10.4 m. Very little recent dune formation can be observed in front of the coastal dunes today.

As on the Havsand/Juvresand on Rømø, which can be regarded as an equivalent of the Kniepsand, major dune complexes have not developed. Between Nebeler Strandweg path and Quermarkenfeuer lighthouse primary dunes have formed reaching heights of around 1.5 m above Kniepsand level (fig. 273). The dune area, however, is separated from the main dune core by a broad stretch of flat beach sand.

Attempts to enhance dune formation have been undertaken at, and south of, the Nebeler Strandweg. A series of brush fences were installed at right angles to the dune margin at the Nebeler Strand-

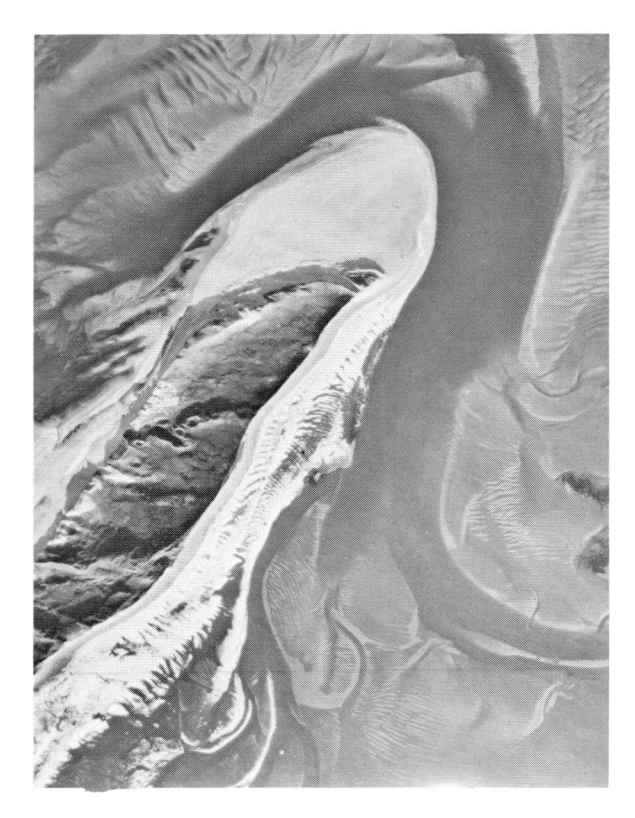

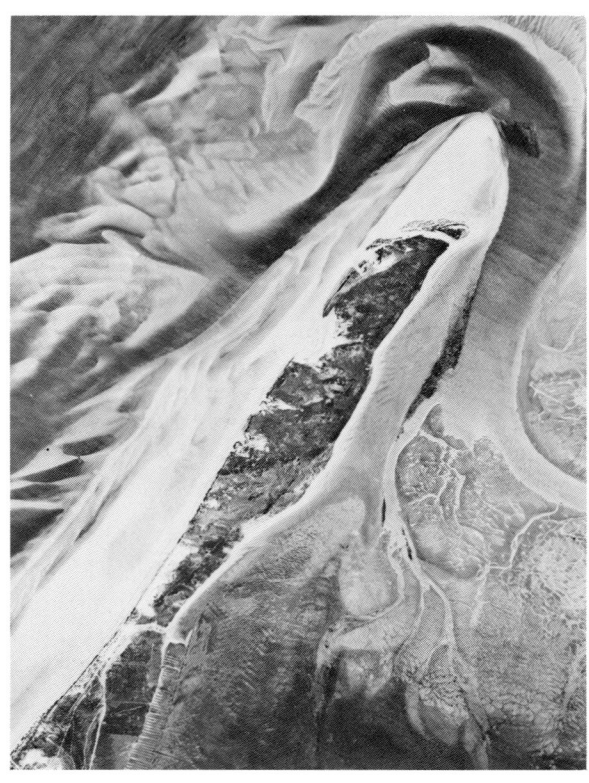

Figure 274: Aerial photographs of Amrum-Odde, the northern spit of Amrum; a: 1978, b: 1985 (Aerial photographs of 7.12.78 and 5.7.1985, freigegeben durch das Landesvermessungsamt Schleswig-Holstein unter den Nummern SH 510/79 und SH 510/85, Bildflug Schutzgebiete Westküste, Amrum, Str.3/400 und Bildflug Nordfriesland, Str.9/53, vervielfältigt mit Genehmigung des Landesvermessungsamtes Schleswig-Holstein vom 3.12.1985).

weg, but dune formation has only occurred along the eastern 100 metres of the fences, and even these dunes are still rather weak. Aeolian sand has also accumulated farther to the west, but the vegetation required for dune formation is lacking here. A brush fence was constructed obliquely across the Kniepsand further to the south, but this was a total failure.

The Kniepsand terminates with a relatively steep seaward margin. Seawards of this margin there are at least two coast-parallel ridge and runnel systems which emerge at low tide. They lie considerably deeper than the Kniepsand surface.

The southern spit of the Kniepsand like similar features on the ends of other islands is relatively unstable. The southward directed sand transport leads to the formation of flood spits which curve northwards towards the Wittdün Peninsula (fig. 278). These spits are finally levelled and destroyed when they start to fill the protected lagoon in front of the sea wall. This cyclic process has a return interval of about 30 years, the flood spits migrating at a rate of approximately 26.7 m/year (Knop, 1963; Giller, 1981: 52).

At the same time the Kniepsand extends northwards, thus protecting parts of the Amrum-Odde. This northward migration has to be seen in relation to the long term morphological changes in the Vortrapptief inlet, and possibly also the future development of the Hörnum Peninsula on Sylt. The Kniepsand is part of the Vortrapptief ebb delta, a marginal flood channel effecting its northward extension. This system, however, has to be regarded as rather unstable, and future developments may either lead to a further extension of the broad beach towards the north or to westward migration of the flood channel and subsequent erosion of part of the Kniepsand.

Like the beach barriers off St.Peter-Ording, the Kniepsand has recently become both higher and smaller. Measurements by the Amt für Land- und Wasserwirtschaft in Husum have demonstrated, that between 1948 and 1980 the gradient of the seaward slope steepened so that the area above

MTHw (ca. NN + 1.1 m) has expanded considerably, whilst the area below MTHw has decreased in size (Giller, 1981).

Amrum-Odde

The northeastern extension of the old dune systems of Amrum are found on Amrum-Odde. This original dune ridge was breached north of Norddorf during the storm surge of 3./4.2.1825. Early attempts to repair this breach by enhancing new dune development failed, the newly created dunes being completely destroyed during the storm surge of 8.1.1839. A second attempt was destroyed during the storm surge of 20./21.10.1845. In both cases the marsh meadows behind the dunes were covered and spoiled by dune sand (Müller & Fischer, 1937 a: 196).

The continued retreat of the west coast of the Amrum-Odde and the widening of the breach between the dune areas threatened neighbouring Föhr, and in 1895 the first four light pile groynes with stone heads were constructed in front of the so-called Risum Gap. Considerable coastal retreat during the storm surge of December 1895 necessitated the construction of six additional groynes (1897), and six lighter groynes between the large ones (1898/99). Despite all these efforts it proved impossible to promote a broader beach and initiate new dune formation. The groynes had to be extended landwards in 1906 after further erosion. Finally, between 1914-17, the Risum Gap was closed by a dyke, all other schemes having failed (Müller & Fischer, 1937 a: 101).

The threat of the complete erosion of the Amrum-Odde peninsula from the seaward side was demonstrated by Gripp (1968: 114). Since then the situation has considerably deteriorated (figs. 274 a and b and fig. 275). The dune remnants on Amrum-Odde are currently being eroded from both sides, a small creek having extended around the northern end of the island endangering the dune area from the Wadden Sea side.

The Wadden Sea side

The marsh areas north of Norddorf (78 ha) and south of Steenodde (38 ha) were dyked between 1933/34 and 1933/35 respectively during a works programme of the 'Freiwilliger Arbeitsdienst'. These areas were originally only protected by dune ridges in the west, the eastern side remaining unprotected. Dyking was out of the question for economic reasons (Müller & Fischer, 1937 a; 166).

There is no natural accretion anywhere on the Wadden Sea side of Amrum. North of Norddorf massive reclamation works have been undertaken (fig. 276). Additional brushwood groyne fields east of Nebel have been abandoned, only the wooden poles remaining from this futile land reclamation scheme (fig. 277).

Erosion is dominant on the Wadden Sea side of the island. An almost 1 m high cliff has developed south of Boragh along the salt marsh margin. Salt marsh erosion is occurring north of Nebel in association with the formation of a sandy beach ridge which migrates inland as the cliff retreats.

Between Steenodde and Nebel the Pleistocene cliff of Ual anj is almost completely covered with slope material because of its minimal rate of retreat. Periglacial cryoturbation structures and ice wedge casts can be seen in the upper part of the cliff, whilst on the surf platform in front of the cliff periglacial structures (rings and stripes) are exposed. The western end of the cliff has been recently secured by artificial sand accumulation.

The dune spit of Wittdün

The southern dune spit of Amrum was completely uninhabited until the foundation of the Wittdün seaside resort about 1890. Whilst strong morphological changes in this area had previously been accepted as inevitable events, following the foundation of Wittdün, coastal protection measures became necessary to secure the built-up area. The dune margin retreated by 3 – 6 m during the storm surge of 5./6.11.1911, necessitating the evacuation of the 'Nordseehallen' because of the danger of collapse.

Several houses were immediately threatened. A sea wall was planned, and in 1914 a 900 m long wall was built. The final backfill of the wall was left incomplete because of the outbreak of the First World War. The storm surges of 13./15.9.1914 and 12.11.1914 immediately caused considerable damage to the wall which had to be repaired in 1915/16, when the original four groynes were supplemented by five additional structures (Müller & Fischer, 1937 a: 130).

It now became obvious, however, that the hitherto unprotected southeastern end of the Wittdün Peninsula was threatened by lee-side erosion. Plans were drawn up to protect the dune spit, but before they could be put into practice in 1918-20 the whole 300 m long southeastern spit was destroyed by erosion (Müller & Fischer, 1937 a: 145).

In the meantime the houses on the Wadden Sea

Figure 275: The steeply sloping, narrow beach of Amrum-Odde, an indicator of coastal erosion (1980).

Figure 276: Land reclamation along the east coast of Amrum, east of Norddorf (1980).

side of the Wittdün Peninsula were endangered by coastal erosion from that side. This led to the construction of a 258.5 m long sea wall in 1921/22 (Müller & Fischer, 1937 a: 156).

The situation has stabilized today (fig. 278). The coastal protection structures extend into areas with a positive sand balance, so that Wittdün is now not directly endangered. Further developments, however, will largely depend on the future dynamics of the Kniepsand.

7.2.3 Föhr

The Isle of Föhr does not form a part of the barrier islands but lies in a more backward and protected position between Amrum and the mainland. The island consists of a number of low Pleistocene cores, which reach a maximum height of 13.2 m south of Oevenum, in the south, and extensive marsh areas in the north.

Dissection of the tidal flats

The tidal flats between Amrum and Föhr are high enough to walk across. The water divide between Föhr and the mainland as well as between Amrum and Föhr has shifted to a more northerly position. Continual displacement and simultaneous dissection of the tidal flats continue today. The tidal

Figure 277: Coastal erosion at Nebel on the Wadden Sea side of Amrum; the brushwood groynes have been abandoned (1980).

flats have been concurrently lowered, especially to the south of the water divide (Knop, 1963: 27), and, in particular, the branches of the Norderaue inlet have migrated far northwards (fig. 279). The recent extension of the tidal creek between Amrum and Föhr, which runs directly past the coast of Föhr, has been partially caused by dredging.

The cliff coast in the south

The south coast of Föhr is particularly prone to erosion (figs. 215 and 280). Huge waves can develop during storm surges in the broad Norderaue inlet and cliffs have developed in the Pleistocene areas of Föhr at Utersum and Goting. The cliff at Utersum was protected by a revetment in 1937 following erosion of the 'Nordseekurheim' area (Müller & Fischer, 1937 b: 255), but in the south erosion of the Goting Cliff continued.

During the last 100 years the cliff has retreated at a rate of over 1 m/year (fig. 281). The 'drowned forest' south of Goting, which is situated at a depth of 3 m, testifies this continued erosion (Müller & Fischer, 1937 b: 7). A sand dam was built on the tidal flats in 1975 to protect the summer house settlement north of the cliff (height: NN + 2.5 m). Despite this an average retreat rate of 4 m was registered at the Goting Cliff during the storm surges of 3./4. and 20./21.1.1976 (Zitscher, Scherenberg & Carow, 1979: 91).

Thrust sand and clay layers are exposed in the cliff. The micro-relief of the cliff face is determined by these varying lithologies, the clay layers, which are more resistant, forming seaward protrusions along the cliff (fig. 282).

The salt marshes in the north

The undyked forelands in the north, which in places extend to widths greater than 500 m, give the impression of predominant accretion, as suggested by Jessen (1914: 250). Map comparison, however, reveals that the development of the north coast of Föhr is closely related to the sand transport pattern in the Hörnumtief inlet, and that the current sand budget is, at best, only balanced.

The recent marsh areas are bounded on the seaward side by beach ridges of coarse gravel (fig. 283), which form barrier-like spits around the western parts of the forelands. These forelands have been eroded in the west as a result of the eastward-directed sand transport in the Hörnumtief, whilst at the same time the spits have migrated eastwards. New salt marshes have developed under the protection of these spits.

The mechanism forming these spits is comparable to the large flood spits on Amrum (Wittdün Peninsula and Amrum-Odde), despite the fact that the Föhr spits are much smaller. The two spits north of Oldsum migrated 200–250 m eastwards between 1878 and 1963. The end of the eastern spit migrated 400 m to the east in the same period, whilst the salt marshes under its protection extended by 300 m (fig. 284). Since then the total fore-

Figure 278: The southeastern spit of Amrum with Wittdün and the Kniepsand to the south (Aerial photograph of 7.12.1978, freigegeben durch das Landesvermessungsamt Schleswig-Holstein unter der Nr. SH 510/79, Bildflug Schutzgebiete Westküste, Amrum, Str.4/245, vervielfältigt mit Genehmigung des Landesvermessungsamtes Schleswig-Holstein vom 3.12.85).

land area has not altered significantly because growth in the east has been balanced by erosion in the west.

Similar sand spits also occur along the more easterly foreland off Ostlandföhr. A land reclamation commission which visited the locality in 1861 referred to a 'regularly built sand groyne, created by nature' (Müller & Fischer, 1937 b: 121). This spit has migrated in a similar way to those north of Oldsum, as can be seen by comparison of recent maps with those of 1882 and 1936 published by Müller and Fischer (pp. 134 and 151).

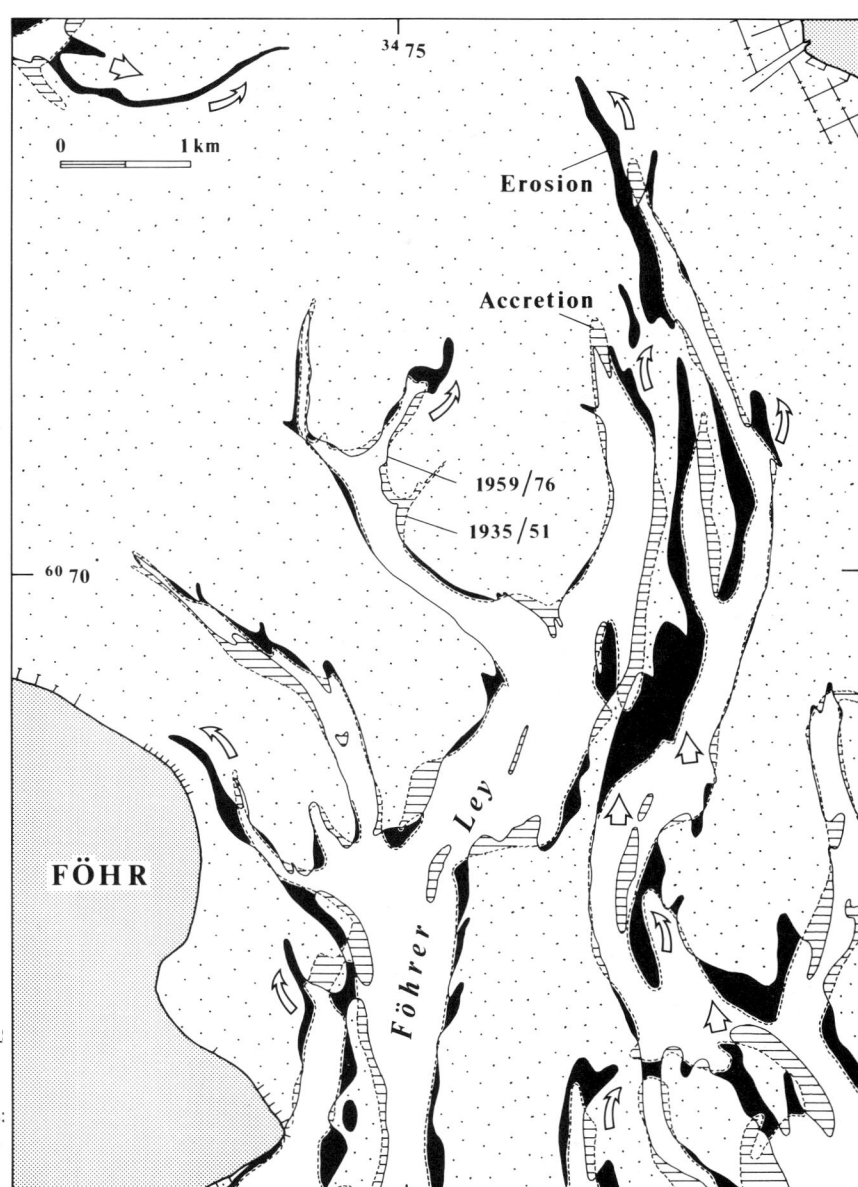

Figure 279: Development of the tidal creeks in the Föhrer Schulter area, northeast of Föhr; dissection of the tidal flats continues. Sources: Map in Müller & Fischer, 1955 and Küstenkarte 1:25,000, Sheet 1216 K Föhr of 1978.

Coastal protection

On Föhr all the settlements were situated on the high safe Pleistocene areas, creating little demand for dyke construction. Therefore the first sea dyke on Föhr was not built before 1492, and its improvement to resist storm surges had to wait until the 19th century (Müller & Fischer, 1937 b: 63). Similarly, there was little interest in the reclamation of additional forelands. Plans to dyke the foreland at Kuham (south of Näshörn) were made as early as 1796, but they did not materialize and the erosion of the forelands continued. The narrow strip of salt marsh at Osterkurzmaße was completely lost during the 1834 storm surge. Groynes were at last built in 1838 to protect Näshörn and Kuham (Müller & Fischer, 1937 b: 85).

The first sea wall, 280 m long, was built at Wyk in 1904. It was a light structure without any basal protection. Incipient lee-side erosion caused its northward extension over another 220 m in 1908. A private initiative led to another sea wall being constructed along the south coast of Boldixum,

Figure 280: Goting Cliff, south coast of Föhr (1981).

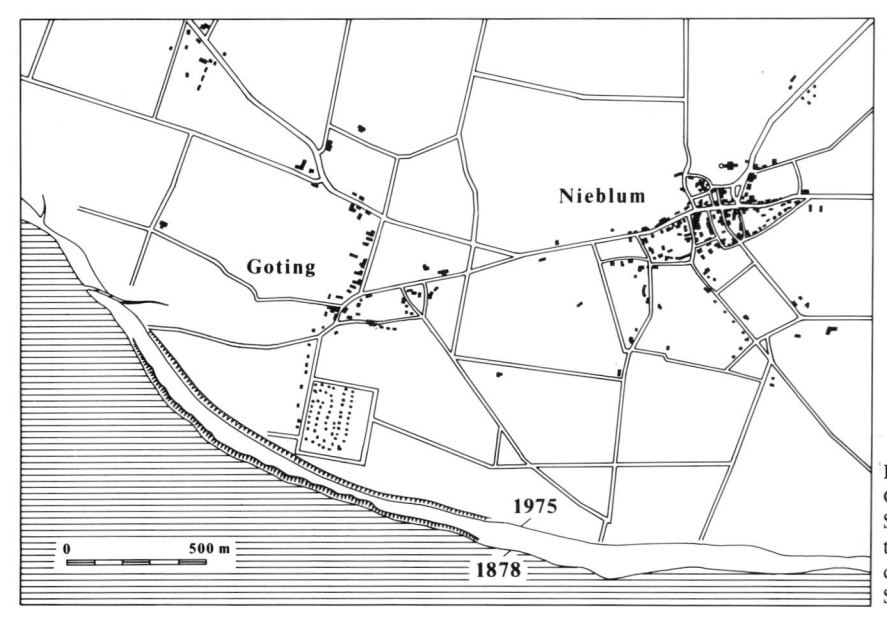

Figure 281: Retreat of the Goting Cliff between 1878 and 1975. Sources: First edition of the Meßtischblatt 1:25,000, Sheet Nieblum of 1880 and Küstenkarte 1:25,000, Sheet 1316 K Amrum of 1977.

west of Wyk during the summer of 1909. This 162.5 m long wall was quickly destroyed over a length of 45 m during a storm surge on 3.12.1909. The walls were subsequently extended, so that by 1929 the structure had reached an overall length of 2.7 km, protecting the entire southeastern extremity of Föhr (Müller & Fischer, 1937 b: 238). There have been no recent problems on Föhr because of its relatively protected position.

7.2.4 Halligen

The Halligen of North Frisia (figs. 285 and 286) are undyked marsh islands which were formed by the destruction of vast marshlands by storm surges during the Middle Ages. Some of the islands thus created were able to be dyked (Nordstrand and Pellworm), whilst others were reincorporated into the mainland (e.g. Dagebüll). The complicated coastal development and dyking history of the North Frisian mainland will not be dealt with

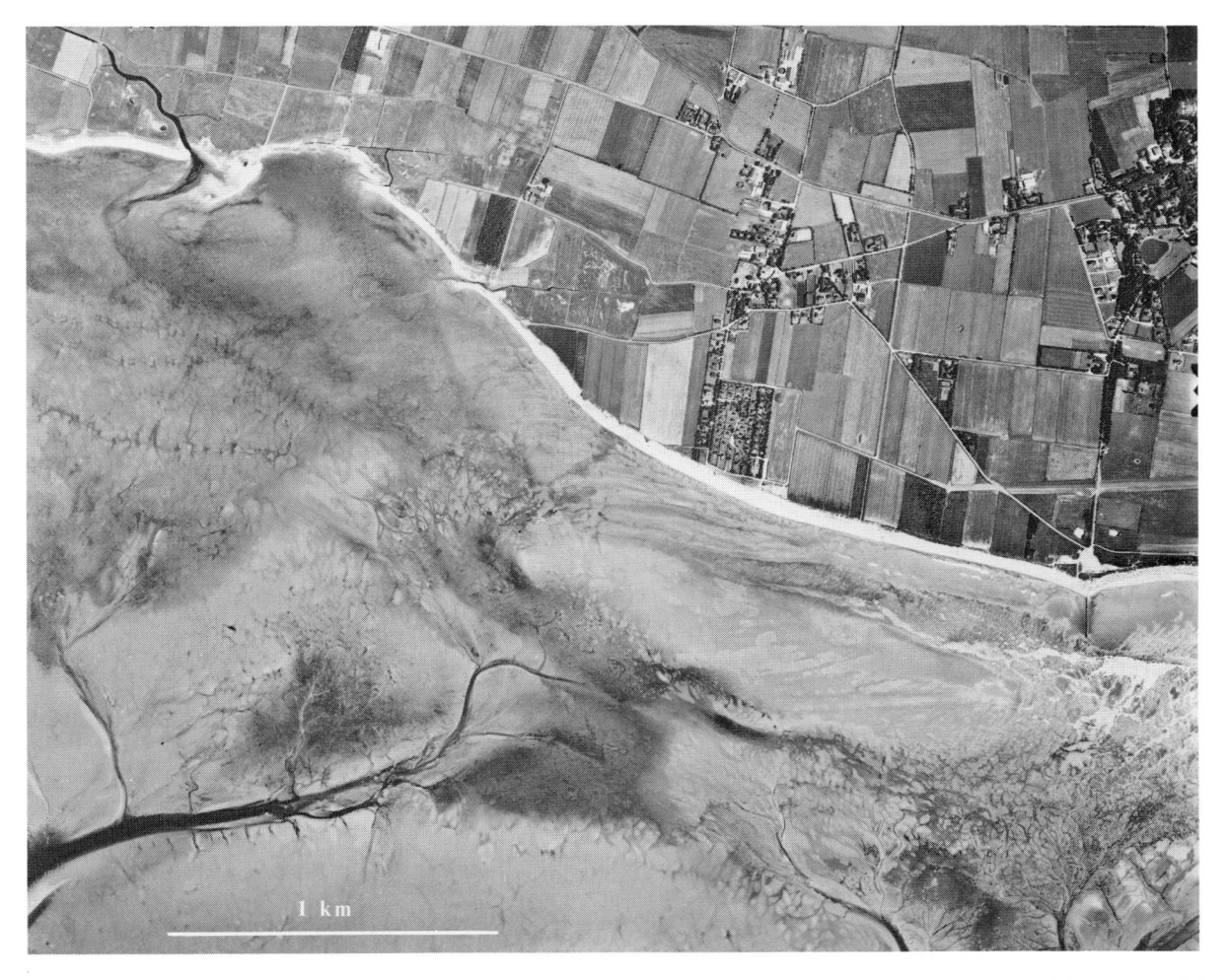

Figure 282: The Goting Cliff area; south of the cliff remnants of the protective sand dam can be seen on the tidal flats (Aerial photograph of July 1985, freigegeben durch das Landesvermessungsamt Schleswig-Holstein unter der Nr. SH 510/85, Bildflug Nordfriesland, Str.10/18, vervielfältigt mit Genehmigung des Landesvermessungsamtes Schleswig-Holstein vom 3.12.1985).

Figure 283: Beach ridge on the spit at Dunsum, north coast of Föhr; the ridge forms a protective barrier behind which salt marsh has developed. The sea dyke can be seen in the background (1981).

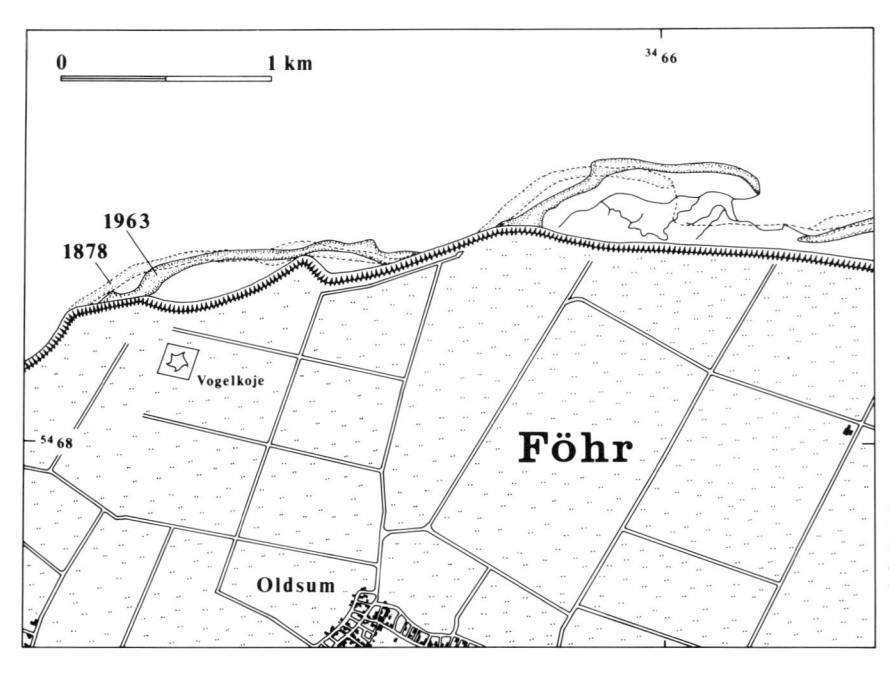

Figure 284: Spit migration along the north coast of Föhr between 1878 and 1963. Sources: First edition of the Meßtischblatt 1:25,000, Sheet Borgsum of 1880 and Küstenkarte 1:25,000, Sheet 1216 K Föhr (1978).

here; it has been discussed in detail by Fischer (1955).

The Halligen are separated from the open sea by only a series of high shoals. Those so-called 'Außensände' (Outer Sands) Japsand, Norderoogsand and Süderoogsand are submerged during every higher than normal tide. They are completely barren, and dune formation never continues beyond the primary dune stage. Like Koresand and Peter Meyer's Sand in the Mandø Gap, the Outer Sands are very mobile forms, shifting landwards at a rate of about 20 m/year (Taubert, 1982: 45). If their migration continues at the present rate, Japsand will have reached Hooge in about 100 years, Norderoogsand will abut Norderoog already in about 50 years. The future development, however, will depend largely on sea-level change and on morphological changes in the Halligen area.

The term 'Hallig'

The term 'Hallig' has undergone a considerable change in meaning during the course of history. In King Valdemar's Jordebog (1231), the present Hallig Oland is mentioned and referred to as an 'island'. The word 'Hallig' occurs for the first time in the land register of the bishop of Schleswig (1509). In this register, however, what we now call 'Halligen' are referred to as islands (Groden = Gröde, Oland, Nortmersk = Nordmarsch), but some 'halgen' near Ofkebul are also mentioned. The latter were apparently not islands but undyked salt marshes along the mainland (Müller, 1917 a: 1).

In the report 'Ein korte Beschrivinge des Lendlins Nordstrandes und deßer Gelegenheit' (1597) by the Nordstrand pastor Johann Petreus, the term 'hallig' is used in a general sense to refer to all undyked foreland, including the undyked islands. This term, however, was not in common usage in the whole Wadden Sea area but only along the west coast of Schleswig (Müller, 1917 a: 2). The word 'Hallig' was later used for all undyked marsh islands; Müller (1917 a: 6), for instance, included islands like Helmsand and Trischen in the Halligen, whilst excluding undyked foreland.

In contrast to Müller (1917), the term 'Hallig' is used here as a genetic term to denote all the undyked marsh islands of the Wadden Sea which are neither formed by recent accretion (Helmsand) nor by emerging shoals (Trischen). The Halligen are remnants of old marsh areas inundated by storm surges. Though in some cases low summer dykes have been erected (e.g. on Hooge), the islands are not protected by sea dykes.

Age of the Halligen

The age of the Halligen is uncertain. The oldest source concerning the Schleswig-Holstein and

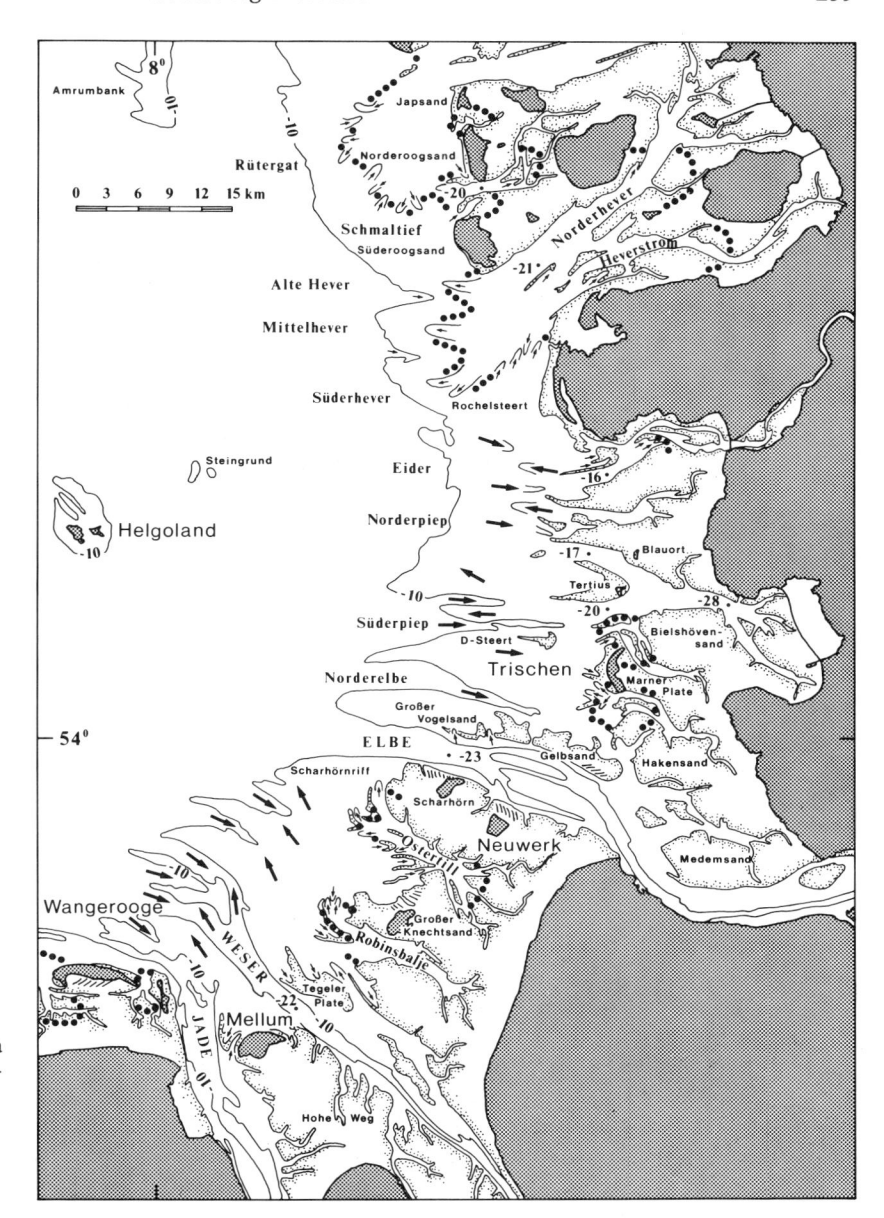

Figure 285: Map of the Wadden Sea area between the Halligen and Wangerooge. The maximum depths of the tidal inlets are given in metres. Dotted lines: boundaries of the ebb and flood deltas. Arrows: distinct ebb and flood channels.

Danish islands, 'Kong Valdemar den Andens Jordebog' (1231), records the following North Sea islands:
1. Fanø
2. Mannö (= Mandø)
3. Rumö (= Rømø)
4. Hiortsand (= Jordsand)
5. Syld (= Sylt)
6. Ambrum (= Amrum)
7. Föör (= Föhr)
8. Aland (= Oland)
9. Gaestaenacka
10. Hwaelae minor
11. Hwaelae major
12. Haefre (= Wester- and Osterhever)
13. Holm (= Tating, St.Peter and Ording)
14. Haelgheland (= Helgoland)

It is remarkable that mention of Nordstrand or Strand is lacking (Müller, 1917 a: 167).

The destruction of North Frisia during the great storm surges of the Middle Ages and early 17th century resulted in the formation of a large number of minor and small marsh islands, only the largest of which have been able to resist erosion

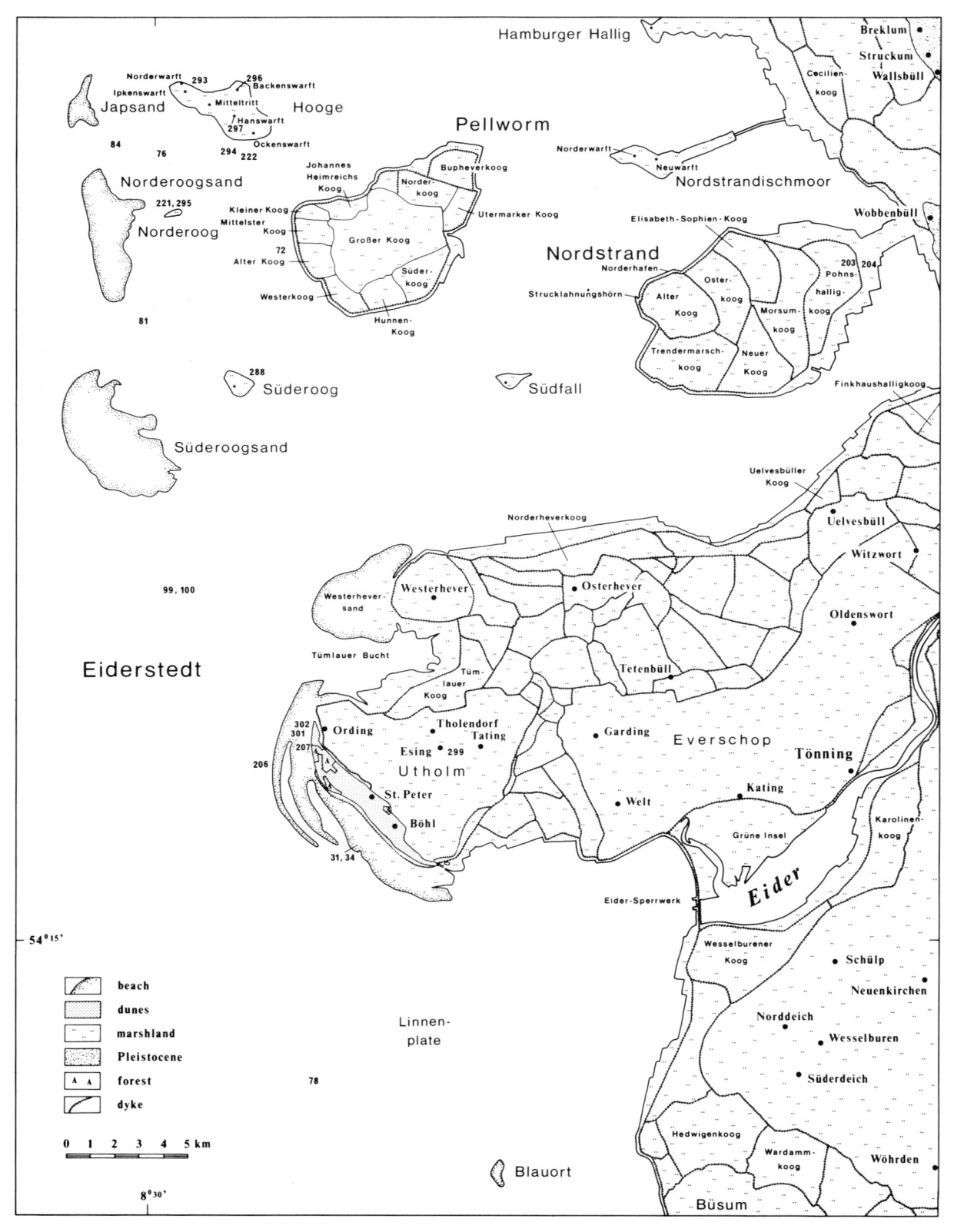

Figure 286: Location map of the Halligen and of the Eiderstedt Peninsula. The numbers refer to figures, those prefixed by 'pl' refer to plates.

Figure 287: The apparently flat surface of Hallig Langeneß with a dwelling mound in the background (1984).

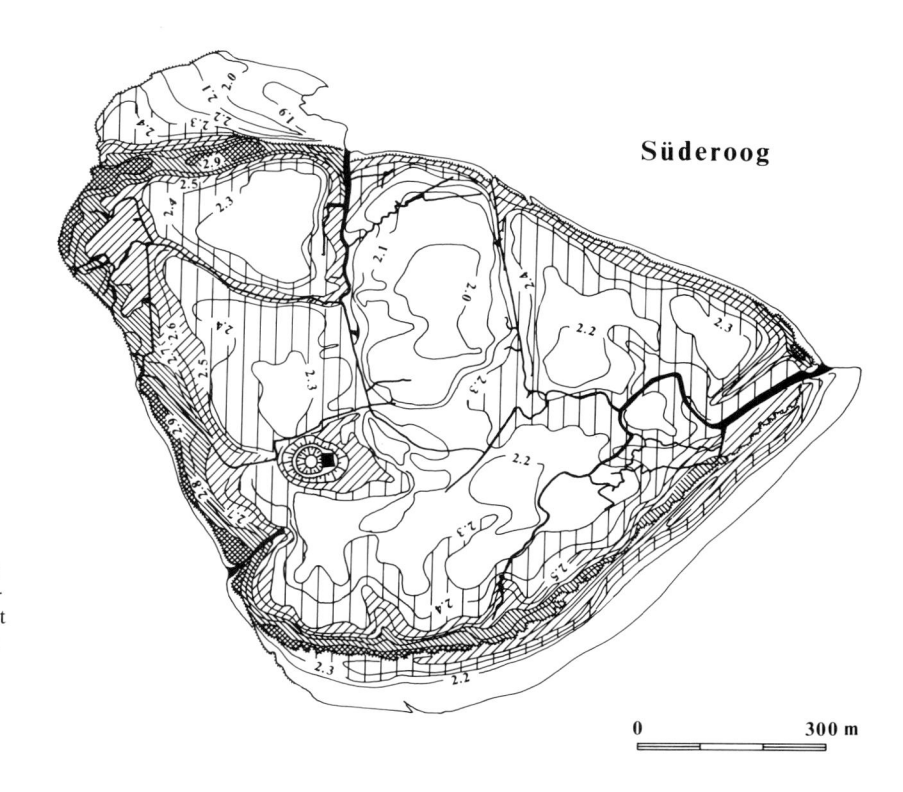

Figure 288: Contour map of Hallig Süderoog, showing the principal relief of a Hallig with beach ridges at the coasts and levées along the salt marsh creeks, encircling closed depressions. Source: Deutsche Grundkarte 1:5,000; Sheet 7036 Süderoog.

and survive up to the present day. The exact number of the original Halligen is unknown. At least 29 drowned Halligen are recorded in the literature and on old maps. Müller (1917 a: 243) assumes that the original number must have been close to 50. Most of these disappeared during the 17th and 18th centuries. Apart from the present Halligen, during the 19th century there were only two remaining Hallig remnants, the exact positions of which have been mapped, the Hainshallig east of Hooge and the Beenshallig south of Gröde. The Hainshallig was destroyed in 1835 or 1836, and of

Figure 289: Hilligenley Warft on Langeneß, protected by strong revetment (1984).

the Beenshallig there was only 'a barren reef of clay earth' left in 1892 (Traeger, 1892: 242) which completely disappeared within a few years (Müller, 1917 a: 250).

The relief of the Halligen

The Halligen surface gives the impression of an absolutely flat marsh plain, the only relief being provided by the Warften (dwelling mounds) (fig. 287). This impression, however, is somewhat deceptive. Detailed surveying carried out for the Deutsche Grundkarte 1:5,000, on which some Halligen are depicted with contours at decimetre intervals, indicate that there is in fact a distinct microrelief.

Those Halligen surveyed in this way (Langeneß, Oland, Gröde, Nordstrandischmoor, Süderoog and Südfall) all exhibit the same relief, of which Hallig Süderoog is a typical example (fig. 288). A beach ridge several decimetres high is formed on the seaward side which under natural conditions migrates eastwards at the same rate as the erosion of the Hallig margin (cf. Traeger, 1892: 245, fig. 3). This beach ridge is supplemented by two, usually lower, branches along the south and north coasts.

As already recognized by Ordemann (1912), the existing Halligen no longer represent fragments of the medieval marsh surface but considerably higher surfaces as a result of later accretion. Accumulation occurs in the 'Landunter' (= land under) periods when the Halligen are flooded during storm surges (plate 22). Accumulation is greatest along the salt marsh creeks which all have developed low levées, the intervening marshes forming shallow closed depressions (fig. 288).

Erosion in the Halligen area

The Halligen area is largely a landscape of net active erosion. This can be illustrated by the fact that the land area of the present 10 Halligen was reduced by about 80 % from ca. 10,000 ha to 2,000 ha between 1650 and today. The Fallstief, a previously unimportant creek, evolved into the large Norderhever inlet (fig. 285). Between Hallig Südfall and Pellworm its depth has increased tenfold and its width at low water sevenfold; its cross profile is now 40 times larger than 300 years ago. This development had undoubtedly been influenced by the direction of the near-coastal tidal wave parallel with the axis of the Norderhever inlet (Nommensen, 1982: 119). Wind-driven drift currents also largely contribute to the intense erosion. About 40,000,000 m^3 of water per tide drift from the Norderhever catchment area northwards into the Süderaue tidal basin through which it drains (Wißmann, 1981: 89).

Though the Hoogeloch inlet between Hooge and Norderoog is small, there is a considerable exchange of water with the adjoining Rummelloch catchment area to the east. This contributes significantly to the lowering of the tidal flats between Norderoog and Hooge (Kemstedt, 1981: 82).

The Amt für Land- und Wasserwirtschaft in Husum has evaluated 1:10,000 maps drawn in 1935

Figure 290: Bank protection east of the very exposed Norderhörn Warft, north coast of Hallig Langeneß (1981).

and 1975 of the tidal flat topography with contour lines at decimetre intervals on the flats and one metre intervals in the channels. These maps indicate an enormous and continuing increase in the cross section of the Norderhever. The western part of the Holmer Fähre channel is another sensitive area. Particularly strong erosion occurs north of Norderhafen on Nordstrand where a short cut has developed between Fuhle Schlot and Holmer Fähre, lowering major tidal flat areas north of Nordstrand between Strucklahnungshörn and the northern edge of the Elisabeth-Sophien-Koog. The Butterloch and Strand channels are also areas of increased erosion (Nommensen, 1982: 120).

Major areas of erosion are also found near the Halligen Gröde and Habel. In particular, exposed *Cerastoderma edule* shells, old clay and peat beds together with decimetre-deep hollows on the flat surface east of Gröde suggest large-scale erosion in the lee of the Hallig. The strong hydrodynamic stress in the area is demonstrated by the relatively high content of sediment coarser than 0.2 mm (partly over 5 %) (Nommensen, 1982: 122).

Today cultural remains are completely absent southwest of Habel and along the margin of the Rocheley channel, even at extreme low water. In this area until a few years ago remnants of the old Süderwarf, the churchyard and tree stumps could be found (cf. Bantelmann, 1966). Their destruction demonstrates the extent of recent erosion (Nommensen, 1982: 123).

In contrast to these areas of erosion, there is some accretion in the inner Nordstrander Bucht, seaward of the brushwood groynes of the Hauke-Haien-Koog, Ockholm-Koog and Sönke-Nissen-Koog, as well as on both sides of the embankments of the Hamburger Hallig and Nordstrandischmoor (Nommensen, 1982: 120).

In the Nordstrander Bucht, however, accumulation is restricted to marginal areas on both sides of the Holmer Fähre channel and in front of the land reclamation works. Signs of erosion can be seen north of the Holmer Fähre channel, where there are numerous hollows and exposed *Mya arenaria* shells in life position. Erosion also prevails north of the Pohnshalligkoog on the other side of the Holmer Fähre (Nommensen, 1982: 123).

In order to halt this erosion, the construction of an embankment from the mainland north of the Hamburger Hallig along the water divide to Pellworm was planned many years ago, the first survey of this course being undertaken in 1912. Only now is there a real possibility of its completion. Its main purpose is to inhibit the drift of large water masses between the tidal basins. The dyking of the Nordstrander Bucht at the same time will considerably reduce the drainage area of the Norderhever. The dyke is at present under construction (cf. plate 18).

Langeneß

Langeneß has an area of about 917 ha and there are 20 Warften (dwelling mounds) (Becker, 1976: 111). The present Hallig Langeneß was formed by the connection of three single Halligen, namely

Langeneß,
Nordmarsch and
Butwehl.

Old sources indicate that Nordmarsch was formerly connected to the Isle of Föhr or at least separated from it by little more than a passable minor creek. In the Nieblum cemetery on Föhr the old graves of Nordmarsch inhabitants have been found, and there was also a road called 'Nordmarscher Kirchenstieg' (Nordmarsch church path) on Föhr. Nordmarsch belonged to the parish of Nieblum until 1599 when its own church was built (Müller, 1917 b: 16).

Reliable reports concerning the form of the Hallig and the extent of land losses are no older than the 18th century. As on the other Halligen, the storm surges of 1717 (Christmas Flood) and 1720 (Sylvester Flood) caused heavy damage. In 1732 the Hallig church had to be relocated into the interior of the Hallig. At the beginning of the 19th century the storm surges of 1.12.1821 and of 3./4.2.1825 caused more heavy damage. The latter was called the 'Hallig Flood' because of its devastating effect on the small, undyked islands. Poverty increased and many inhabitants emigrated to the safer islands. Nordmarsch was combined with Langeneß into one parish in 1838, and the last church on Nordmarsch was demolished before 1840.

During the spring of 1847 Nordmarsch was connected by an embankment with neighbouring Langeneß, from which it had been separated 'from time immemorial' by a 25 m wide creek which was 6 m deep in the centre (Müller, 1917 b: 33). In 1868/69 the 30 m wide channel between Langeneß and Butwehl was also dammed, but against the will of the inhabitants of the larger Langeneß who saw no advantage in becoming connected with Butwehl (Müller, 1917 b: 90f.).

With the continuation of coastal retreat, the first revetment was built in 1872 in front of the Hilligenley Warft (fig. 289). A further extension across the whole western end of Langeneß was not completed, however, because of the resistance of the local population, who were vehemently opposed to abandoning Hallig land in order to permit the shortest possible course for the planned bank protection (Traeger, 1892: 328).

The storm surge of 14./15.10.1881 again caused coastal retreat of over 2 m on the western and northern side of the Hallig, and after another storm surge on 28.10.1884 the situation of the Peterswarft in the northwest became critical. The inhabitants attempted to secure their Warft by constructing a bulkhead and by fixing the slopes of the dwelling mound. By 1894, however, the Warft had to be abandoned and the Neu-Peterswarft built in the interior of Langeneß, the youngest of the Hallig Warften (Müller, 1917 b: 34). The old Peterswarft was later included in a coastal protection scheme for Langeneß and chosen as the site for a lighthouse erected in 1902.

Before the 17th century the Halligen Langeneß and Gröde were not separated by deep water, permitting the inhabitants of Butwehl to attend the church on Gröde. This practice, however, had to be abandoned in 1681 (Müller, 1917 b: 63; Hansen, 1814: 10).

Nordmarsch and Langeneß alone lost an area of 110 ha between 1877 and 1897 (Müller, 1917 b: 100). This was partly caused by the complete failure to close the channels as they developed, even if belated attempts were made. There was little initiative amongst the inhabitants towards coastal protection because there was no private land ownership on the Halligen, so that all losses were borne by the community. According to Richardsen (1902: 64) this attitude was especially prevalent on Langeneß-Nordmarsch.

The state finally intervened in order to maintain the Halligen as a protective bulwark in front of the mainland coast. A 3,500 m long stone bank protection was built between 1901–1904 extending from the Rixwarft in the southwest via the abandoned Warften Hallge (left after 1825) and Alte Peterswarft to the northeast. At the same time a system of brushwood groynes was built to initiate accretion and a number of salt marsh creeks were dammed (Müller, 1917 b: 105).

The Hallig Langeneß (including former Nordmarsch and Butwehl) is currently completely surrounded by bank protection (fig. 290) supplemented by land reclamation fields in the east. Drainage of the central parts of the Hallig, especially after storm surges, has been further improved by construction of a new sluice near the Treuberg Warft (Riecken, 1981: 72).

The stabilization of the Hallig margins during the 20th century has completely halted coastal retreat though erosion of the tidal flat surface continues. The traces of medieval salt peat digging south of the Hallig (fig. 223; cf. Bantelmann, 1939) are being increasingly destroyed. König (1972: 40 and figs. 10 a and b) has compared aerial photographs of 1936 and 1958, demonstrating the degradation of the old land surface. Aerial photographs of the Landesvermessungsamt Schleswig-

Figure 291: The embankment connecting Langeneß, in the southwest, with Oland in the northeast. Construction of the embankment has enhanced mud accumulation in its lee. On the windward side a shallow creek has developed, endangering the embankment (Aerial photograph of 28.5.1985, freigegeben durch das Landesvermessungsamt Schleswig-Holstein unter der Nr. SH 510/85, Bildflug Nordfriesland, Str.ll/28l, vervielfältigt mit Genehmigung des Landesvermessungsamtes Schleswig-Holstein vom 3.12.85).

Holstein of 1985 reveal that erosion has continued since 1958.

Oland

Hallig Oland covers an area of about 117 ha. Fifteen houses exist on the single dwelling mound, together with the Hallig church and the school (Becker, 1976: 110). Oland is the only Hallig mentioned in King Valdemar's Jordebog (1231) under its present name, so that it can be regarded as the oldest Hallig. Nineteenth century sources indicate that in former times (in the 15th century) summer dykes existed on the island, as on Hooge and Gröde (Müller, 1917 b: 117).

Oland's history, like that of the other Halligen,

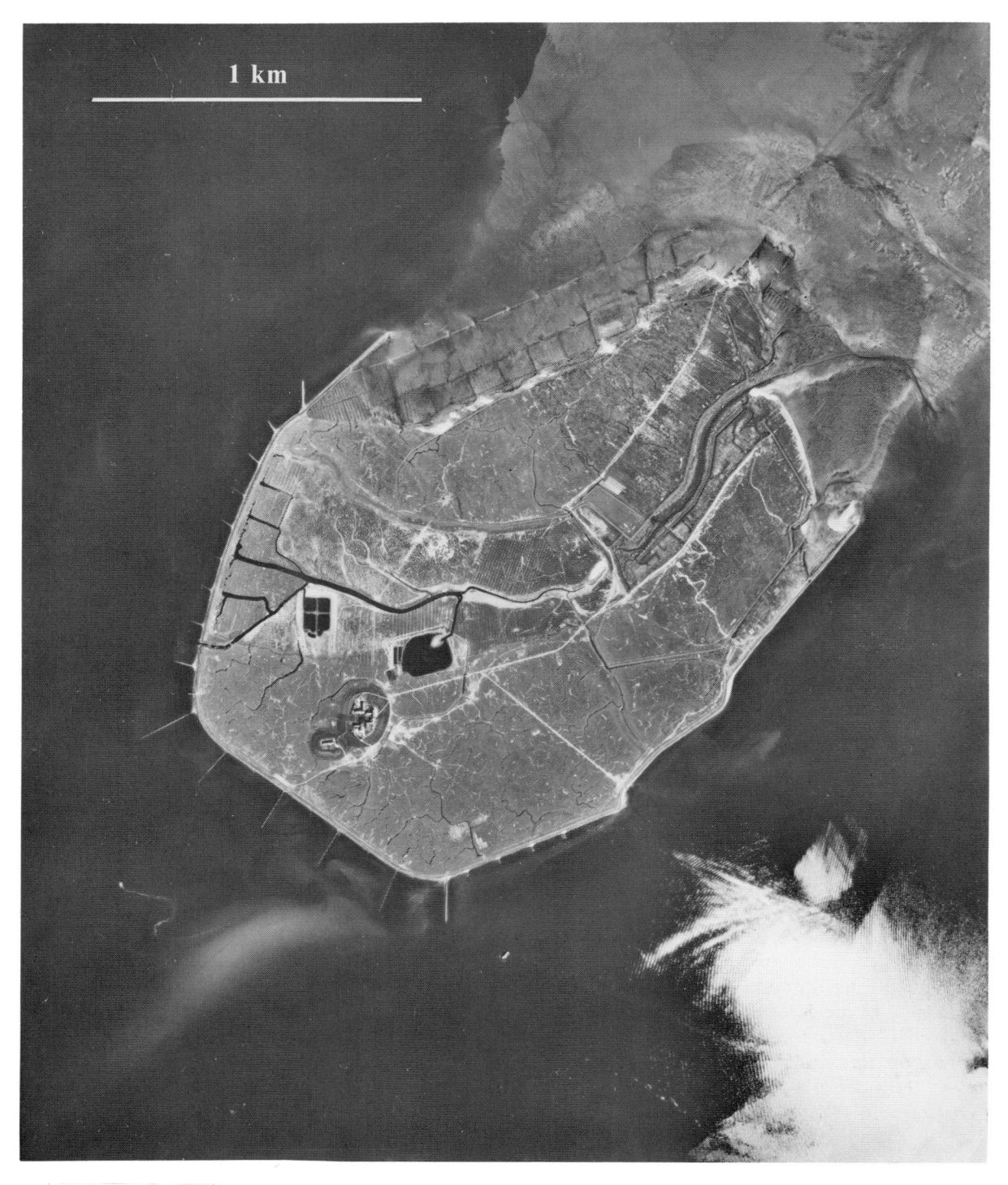

Figure 292: Hallig Gröde. The Hallig was formed by the dyking of two separate Halligen, Appelland in the north and Gröde in the south (Aerial photograph of 4.6.1985, freigegeben durch das Landesvermessungsamt Schleswig-Holstein unter der Nr. SH 510/85, Bildflug Nordfriesland, Str.12/238, vervielfältigt mit Genehmigung des Landesvermessungsamtes Schleswig-Holstein vom 3.12.85).

is a catalogue of land losses. The relocation of the church into the interior had to be considered in 1797, and this was undertaken after renewed land losses in 1824. The second Warft on Oland, the Piepe, had to be abandoned after 1840 (Müller, 1917 b: 120).

The heavily eroded western end of Oland, where about 240 m had been lost since 1750 (about 2 m/year), was finally stabilized by a stone dyke in 1896. In addition 10 pole groynes and 8 brushwood groynes were built and work commenced on the embankment connecting Oland with the mainland at Fahretoft. The embankment was completed on 12.10.1899. Between 1897-99 an embankment from Oland to Langeneß was also built. Under the protection of both embankments accretion occurred at the eastern ends of both Halligen. Together with the other coastal protection measures this led to a halt, or at least a considerable reduction, in coastal erosion (Müller, 1917 b: 130).

The storm surge of 30.12.1904 caused several breaches in the embankment between Oland and the mainland, but these were quickly repaired, as was the damage caused by the northwest storm of 28.9.1914 (Müller, 1917 b: 146). However, the embankment fell into disrepair during the course of the First World War and the storm surge of 16.2.1916 caused heavy damage (Becker, 1976: 110). The dam was rebuilt between 1925-27 along a new course between Oland and Dagebüll (length: 4,872 m) and equipped with a railway for direct transport to the Hallig. The embankment reached a height of 0.5 m above NN. The 3 km long embankment between Oland and Langeneß was also rebuilt in 1928 and equipped with a railway (Riecken, 1981: 39).

Recent accretion can be observed along the embankments between Oland and Langeneß and between Oland and the mainland (fig. 291). Young salt marshes have developed east of Oland and Langeneß, in the lee of the islands and the embankment and a foreland of over 180 ha has formed on the mainland in the protected bight between the embankment, Dagebüll and Fahretoft (Heydemann, 1981: 40).

Gröde

At present the Hallig Gröde covers an area of about 278 ha. There are only two dwelling mounds on Gröde, on one of which is sited a lodging house of the Marschenbauamt (Becker, 1976: 112). In earlier times Gröde must have been much larger bearing in mind that Butwehl was part of its parish. Oak and birch trunks, all of which had fallen to the southeast, were found southeast of Gröde and southwest of Habel in the tidal flats. As on Hooge, grain was grown in the summer polders on Gröde. About 1750 the remnants of two such polders remained, but grain production had been discontinued (Hansen, 1814: 15).

In 1625 the Gröde church was lost in the so-called 'Ice Flood'. The rebuilt church had to be replaced again after the storm surge of 1634, and between 1756 and 1759 the Knutz-Warf (= Knudtswarft) also had to be relocated. In 1779 the church had to be relocated for a third time, and the Warft Neuherst had to be abandoned after 1790. Shortly after 1860 the Warft Alte Herst also had to be abandoned (Müller, 1917 b: 155).

Neighbouring Appelland was inhabited until the mid-19th century, but after heavy land losses the last Warft was abandoned in 1860 (Müller, 1917 b: 174). The channel between Gröde and Appelland increasingly silted up during the 19th century. As the protection of Gröde and Appelland was only considered necessary as a bulwark to protect the mainland coast, the shortest possible course was chosen for the bank protection finally built between 1899 and 1902 which dammed the channel between the two Halligen. The revetment had an overall length of 2,400 m; it cut off 340 m of the western end of Appelland and 380 m of the western end of Gröde. These areas were quickly eroded, changing the characteristic form of the two elongated, west-east-striking Halligen into a single, almost round island (Müller, 1917 b: 195).

In 1926/27 the revetment was extended to its present length of 3,277 m to protect most of the Hallig margin (Riecken, 1981: 40). The Knudtswarft was additionally reinforced in 1967/68 by a ring dyke which was further raised in 1978 (Riecken, 1981: 72). Whilst the creek between Gröde and Appelland quickly silted up following the construction of a dyke in the middle of the double island in 1930, coastal erosion continued in the east. König (1972: 42) has demonstrated that between 1935 and 1968 about 30 - 50 m of salt marsh were lost. The light grey beach ridge visible on the aerial photograph (fig. 292) demonstrates that erosion is continuing today. Between 1968 and 1985 salt marsh erosion was about 10–15 m. It is important to note that this process is not limited to the unprotected part of the east coast but that it also extends into the brushwood groyne fields in the north. Here the margin of the salt marsh has been eroded in some places back to the foot of the summer dyke of 1952.

Habel

There are very few reliable old sources concerning the Hallig Habel. A plan in 1711 to connect Habel with the neighbouring Hallig Herst never materialized and Herst was subsequently completely eroded. Only two dwelling mounds were left on Habel in 1825. The Norderwarft was heavily damaged during the storm surge of 3./4.2.1825 and had to be abandoned after 1866. The Süderwarft was severely damaged during the storm surge of 14./15.10.1881 but was saved. The heavy land losses of Habel during the 19th century have been recorded by Hansen (1894). Habel was bought by the state in 1905 (Müller, 1917 b: 181).

Erosion continued unhindered by coastal protection measures so that the last inhabitant left the island in 1923 when the remaining area had become uneconomic. It was not until 1934 that coastal protection began (Riecken, 1981: 42), and the island had by then decreased to its present size of 4 ha. The gradient of the Warft slope was decreased after the storm surge of 1962 and the house was protected by a semi-circular dyke. Despite these measures both the Warft and house were severely damaged during the storm surge of 24.11.1981. The costs for the reconstruction of both were estimated at DM 600,000, a considerable expense for a tiny uninhabited island (Riecken, 1981: 9).

Hamburger Hallig

At the beginning of the 17th century the Hamburger Hallig still formed part of the foreland of the old Isle of Nordstrand. On 23.3.1624 the dyking rights for this foreland were given to the Amsinck brothers of Hamburg by Duke Friedrich, and they started to dyke their property during the same year.

Although the severe storm surge of 1634 was responsible for the final separation of Pellworm from Nordstrand, the Amsinck-Koog suffered less than its surroundings. On Meyer's map of 1649 the area is depicted as 'dry land', separated from the rest of Nordstrand by an abandoned polder area which soon suffered erosion. The proprietors remained on their land and attempted to reclaim further areas, but after the death of Arnold Amsinck in 1658 the polder had to be abandoned, and the Hamburger land evolved into a Hallig.

The Hamburger Hallig only re-awakened interest after the land reclamation works on the mainland coast were intensified after 1847. The Hamburger Hallig was the first Hallig to be connected to the mainland. The first brushwood groyne connection was built in 1859/60, but this was soon destroyed by ground failure in its central sections. Another brushwood groyne was built 1.4 km further to the south in 1866/67. In the meantime extensive accretion had occurred on both sides of the abandoned groyne fragments in the north. A stronger connection with the mainland was built in 1874/75. In 1877 the state bought the Hamburger Hallig; as a result the western end was secured by a 1,600 m long stone bank protection (1880-83), the connection to the mainland was reinforced (1880/81) and the northeastern end of the Hallig was strengthened by a 190 m long bank protection (1886/87) (Müller, 1917 b).

Today the Hamburger Hallig is effectively a part of the mainland with which it is connected by a broad stretch of foreland. It was originally planned to include this area in the Nordstrander Bucht dyking project (Ministerium für Ernährung, Landwirtschaft und Forsten, 1977: Anl. 9) but this suggestion was abandoned because of ecological objections.

Hooge

Hallig Hooge covers an area of about 594 ha and supports 10 inhabited dwelling mounds (Becker, 1976: 107). The Halligen Hooge and Norderoog are probably remnants of the former large island of Hwaelae minor mentioned in King Valdemar's Jordebog (1231). Müller (1917 b: 268) concurs with Geerz (1859; *op. cit.* in Müller, 1917 b: 269) in regarding 'Hwaelae' as the equivalent of 'Waeloe' (beach island). Hansen (1894 a: 66; *op. cit.* in Müller, 1917 b: 269) has pointed out that 'ae' was repeatedly written instead of 'oe' in Valdemar's Jordebog, for example in the Baltic Sea the islands Omae (Omø) and Sprowae (Sprogø).

Pellworm, when still separated from Nordstrand during the early Middle Ages, may also have formed part of Hwaelae minor. A close relationship between Pellworm and Hooge is suggested by the fact that some of the Hooge taxes had to be paid to the old Pellworm church. In addition, an old undated document in the Hooge parish archive states: 'It is well known that Hooge was connected with Pellworm...' (Müller, 1917 b: 269).

In the Petreus report (1597) it is stated that at Ipke Sandt, Amrhum and Buhrr great sand dunes extended far out to sea. The location of Ipke Sandt is unknown. From the existence of an Ipkens Warft on Hooge, Müller (1917 b: 270) suggests that the sand was possibly situated west of Hooge, in the

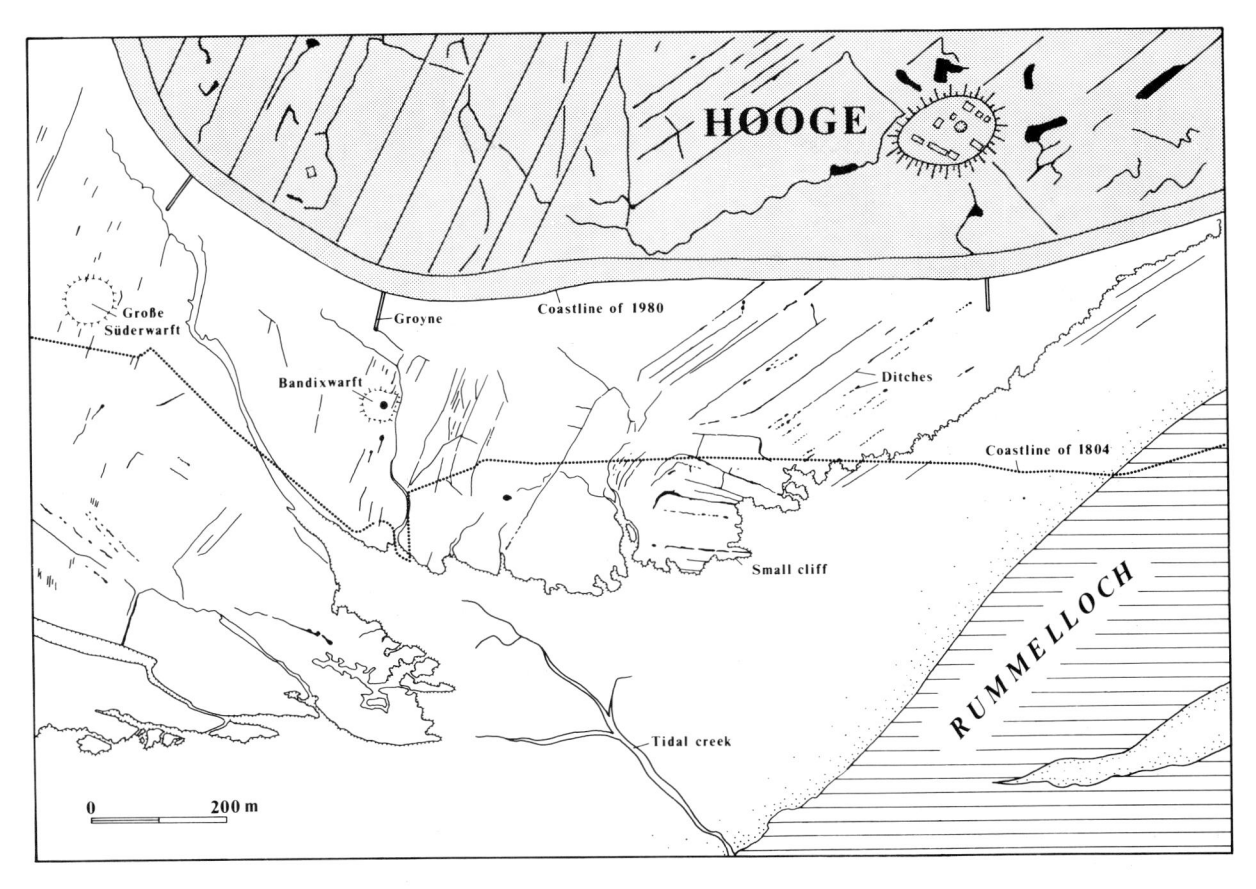

Figure 293: The old marsh surface exposed along the south coast of Hallig Hooge. Source: Aerial photograph of the Landesvermessungsamt Schleswig-Holstein of 1980.

area of the present Japsand and Norderoogsand, in which case it might have formed part of a formerly more complete dune barrier seawards of the Halligen.

It has to be stated, however, that this argument is very weak. The name Ipke was widespread on the Halligen and as a result of the patronymic name system it could have occured anywhere. It was only on 8.11.1771 that a state decree ordered the father's name system to be replaced by family names. This was only enforced from the early l9th century following another decree in 1812 (Panten, 1979).

A separate parish was established on Hooge following the widening of the tidal creek between Hooge and Pellworm which terminated the connection between them. The church on Hooge was built between 1624-37 and though it never had to be relocated, the Hallig suffered severe land losses, seven Warfts being lost between 1593 and 1653 alone (Müller, 1917 b: 271).

Despite these continuous land losses Hooge experienced a period of greatly increased wealth during the l8th century and the population increased to about 700. Many of the men worked as sailors and agriculture was neglected. Grain production in the summer polders on Hooge was finally abandoned following the 'Ice Flood' of 1720; these polders were never reconstructed (Müller, 1917 b: 275). The polder area is still visible in the field from its deviating drainage pattern. Similar polders existed on the other Halligen but all traces of them have been lost.

Remnants of the Warfts which were lost in the l9th century can be found on the tidal flats south of Hooge (fig. 293). The last house on the Sievertswarft was demolished in 1810 and the Große Süderwarft was also lost early in the 19th century. The Kleine Süderwarft (3 houses) and the Bandixwarft (2 houses) were destroyed by the storm surge of 1825 (Schulz, 1946: 17).

The golden age of the Hallig declined during the

Figure 294: Low summer dyke on Hallig Hooge; accumulation of shell fragments in the foreground (1981).

19th century, continuing coastal retreat forcing many of the inhabitants to leave. Coastal protection measures were not undertaken before the 20th century, the outermost Halligen until then being regarded as useless as a protection for the mainland coast and of too little value to be protected for their own sake.

The first experimental stretch of summer dyke with a stone pavement was built in the west in 1905, followed by the construction of 5 brushwood groynes in the southeast in 1906. The present summer dyke on Hooge was built in 1911. It is 11.1 km long and has a height of NN + 1.70–2.20 m. A 4,300 m long section on the southwestern, western and northern side has been protected by a stone pavement (fig. 294), and the open salt marsh creek mouths have been blocked by sluices (Müller, 1917 b: 290).

The construction of the summer dyke improved the agricultural potential of the vegetated area, the number of floods being reduced to 2–3 per year. This caused the intensified leaching of the salts by rain water percolation, and the flora changed to freshwater species. In 1914 only 190 cattle could graze on the island, but this had increased to 485 by 1925 (Riecken, 1981: 43). Further improvements were effected by the reinforcement of the dyke in 1927/28 (Schulz, 1946: 9).

A number of deleterious consequences of the dyking emerged in time, however. The reduction in the number of floods caused a decrease in soil fertility, and productivity soon declined. Attempts to grow oats on Hooge (1936) failed. The common land ownership on Hooge was transferred into private ownership in 1941 in order to stimulate more intensive agriculture on the island (Riecken, 1981: 43).

Nordstrandischmoor

Four dwelling mounds exist on Nordstrandischmoor, which at present covers an area of about 180 ha (Becker, 1976: 108). Like the Hamburger Hallig, Nordstrandischmoor is a remnant of the old Isle of Nordstrand which was destroyed in 1634. The so-called 'Wüstes Moor' (Deserted Moor) was an uninhabited swamp area unaffected by peat production. The inhabitants of the neighbouring polders found refuge in this area after dyke breaches. On discovering that their polders could not be reclaimed after the 1634 storm flood, some of the homeless settled on the 'Wüstes Moor', turning it into a Hallig. Sheep grazing provided a small source of income (Müller, 1917 b: 309).

The storm surges of the 18th and 19th centuries resulted in considerable land losses. A land reclamation commission in 1860 and 1861 reported that the island was suffering strong erosion in the west but that recent accretion had occurred in the east. The commission recommended that salt marsh formation should be enhanced by the construction of brushwood groynes and a groyne connecting Nordstrandischmoor with the mainland (Müller, 1917 b: 323).

Decades passed, however, before the protection of the Hallig became a reality. The storm surge of

Figure 295: Bank protection and brushwood groynes on Hallig Norderoog (1981).

Figure 296: Water reservoir (Fething) on Hallig Hooge, Backenswarft (1981).

14./15.10.1881 caused severe damage to the houses and dwelling mounds. In 1890 the four Warften of Nordstrandischmoor were inhabited by 27 people but by 1917 the numbers had decreased to 3 families living on just two dwelling mounds. Of the planned coastal protection structures, by 1917 only minor experimental sections had been completed (Müller, 1917 b: 332).

As late as 1933 a 5.4 km long sheet-pile groyne was built from Cecilienkoog to Nordstrandischmoor. The groyne was almost completely destroyed during the Second World War through lack of maintenance. Only the steel sheet-pile wall remained during the 1950's, heavily corroded by salt water (Riecken, 1981: 40).

The groyne was subsequently rebuilt. By the early 1970's a broad foreland had developed in front of the Cecilienkoog on the mainland coast, but at about the same time the state decided that subsidies should only continue for the maintenance of a 400 m broad foreland in front of the sea dykes. Major portions of the reclamation area therefore had to be abandoned, but the overall foreland area could be maintained (Andresen, 1980).

The area is currently affected by the construction of the Nordstrander Bucht polder. The new dyke will halve the distance between Nordstrandischmoor and the mainland (now 7.5 km) and will undoubtedly enhance accretion along the groyne.

Norderoog

Norderoog covers an area of about 11 ha (Becker, 1976: 103). This Hallig was first mentioned by Petreus in 1597 under the name of 'Norder Oug'. He reports that the island was inhabited. The effects of the 1634 storm surge on the island are unknown, Norderoog not being mentioned in Heimreich's (1819) account of the disaster. Archive material indicates that the island was still inhabited soon after this storm surge (Müller, 1917 b: 339).

Eighteenth century sources report that Norderoog was inhabited by only one family which lived by raising cattle and hunting seal (about 130 seals per year). The Hallig is said to have been surrounded by 'mud beds' which extended westwards to the so-called 'Englischer Sand' (Müller, 1917 b: 340), possibly the present Norderoogsand.

The Warft on Norderoog was damaged and the house on it almost destroyed during the 1825 storm surge, making the Hallig uninhabitable for a number of years. Two new houses were built in 1830 and 1834, but these were both lost, along with the Warft, during the storm surges after 1850 (Müller, 1917 b: 341).

Traeger (1892: 304) reports that sand approaching the Hallig from Norderoogsand encircled the small island on three sides, forming a low beach ridge with dune vegetation. Of all the Halligen beach ridges, these seem to have been the most well developed because of a positive sand supply. This positive tendency is also shown by the fact that the area of Norderoog remained almost constant between 1873 (22.7 ha) and 1912 (22.2 ha), when the first triangulations were made (Knop, 1963: Tab. 3).

The coastal retreat of Norderoog between 1927 and 1958 has been described by König (1972). Norderoog is still endangered by coastal erosion despite recent coastal protection works (figs. 211 and 295). Further developments in the tidal creek systems surrounding the island will be of fundamental importance for its survival. The small channel described by Knop (1963) between Norderoogsand and Norderoog appears to have remained the same size, but incision of the tidal flats south of Norderoog has intensified.

Süderoog

The Hallig Süderoog (fig. 288) covers an area of approximately 60 ha, and, like Norderoog, only had one house in 1597 according to Petreus. The Hallig was also called 'Boy Ocksens Halg'. The storm surge of 1634 severely affected the Hallig; 10 people drowned and two houses were completely washed away. Part of the Hallig was protected by low summer dykes when it was mapped in 1711. A report of 1867 mentions the unusually 'regular conduct of agriculture' as well as 'purposefully constructed and well kept dams' which should prevent expansion of the salt marsh creeks. Such features were not considered characteristic of the Halligen. Süderoog, however, was later eroded on all sides (Müller, 1917 b: 358) and because of its small size and great distance from the mainland it was not considered for any coastal protection projects (Müller, 1917 b: 363).

Südfall

At present Südfall covers an area of 56 ha. According to Petreus only one house existed on the Hallig in 1597. By 1634, however, the island population had grown, and there were plans to connect the island by dykes with Nordstrand. According to Heimreich (1819, Vol.II: 150), 46 people were killed during the storm surge of 1634, but the island was soon repopulated. There were still 7 houses on Südfall in 1781, but one of the two Warften, the Süderwarft, was destroyed during the 1821 storm surge, and a new Warft, the Osterwarft, was established in the interior of the Hallig. The storm surge of November 1824 caused more heavy damage to the island.

König (1972: 43), using aerial photographs, has documented the development of Südfall between 1936 and 1958. Only the western coast of the Hallig was protected, the eastern coast being further eroded. The revetment in the west was heavily damaged during the Second World War and had to be rebuilt in 1954 after a few more metres were lost by erosion.

Recent coastal protection measures

The dwelling mounds on the Halligen were originally relatively small, steep terps. Time and again it was demonstrated that the steep slopes increased the susceptibility of storm surge damage. The Warften also were too low to protect their inhabitants effectively against the extremely high water

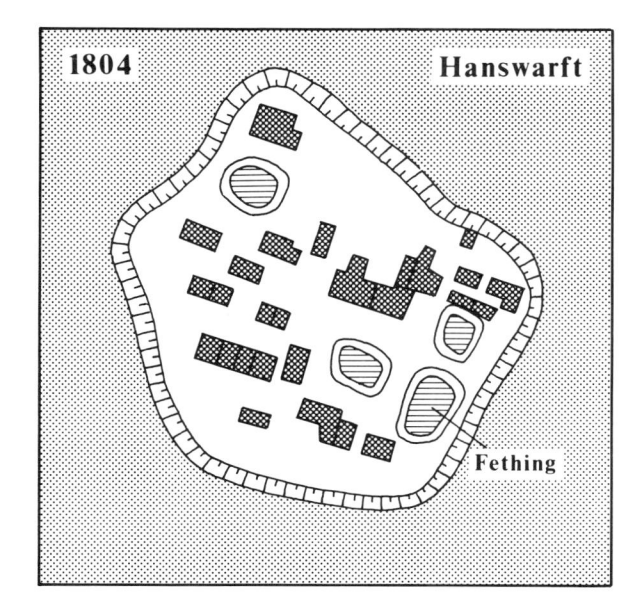

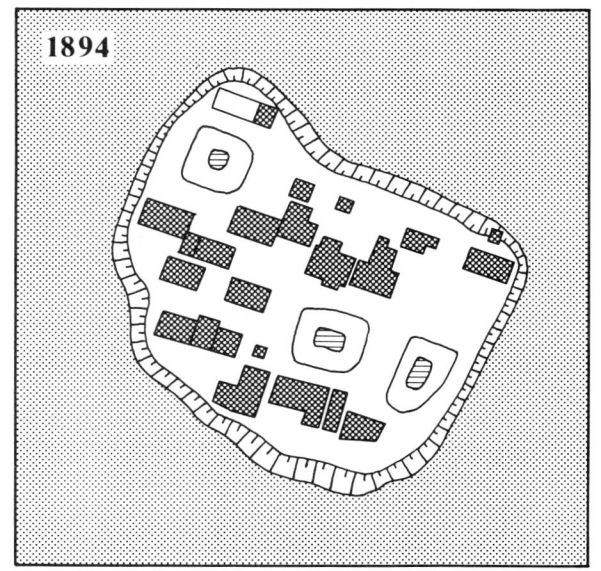

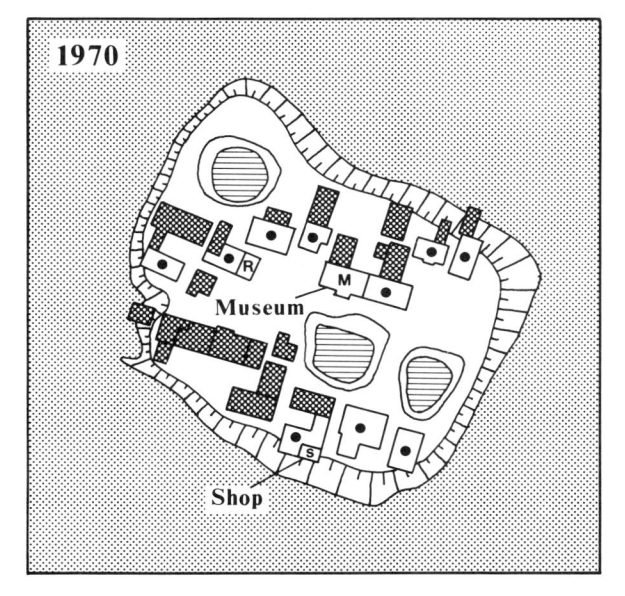

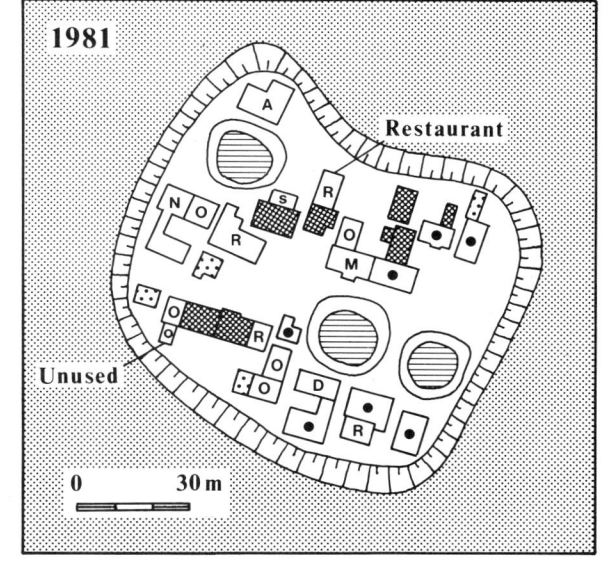

Figure 297: Changing use of the houses on the Hanswarft, Hooge. In the l9th and early 20th centuries all the buildings were used for agricultural purposes. Today tourism and associated activities dominate the Warft, and vacant buildings are awaiting change into guest houses or summer houses. Shaded = farmhouse; single dot = boarding house; N = environmental protection; A = administration; O = vacant; R = restaurant; dotted = summer house; M = museum; S = shop; D = depot. Sources: Traeger (1896), Müller (1917 b) and Riecken (1981).

levels that were experienced during storm surges. The Warft on Oland was the first to be raised, in 1939, and in 1953/54 the Norderwarft on Hooge was also raised and its slope gradient reduced (Riecken, 1981: 47).

The justification for such measures became obvious after the Holland storm surge of 1953. In a report of the 'Arbeitsgruppe Küstenschutz des Küstenausschusses Nord- und Ostsee' (1957: 132) several possible solutions were proposed. The best solution, the raising of all the Warften, was considered out of the question because of the high costs involved. Proposals were made instead to reduce the slope gradients of the mounds and to in-

Figure 298: Destruction of the old salt marsh drainage system on Hallig Langeneß; the salt marsh stratification is exposed in the ditch (1981).

stall storm-surge shelters. The 'Programme for the protection of the Hallig people and for the improvement of their conditions' was started in 1961.

The first developments took place in 1961. Apart from the reconstruction of a house on Oland and the construction of a new house on Langeneß, the three endangered farms on Nordstrandischmoor were regarded as the most urgent cases. The shelters installed in the new houses rest on four concrete piles sunk 4 m deep into the Hallig ground. The base of the shelters is situated about 2 m above Warft level, i.e. 6–6.5 m above NN (Riecken, 1981: 49).

The severe storm surge of 16./17.2.1962 emphasized the urgency required in the restoration of the Halligen; the programme, originally estimated to take 10 years, was completed within 5 years. The estimated costs increased because some of the houses which had originally been planned for modification had to be completely rebuilt after their destruction during the 1962 flood. Seventy-four individual projects were completed at a total cost of 12,700,000 DM, of which 6,400,000 was supplied by the Hallig people. All the Warft slopes had been modified by 1968 (Riecken, 1981: 62).

Structural change

A number of measures were undertaken simultaneously to improve the infrastructure of the Halligen. With the single exception of Süderoog, all the inhabited Halligen were connected to mainland electricity and water supply systems (fig. 296) between 1954 and 1976. New landing places were built to improve traffic connections (Gröde, Hooge and Langeneß: 1963, Südfall: 1968; Nordstrandischmoor: 1975), and a regular boat connection between Hooge, Langeneß and the mainland (Schlüttsiel) began in 1960. The roads and tracks on the Halligen were also improved (Riecken, 1981).

These developments initiated an enormous

structural change on the Halligen from about 1965, dominated by the increasing influence of tourism. There is a growing tendency for the islanders to sell their houses to affluent people from the mainland. This has resulted in a population decline on Hallig Hooge between 1970 and 1983 from 191 to only 119 inhabitants, a reduction of over 30 % (Schirrmacher, 1983: 167). At present (1983) there are 51 islander households as opposed to 11 holiday apartments or summer houses (almost 20 %; Schirrmacher, 1983: 113) on Hooge. The income from tourism far exceeds the potential earnings from agriculture, and growing numbers of islanders are turning to this new source of wealth (fig. 297). The destruction of the traditional Hallig landscape by modern agricultural programmes, such as have occurred on Langeneß (fig. 298), should be avoided in the future.

7.2.5 Pellworm

Like its neighbour Nordstrand, the Isle of Pellworm is a remnant of the former large Isle of Strand ('Old Nordstrand') which was destroyed during the catastrophic storm surges of 1362 and 1634. Pellworm was initially separated from Nordstrand during the 'Große Mandränke' of 1362, and the islands were only reconnected by the dyking of the Norder Neuen Koog in 1550/51 after 200 years of incessant labour (Müller & Fischer, 1936 a: 48)

Following this dyking the Isle of Old Nordstrand took on a horseshoe-shaped configuration as depicted in a number of old maps. Further dyking projects were of only limited success. The second 'Mandränke' of 11./12.10.1634 led to the final destruction of the large island, Pellworm being at last separated from Nordstrand. Nordstrandischmoor and Hamburger Hallig were also separated from Nordstrand at the same time (Müller & Fischer, 1936 c).

Despite the enormous damage inflicted on Pellworm by these repeated inundations, re-dyking commenced almost immediately. By as early as 1635-37 the dykes of the Großer Koog, Kleiner and Mittelkoog, Alter Koog and Wester Neuer Koog (= Johannes-Heimreich-Koog) were reconstructed. The Norder Neuer Koog could be reclaimed in 1657, the Westerkoog in 1644, the Hunnen-, Süderkoog and Hensebek-Koog in 1672, and finally the Großer Norderkoog in 1687.

Further reclamations were not possible for a long period. It was only in 1790 that the small 2.5 ha Ostersiels-Koog was dyked, followed in 1938 by the dyking of the Buphever-Koog, bringing Pellworm up to its present size (Quedens, 1982). The Buphever-Koog was regarded as the first step towards the completion of an embankment connecting Pellworm with the mainland, but it took another five decades before this project reached fruition.

The Pellworm economy has always been based on agriculture. The large distance to markets and the relatively inaccessible connection with the mainland, however, have proved increasingly adverse. Being a marsh island without a potential recreational beach, Pellworm is unable to attract many tourists, but some improvements may accrue with improved links.

7.2.6 Nordstrand

In contrast to Pellworm the surviving population on Nordstrand after the storm surge disaster of 1634 was insufficient to restore the dykes. At first it proved impossible to attract other 'Interessenten' in re-dyking projects, and the island deteriorated over a period of 17 years, when many of the inhabitants emigrated.

Dutch 'Partizipanten', who were endowed with excessive privileges by the Duke, finally started re-dyking in 1654. Quedens (1984: 23) has described the enormous problems faced by the local population who suddenly found themselves deprived of almost all rights. Resistance was so great that the participants had to hire Dutch dyke workers in order to proceed with their projects. Nordstrand became a Dutch colony *de facto*; Dutch became the main language used, and a catholic church was established on the island.

After a series of successful dykings, which were very profitable for the participants, they ran into bad luck during the middle of the 18th century. The participants had invested enormous sums in the dyking of the Christians-Koog between 1737-39. The project became far more expensive than anticipated, but large profits were expected. The dykes were breached, however, during a storm surge on 11.9.1751, and the debtors could not raise the additional sums required for their repair. The final fate of the project was sealed in 1756 when the remnants of the dyke were levelled by another storm surge. Some of the participants went bankrupt, leading finally to a complete withdrawl of the Dutch from the island (Quedens, 1984: 29).

Part of the abandoned Christians-Koog was re-dyked in 1770/71 by Count Desmercieres and given the name 'Elisabeth-Sophien-Koog'. This polder, which was especially well dyked, was the only one on Nordstrand which withstood, almost unaffected, the storm surge of 1825; the other polders (Trendermarschkoog, Oster-Koog, Alter Koog and Neuer Koog) were all seriously damaged. The old dykes were finally modernized after the flood. This had previously been demanded by the dyke inspector but rejected by the local population.

The Morsum-Koog, which was dyked in 1866 two years after the German-Danish War, was the last privately dyked polder on the island. Coastal protection and land reclamation were declared state concerns in 1871. The first dyke stone pavements on Nordstrand, covering 3 km, were completed in 1879, the state contributing 70 % of the costs (Quedens, 1984: 39).

A connection was built between Nordstrand and the small Pohnshallig island to the east at the same time as the dyking of the Morsum-Koog (cf. section 5.3.5). Rapid accretion began, and the dyking of another polder soon became possible. The plans had to be shelved at the beginning of the First World War, but work started in 1919. A number of problems, including storm surge damage in 1923, delayed the completion of the dyke until 1925 (Quedens, 1984: 40).

In 1935 Nordstrand was connected with the mainland by a causeway. Extensive land reclamation works were undertaken on both sides of this embankment, and within decades a large foreland had developed. The northern part of it will be incorporated in the current dyking of the Nordstrander Bucht.

The dyking project was controversial mainly because of its anticipated ecological effects; the most likely consequences have been analysed in a series of reports, a shortened version of which has been published by the Schleswig-Holstein Minister of Food, Agriculture and Forests (Minister für Erährung, Landwirtschaft und Forsten, 1981). The hydrological consequences were investigated by Partenscky & Schwarze (1981 a, b) and Partenscky & Dieckmann (1981), the geological context by Köster (1981) and the morphological situation by Higelke (1981). These specialists rejected the original major plan, favouring the alternative so-called 'small solution' consisting of a dyke from the Sönke-Nissen-Koog to Nordstrand (Flessner, 1981: 8), ensuring maximum coastal protection at minimum environmental cost.

7.2.7 EIDERSTEDT

The Eiderstedt Peninsula is the only area between Blåvandshuk and Den Helder where the barrier coast forms part of the mainland. The fossil beach ridges of Garding (fig. 299) and Tating demonstrate that the area formed an early extension of the mainland. The present beach barrier system at St.Peter–Ording, however, cannot be regarded as a continuation of the old system, and has developed more recently under completely different circumstances (Menke, 1984: 56).

Historical development

The first settlement on Eiderstedt, on the elevated beach barriers, started during the first centuries A.D. The settling of the flat marsh followed at about 100 A.D. and continued uninterrupted until the 5th century. Subsequent evidence is lacking, possibly explained by the emigration of Angles and Saxons into England. Traces of renewed settlement date only from the 8th century and the succeeding Viking Age (9th – 11th century). The original settlements on the flat marsh surface were later altered into dwelling mounds (Bantelmann, 1976: 9).

Progressively extensive marine transgressions subdivided Eiderstedt into two islands, Everschop and Utholm (fig. 300), which were separated from each other by the Süderhever tidal channel. The dyking of a number of small polders in the Süderhever area enabled the reconnection of Everschop and Utholm in 1231.

The heavy storm surges of the 14th century which destroyed North Frisia also affected Eiderstedt. The Isle of Westerhever was separated from Utholm in 1361 (Rohde, 1976: 75) and had to be temporarily abandoned (Fischer, 1956: 76).

The existing Tümlauer Bucht is a remnant of the former Fallstief, a tidal stream which was connected with the Süderhever and which had severed Westerhever from Utholm. The Fallstief was dammed following heavy damage during the All Saints Flood of 1.11.1436. Further dykings connected Westerhever with the large Isle of Everschop (Fischer, 1956: 76). It was not until 1489 that it was possible to reconnect Eiderstedt with the mainland by dyking the Dammkoog (Fischer, 1956: 66).

The second 'Mandränke' of 1634 caused permanent land losses especially in the Lundenbergharde area of northeastern Eiderstedt.

Land reclamation work intensified during the

Figure 299: The Haferacker sand pit at Esing on Eiderstedt exposing the fossil Garding beach barrier (1984).

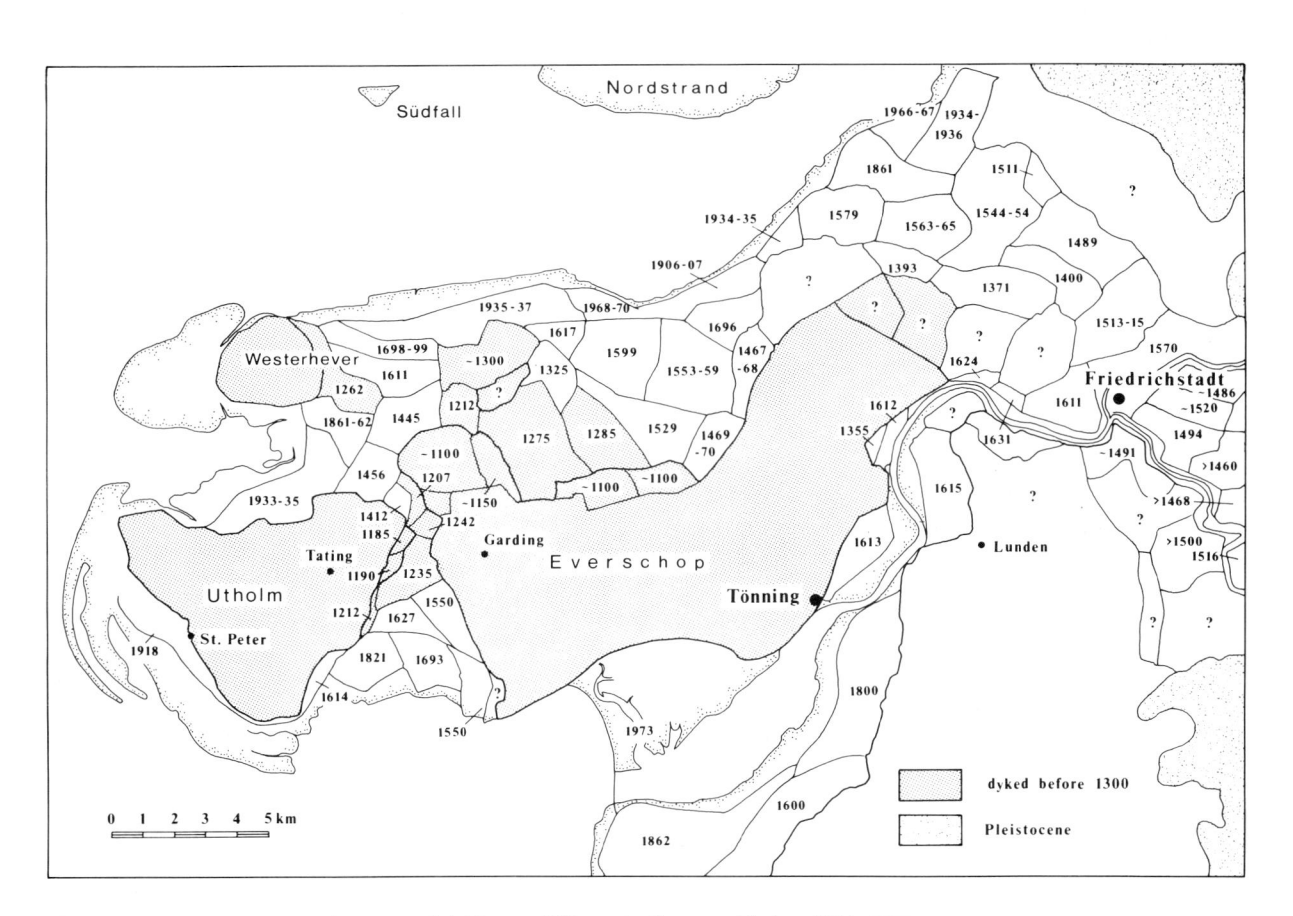

Figure 300: The historical development of dyking on Eiderstedt. Sources: Fischer, 1956, 1957, 1958.

Figure 301: Shifting sand on the recent beach barrier at St.Peter-Ording, Eiderstedt Peninsula (1981).

Figure 302: Densely vegetated lagoon between the beach barrier and the coastal dunes at St.Peter-Ording. Desiccation cracks developed in the mud deposits during the ebb; the whole area is now flooded again by the rising tide (1981).

20th century. The Tümlauer Koog (585 ha) was dyked between 1933-35, the Ülvesbüller Koog (105 ha) between 1934/35, the Finkhaushalligkoog (470 ha) between 1934-36 and the Norderheverkoog (605 ha) between 1935-37.

The dunes

The age of the Eiderstedt dunes is unknown. Some indication of their age can be deduced from the relationship of the dunes to the dykes. The first dykes on Utholm were constructed in the west, and at present the dyke is situated behind the dunes. The dyke of the old Ording Koog also dissects the dune area, suggesting that it too might be older than the dunes. Dune formation therefore seems to have begun only after the storm surge of 1362 when the polders were destroyed.

However, it is known that the dunes were in existence at the time of the 1634 storm surge (Jensen, 1933: 60). In 1725 the church at Ording had to be relocated over 1 km to the east because it had become engulfed in dune sand. The old church was subsequently completely covered by the migrating dunes; its remains later appeared on the beach and were completely destroyed during the storm surge of 1788 (Degn & Muuss, 1979: 174)'

The dune ridge was breached at St.Peter and the polder flooded during the storm surge of 1825. Since then the dunes have stopped migrating (Jen-

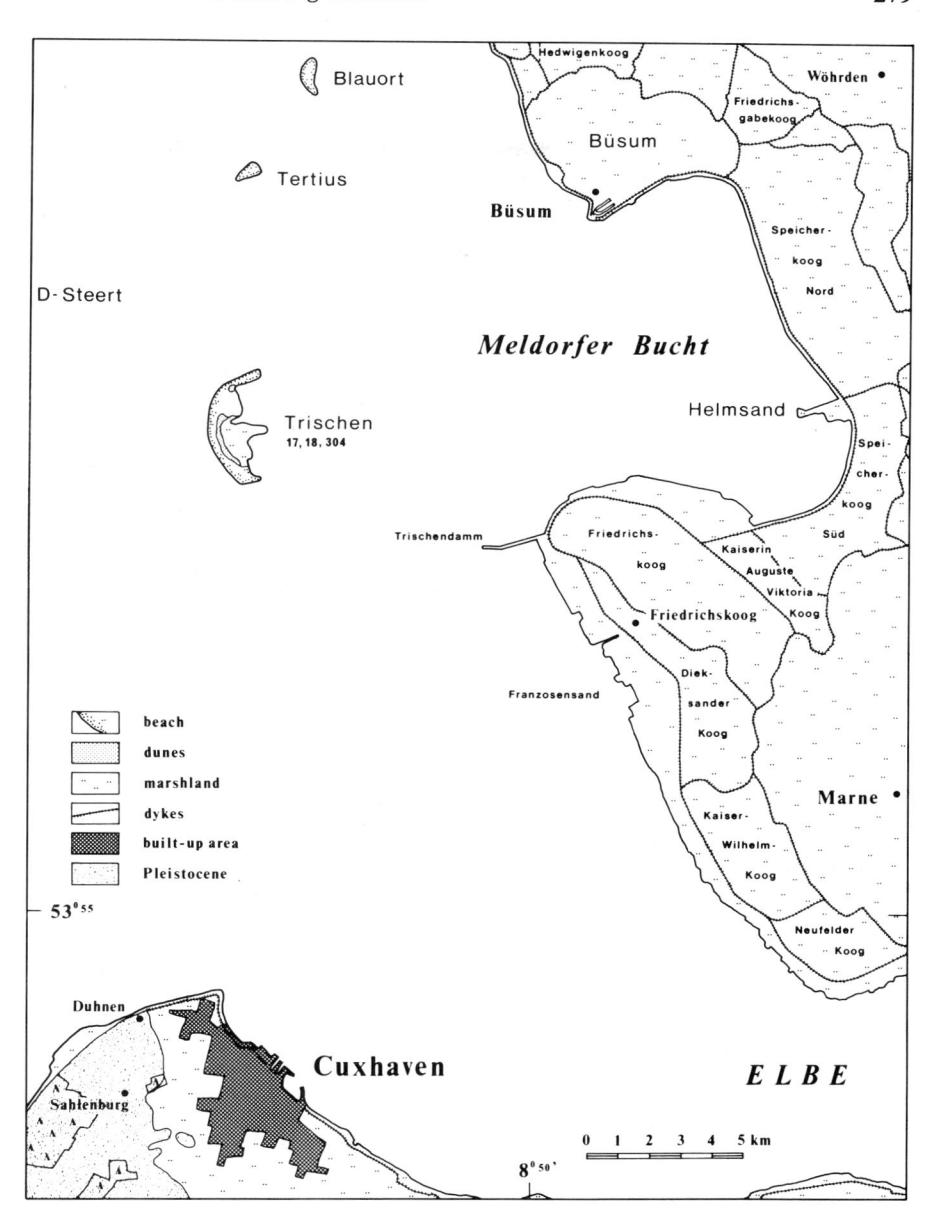

Figure 303: Location map of the area between Blauort and Cuxhaven. The numbers refer to figures, those prefixed by 'pl' refer to plates.

sen, 1933: 60) and are now largely covered with pine forest.

The beach barriers

At present the sand budget of Eiderstedt is remarkably positive. The aerial photograph of St.Peter-Ording (fig. 206) and the satellite image (plate 18) both show the system of broad, flat beach barriers that are accumulating along the coast. Whilst the barriers migrate slowly eastwards under the influence of wind (fig. 301) and waves, the sand transport on them is directed mainly southwards. This was first demonstrated by Jensen (1933: 104) who detected a progressively southwards fining of sand based on grain-size analyses of samples from the 'beach flat' and the seaward 'sand bank'.

Elongated lagoons have developed under the protection of these barriers; these gradually silt up and become vegetated (fig. 302). Eventually the barriers transgress the lagoonal sediments leading to a further elevation of the ground.

The sand balance has not always been this positive. After about 1610 the old beach barrier extended as far north as the present Rochelsand (north of

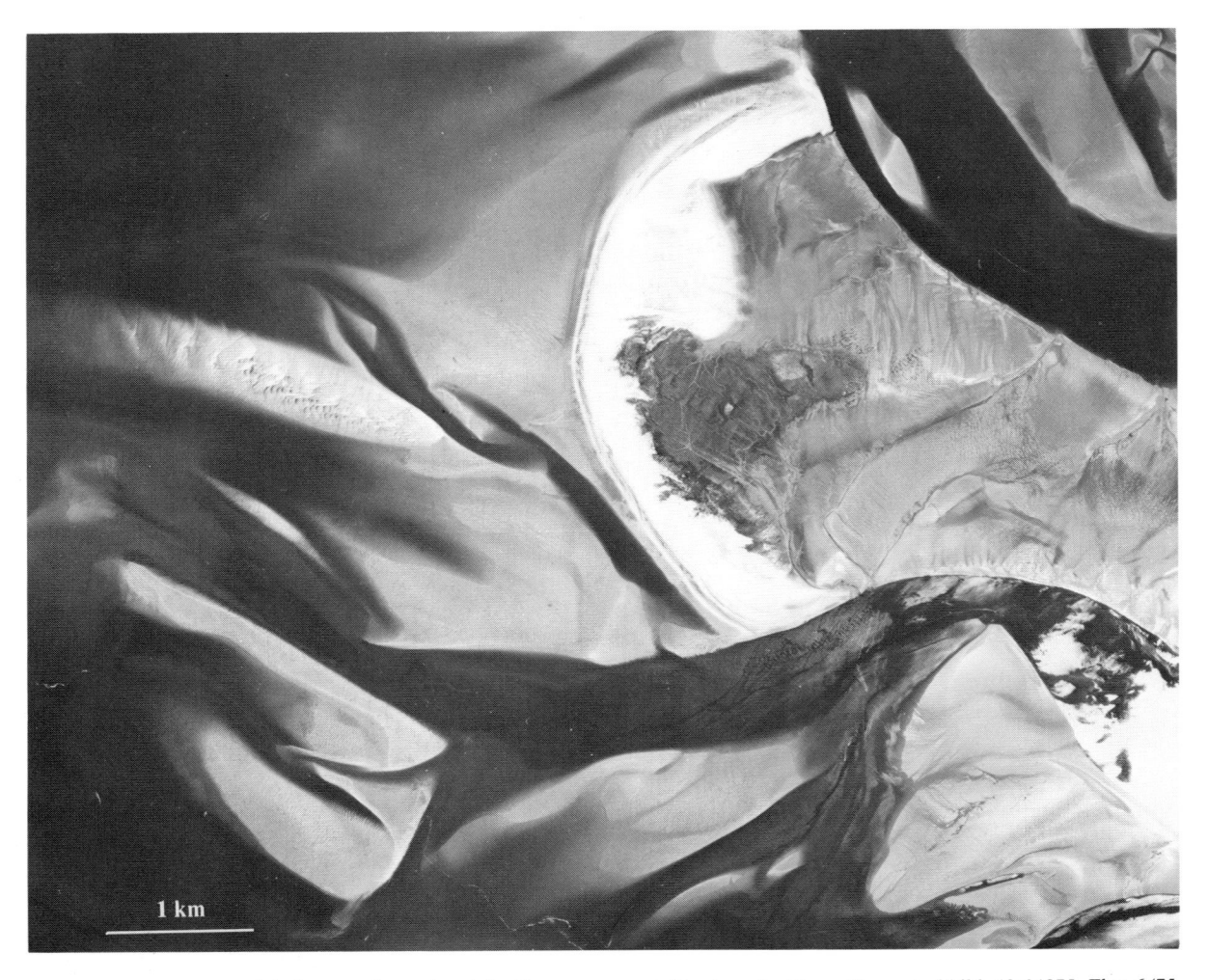

Figure 304: Trischen (Aerial photograph produced for Vermessungsamt Hamburg by Firma Hansa Luftbild, 12.6.1975, Flug 6/75, Str.10/68, freigegeben durch Regierungspräsident Münster unter Nr.1426/75).

the Tümlauer Bucht), but it was subsequently eroded from the north and migrated eastwards leading to some coastal retreat (Degn & Muuss, 1979: 174).

7.2.8 THE MOBILE ISLANDS OFF DITHMARSCHEN

The coast of Dithmarschen between the Eiderstedt Peninsula and the Elbe Estuary (fig. 303) is an area of strong sediment displacement expressed in rapid channel and bar migration (cf. section 5.2.3) and in the formation and destruction of shoals and islands. The surface morphology of the area has been described by Wrage (1930, 1958), and the channel migrations have been traced by Bahr (1963) and Higelke (1978). Lang (1975), who has studied the morphological development of the area from the middle of the 16th century until the present, using old maps and documents, has provided a thorough description of the histories of the islands and sands.

Trischen

Most is known about the development of the Isle of Trischen. Old documents from the 16th century

Plate 22: 'Landunter' on Hallig Langeneß, looking towards the Rixwarft (1981).
Plate 23: Salt marsh margin under erosion at eastern Borkum. The eroded sand is thrown onto the salt marsh forming a low sandy beach ridge whilst the fines are removed (1984).

PLATES 22-23

PLATES 24-26

give no indication of the island under its present name or any other; it can be assumed that the island was not in existence at this time.

Trischen is first mentioned in a court document of 1610, and has appeared regularly in documents since then. Even by 1689, when the island was first described in some detail it was referred to as 'a very low sand'. Trischen was first mapped in 1721, if only in a strongly distorted form. Strong salt marsh formation on its landward side was recorded in 1735, but this area was not used for hay production. Maps from the middle of the 18th century fail to record any vegetation on the island, suggesting that this recent salt marsh was soon covered over by sea sand (Lang, 1975: 52).

By the end of the 18th century, however, some vegetation seems to have developed in the east of the island as indicated by some comments written on a map of 1797. These old sources also indicate that Trischen had always been very mobile. The development of vegetation was not mentioned again in documents and maps until the middle of the 19th century (Lang, 1975: 56).

Accretion was recorded as occurring in 1854 in the form of small vegetated 'islands' only 1 m^2 in size. These 'islands' subsequently grew, and in 1868 land reclamation projects were initiated to promote this accretion. The salt marsh area had expanded to 16 ha by 1872, the *Salicornia* areas to as much as 47 ha. A grass meadow of 66 ha had formed by 1884. From 1885 onwards low dunes started to form in the west, less than 1.5 m high, under the protection of which the meadow grew to a size of 103 ha by 1894.

Dune formation increased further. By 1898 the dunes had reached a height of 6 m and formed a 1.2 km long and 50–100 m wide belt in the western part of the young island. The vegetation was bounded in the west by a sand flat 1 km wide and 7 km long including the dunes (Lang, 1975: 58).

The early agricultural use of this vegetated area had to be abandoned after the severe winter storms of 1898/99 when the dune ridge was breached and a large part of the area covered with sand. However, accretion continued in the east, land reclamation works beginning again in 1902. By 1907 a vegetated area of 80 ha had redeveloped. This increase then stagnated when fresh accretion in the east was balanced by erosion and sand coverage, especially in the north (Lang, 1975: 59).

Although it was known at this time that deeper channels were approaching Trischen in the west, there were plans laid down after the First World War to dyke the salt marsh in order to permit regular farming on the island. This project was started by private initiative in 1922 and completed in 1925, encircling a polder of 89 ha. However, dune erosion continued in the west (Lang, 1975: 60).

When the state took over the responsibility for coastal protection the dyke was raised and supplemented by groynes and a revetment between 1927 and 1935. The heavy damage caused by the October storms of 1936, however, led to the final decision to abandon the island despite the high investments already incurred. The tenant persevered on the island, but erosion proceeded rapidly. When the dyke of the Trischenkoog was breached in 1942, large parts of the vegetated area were spoiled by sand. Finally in 1943 the tenant had to leave Trischen (Lang, 1975: 61).

Since then the island has migrated further eastwards, but maintained its old shape (fig. 304). It still consists of a sickle-shaped sandy shoal, with a core of low dunes. The vegetated area, which is still extending eastwards, is situated in the lee of the dunes.

Dieksand

Like Trischen the former Isle of Dieksand seems to have formed from an emerging sandy shoal. The island was first referred to as the 'Hochsandt' (high sand), its first description as 'Dieksand' being in 1545. Vegetation started to develop on the island in 1585. Accretion intensified during the following years, and the island became inhabited in 1594. However, Dieksland continued to be displaced in a northeasterly direction, and in 1616 the tenant had to leave, because his dwelling mound, the third on the island, had been destroyed and the pasture covered with sand (Lang, 1975: 64).

A fourth Warft was built on Dieksand before 1695, and after further accretion part of the island was dyked in 1817. This polder did not last long

Plate 24: Old marine clay exposed seaward of the revetment along the western end of Wangerooge (1979).
Plate 25: Detail of the old clay layer showing shells in life position. The shells have yielded a radiocarbon age of 1500 B.P. (1979).
Plate 26: Construction pit in the centre of Wangerooge, showing intercalations of sand and humic layers resulting from phases of flooding and subsequent salt marsh growth (1978).

however, the dykes being destroyed by the storm surge of 1825. In 1853/54 it finally became possible to connect the entire Dieksand, renamed Frederik-VII-Koog (today: Friedrichskoog), with the mainland (Lang, 1975: 73; Fischer, 1957: 260).

Helmsand

The present Isle of Helmsand developed around 1720 as an emerging shoal. Extensive *Salicornia* vegetation was reported growing there about 1730, and by 1737 a 2.6 ha area of vegetation had developed. The island grew rapidly in size and by the first triangulation in 1754 it covered ca. 87 ha, reaching almost 100 ha by 1775 (Lang, 1975: 41).

When it was finally decided in 1865 to build a dwelling mound on the island to allow permanent settlement, the positive tendency had long since reversed. Helmsand had decreased to 75.5 ha by 1818, possibly because of an approaching tidal creek from the east. Only 33.5 ha were left by 1865, dwindling to a mere 19.4 ha by 1881. Only the embankment connecting the island with the mainland, which was completed in 1928, saved it from complete destruction (Lang, 1975: 45).

Büsum

The Isle of Büsum, like the other islands off Dithmarschen, developed from a sand bank. The Büsum parish is first mentioned in a document of 1140. The chronicler Neocorus, who started his work on the island in 1578, reports that Büsum was bordered on the seaward side by dunes. Salt marshes seem to have developed under the protection of these dunes. Though the dunes were later completely eroded, the marsh island itself was saved. The gap between Büsum and the mainland was silted-up to such an extent by 1585 that it could be crossed by a dyke. In 1609 the Wardammkoog (790 ha) between the island and the mainland was dyked. Later dykings further connected Büsum to the mainland. Only the courses of the old dykes give an indication of the shape of the former island (cf. fig. 303) (Fischer, 1957: 56).

Maxqueller

The two neighbouring Maxqueller islands were formed in the southwestern part of the Dithmarschen tidal flats around 1765. The vegetated area had reached such a size by 1782 that it could be tenured for hay production. With further accretion in the east the distance to the mainland decreased, but the islands were eroded in the west. The Maxquellers were connected in 1846 after an attempt in 1834 had failed. By 1857 the remnants of the Großer Maxqueller had become attached to the foreland of the Kronprinzenkoog. The whole area was finally dyked as the Kaiser-Wilhelm-Koog in 1873 (Lang, 1975: 78).

Another salt marsh island, the Franzosensand, developed north of the Maxquellers during the 19th century; it was dyked between 1933 and 1935 after becoming attached to the mainland (Adolf-Hitler-Koog; today: Dieksander Koog) (Lang, 1975: 81).

Extinct islands

A number of islands which once existed on the tidal flats off Dithmarschen were later destroyed by erosion. In most cases the information available about these former islands is rather vague:

Bielshöft and *Hundsand* were neighbouring, sand islands near Büsum, which are mentioned repeatedly from the 16th century onwards. Although their exact positions cannot be reconstructed from the strongly distorted maps, they were probably located WSW of Büsum. The islands seem to have grown in size during the first half of the 17th century. It is likely that small salt marsh areas developed on Bielshöft in the lee of low dunes. The dune ridge was breached early in the 17th century, and reconstruction was regarded as impossible. That the island was inhabited at this time is indicated by the single house drawn on contemporary maps, but this house is missing on maps from 1650 onwards. Bielshöft was still used for grazing, but the island never recovered after the Christmas Flood of 1717. When the island was last mentioned in 1754 it was being used for sheep grazing, on later maps only appearing as a sand bank (Lang, 1975: 23).

Lang has pointed out that *Alt-Helmsand* was not identical with the present Isle of Helmsand, which formed about 1720 (1975: 36). It is probable that Alt-Helmsand was situated 1–2 km further north and also further seawards. Alt-Helmsand was first mentioned during the 14th century. Hay was being produced on the island in 1524, and since at least 1551 the island was used by people from Büsum for cattle grazing. Strong erosion started during the second half of the 16th century. The island ceased to be used for grazing after the All Saints Flood of 1570, and hay production also ceased in 1578. Alt-Helmsand completely disappeared about 1600.

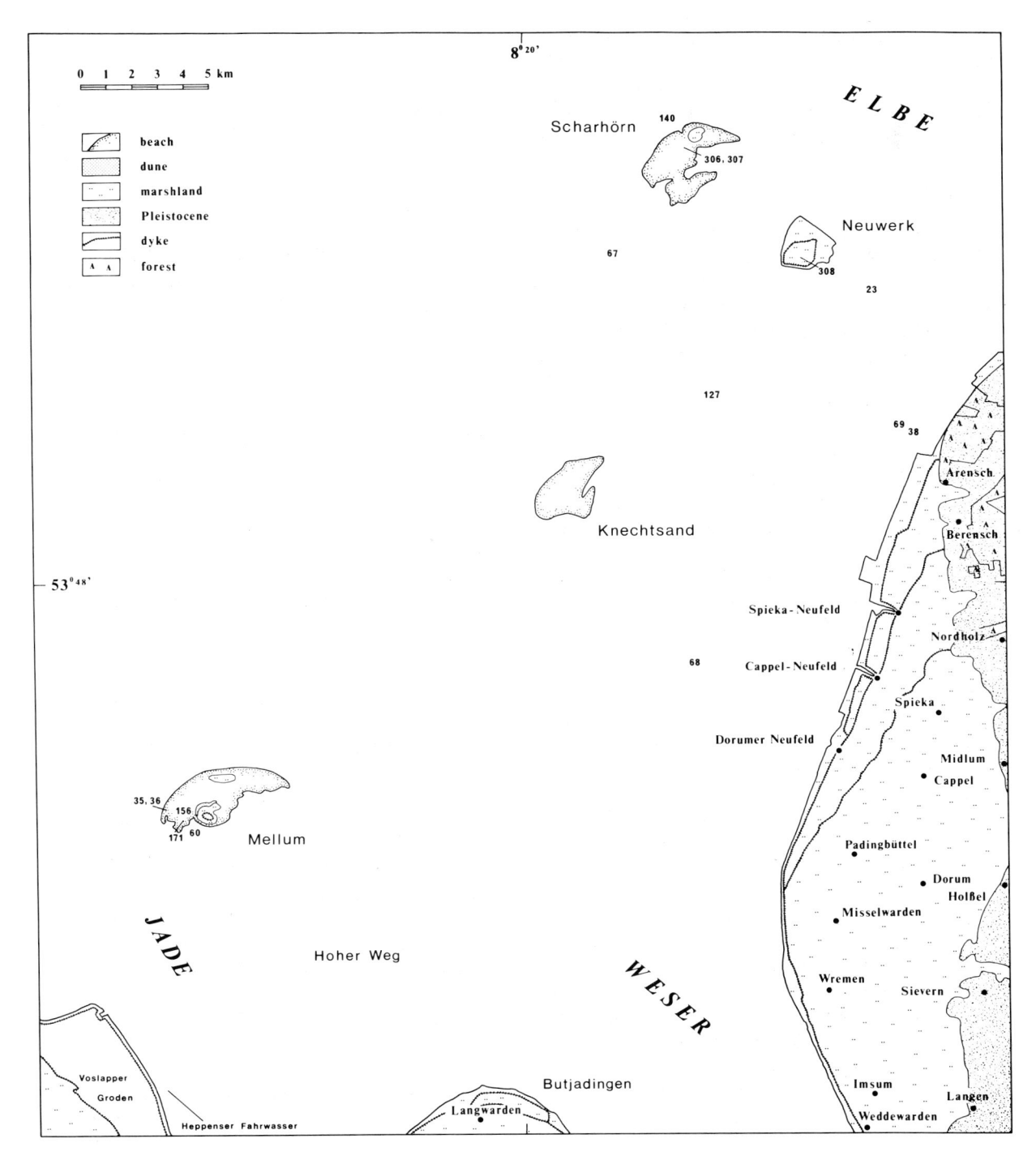

Figure 305: Location map of the area between the Elbe and Jade estuaries. The numbers refer to figures, those prefixed by 'pl' refer to plates.

Tötel was first mentioned in 1568, its size reported to be about 46 ha and 'covered with grass'. By 1596, however, the Isle of Tötel 'had been torn apart by the wild waters' (Lang, 1975: 48).

Lang (1975: 49) points out that the destruction of these islands certainly did not occur as a single sudden event. The islands were unlikely to have been 'torn apart' by a single storm surge; it is more probable that their grassland was covered with sand, which led to vegetational changes, permit-

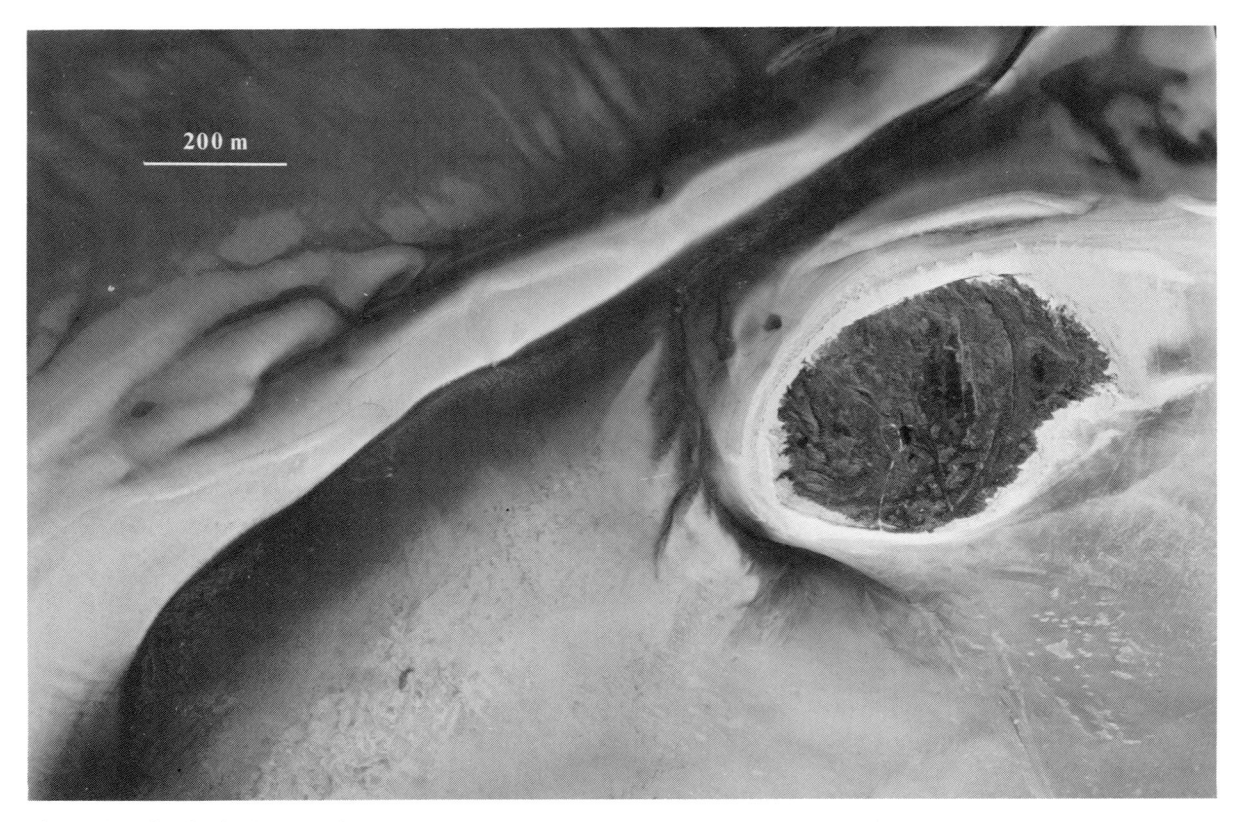

Figure 306: Isle of Scharhörn. Aerial photograph of the Landesvermessungsamt Hamburg of 25.4.1984, Bildflug 210/84, Str.1/668, freigegeben vom Luftamt Hamburg unter Nr. LAH 610/84.

ting the colonisation of different plant communities less resistant to erosion than the original salt marsh vegetation. Thus, after some decades, the islands were finally levelled by continued erosion.

The high sands off Dithmarschen

Blauort

The sand island of Blauort west of Büsum has repeatedly changed its shape in the course of its development. However, since it was first mentioned in 1551 it has never been recorded as having dunes or salt marshes (Lang, 1975: 20). The sand seems to have been largely static between 1841 and 1938, but afterwards it started to migrate eastwards at a rate of about 35 m/year (Wieland, 1972: 133).

Tertius

The Tertius sand island was situated 10 km west of Büsum in the fork between the Norder- and Süderpiep tidal streams. The sand was a relatively immature feature; it was already recorded on a chart of 1900, but situated well below mean tidal high water (MTHw). It also appeared as such on a chart in 1922. On aerial photographs taken in 1938, however, Tertius can be seen as a U-shaped sand island above the level of high water. The U, which was open to the east, surrounded a small muddy tidal flat area. The eastward migration of Tertius has been recorded in aerial photographs taken in 1958 and 1966 by König (1972: 60). As it migrated it decreased in size, the Küstenkarte 1:25,000, Sheet 1818 K Büsum (mapped in 1973), showing Tertius as a small residual island eroded from all sides. On the satellite image of 1984 (plate 18) Tertius ceased to be visible as a sand body above high water.

The migrating barrier off Dithmarschen

Whilst some islands off the coast of Dithmarschen have disappeared, others are just about to emerge (cf. satellite image of 1984, plate 18):

1. A higher sandbank has formed on the Linnenplate, which is no longer flooded during ordinary high tides. It reached a maximum height of

NN + 1.5 m in 1974.

2. Southwest of this sandbank on the Isern Hinnerk another high sandbank has developed, reaching a maximum height of NN + 1.1 m in 1974.

3. The D-Steert, which migrates coastwards northwest of Trischen, had reached a height of NN + 0.8 m in 1976. It is clearly depicted as an extraordinarily high sandbank on the satellite image of 1984.

In contrast to the stable barrier coast of the West and East Frisian Islands and of the North Frisian Wadden Sea, a 'migrating' barrier has developed in Dithmarschen where new islands are continually formed along the seaward margin, older parts of the barrier migrating eastwards, some to be eroded and others joining the mainland coast.

The known facts about the coastal development of this area do not permit any predictive statements to be made with any confidence. This is partly a consequence of differing rates of reaction to changing environmental conditions, so that fundamental changes in the morphodynamic system may only be felt after several decades. This is especially true of the earlier centuries before the production of accurate maps where one is forced to rely on written reports and sketches. With the aid of detailed maps such as the Küstenkarte of the KfKI and with satellite imagery it will perhaps be possible in the future to achieve a better understanding of the processes controlling the immense changes in this part of the Wadden Sea.

7.3 NIEDERSACHSEN

7.3.1 SCHARHÖRN

The tidal flats between Neuwerk and Scharhörn (fig. 305) have been the subject of extensive scientific investigation during the last 25 years. These investigations were stimulated by the proposed construction of a deep-water port for Hamburg in the tidal flat area; Hamburg had acquired the tidal flats in 1961 in order to develop such a port. However, by the end of the planning phase in 1977, a deep-water port for Hamburg was no longer regarded as necessary because:

1. The size of ships had not increased further
2. The Elbe navigation channel had been dredged to a depth of SKN - 13.5 m between 1974-1978 (Keil, 1985: 55)
3. Worldwide economic depression had also affected Hamburg

The project has been shelved for the time being (Naumann & Laucht, 1985: 158).

However, a mass of scientific data has been obtained. Linke (1969, 1970) investigated the geology of the area, whilst Newton & Werner (1969) and Göhren (1971) studied the movement of sediment in the Elbe estuary. Detailed knowledge of the morphodynamics of the area was obtained from these extensive investigations, including the realization that wind-induced drift currents were of prime importance in controlling sediment movement on the open, unprotected tidal flats. The predicted ecological consequences of the deep-water port project have been summarized by Göhren (1976).

From the evaluation of numerous boreholes in the Scharhörn area, Linke (1970) came to the conclusion that the island had formed in its present position as long as 3,500–4,000 years ago. This interpretation was questioned by Göhren (1970: 40) because of the recent morphodynamics of the area. Lang's historical investigations (1970) were only able to clarify the more recent development of the island from the Middle Ages until the present.

Historical development

The name Scharhörn was first mentioned in documents dating from 1466; the sand bank, however, had long existed before this. It had been recognized in 1299 as a dangerous obstacle for the navigation in the Elbe estuary. The usage of the names 'Alt-Scharhörn' and 'Neu-Scharhörn' (Old and New Scharhörn) during the 16th century demonstrates that the name Scharhörn had been used for more than one locality. However, only one 'Scharhörn' is mentioned from the 17th century onwards. Alt-Scharhörn disappeared during the 17th/18th centuries when the Ostertill channel shifted northwards towards the Elbe (Lang, 1970: 111). A beacon was erected on Scharhörn for the first time in 1661. Though this has been repeatedly destroyed or damaged, it has not changed its position since at least 1786. Scharhörn is described as a high sand in a document of 1721, suggesting that the island was no longer flooded during high tide. This early high sand phase is last mentioned in 1784 (Lang, 1970: 124).

After a period of lowering, the shoal emerged once again during the middle of the 19th century. Scharhörn is described as lying about 2.8 m above low water on a Meyer map of 1868 and during the 1:25,000 Prussian mapping in 1878 heights of 4.6

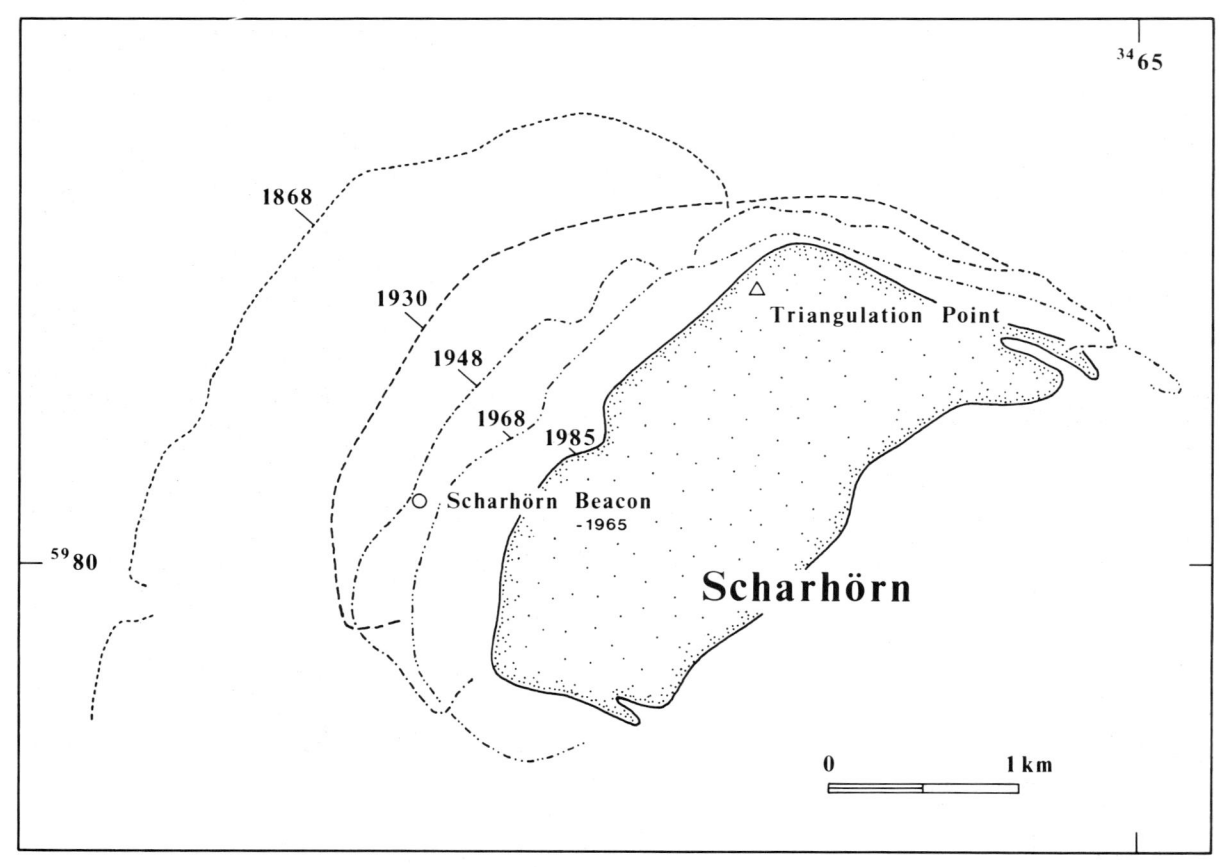

Figure 307: Migration of the Isle of Scharhörn between 1868 and 1982, after Göhren (1971) and Seekarte 1:50,000 Nr.2 Mündungen der Jade und Weser, 7.Ausgabe 1982 XII.

to 5.4 m above NN were measured around the beacon. This higher sand, which emerged during low tide, is last mentioned in 1886. It had disappeared before 1914, as indicated by an old photograph which depicts the Scharhörn beacon rising from wet tidal flats (Lang, 1970: 127).

The present Isle of Scharhörn has migrated northeastwards over the centuries. Along with other morphological changes, the Ostertill channel sanded up after 1900, and its former course is no longer discernible on charts of 1907. Map comparison indicates that the tidal flats around Scharhörn are extremely mobile and that their topography has changed considerably during the last few centuries, very much like the tidal flats off Dithmarschen (Lang, 1970: 116).

In 1927 Scharhörn was situated 0.2–0.5 m above mean tidal high water. From 1927 onwards attempts have been made to create dunes here by sand fencing and the planting of *Ammophila arenaria*, the first appearance of vegetation on the island. By as early as 1929 a storm-flood free area had been created, which grew to 4 ha by 1936. This artificially created island has been maintained ever since (fig. 306), but, having migrated eastwards, it has moved away from the beacon. In 1970 its foundations could be found about 1.95 km west of the dune margin.

Consequences

The constant position of the Scharhörn beacon seems to contradict the recent measured migration of the island (fig. 307). This can be explained, however, if we assume that prior to the formation of the dune island Scharhörn was the name for a procession of sand banks migrating upwards onto the Scharhörn flats, where their sediments were dispersed. This assumption is supported by the fact that the beacon has been reported several times standing in water (1707, 1822 and 1914; Lang, 1970) and on a number of other occasions on high ground.

Niedersachsen

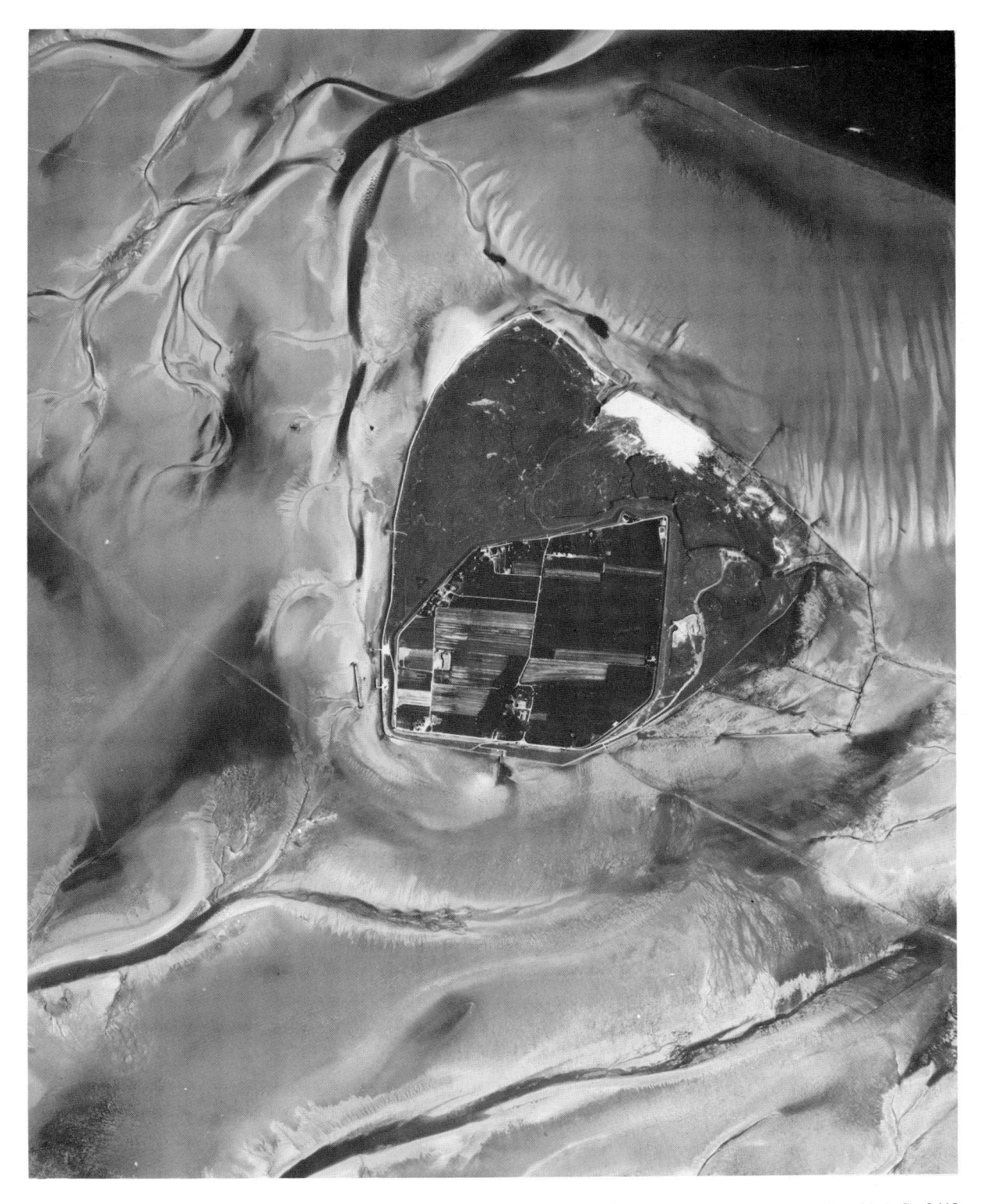

Figure 308: Isle of Neuwerk. Aerial photograph of the Landesvermessungsamt Hamburg of 12.6.1975, Bildflug 6/75, Str.8/45, freigegeben vom Luftamt Hamburg unter Nr. 1426/75.

7.3.2 NEUWERK

Though the Isle of Neuwerk is first mentioned in a document of 1286, it probably formed much earlier. The impressive stone tower on the island, erected by the City of Hamburg in 1310, gives the name to the island (Neuwerk = New Works). The island was dyked between 1556 and 1559, and as the polder area measured during the first primitive survey of Neuwerk in 1574 is almost identical with its present size, it can be assumed that no major changes have occurred since then. Nothing is known, however, about the extent of the undyked salt marshes at that time, although allegedly the island was then much larger (Lang, 1970: 61).

At present Neuwerk is only a marsh island, but it is surrounded by very sandy tidal flats. Old reports mention sanding-up of the island on a number of occasions. A document concerning the storm surge of 1717 reports that the salt marshes were largely covered by sand and the polder fields spoiled by sand having been washed in through the breached dyke. Reinke (1787) shows a broad belt of sand dunes in the north on his map of Neuwerk, and the dunes are also described on a map by Ebels (1854). They also appear on the first edition of the 1:25,000 map of 1878, extending to a height of 2.8–3.3 m above NN. They could still be traced until 1904 when their remnants disappeared under coastal protection structures (Lang, 1970: 68).

Neuwerk is now surrounded by a bank protection. The existence of the island is not threatened as it is situated on high tidal flats a considerable distance from the Elbe and any other important channel. Coastal erosion has been completely halted by the construction of the revetments, and there are even minor signs of accretion in the east (fig. 308).

7.3.3 KNECHTSAND

Blaeuw (1608) records an island called 'Teutel' in the present Knechtsand area. The exact position of the island, however, cannot be reconstructed before 1789. The Knechtsand seems to have migrated first in an easterly and later in a northeasterly direction. Homeier (1969: 78) has provided estimates for the average migration rates during the periods 1789 - 1859 (14.3 m/year) and 1859–1964 (29.5 m/year). During the last few centuries the island has rapidly changed its form and size.

The sand supply for the Knechtsand is provided by the shoals entering the area from the northwest. Their internal structure reflects the landward directed sand transport, with seaward directions only representing a minor component (Wunderlich, 1983: Tafel 1).

The tidal flat areas around Großer Knechtsand are of major importance as a resting area for moulting shelducks (*Tadorna tadorna*) (Goethe, 1983 a: 39). Since 1960 a warden's hut has been manned on the highest part of the Knechtsand, the so-called Turminsel.

Today the Große Knechtsand consists of a sand bank of about 12 km^2 which remains dry during normal high tides. The island and the surrounding tidal flats were used as a training area for the British Royal Air Force between 1952 and 1958. Since 1958 some attempts have been made, with some success, to enhance dune formation on the sand bank, as on Scharhörn.

Kramer (1961: 85) pointed out as early as 1961 that the dunes cannot prevent the migration of the island because of the strong morphodynamics of the area, and attempts to stabilize the island using massive coastal engineering structures remain out of the question.

As on neighbouring Mellum, the area of the Knechtsand which lies above MTHw currently experiences erosion (Wunderlich, 1983: 205). Whilst the primary dunes in 1970 still covered an area of 9 ha, only 3 ha were left in 1980. By 1985 the dunes had completely disappeared (Wunderlich, oral communication). The warden's hut has had to be relocated twice northwards since its construction (Abrahamse & Luitwieler, 1984).

7.3.4 MELLUM

Like Scharhörn and Trischen the Isle of Mellum must be regarded as a young island (Hartung, 1975). It is situated at the outer end of the Hoher Weg tidal flats between the Jade and Weser. The island is mentioned as a high sand bank on a map of 1792 (Schäfer, 1941: 43). It was only in 1903, however, that Schütte (1905: 31) recorded the development of vegetation on Mellum. This had allegedly occurred after 1870 (Schütte, 1924; *op. cit.* in Schäfer, 1941: 44).

The formation of Mellum has to be seen in connection with the shoals approaching from the sea, from the Jade bar, as can be seen on the aerial photograph (fig. 309). This system of sand banks has shifted landwards through the centuries, at a rate of roughly 15 m/year according to Homeier (1974). Homeier assumes that Mellum had previ-

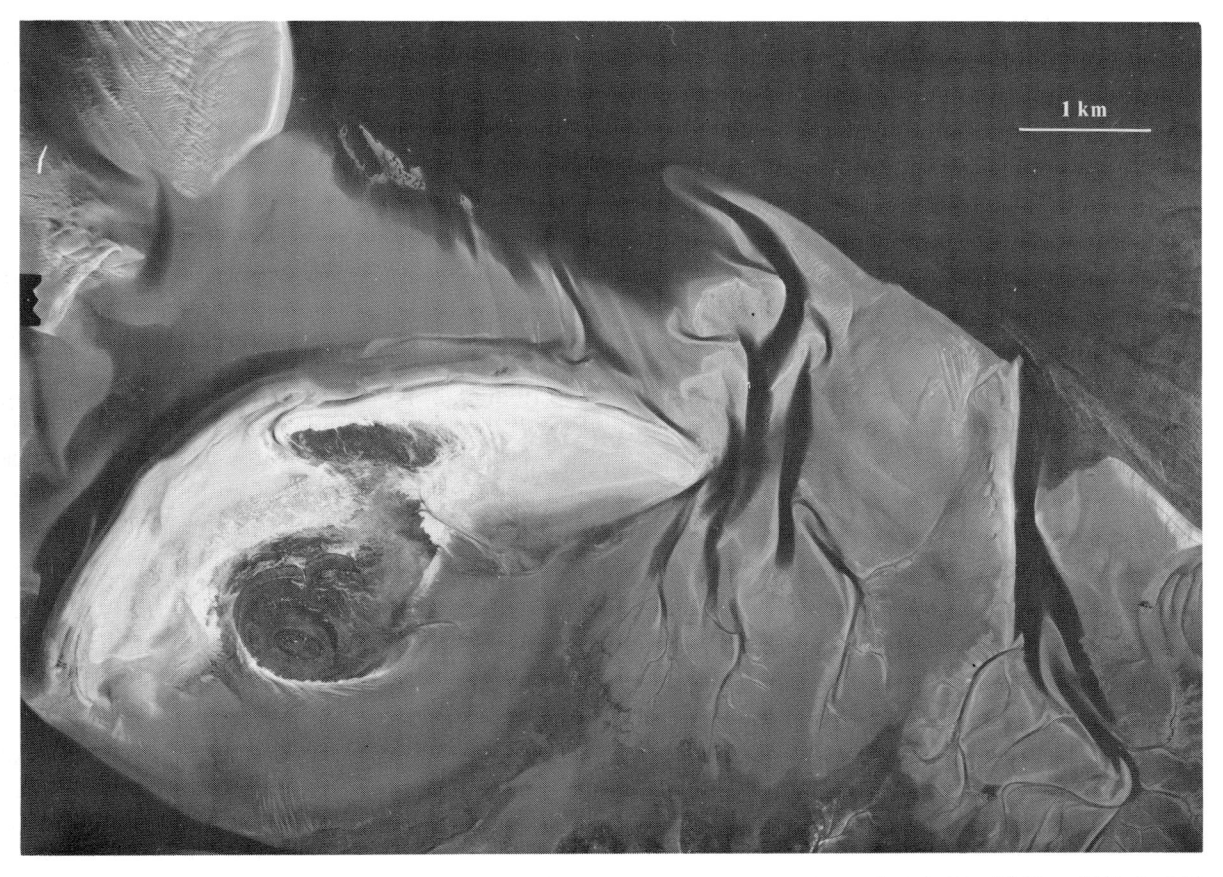

Figure 309: Isle of Mellum. Aerial photograph of the Landesvermessungsamt Hamburg of 12.6.1975, Bildflug 6/75, Str.5/3l, freigegeben vom Luftamt Hamburg unter Nr. 1426/75.

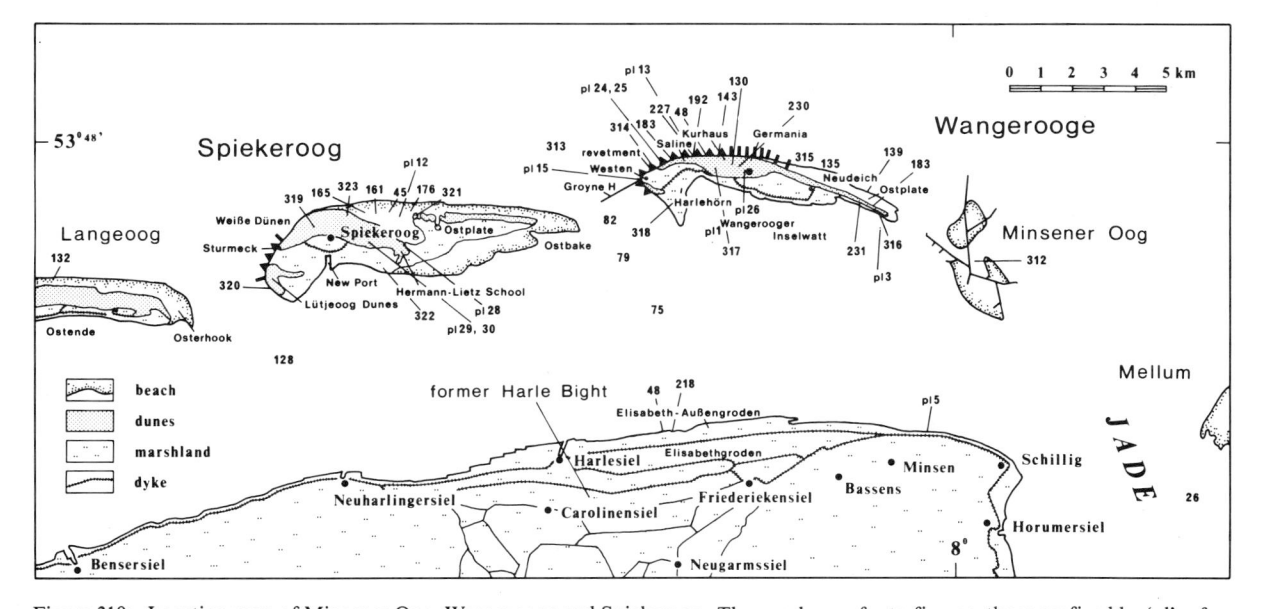

Figure 310: Location map of Minsener Oog, Wangerooge and Spiekeroog. The numbers refer to figures, those prefixed by 'pl' refer to plates.

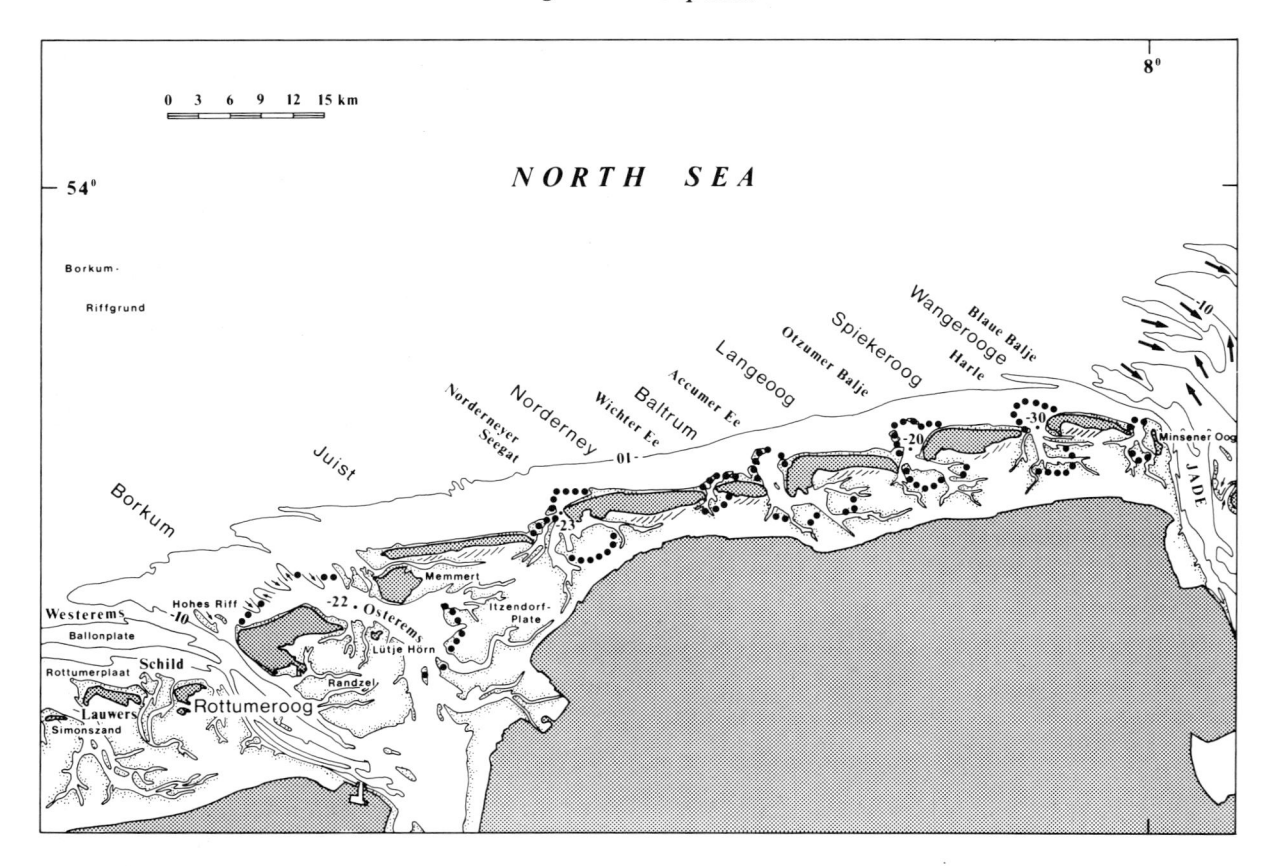

Figure 311: Map of the Wadden Sea area between Jade and Ems. The maximum depths of the tidal inlets are given in metres. Dotted lines: boundaries of the ebb and flood tidal deltas. Arrows: distinct ebb and flood channels.

ously existed as a salt marsh island in its present position. Its striking emergence at the end of the 19th century could have been the result of the convergence of the 'beach island' (sand bank) with the 'marsh island'.

The further development of Mellum between 1908 and 1952 has been documented by Schäfer (1954: 396). During this period it continued to migrate steadily. Between 1962 and 1978 Mellum had migrated an additional 250 m towards the southeast, i.e. approximately 15.6 m/year (Wunderlich, 1979).

7.3.5 Minsener Oog

Minsener Oog is situated between Wangerooge and the Jade (figs. 310 and 311). It was still a sand island about 5 km long around 1650, but since then it has been progressively eroded by the Blaue Balje and the Jade (Homeier, 1974). The present form of Minsener Oog has to be seen in close relation to the engineering of the Jade necessitated by the construction of the Wilhelmshaven naval base. Following the foundation of the port, the navigability of the Jade channel became a major problem, with sands drifting in from the west.

The engineering works in the Jade started on Minsener Oog in 1909. At first a NW-SE striking Hauptdamm (fig. 312) was built with the aim of crossing the Blaue Balje inlet and thus connecting Minsener Oog with neighbouring Wangerooge. However, this proved technically impossible at that time (Schubert, 1970).

The engineering work was completed in 1913, but it was of only limited success since it did not involve dredging. Battleships were unable to cross the Jade bar 2 hours before and 1.5 hours after low water at the beginning of the First World War. Only with the use of numerous dredges has this obstacle been removed and the channel kept open since (Frede, 1938: 40).

The Jade channel regulation work on Minsener Oog was intensified after the First World War. Be-

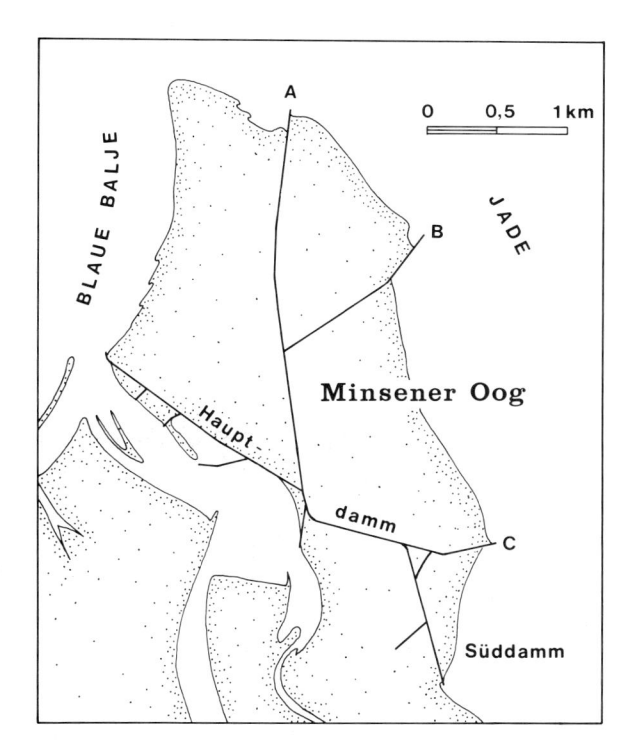

Figure 312: Coastal engineering structures on Minsener Oog, from Ehlers & Mensching (1982). Sources: Schubert (1970) and Abrahamse & Luitwieler (1982 b).

tween 1918 and 1926 Groyne A was extended northwards, the main dam and Groyne C heads were reinforced by fascine mattresses, four additional short groynes were attached to the main dam in order to fend off an approaching tidal creek, and Groyne B was built in order to protect the head of Groyne A. The Süddamm (South Dam) was built in 1936 to prevent the development of a channel between the Blaue Balje and the Jade.

The plan to dam the Blaue Balje (Krüger, 1938) was postponed. After the Second World War, the Allies demolished the Hauptdamm on Minsener Oog in order to promote sand accumulation in the Jade, and access to the island was prohibited to the Germans. When the ban was lifted in 1954 the breach in the Hauptdamm had widened beyond repair, the Blaue Balje inlet having considerably extended its catchment area as a result.

The Jade navigation channel is dredged at present to a depth of 18.5 m below SKN. Continual dredging is necessary because the strong tidal currents tend to fill artificial overdeepened areas. About 8.5 million m³ of sand and mud have to be removed annually (1981 data). Minsener Oog was considered as a possible dumping ground for the dredged sediment when other disposal sites became scarce, and in 1975 a test pumping was a complete success.

An original plan involved the creation of a dump of 10 km² (height: NN + 5 – 6 m) within a period of 10 – 12 years. This would have created an island twice the size of Wangerooge. The plan was abandoned as a result of excessive cost, and between 1980 and 1981 an area of only 220 ha had accumulated beyond the high-water level. The dredged sediment from the Jade is today dumped further seawards (Abrahamse & Luitwieler, 1982 b).

7.3.6 WANGEROOGE

Wangerooge illustrates rather nicely the point that barrier islands are mobile forms which adjust themselves to changing morphodynamic conditions. Layers of marine clay are often exposed on the beach at the western end of the island (plates 24 and 25); these must have formed in a protected position in the lee of the barrier. Repeated radiocarbon dating of the contained shells has yielded ages of around 1,500 years B.P. (Sindowski, 1969: 32; Hanisch, 1980: 224). At that time Wangerooge was situated several hundreds of metres further seawards, and also considerably further to the west.

Historical development

The shape of Wangerooge has changed more than that of any other major barrier island in the Wadden Sea since the middle of the 19th century. The village became close to the western end of the island about 1850 due to the continuing eastward migration of the island. First brushwood groynes were constructed between 1818-34, but they proved to be too light. The narrow belt of coastal dunes in the west and north only provided weak protection. The eastern part of Wangerooge, east of the present village, was still an unvegetated sand flat at this time, flooded during every higher tide.

The coastal dunes at the old Westturm (west tower) and at the lighthouse were breached during a storm surge over Christmas 1854. On New Years Day 1855 a second storm surge inundated the village through the breaches and destroyed 21 of the 73 houses. The complete destruction of the village, now situated unprotected at the western end of Wangerooge, was then simply a question of time

(Schmedes, 1929).

The Oldenburg Government wanted the island to be evacuated in order not to spend excessive money on coastal protection. Every family which moved to the mainland received financial support of 300–350 Taler, and the number of inhabitants fell from 342 in 1855 to 84 in 1861. Many islanders, however, wanted to stay, and after another storm surge on 10.3.1860 the Oldenburg Government finally gave in and agreed to the construction of a new settlement on the eastern part of the island. The relocation took place in 1863 (Jürgens, 1977: 109).

Fixing of the western end

In the meantime the further deterioration of Wangerooge seemed unacceptable. In a secret treaty of 20.7.1853, Prussia had purchased a 334 ha piece of land at the mouth of the Jade Bay for the construction of a naval base. In order to protect the Jade navigation channel it seemed necessary to control the vast sand bodies to the west. The 'Reichsdeich' dyke was built in 1874, connecting the western and central dunes on Wangerooge in order to stabilize the island.

Simultaneously, the first massive groyne was built, protecting the old Westturm. This tower, which had been built between 1597 - 1601 and which had served as a beacon ever since, was then situated outside the dune area on the foreshore. It was finally blasted over Christmas 1914 before falling to marine erosion because it was feared that the British Navy might use it as an orientation point in a possible raid against Wilhelmshaven (Lüders, 1977). Remnants of the outer pavement of the tower are still visible at Groyne B on western Wangerooge.

The coastal protection measures, however, were insufficient to secure the western end of the island for a long period. The Reichsmauer seawall and additional groynes were built (fig. 230) between 1894-97 in order to prevent a breach of the coastal dunes. Dune erosion in the lee of the new wall caused the Oldenburg Government to extend the wall for 500 m from the present Kurhaus to the Hotel Germania and in 1905 another extension of 400 m as far as Groyne R was necessary (Jürgens, 1974).

An additional seawall was erected at the southwestern end of Wangerooge in 1916, but after a series of eleven storm surges during October/November 1917 half of the structure collapsed and had to be rebuilt.

The polder

The small Dorfgroden polder (20 ha) south of the village was dyked in 1902. This dyke reached a height of NN + 4.65 m, while the Westinnengroden (47 ha; dyked 1912) and the Ostinnengroden (ca. 100 ha; dyked 1923-25) dykes reached heights of 4.80 m (Borchers, 1929). After the 1962 storm surge, in which the Dorfgroden dyke was breached, the dykes on Wangerooge were heightened and reinforced. The dyke was extended westwards into the dune area west of the railway station at the same time.

Largely following the course of the old Westinnengroden dyke the new Westgroden dyke was built in 1976. It was extended eastwards to the Dorfgroden, combining both polders into a larger one. This dyke was engineered to resemble the dune landscape it protects. The low degree of exposure meant that a sand dyke without a clay cover sufficed, permitting a dune-like vegetation to develop.

Groyne H

The Harle inlet between Spiekeroog and Wangerooge differs from all the other inlets of the Wadden Sea because it is confined by a groyne over 1 km long, the Wangerooge Groyne H (fig. 313). This groyne was built between 1938–1941 in order to fend off the threatening approach of the Dove Harle, a side tributary of the Harle inlet.

This aim was achieved despite the fact that the groyne was never built as planned to the height of mean tidal high water level (MTHw) because of the war (Lüders, 1952: 26). The western head of Wangerooge was protected, but the adverse consequences caused by the groyne soon became evident (cf. Hanisch, 1981):

1. The inlet trunk was incised to a depth of NN−30.2 m at the groyne head. Comparable depths are found in none of the other East Frisian inlets. The maximum depth in the Oster-Ems between Borkum and Juist is NN−22.4 m, in the Norderneyer Seegat NN−22.7 m. All the other inlets are shallower.

2. The groyne soon started to sag because of lateral undercutting and marginal scour hollow formation, causing it to become increasingly submerged.

3. Maps presented by Homeier (1973: Anl. 3 and 23) demonstrate that the depth of the Strandbalje channel (fig. 313) has increased considerably since the construction of Groyne H. The strong coast-

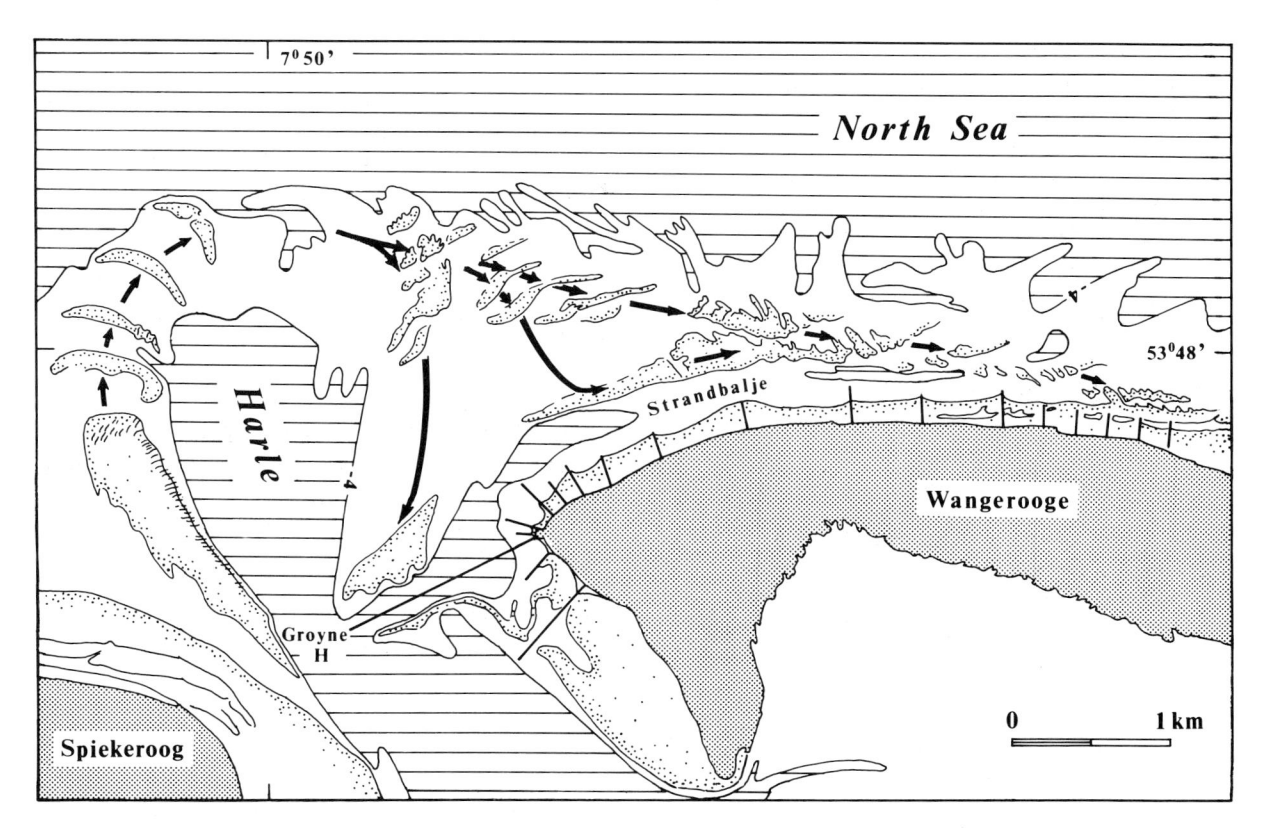

Figure 313: Harle ebb delta and routes of resultant sand migration. The true sand migration, as exemplified for the Wichter Ee in section 5.2.4, is predominantly at right angles to the course of the ebb delta arc of shoals. Hatched area: below NN - 4 m. After Ehlers & Mensching (1983). Sources: Küstenkarte 1:25,000, Sheets 2212 K Wangerooge (1980) and 2112 K Jademündung (1977); Aerial photograph mosaic of 23.6.1973 from Luck & Witte (1974), and Hanisch (1981: fig. 8).

parallel currents in the Strandbalje generally prevent beachward sand transport, so that the sand approaching Wangerooge via the ebb delta shoals is transported much further eastward. Most sand only reaches the beach east of the Wangerooge settlement.

During the 1960's, the Strandriff shoal only migrated once through the Strandbalje onto the Wangerooge beach, giving rise to a temporary standstill in the lowering of the beach (Homeier, 1973: 37). The resultant sand movement in the Harle ebb delta fans out into three possible directions at its eastern end (fig. 313). At present only the easternmost route leads to the beach on Wangerooge; the western route, under the influence of Groyne H, leads back into the Harle, and the sand of the central route is diverted eastwards until it joins the eastern route.

The western and central routes were more strongly developed before the construction of Groyne H. Homeier (1973, Anl. 15) has reconstructed the positions of the ebb-delta shoals for 10 different stages between 1869 and 1972; the easterly route dominated in three out of four cases after the construction of Groyne H (only exception: 1960), whilst previously the other two routes had prevailed.

Additional structures

The storm surge of 17.2.1962 caused considerable damage on Wangerooge. The seawall in the west was largely destroyed and had to be replaced by a modern revetment between groynes U and G (plate 15). The foundations of the old seawall were largely incorporated into the new structure. Only at its eastern end, where erosion had been extensive, the remnants of the wall lie seawards of the revetment on the beach.

After a few years lee-side erosion necessitated the extension of the revetment eastwards almost as far as Groyne J using semi-high bank protection.

Figure 314: Dune erosion on top of the semi-high revetment (1978).

Figure 315: Recent dunes forming east of the Wangerooge settlement (1979).

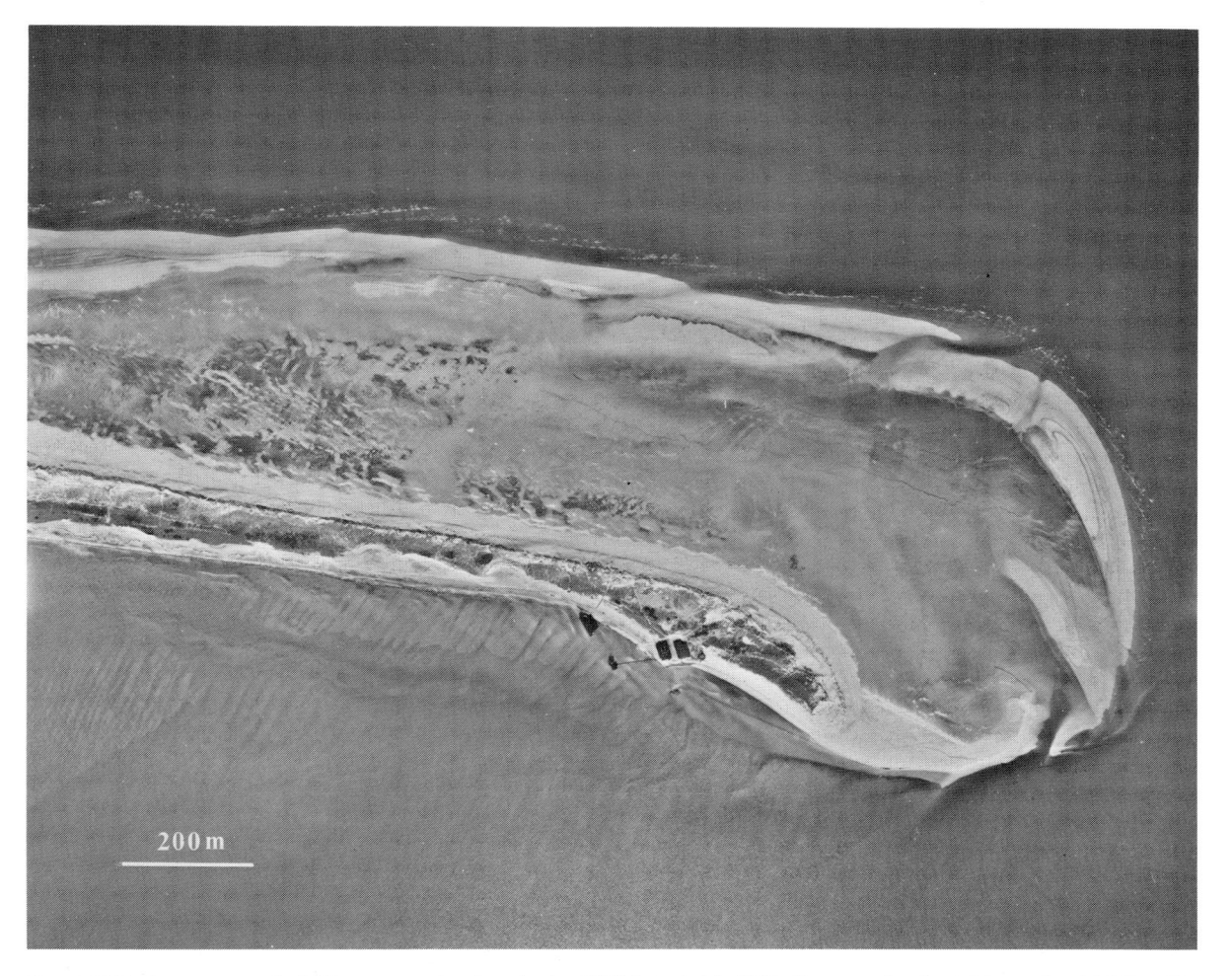

Figure 316: The eastern end of Wangerooge showing the 'stuifdijk' started in 1904. Dune erosion is visible on the tidal flat side. Aerial photograph of the Niedersächsisches Landesverwaltungsamt, Landesvermessung, of 3.5.1972, Bildflug Wangerooge (804), freigegeben am 2.9.1972 unter Nr. 1188/72 durch Regierungspräsident Darmstadt.

This structure, which was only intended as a provisional measure, was severely damaged during the 1973 storm surges and was unable to prevent dune retreat. The dune margin was eroded by another 20 m (fig. 314), and lee-side erosion started at the eastern end of the new structure.

The series of storm surges in autumn 1973, especially those of 19.11. and 6.12., caused additional damage on Wangerooge. The Südwestmauer in the southwest suffered considerably, and 15 m of dune were lost in the lee of the Krügermauer seawall. During another storm surge on 14.12.1973 the dune cliff failed west of the Saline in the lee of the half-height revetment, exposing part of the Reichsmauer which had been completely buried since 1922.

Additional coastal protection measures became necessary. In 1974 the half-height bank protection was reinforced with a 25 cm thick layer of asphalt concrete, giving the structure a crest height of MTHw + 3.30 m.

During the autumn of 1979 a 380 m long revetment had to built in front of the new Kurhaus. This revetment has a height of 8.50 m and a 1:3 slope. Its construction became necessary because the Reichsmauer no longer sufficiently protected the Kurhaus, which had opened in 1973. Unnecessarily, it was built 50 m seawards of the other houses along the Wangerooge beach.

Beach replenishment became necessary in front of the village in 1982. Sand was supplied to the western end of Wangerooge in 1984, and in

Figure 317: Bomb crater landscape west of the Wangerooge settlement. Aerial photograph of the Niedersächsisches Landesverwaltungsamt, Landesvermessung, of 3.5.1972, Bildflug Wangerooge (804), freigegeben durch Regierungspräsident Darmstadt unter Nr. 1188/72.

1984/85 a new revetment was built connecting the existing revetments in the west and in front of the village. Currently over 50 % of the seaward coastline of Wangerooge is strengthened by sea-walls, revetments and groynes; the beach in front of these structures has been lost, where it was not supplied artificially by beach replenishment.

The dunes

Dune formation has only occurred at two places on Wangerooge in recent years:
1. A dune ridge several metres high has developed along the brush fences built between the Neuer Westturm (new west tower) and the quay.
2. New dunes have developed in front of the dune ridge on the eastern end of Wangerooge between the end of the seawall and about 400 m east of the eastern end of the Ostinnengroden polder (fig. 315) (cf. Hempel, 1980).

The dunes on Wangerooge no longer constitute natural landforms, having been considerably reshaped by human interference. The coastal dunes along the beach only combined to form a single dune ridge with the assistance of coastal protection measures at the end of the 19th century (Sindowski, 1969). The dune ridge on eastern Wangerooge was built as a stuifdijk (dune dyke) after 1894 (fig. 316). The eastern end of the island had originally been a sand flat like the Ostplate of Spiekeroog (Krüger, 1929: 218). All the dunes on the island have been artificially stabilized. As on most of the other islands, the coastal dunes are now fenced with barbed wire; specially constructed wooden footpaths have been installed for the use of tourists.

A great number of bunkers were built in the dunes during the First and Second World Wars. After the Second World War they were demolished and covered by sand, but most of them can still be seen. Other military installations include earth walls, an abandoned firing range and the founda-

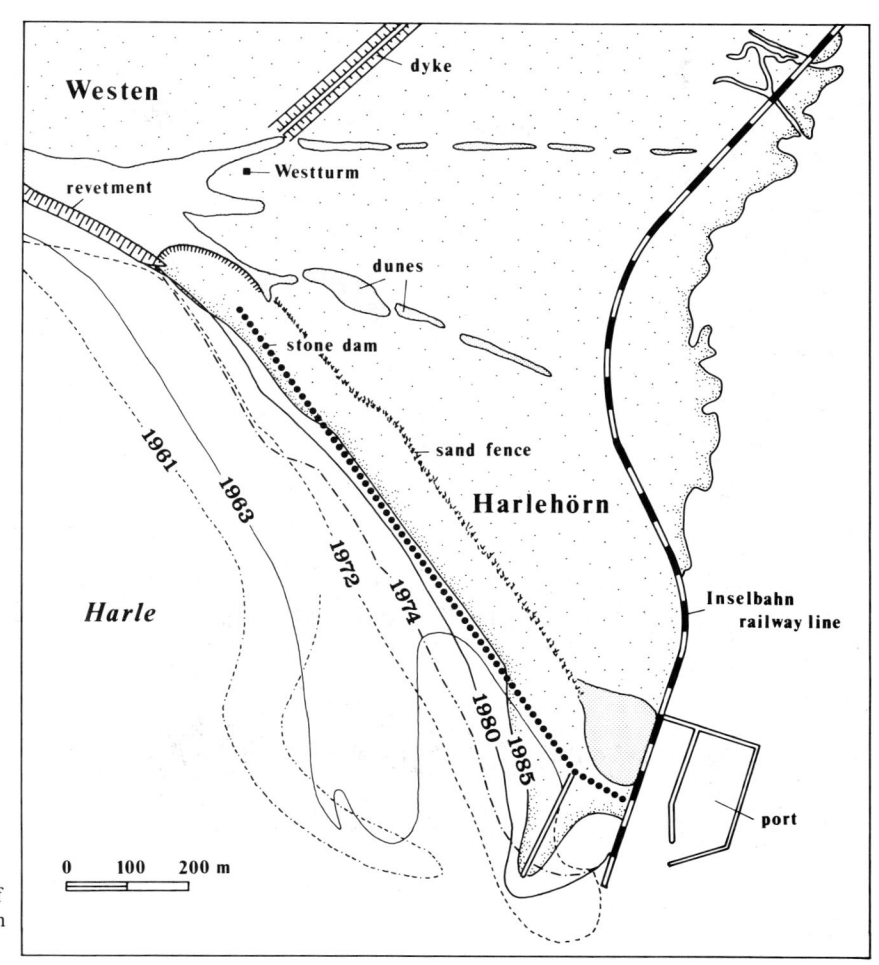

Figure 318: Recent development of the Harlehörn peninsula (based on Ehlers & Mensching, 1982).

tions of barracks. American bombers attacked Wangerooge shortly before the end of the war (on 25.4.1945), turning the island into the crater landscape visible today (fig. 317) (Ehlers & Mensching, 1983: 507).

The non-dune areas of Wangerooge also consist largely of sand (cf. Sindowski, 1969). Plate 26 shows an excavation for a construction pit in the centre of the village. The sand beds with intercalated humic layers reveal how the sand flats south of the dunes in central Wangerooge were formed. The sand represents phases of intensive flooding when dune and beach sand was washed onto the salt marshes, whilst the humic layers represent phases of salt marsh growth during calmer weather conditions.

Harlehörn

The Harlehörn peninsula south of the western end of Wangerooge originally formed part of the Harle tidal delta. It has migrated round the western end of the island following the stabilization of the dunes at the end of the 19th century (cf. fig. 130). The eastward retreat of the Harlehörn has continued in recent years (Ehlers & Mensching, 1982: 35). Aerial photographs of 1961 still show a 250 m wide sand beach west of the stone dam, but by 1974 this had shrunk to only 100 m and only 50 m were left by 1976. By 1980 it had decreased to 20 m. In 1961 a broad strip of salt marsh on the Wadden Sea side of the Harlehörn extended southwards almost as far as the quay, but by 1974 most of it had been covered with sand as a result of the 1973 storm surges.

Today the situation has slightly changed. The dune ridge on the Harlehörn has grown to a height of over 3 m, which will prevent further washovers on the peninsula. Erosion on the western coast continued up until 1984 (fig. 318), but it seems

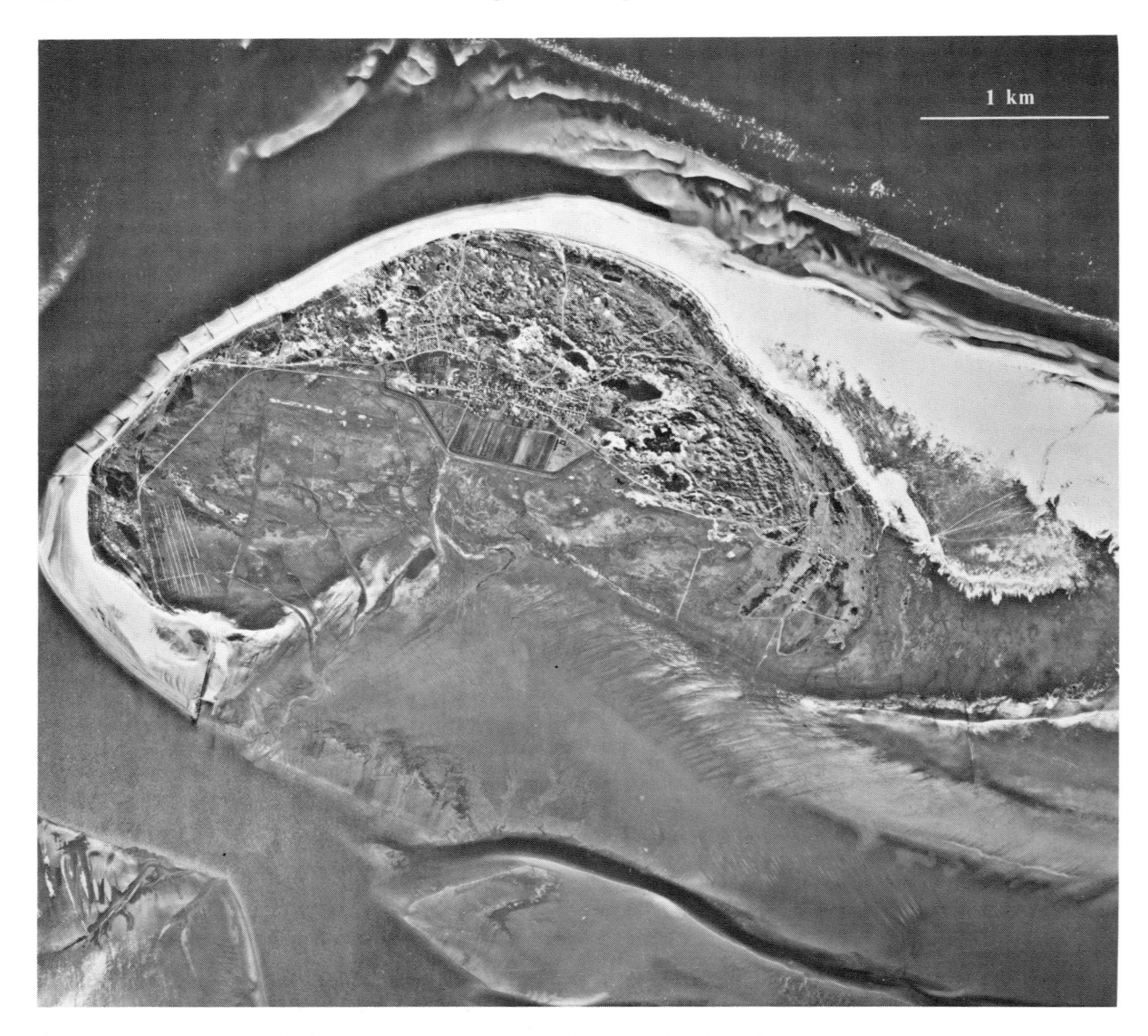

Figure 319: The dune core of Spiekeroog. Aerial photograph of the Niedersächsisches Landesverwaltungsamt, Landesvermessung, of 17.4.1974, Bildflug Ostfriesische Inseln (1086), Str.3/063, freigegeben durch NLVA–Abt. Landesvermessung–Hannover unter Nr. 34/74/1086.

possible that the sand supply for the Harlehörn beach may increase again as a result of the recent changes in the Harle Inlet morphology (cf. section 5.1.2).

7.3.7 SPIEKEROOG

The Isle of Spiekeroog is at present experiencing considerable change. Whilst the northwestern end of the island has been slightly eroded over the last few centuries, Spiekeroog's total area has nevertheless increased. According to Roterberg (1983: 78) Spiekeroog now covers an area of 17.48 km^2.

The island consists of an old dune core in the west and an elongated eastern sand flat. The dune core forms a broad, u-shaped dune belt which is open to the south. The village has been built along the southern margin of the dunes. The Wester-, Süder and Ostergroen salt marshes have developed under the protection of the dune core; a minor part of this area has been dyked south of the village (fig. 319).

Figure 320: Dune erosion at the southwestern end of Spiekeroog, Lütjeoog Dunes (1984).

Figure 321: Formation of young dunes with *Ammophila arenaria* on the Ostplate. View towards west; background: the dune core of Spiekeroog; left: young salt marshes forming behind the dunes (1984).

Historical development

Like most of the other islands Spiekeroog has suffered considerably from coastal erosion during the historical period. The 'All Saints Flood' of 1.11.1570 caused heavy dune losses and necessitated the relocation of the village from the western end of the island to its present site (Meyer-Deepen & Meijering, 1983: 39).

However, the storm surge of 1825 caused little damage on Spiekeroog, though some coastal dunes were lost and the pasture areas were covered with sand. In order to prevent such problems in the future, the Hannover Government built a dyke connecting the main dune core with the dunes on the southwestern peninsula. This dyke, however, was breached during the severe storm surges of 1854/55, leading to the destruction of the old village on neighbouring Wangerooge. The first massive coastal protection structures were therefore erected in the northwest in 1873. Additionally, a stronger dyke, the so-called 'Richel Dyke', was built south of the village between 1883/84, protecting an area of 11 ha (Möbius, 1931: 9).

The severe storm surge of 13.3.1906 damaged the revetment, and the dyke surrounding the village

Figure 322: Young salt marsh formation on the Wadden Sea side of Spiekeroog, south of the Hermann-Lietz School (1984).

was breached in several places. The village was flooded, and once again gardens and pastures were covered with sand.

As erosion in the northwest continued the wide beach in front of the coastal protection structures was lost within a few decades. In 1931 the pleasure beach therefore had to be relocated towards the east to its present position, and in 1936 a steel sheet-pile bulkhead was built in the northwest to stem further erosion. This was heavily damaged during the storm surge of 18.10.1936, which also destroyed the recently completed sand dykes around the Hermann-Lietz School.

The storm surge of 17.2.1962 damaged the revetment on western Spiekeroog in four places and caused heavy lee-side erosion (Janssen, 1962); it also breached the dykes and flooded major parts of the village. As a result the new Süddeich dyke was built (1,630 m long, 5.7 – 6.0 m high) in 1968. The traffic connection with the mainland was improved with the construction of a new port south of the village, replacing the old quay in the southwest. The new port was opened in 1981.

Though storm surges have repeatedly struck Spiekeroog, recent land losses have been moderate. Erosion in the northwest has continued, but at a rather slow rate. This is because the main ebb channel of the Otzumer Balje adjoins the coast of Langeoog, leaving Spiekeroog in a more protected position. The revetment had to be extended by tetrapods northwestwards in 1975. Fig. 320 shows dunes at the southwestern end of Spiekeroog experiencing erosion. As the settlement has not been extended to the beach, unlike Wangerooge, further massive protection measures will be unnecessary during the foreseeable future.

The Lütjeoog Dunes

The Lütjeoog Dunes started to form during the 17th century on a sand flat which was part of the Otzumer Balje flood delta (Meyer-Deepen & Meijering, 1983: 12). The dune core was still represented as a separate island in the Wadden Sea on Horst's map of 1758, and called 'Klein-Oog' (Backhaus, 1943). The Süderdünen, the southern branch of the small dune arc, started to form about 1841/42. When Lütjeoog and Spiekeroog grew together to form a single island both dune cores were separated only by a minor tidal creek in the salt marshes, and the remaining gap was closed in 1873.

At the same time the island extended towards the east. Large amounts of sand were deflated from the originally unvegetated Ostplate area, resulting in the expansion of the main dune core to the southeast.

The growth of the Ostplate

Spiekeroog and Wangerooge were originally separated by two tidal inlets; the Harle in the west and the Rote Balje further to the east. The catchment areas of both inlets decreased with the increasing

Figure 323: Old dune area on Spiekeroog, northeastern part of the dune core (1984).

reclamation of the Harlebucht. As the Schillbalje extended its catchment area to the east, the importance of the western inlet diminished. The Rote Balje was then renamed 'Harle', and the original Harle became the 'Alte Harle' (Old Harle). In 1960 the latter only formed a very small inlet remnant directly west of the Harle (fig. 82), and by 1972 it had completely disappeared.

In this way the Ostplate of Spiekeroog could extend further and further to the east. This development continues at present, because the western part of Groyne H on Wangerooge, which is over 1 km long and keeps the Harle in its position, sagged considerably allowing the inlet trunk to shift eastwards. Only the easternmost 400 m of Groyne H remain intact. Around 1950 primary dunes began to form on the broad Ostplate sand flat. These have now developed into a belt about 3 m high (fig. 321, plate 28) extending from the old island core eastwards beyond the Ostbake beacon. This dune belt, which is about 20 m wide, continues to grow in height and extent. It is intersected by a number of washover areas, the largest of which occurs directly east of the old dune core (fig. 319, plate 12). The Ostplate was declared a nature reserve in 1970.

Protected by the dunes, a broad salt marsh area has developed on the Wadden Sea side of the island. The development of dunes on the Ostplate has cut off the sand supply for the southeastern end of the old dunes. However, the expansion of the dune belt has added so much to the protection of the south coast of Spiekeroog that young salt marsh is now forming south of the old dune core east of the village (fig. 322). As on Baltrum, the old erosive cliff along the salt marsh margin can still be identified as a terrace edge in the salt marshes, and in a number of places old salt-marsh creeks terminate there.

The 'Green Island'

Spiekeroog is currently advertised as the 'Green Island', with old trees in the village as well as the small forests in the dune valleys. Of all the islands, Spiekeroog has the most rigorous dune protection measures, keeping the old dunes vegetated (fig. 323, plates 29 and 30). There are no active blowouts; where they do form they are immediately repaired with brushwood and hay. Numerous signposts remind tourists not to leave the footpaths in the dune area, and dune destruction by trampling is extremely low.

The woodland areas are obviously considerably endangered by inundation. A storm surge during the 1928/29 winter, for example, led to the intrusion of salt water into one of the dune valleys, causing the loss of many trees (Dhonau, 1931: 32).

Rabbits, which are a major factor in promoting blowouts on some islands (e.g. on Norderney), are extinct on Spiekeroog. Rabbits were originally introduced to the islands for hunting, but when their

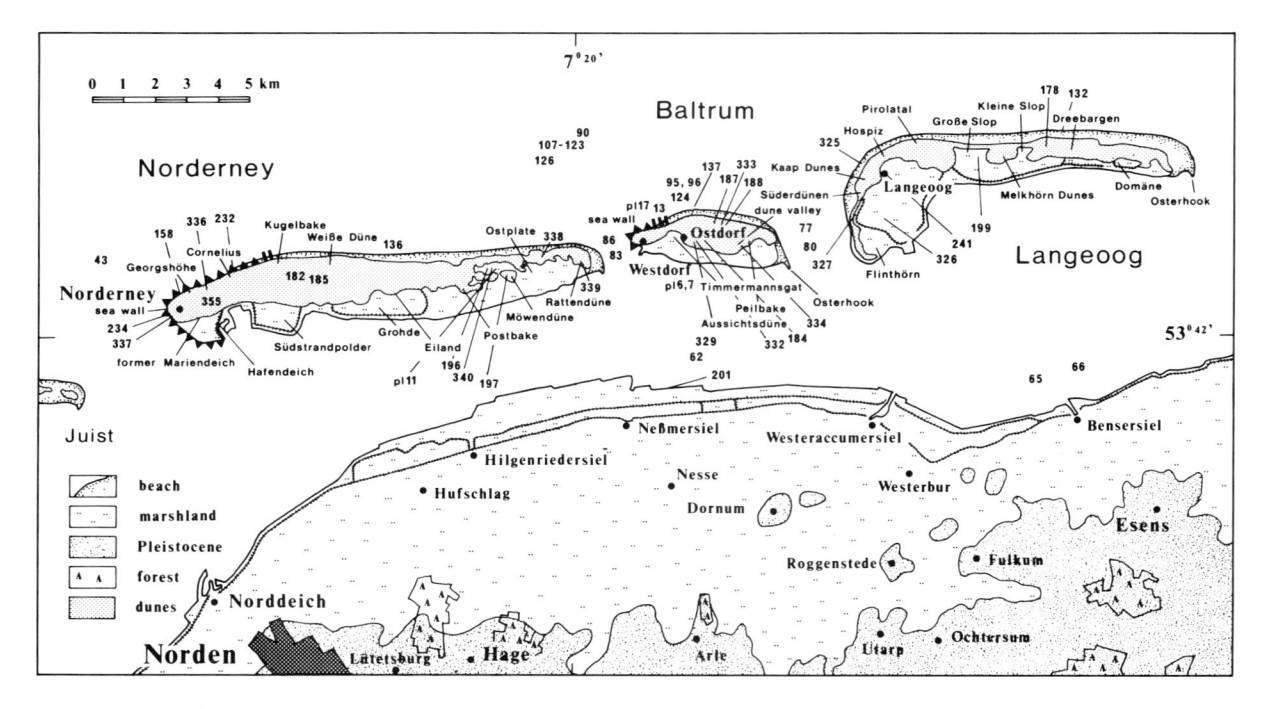

Figure 324: Location map of Langeoog, Baltrum and Norderney. The numbers refer to figures, those prefixed by 'pl' refer to plates.

destructive role became obvious they were eliminated and replaced by hares, which, unlike rabbits, live on the ground surface. The last rabbit on Spiekeroog was killed in 1889 (Meyer-Deepen & Meijering, 1979: 97).

7.3.8 LANGEOOG

The present Isle of Langeoog (fig. 324) formed by the merging of three dune cores, which were still separated by washover areas as late as 1891 when the first 1:25,000 scale map was made. This map also gives some indications of the positive sand budget which has characterized Langeoog throughout the late 19th century. A new coast-parallel dune ridge had formed in the south in front of the Kaap Dunes. This terminates in the south in a dune fan, the so-called Süderdünen. Young dunes had also started to form seawards of the Herrenhus Dunes, the Melkhörn Dunes and the dune core at the eastern end of Langeoog, reaching a maximum height of 6.3 m off Dreebargen in 1891. All those recent dunes have developed into impressive dune ridges today. The situation had not always been so favourable, however; Langeoog is the only major North Sea barrier island which has been uninhabited for a number of years in recent times.

Geological history

Like the other East Frisian Islands, Langeoog is a sand island which evolved relatively late in the Holocene. Geological mapping has revealed an old island core under western Langeoog which can be subdivided lithostratigraphically down to a depth of about NN - 0.5 m. Age determinations have permitted the following reconstruction:

1. A sandy shoal developed in the present area of Langeoog, finally emerging beyond mean high water level between the 8th and 2nd centuries B.C.
2. The first dunes formed during the 1st century B.C., providing protection for salt marsh development, which has been dated palynologically.
3. Extensive re-deposition of sediments occurred after the turn of the millenium, leaving the island almost unvegetated for a number of centuries.
4. Salt marsh development only re-started during the 13th century (radiocarbon-dated).
5. Extensive dune formation occurred from the 13th to the 17th century (Barckhausen, 1969: 273).

Figure 325: The wide northwestern beach of Langeoog (1980).

Historical development

Dune erosion during the 1921-23 storm surges exposed the settlement remnants and wells of an old village on the northwestern beach off the Hospiz. An English copper coin was discovered, which, having been minted in Canterbury between 1307 and 1327, helped to date the settlement (Tongers, 1975: 41). This corresponds well with the first mention of 'Langeooch' in a document of 1398, proving that the island was inhabited during the 14th century (Backhaus, 1943: 76).

Langeoog was originally protected from the sea by a continuous dune ridge, which is indicated on a map as late as 1585 (Backhaus, 1943: 76). Subsequently the dunes were repeatedly breached, but the age of the formation of the large washover areas is unknown.

Langeoog was one of the poorest islands, and provided unfavourable conditions for settlement. In 1660 only 62 people lived on the island. Shifting sand was a serious problem and the village had to be relocated several times during the 17th century alone (Tongers, 1975: 50; Backhaus, 1943: 76).

There was no dune protection on Langeoog prior to 1692. Official suggestions to plant *Ammophila* and keep the cattle out of the dunes were rejected. Only after 1700 were Dutch dune workers, so-called 'Dünemeyers', brought to the island to initiate the construction of bush fences, plant 'Helm' (*Ammophila arenaria*) and thus stem deflation. At this time the grazing of cattle, mowing of grass and even walking were strictly forbidden in the dunes. Despite these measures the dunes were not fixed for a considerable time (Tongers, 1975: 51).

The Christmas Flood of 1717 caused a serious breach in the dunes in the area of the present Große Schlopp, causing the destruction of the church and vicarage. Only four families lived on Langeoog at this time, and the remaining two families left in 1721 when the further use of the island seemed impossible. The island remained uninhabited until 1723. A few families from Helgoland subsequently settled on the island, only to leave again after a few years. However, other families from the mainland settled there, a new village developed, and in 1740 an additional settlement was founded on the eastern part of the island in the present Domäne area. By 1765 there were 28 people living on Langeoog.

The Kleine Schlopp washover area was closed around 1890 by natural dune development. An initial attempt to close the Große Schlopp using an artificial dune ridge failed in 1894. In 1902 a sand dam was built, but this too was destroyed by the floods. The Große Schlopp was finally dammed in 1906 (Backhaus, 1943: 81), once again forming a continuous dune barrier.

Dune formation and erosion in the northwest

The positive development which followed the severe land losses of the early 18th century started

Figure 326: Aerial photograph of the western end of Langeoog, showing the eastern end of the Accumer Ee ebb delta, Langeoog settlement and the Flinthörn. Aerial photograph of the Niedersächsisches Landesverwaltungsamt, Landesvermessung, of 29.5.1966, Bildflug Ostfriesische Küste (419), Str.2/294, freigegeben durch Regierungspräsident Münster/W. unter Nr. 2316/66.

with the attachment of a series of shoals from the ebb delta between 1815 and 1828. The sand was probably derived from the serious land losses occurring along the western end of Baltrum at that time. A new dune ridge on northwestern Langeoog formed as a result, but this positive trend was halted shortly after 1900 (Homeier & Luck, 1971: 11) and the situation has remained relatively stable ever since. During the years 1891, 1921 and 1936-38 there were dune losses at the northwest point, but these were partly replenished with the arrival of new shoals (Backhaus, 1943: 80).

Two periods of erosion affected the northwestern beach of Langeoog after the Second World War, both preceding the arrival of new shoals. A section about 1 km long west of the water tower started to retreat after 1945, and between 1947 and 1955 the dunes were eroded at this point by a maximum of about 150 m (ca. 20 m/year). The dunes were rapidly replenished with the attachment of the new shoal.

The second phase of erosion began after 1960 north of the Pirolatal dune valley, where a section of 2 km retreated by 13 m per annum. In 1971/72

Figure 327: Recent dunes protecting the old dyke of the Flinthörn peninsula (1980).

a 2,500 m long revetment of flexible sand-filled plastic tubes was erected, subdivided by transverse tubes into several compartments, in order to prevent the complete destruction of the coastal dune ridge and to avoid the flooding of the Pirolatal. The compartments between the tubes were replenished with 550,000 m^3 of sand dredged from the Große Schlopp area; the resulting basin can be seen on the aerial photograph (fig. 199). This relatively light measure enabled the dune foot to be protected until the next shoal arrived, thus avoiding the construction of a massive revetment (Lüders, Führböter & Rodloff, 1972; Luck, 1974).

Since the first mapping of the Langeoog 1:25,000 map sheet in 1891, more than 50 m of dune have been lost in the northwest. In contrast with the other East Frisian Islands (only exception: Spiekeroog), the built-up area has not extended to the beach, so that massive coastal protection structures have not been necessary (fig. 325).

At present the coastal dunes form a continuous dune ridge, which, following the closure of the Große and Kleine Schlopp washovers, protect the central part of the island against flooding from the sea. This youngest dune ridge is over 20 m high in places and therefore much higher than the older dune core of eastern Langeoog which averages about 14 m. Between the young dunes and the old dune core of the eastern dunes, a 100–200 m wide dune valley has formed, barred from inundations by low dykes built at both ends. The entire eastern end of Langeoog has been declared a nature reserve.

The Flinthörn

The flood spit of Flinthörn started to extend southwestwards at the beginning of the 19th century. The first dunes formed here between 1825 and 1841 (Homeier, 1963 b: 17). During the last hundred years the Flinthörn spit has migrated further around the western end of Langeoog and has expanded southwards towards the Langeooger Balje. The low dune landscape of the Süderdünen was largely destroyed during the construction of the airfield. The military airfield was demolished after the Second World War and never reconstructed (fig. 326). A new private airfield has been opened east of the old one. A continuous dune ridge has developed seawards of the old airfield, reaching a current height of 10 - 14 m, protecting a dyke built around the old airfield area in 1939/40 (fig. 327).

7.3.9 BALTRUM

The Isle of Baltrum is the smallest of the inhabited East Frisian Islands. During the 19th century it consisted of two dune cores separated by the Timmermannsgat washover area. Though this breach

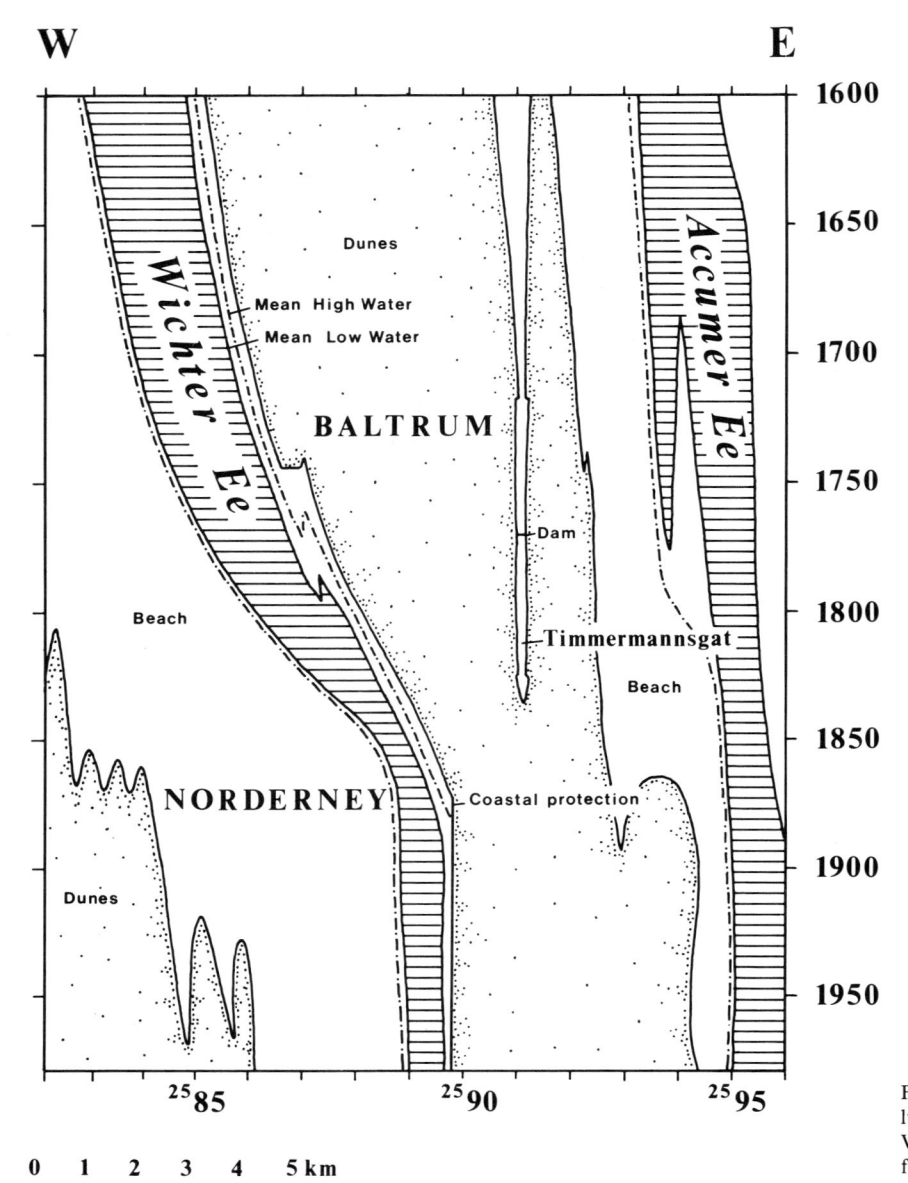

Figure 328: Time-space diagram illustrating the displacement of the Wichter Ee and Accumer Ee inlets from 1600 to the present (after Luck, 1975).

was closed in 1840 its former position can still be seen in the field between the Ostdorf and Westdorf settlements.

Coastal protection

The existing settlement is not the oldest on the island. Dune erosion in 1738 at the margin of the Wichter Ee uncovered the remains of an older village which must have been abandoned as a result of shifting sand prior to 1650. It was situated about 2 km west of the present Westdorf village (Backhaus, 1943: 66).

The island has decreased in size during the last few centuries. Whilst the Wichter Ee shifted rapidly eastwards, the displacement of the Accumer Ee took place at a much slower rate, with the result that Baltrum between them was strongly eroded. The time-space diagram (fig. 328) also shows that the inlets grew progressively narrower, and that the dune areas on both Norderney and Baltrum expanded during the middle of the 19th century.

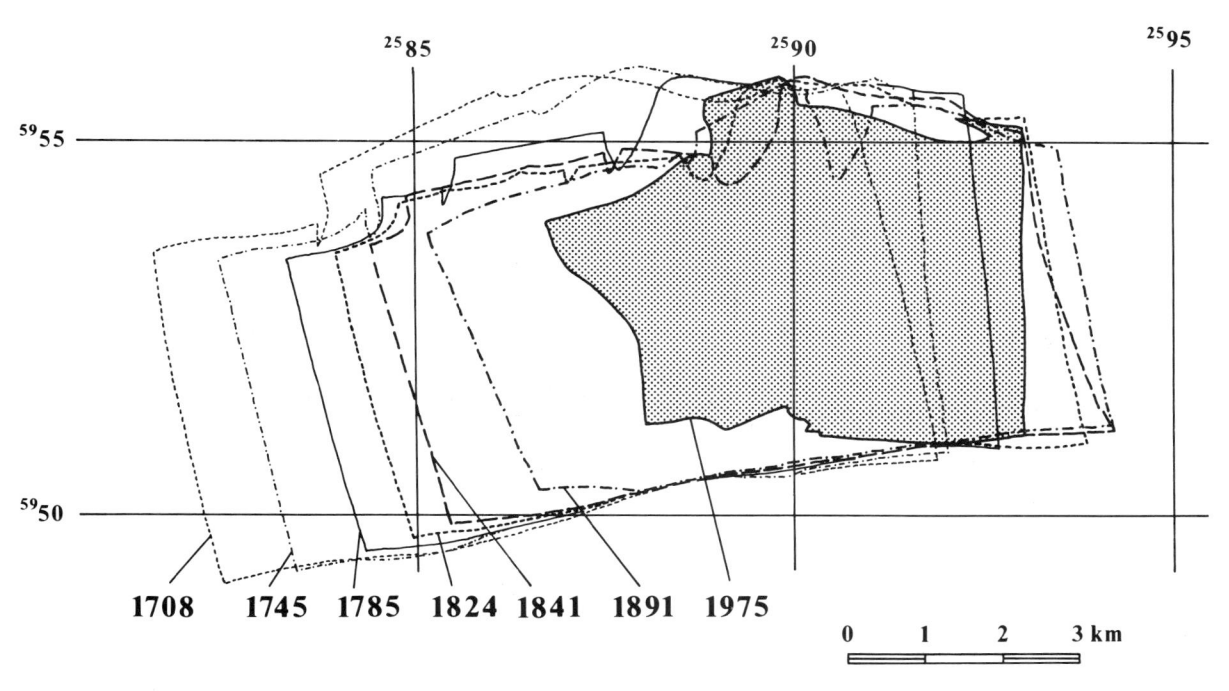

Figure 329: The changing catchment area of the Wichter Ee inlet (after Luck, 1975).

The decreasing size of Baltrum also had an influence on the shape of the Wichter Ee catchment area (fig. 329). Whilst both water divides limiting the Wichter Ee drainage basin shifted eastwards during the 18th century, albeit at different rates, the eastern water divide stagnated from 1824 onwards; it even migrated westwards after 1891, further decreasing the Wichter Ee catchment area at the cost of the Accumer Ee. If this tendency continues it will ultimately lead to the complete closure of the Wichter Ee inlet.

The migration of the Wichter Ee, however, continued to play an important role in shaping Baltrum. Continued coastal erosion at the western end of the island released considerable quantities of sand, which invaded the settlement in the form of migrating dunes. The pastor's home was partly covered by sand in 1793 and completely covered in 1797, necessitating its reconstruction in 1800. At the same time the old church was destroyed by coastal erosion in 1797. The relocation of the village, which had become urgent, was largely complete by 1825, and of the original 25 houses, only seven were left in the west (Backhaus, 1943: 68).

According to a contemporary report of the 3./5.2.1825 storm surge the magnitude of the damage on Baltrum was only comparable to the destruction on the Halligen. The remaining Westdorf houses were either destroyed or so heavily damaged that they had to be abandoned. The Timmermannsgat, which had formed about 1700, was widened into a breach of 680 m. Five houses were damaged in the new Mitteldorf village in 1825, and only a single house survived intact in the small Ostdorf settlement, which had been built in 1820 (Backhaus, 1943: 68). In 1826 the Timmermannsgat continued to flood during every tide that was higher than normal. Its closure was regarded as essential in order to protect the island from further deterioration. However, the breach was only closed in 1840. The continuation of dune erosion in the west threatened to destroy Mitteldorf (now called Westdorf), so the first coastal protection structures were built in 1873. As on the other islands, it soon became obvious that the original structures were far too weak to protect the island, therefore a long series of coastal engineering countermeasures began:

1873: Construction of the first groyne on Baltrum. Continuing beach erosion restricted its length to only 135 m instead of its planned length of 180 m. It was severely damaged during the storm surge of December 1873.

1874: Groynes B and C (165 m and 135 m long)

Figure 330: Oblique aerial photograph showing the Wichter Ee ebb tidal delta and the coastal protection structures at the western end of Baltrum (seawalls, revetments and groynes). Aerial photograph of F.Böker, Hannover, of 14.6.1984, freigegeben Brg. Nr. 5765/591.

were built together with a 340 m long seawall.

1875: Reconstruction of the damaged groyne A (125 m long) south of its old position. Extension of the seawall by another 200 m.

1876: The approaching Wichter Ee undercut the groyne heads. The heads of groynes B and C were then reinforced by fascine mattresses and the 225 m long groyne D was built.

1877: Smaller middle groynes were constructed between groynes A, B and C because of continued beach erosion. The 125 m long groyne E was built in the same year.

1878/79: The seawall was breached in two places and had to be repaired. The groyne heads of groynes A, B, C and D had to be reinforced and groynes A and B supplemented by additional berms.

1880: Groynes A and B were strengthened with a third berm, groyne C with a second berm. Continued beach erosion caused the construction of groyne D_1, and the seawall was converted into a palisade revetment.

1883: The palisade revetment was severely damaged, and heavy dune erosion in the lee of the structure led to its being extended by another 300 m eastwards. The 170 m long groynes F and G were built.

1884: The continuing approach of the Wichter Ee was countered by the sinking of ships' hulls, of 15–17 m length, which protected groynes A, B and C. A storm surge in October led to renewed leeside erosion east of groyne G.

1885/86: The seawall was extended by another 800 m, and groynes H, J and K built.

1887/88: After lee erosion along the southwestern end of the revetment, it was extended by another 450 m and groynes L and M were constructed.

Figure 331: The extension of the Baltrum harbour jetty has enhanced accretion on the Wadden Sea facing coast (1980).

Further improvements and repairs followed. Fülscher (1905: 543) commented that between 1873-1903 about 2,800,000 Mark were spent on coastal protection for Baltrum. If the alternative of relocating the entire village of 30 houses to the safe eastern end of the island had been adopted instead, the costs incurred would have been below 150,000 Mark. If this was considered inacceptable or undesirable, it would have been wiser to have built a single strong coastal protection structure instead of this piecemeal solution. Fülscher (1905: 560) estimated that at least 600,000 Mark could have been saved in this way.

Today the western end of Baltrum is protected by numerous groynes and a seawall, which is built along the lines of the Norderneyer profile in the west, and the steeper Juister profile in the northeast (fig. 330). The steep Juister profile causes the release of enormous forces when waves break against the wall (plate 17). The storm surge of 17.2.1962 caused heavy damage in the west, the wall being destroyed over a length of 175 m. In addition, the seawall was breached in two places in the north (Janssen, 1962). A major part of the wall is therefore now protected with a modern revetment.

Positive sand budget in the south and east

The revetment extends eastwards into an area of positive sand budget where currently no leeside erosion has occurred. A second revetment was built between its western end and the quay where, between groynes D and M, the last palisade revetment remains seaward of the modern structure.

The enlargement of the port and extension of the jetty have led to better protection of the landward side of the island against storms from the west, resulting in increased accretion which has continued until the present day (fig. 331). The old brushwood groynes, which were originally planned to promote accretion, are situated today far inland in the salt marshes. Islands of *Spartina* have expanded relatively far out onto the tidal flats.

Two major dune ridges, which still experience active development, are situated in the northeastern part of the island. The inner dune ridge extends from the pleasure beach eastwards as far as the Peilbake beacon, the outer dune ridge following seaward at a distance of 200–300 m. Whilst the inner dune ridge is largely vegetated (fig. 332) and only modified by a number of blowouts, the outer dune ridge comprises recent 'white dunes' which are only sparsely vegetated with *Ammophila arenaria*. The grey and black dunes of southwestern Baltrum (fig. 333) are separated from the younger dunes of the northeast by a broad dune valley. Deflation has lowered the ground here as far as the groundwater table, leading to the development of aquatic vegetation (fig. 334).

The outer dune ridge has been breached in two places in the east. These breaches have been repaired using bush fences and are now almost completely closed again. The outermost dune ridge, especially at its eastern end, is in a state of continuous growth and renewal, a result of the strongly positive sand budget at this end of the island. A major primary dune field occurs at the very eastern end of the island, the Osterhook.

Figure 332: The inner dune ridge in the eastern part of Baltrum (1980).

Figure 333: Vegetated grey dunes east of the Westdorf village of Baltrum, looking from the Aussichtsdüne towards the west; the dune vegetation is protected by fences, and all footpaths are paved (1984).

7.3.10 NORDERNEY

Like most of the East Frisian Islands, Norderney consists of a relatively old dune core in the west and a sand flat in the east occupied by younger dunes. The settlement on Norderney is situated at the western end of the island (fig. 336). The island has experienced strong eastward displacement, while its western end remained stable (Luck, 1975: 50).

Historical development

Maps dating from the 16th century show an island situated between Juist and Norderney, the Isle of Buise, which does not exist today. Historical sources indicate that until the catastrophic storm surges at the end of the 14th century, Buise and Norderney had constituted a single island. Sometime before 1398 the dunes were breached, and a new tidal inlet formed. The dating of this event is unknown, but it may have been connected with the vast extension of the Leybucht in 1373 (Homeier, 1964: 12). Buise was subsequently completely eroded by the Buisetief and Norderneyer Seegat inlets, which at present form the double inlet between Juist and Norderney. The island disappeared completely about 1690.

The earliest reports concerning Norderney (since 1650) are favourable, the positive sand budget at this time resulting from the destruction of Buise. As a result the Christmas Storm Surge of 1717 caused little damage on the island. First indications of coastal erosion, however, are mentioned in reports written in 1773 and 1804 (Backhaus, 1943: 53). Since that time a negative sand budget has continued in the west. After a few decades, following the development of the island as a sea-side resort, Norderney became the first island to be protected by massive coastal engineering structures (fig. 337):

1. In 1846 the first two 400 m long brush groynes were built at the western end of the island. One was quickly destroyed during a storm surge in December 1847.

2. Further dune retreat prompted the construction of a 950 m long seawall in 1857/58, again leading to increased erosion in front of the structure.

3. The first five stone groynes were built between 1861 and 1863 to counter this erosion and stabilize the beach.

4. The seawall between the western end of Norderney and the Georgshöhe dune was built between 1874 and 1877. Five additional groynes were installed at the same time to protect the beach seawards of the wall. The groynes halted the eastward migration of the Norderneyer Seegat inlet, causing further incision from SKN-16 m to -19 m.

5. This necessitated further reinforcement of the groyne heads, installed between 1897–1900.

6. Dune erosion increased in the lee of the seawall, necessitating further extensions to the wall towards the east as far as the present 'Cafe Cornelius' between 1913/14.

7. The persistence of leeside erosion made repeated extensions of the wall necessary. In 1938 the wall was further extended by a revetment beyond the eastern end of the settlement.

8. After the Second World War the wall and groynes were in bad repair, and the beach in the west had been considerably lowered by erosion. Storm surges between January and March 1949 destroyed the seawall over a distance of 450 m. A proposal to counter erosion by building a dyke

Figure 334: The dune valley of eastern Baltrum, reduced to groundwater level by deflation (1980).

Figure 335: Norderneyer Seegat inlet and the western end of Norderney showing the town and coastal protection structures. Aerial photograph of the Niedersächsisches Landesverwaltungsamt, Landesvermessung, of 17.4.1974, Bildflug Ostfriesische Inseln (1086), Str.3/039, freigegeben durch NLVA–Abt. Landesvermessung – Hannover unter Nr. 34/74/1086.

Plate 27: Primary dunes on the Westerstrand sand flat of Schiermonnikoog (1985).
Plate 28: 'White dunes' forming on the Ostplate of Spiekeroog around colonies of *Ammophila arenaria* (1984).
Plate 29: 'Grey dune' on Spiekeroog, windward slope vegetated with *Empetrum nigrum*, leeside covered with *Corynephorus canescens* (1984).
Plate 30: Close-up view of the *Corynephorus* area of plate 29, demonstrating the open spaces between the plants which make the grey dune slopes susceptible to erosion (1984)

27	28
29	30

PLATES 27-30

PLATE 31

Figure 336: Oblique aerial photograph of Norderney from the west showing the coastal protection structures. The hotels on the front are protected by a sand dam, built following the flooding in 1962 caused by the February storm surge. Aerial photograph of F.Böker, Hannover, of 16.6.1984, freigegeben Brg. Nr. 5765/592.

from Juist to the mainland (Thilo & Kurzak, 1952) was turned down in favour of the less extreme measures of firstly beach replenishment in 1951/52 (Kramer, 1959), and secondly by further extensions to the revetment and the groynes, the latter as far as the Kugelbake beacon. These improved the situation, and in 1953 the western end of Norderney was finally protected by 6,000 m of seawalls and revetments, and 32 groynes.

9. Erosion in the lee of the structures has continued, leading to severe dune losses east of the Norderney settlement. The storm surges of 4.1. and 7.2.1981 breached the coastal dune ridge between Kugelbake and Weiße Düne. Sand-filled plastic tubes like those used previously on Langeoog were installed to stem the erosion, but the sediment budget has remained negative.

The present situation

The northwestern shore of Norderney currently gives the impression of a step-like retreating coastline from west to east with a narrow beach and numerous coastal protection structures (figs. 335 and 336). The situation has considerably worsened during the last 30 years, the main reason being the

Plate 31: Satellite image of the Niedersachsen coast between the Weser estuary and Norderney; Landsat-5 TM image of 22.8.1984, Track 197, Frame 023 (DFVLR).

Figure 337: Narrow beach at the western end of Norderney (1982).

increased frequency of storm surges:

Number of storm surges per year

Period	Light	Severe	Very severe
1951–60	7.7	1.5	0.1
1961–70	10.7	1.4	0.7
1971–80	9.3	1.7	0.8
1981	17	3	0
1982	17	0	2
1983	31	5	1*
1984	16	2	0

*Only period 11.11.1983 – 8.2.1984.

Repeated beach replenishments have been necessary in order to counter the negative sand balance (cf. section 5.3.7). Since the first beach replenishment in 1951/52 this procedure has been repeated six times, most recently in 1982 and 1984.

The heavily defended western end of the island is still suffering from a shortage of sand. The recreational beach is restricted to narrow zones northeast and southwest of the town (fig. 337).

At the Weiße Düne about 5 km east of the western end of Norderney the situation changes over a distance of only a few hundred metres. West of the Weiße Düne can be seen the strong, continuously deepening breach in the coastal dunes due to the lee-side effect of the groynes, but beach width increases rapidly to the east. This is the area where the shoals of the Norderneyer Seegat ebb delta attach themselves to the Norderney shore.

The polders

A recurved spit started to form south of the town during the 19th century as a result of the stabilization of the western end of the island. The Mariendeich dyke was constructed in 1840 forming a small polder to protect the town, and a port was built in 1880 behind the flood spit. Three years later the polder was enlarged as far south as the Hafendeich. The other major polder on Norderney, the Grohde Polder (167 ha) in the centre of the island, was dyked between 1922-26. The Südstrandpolder is the youngest of the Norderney polders, dyked during the Second World War in order to reclaim land for an airfield. This plan was abandoned, and in 1962 the wetlands of the polder were turned into a nature reserve.

Development of the Ostplate

The eastern part of the island enjoys an obviously positive sand balance. The eastern end was still a largely unvegetated sand flat during the early 19th century. Dune formation was initiated during the middle of the century. By 1866 a continuous dune belt extended as far as the Eiland area, and by 1892 it reached the Postbake (post beacon). Today the Ostplate is covered with dunes almost to its outermost extremity. The backshore, which supplies the sand for additional dune formation, is several hundred metres wide.

The Norderney Ostplate is one of the few relatively natural areas in the East Frisian Islands. No artificial dykes or dune ridges have been constructed, and during exceptionally high tides the area is largely inundated, like the eastern end of Spiekeroog. The dunes are concentrated as several elliptical dune cores interrupted by washover areas (fig. 338 and plate 11).

During strong storm surges the recent dunes can be heavily eroded or even eliminated, especially in the easternmost part of the area. A typical example of this is the Rattendüne complex. This dune, which was originally several metres high, was completely destroyed during the 1962 storm surge, an unvegetated area in the salt marshes being the only

Figure 338: Oblique aerial photograph of a washover area on the Ostplate of Norderney, also showing the wide beach and the recent salt marshes which have developed since 1860 under the protection of the recent dunes. Aerial photograph of F.Böker, Hannover, of 16.6.1984, freigegeben Brg. Nr. 5765/593.

Figure 339: Oblique aerial photograph showing path systems and eroded recent dunes on the Norderney Ostplate. Aerial photograph of F.Böker, Hannover, of 16.6.1984, freigegeben Brg. Nr. 5765/594.

testament to its former position (cf. fig. 339). Eroded dunes remain visible as light patches in the salt marshes for rather a long time. Following the removal of the topsoil, vegetation has great difficulty in re-colonising such areas. The slightly elevated sand mounds are also favoured by rabbits, which occur in vast numbers on Norderney in contrast to some of the other islands. The open sand areas inhibit deflation which would lead to further levelling.

Vegetation is also destroyed along the Ostplate footpaths, which soon present axes of erosion penetrating down to the groundwater table. When a path becomes excessively muddy, new parallel paths are established, so that after a few decades a system of sub-parallel paths develops several tens of metres wide (figs. 339 und 340). As can be seen on fig. 339, the uncontrolled use of off-road vehicles by the state authorities (no private cars are allowed here) plays a major role in the initiation of such path systems.

Figure 340: Path system on the Norderney Ostplate with the Möwendüne dune core in the background and recent dunes on the left (1984).

7.3.11 JUIST

Juist is the longest of the East Frisian Islands with an overall length of 17 km (fig. 341). Its elongated form contrasts with the more bulbous shapes of the other islands and probably results from the detachment of its western flood spit at an early stage to form the present Isle of Memmert (Gaye & Walther, 1935: 556). Three dune cores (Bill, Haiddünen, Kalfamerdünen), which were formerly separated by washover areas, have merged with human assistance. The island has remained very stable over the last 100 years (fig. 342), making massive coastal protection structures unnecessary.

Historical development

Like most of the other East Frisian Islands, Juist was first mentioned in 1398. Around 1568 it still had one continuous dune ridge, but by 1585 the coastal dunes on both sides of the church had disappeared. This old church had been erected about 1300 north of the now central parts of the Hammer Lake. It collapsed in 1660 as a result of coastal erosion and was re-built at the northwestern end of the Ostland dunes (Backhaus, 1943: 32).

The foundations of the old church were preserved in the sand and were discovered by Leege in 1900 about 400 m north of the low water line. Therefore, since the construction of the church the island had migrated about 600 m southwards, i.e. about 1 m/year (Luck & Stephan, 1983: 15). Early efforts to protect the dunes were assisted by 'Dünemeyers' (dune workers) imported from the Netherlands at the beginning of the 18th century (Backhaus, 1943: 32).

The breach in the coastal dunes in the Hammer area which had been closed in 1690 was re-opened and considerably widened during the Christmas Flood of 1717, lowering the height of the island at this point almost to beach level. A second breach, the so-called Schweinshammer, began to open in the Haiddünen at the beginning of the 18th century. An attempt to close the Hammer gap with brush fences failed, and during the storm surge of 1743 the gap was widened to between 120–240 m. Continued erosion until 1780 led to the abandonment of Westdorf village and subsequent resettlement in the Osterloog, the present site of Juist village; the island church had already been moved again to the new site in 1742. The final sealing of the Schweinshammer using sand dykes was achieved in 1743 (Backhaus, 1943: 42).

By 1866-77 it had become possible to construct a dam between the Bill Dunes and the main dune core of the Haiddünen, thus closing the Hammer washover area (Backhaus, 1943: 47). The Hallohmsglopp washover between Haiddünen and Kalfamerdünen dune cores was closed at about the same time (1885-1890) (Windberg, 1931: 53).

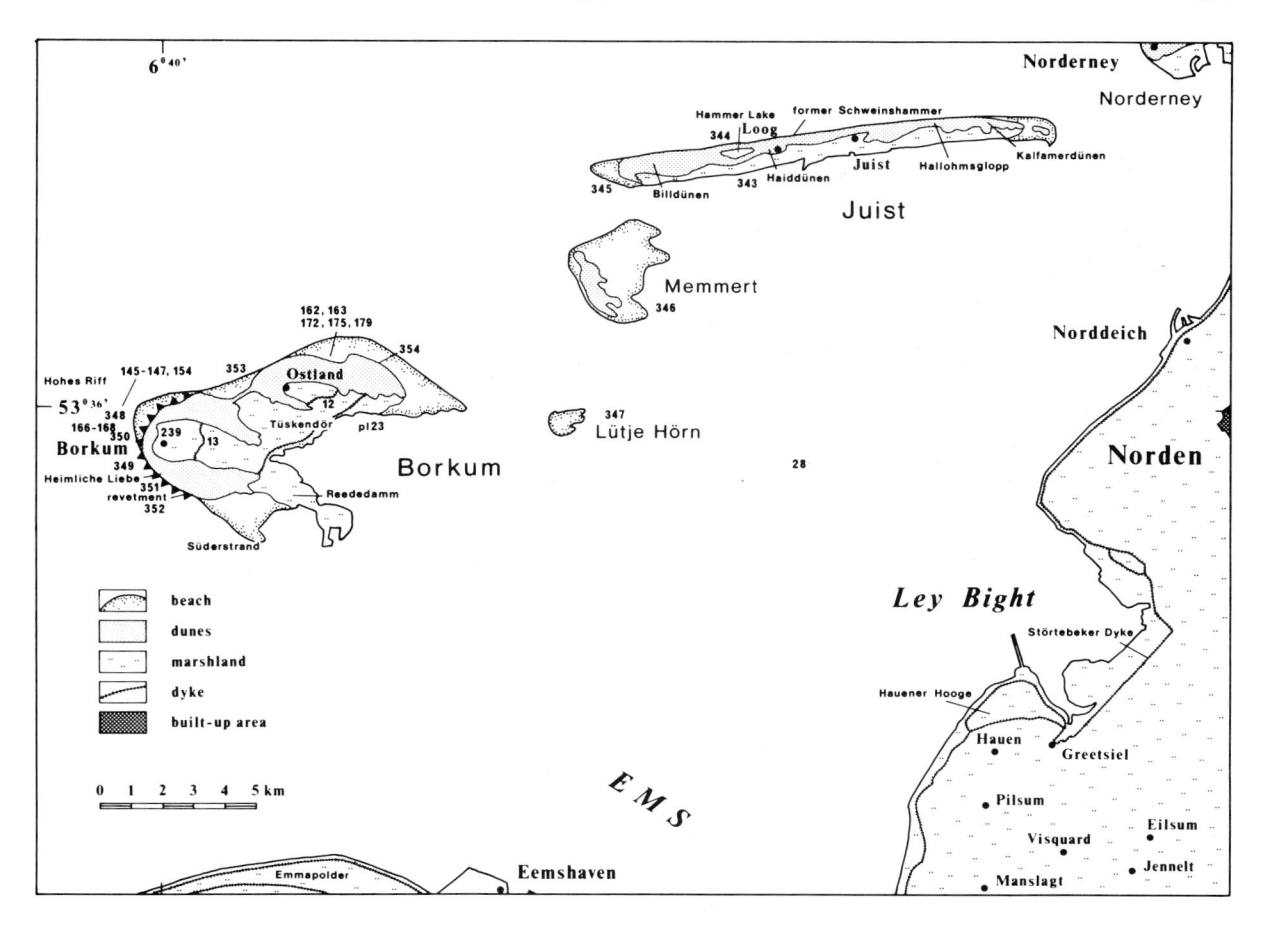

Figure 341: Location map of the area between Juist and Borkum. The numbers refer to figures, those prefixed by 'pl' refer to plates.

Positive sand balance

Three generations of dunes can be identified on Juist (Backhaus, 1943: 47). The oldest dunes formed before 1717, the second belt from then until about 1800. The latter belt was badly damaged during the storm surge of 1825 but recovered during the subsequent decades. The youngest coastal dunes have developed since 1880 (Windberg, 1931: 59). Between 1840–1900 the dune core of Juist extended 2,600 m to the east demonstrating that, as on most of the other islands, a period of positive sand balance has been prevalent since the middle of the 19th century.

The salt marshes of the Bill were dyked in 1875. At the end of the 19th century Juist village was well-protected behind the dunes (c.f. map in Scherz, 1900), but with the onset of tourism new hotels have been built much closer to the beach.

A short period of intensive erosion began in 1901, leading in 1913 to the construction of a 1,400 m long seawall and seven groynes in front of the village. This erosive period, however, did not persist, and the structures were soon completely engulfed by recent dunes (Backhaus, 1943: 47). They have not re-appeared since.

In 1927 the Hammer Bucht was closed by a dune ridge created with sand fences on the seaward side of the Hammer Lake (figs. 343 and 344) (Backhaus, 1943).

During the last 300 years Juist has not been eroded in the west but has grown considerably towards the east (Homeier, 1964). A sufficient sand supply has made the island one of the most stable barrier islands in the Wadden Sea. Luck & Stephan (1983), however, have concluded that the positive sand budget enjoyed on the north coast will not persist for a long time. During the last 20 years major morphological changes in the Oster-Ems, possibly linked with the current intensive rise in

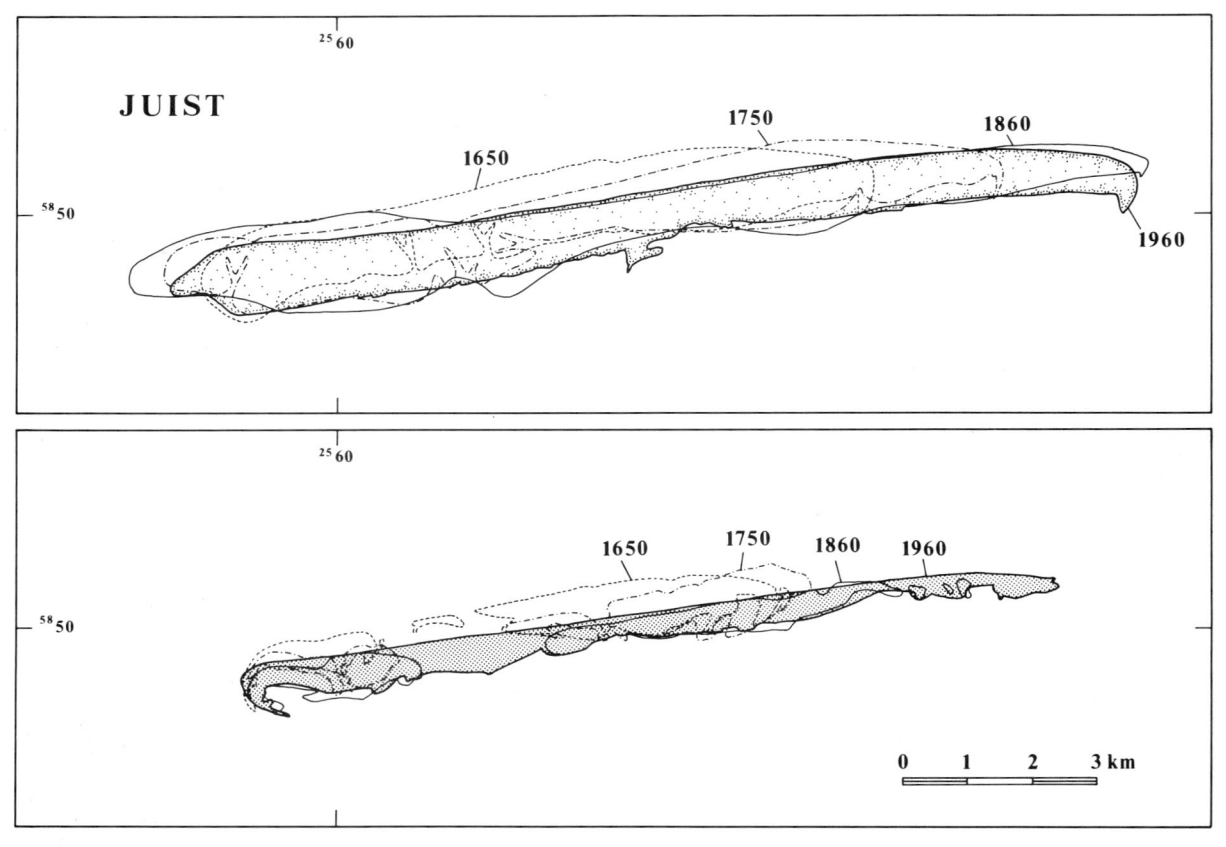

Figure 342: Historical development of the coastline of Juist (above) and the dune areas (below) from 1650 to 1960 (after Luck & Stephan, 1983).

sea level, have initiated strong erosion of the westernmost dunes.

As early as 1931 Windberg (1931: 61) had pointed out that the shoals of the Oster-Ems ebb delta had increasingly approached the western end of the island since 1851, and this trend has continued. Fresh sand no longer reaches the island at its western end, but a little further to the south (fig. 345), so that the development of a recurved spit, such as occurs on the other East Frisian Islands, seems to be imminent.

7.3.12 Memmert

The Isle of Memmert was recorded as a sand bank on maps as early as the 16th century. A description of 1650 states that Memmert was no longer flooded during ordinary high tides and that Helm (*Ammophila arenaria*) had colonized some small primary dunes. It was only at the end of the 19th century, however, that salt marsh formed under the protection of these dunes. The first salt marsh

Figure 343: The Hammer Lake on Juist, looking towards the south (1982).

Figure 344: The recent dune ridge separating Hammer Lake from the beach (1982).

Figure 345: The western end of Juist with a small secondary inlet on the sand flat (cf. section 5.2.1), the mouth of the Juister Balje and the Isle of Memmert to the south. Aerial photograph of the Niedersächsisches Landesverwaltungsamt, Landesvermessung, of 17.4.1974, Bildflug Ostfriesische Inseln (1086), Str.3/027, freigegeben durch NLVA−Abt. Landesvermessung−Hannover unter Nr. 34/74/1086.

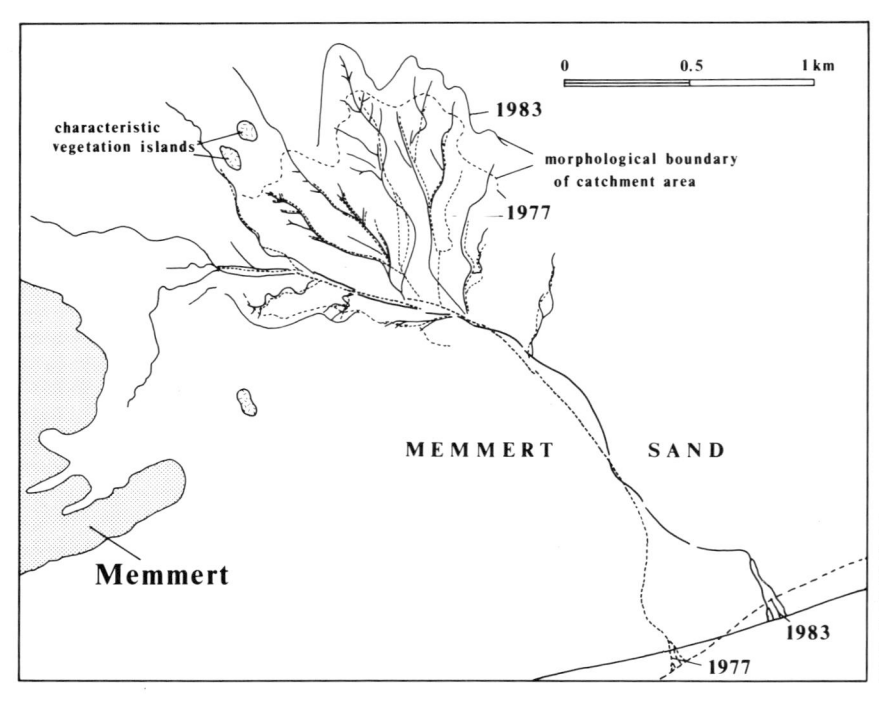

Figure 346: Changes in a dendritic drainage system developing in the lee of the Isle of Memmert, between 1977 and 1983; based on aerial photographs of the Niedersächsisches Landesverwaltungsamt, Landesvermessung.

plants were recorded in 1891 (Leege, 1935: 23).

Memmert was declared a bird sanctuary in 1907 and in 1908 the two main dune groups were connected by a sand dyke created with brush fences. The dunes were immediately exposed to erosion, but the dyke prevented further washovers and salt marsh growth intensified in the east (Homeier, 1968: 28). However, dune erosion continued in the west (Luck, 1968). At present this does not seriously endanger the island because the point of attachment of the Oster-Ems ebb-delta shoals has shifted increasingly southwards. This favours the Kachelotplate sand bank off Memmert (cf. Luck & Stephan, 1983: 22) which may indicate the onset of a period of increased sand supply for the island.

Simultaneously with the growth of Memmert, the sand flat behind it has become increasingly dissected. Three major dendritic drainage systems have developed on the flat. The recent growth of the southwestern system is shown in fig. 346. The development of this system can be traced back to 1937 using old aerial photographs (cf. Luck & Stephan, 1983: Anl. 17).

7.3.13 Lütje Hörn

Lütje Hörn is a small sand island southeast of Borkum, characterized by a few scantily vegetated primary dunes (Leege, 1935: 30). Like Memmert the Isle of Lütje Hörn forms part of the flood delta of the Oster-Ems. The island was recorded as early as 1585 on a map by Haeyen under the name of 'Hooghe Hörn'. A first approximation of its position can be determined using the map of Faber (1642); at this time Lütje Hörn was situated roughly 3.2 km northwest of its present position (Homeier, 1963 a). The island therefore seems to have migrated at a rate of about 10 m/year, which corresponds closely to its current migration rate (1957–1977: 11 m/year; fig. 347).

7.3.14 Borkum

Borkum is the westernmost East Frisian Island and, covering 32 km^2, it is also the largest. The island is situated between the Wester-Ems and Oster-Ems inlets which are both connected by a broad and shallow channel. The island consists of two dune arcs which are open to the east. Until the mid-19th century both arcs were separated from each other by a washover area which was regularly flooded during the higher tides. The gap between the two dune arcs was closed about 1860.

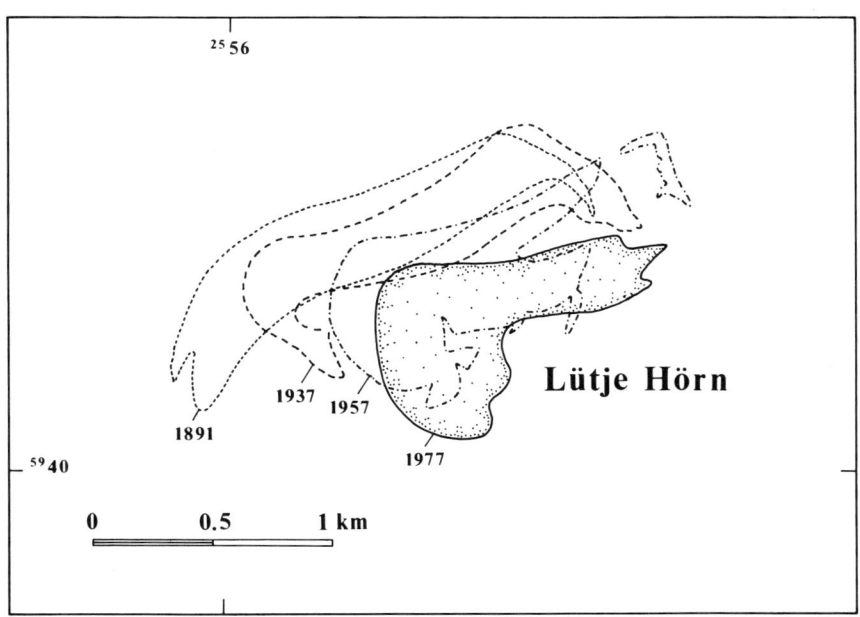

Figure 347: Migration of the Isle of Lütje Hörn between 1891 and 1977. Sources: Homeier, 1963 a and Küstenkarte 1:25,000 Sheet 2406 K Borkum (1977).

Historical development

Backhaus (1943: 34) believed that Borkum and Juist had originally been part of a larger island called Bant which had allegedly become divided between 1100 and 1398. Evaluation of historical data, however, has shown that this was not the case. Homeier (1965: 38) demonstrated that Borkum and Juist have always been separate islands, and that Bant had only been a hallig-like marsh island situated between Borkum and the mainland, east of the Oster-Ems. In 1470 Bant still covered an area of about 17 km^2. Salt was produced from peat on the island, but the peat digging apparently furthered erosion (Marschalleck, 1973) and the island disappeared completely about 1780 (Pötter, 1982: 55).

Borkum already conformed to its present shape when it was first described in 1650 and 1657 (Backhaus, 1943; Homeier, 1979 b: 16). Through the centuries the island has been slightly eroded in the northwest and slightly extended in the east. In 1789 remnants of old wells were found, together with urns from 'pre-Christian times' on the beach 800 m from the present seawall. Using these data a maximum retreat rate of the dune foot between 800 and 1650 A.D. of 2.1 to 2.3 m/year can be calculated. This rate accords well with the retreat rate of 2.2 to 2.4 m/year measured from 1650 until the construction of the seawall (Homeier, 1979 b: 13).

Erosion in the west

The western dune arc of Borkum has not been completely preserved, suffering partial erosion in the northwest. The first groynes were built here in 1869 when the beach was still over 200 m wide, enabling uncomplicated construction of the groynes on dry sand during low water. Attempts to curb coastal retreat had been unsuccessful before the construction of a seawall (in 1875) which had reached an overall length of 3,870 m by 1906. The wall was unable to withstand the winter storms for long, however, because of its steep profile. During the storm floods of October 1936 a major section of the wall was destroyed (Wegmann, 1982) and in the storm surge of 21.12.1954 another 100 m collapsed. A major section of the wall (1,870 m) was later rebuilt with a more gently sloping profile (Wasser- und Schiffahrtsdirektion Aurich, 1968). The danger lessened, however, with the approach of another major shoal from the Hohes Riff area.

The sand supply for the beaches on Borkum is controlled by the migrating pattern of the Hohes Riff shoals. The Hohes Riff is part of the rather incomplete ebb delta of the Wester-Ems, the recent morphological changes of which have been discussed by Samu (1972, 1982). At present the Hohes Riff consists of two parallel striking shoals slowly migrating in the direction of their long axes; in this way they move towards the beaches on Borkum

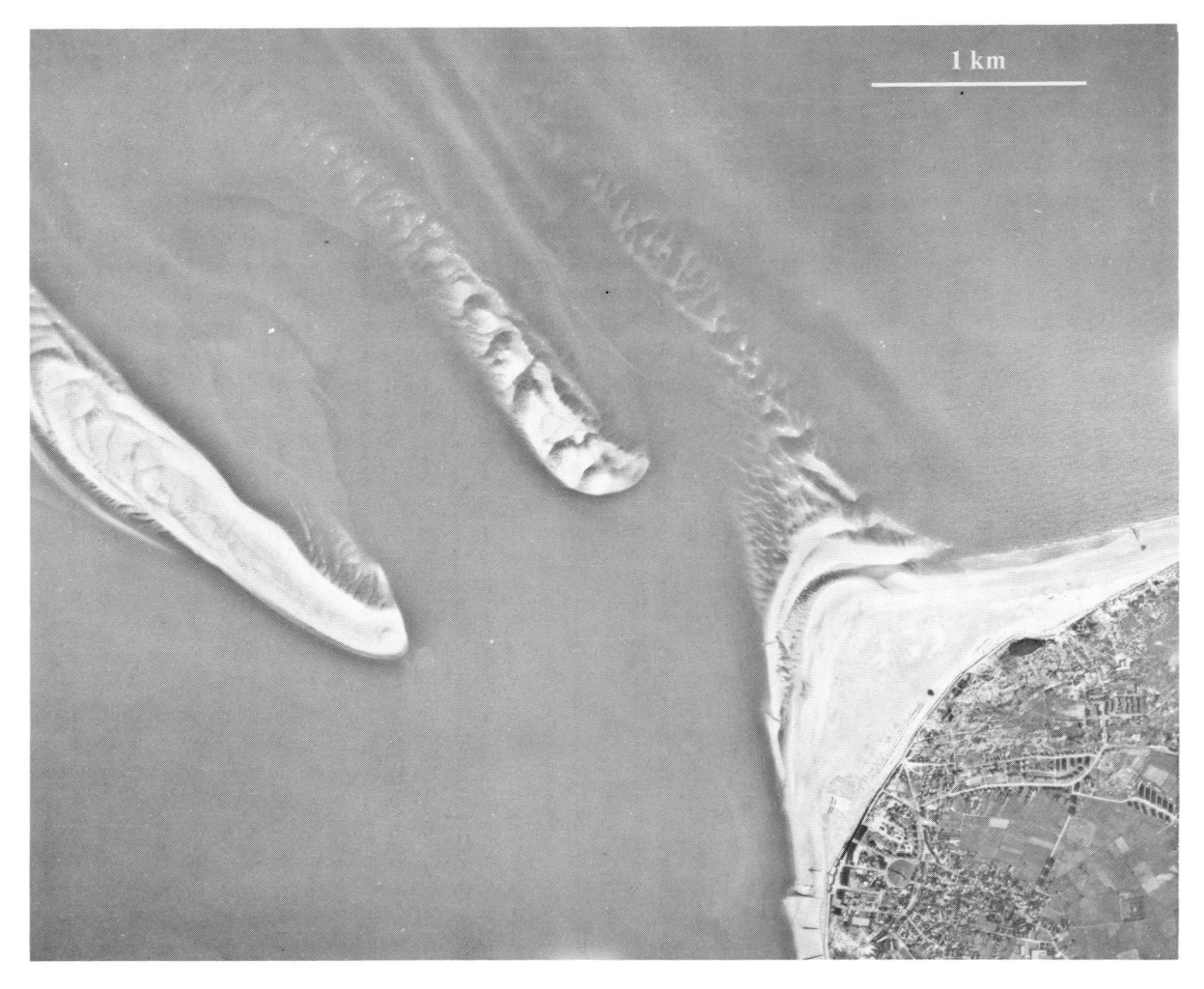

Figure 348: Shoals of the Hohes Riff approaching the northwestern end of Borkum. Aerial photograph of the Niedersächsisches Landesverwaltungsamt, Landesvermessung, of 20.8.1966, Bildflug Ostfriesische Küste (419), Str.6/063, freigegeben durch Reg.Präsident Münster/W. unter Nr.2316/66.

(fig. 348) at a rate of about 50 m/year (cf. Luck & Witte, 1974), much slower than the shoals in the ebb deltas of the other East Frisian Islands.

The Hohes Riff consists of what can be termed 'megashoals' (cf. section 5.2.2). The aerial photograph shows that they are covered with megaripples which migrate much faster than the shoals themselves. These migration rates, however, have not so far been measured.

Homeier (1971: 21) and Homeier & Luck (1977: 85) have summarized beach development on Borkum as follows:

1. Long term beach development is controlled by cycles of accumulation and erosion which can be observed over long periods of time. Each cycle starts with the attachment of a major shoal from the Hohes Riff on to the northwestern beach, leading to an instantaneous improvement of the sand balance at this point. Two such events have been recorded, the first towards the end of the 17th century, the second in the middle of the 20th century. The size of the 17th century shoal is unknown, but appears to have been considerable.

2. The resultant zone of excessive beach width created by the attachment of the shoal subsequently divides into two beach maxima, migrating northeastwards and southeastwards respectively at almost equal rates of slightly less than 25 m/year.

Figure 349: The area with negative sand balance south of the point of attachment of the shoals, protected by a seawall, groynes and a revetment. In the northeast the old dyke of the Borkum polder can be seen, several ponds documenting its repeated breaching. Aerial photograph of the Niedersächsisches Landesverwaltungsamt, Landesvermessung, of 18.6.1983, Bildflug Borkum (1979), Str.3/056, freigegeben durch NLVA–Abt. Landesvermessung–Hannover unter Nr. 49/83/1979.

Figure 350: The boundary between positive and negative sand balance at the western end of Borkum (cf. fig. 349). The hotels built directly on the beach around 1890-1900 are in the background (1984).

Figure 351: New dune ridge formed on top of the revetment south of the 'Heimliche Liebe' restaurant (1984).

3. New dunes form under the influence of both migrating sand bodies.

4. After these maxima the cycle returns to its starting point. Beach minima are reached at distances of 3.5 km on the northern and 2.5 km on the southwestern beaches. The minima migrate at the same rates as the maxima.

This entire process has caused Borkum to migrate slowly eastwards, erosion prevailing in the west and accumulation in the east.

The boundary between the areas of positive and negative sand balance is situated at present along the recreational beach (figs. 349 and 350). An extremely wide beach occurs north of this boundary, whilst south of it the seawall directly faces the sea.

Figure 352: The exposed revetment base in the southeast (1984).

Additional coastal protection works in the southwest

Erosion started to intensify south of the seawall after 1950, threatening the adjoining dune area. Two World War II bunkers, which had originally been sited within the dunes, soon lay 50 m outside on the beach. Overall, about 100 m of dune terrain has been lost (Homann, ca. 1973: 68). To offset erosion south of the 'Heimliche Liebe' restaurant a 600 m long revetment was built in two stages (1956 and 1962/63), later supplemented by strong concrete groynes. The formation of a new coastal dune ridge was also encouraged by brush fences (fig. 351).

Erosion has intensified at the southeastern end of the revetment, and the remaining beach in front has been lowered, exposing the base of the structure. No sand deposition occurs here (fig. 352). A vast, largely unvegetated sand flat is situated southeast of the revetment, representing the marginal part of the Wester-Ems flood delta. Groups of primary dunes have formed on the northern part of this sand flat, intersected by a narrow washover area. Some of the sand lost at the western end of the island migrates into this area.

The polder

Agriculture has never played more than a minor role on Borkum because of poor soils. During the climax of the whaling period around 1750, however, population increased considerably and there was great demand for grain. This led to the settlement (1752) and dyking of the Ostland area, but most of the dyked areas in the polders of eastern and western Borkum were used for cattle grazing and hay production (Meier, 1863: 45; Wegmann, 1982). Farming has become easier since the construction of a modern sea dyke in 1932. This was breached at Ostland during the storm surge of 17.2.1962 over a length of 100 m (fig. 12).

The dykes on Borkum were last raised in 1977/78. The 20 ha Tüskendörsee lake south of the airport is not a natural feature but was formed as a result of sand pumping for dyke construction. The Reededamm, the embankment connecting the island with the ferry terminal, had also been badly damaged during the storm surge of 1962. It was recently protected against storm surges by the construction of a 2,550 m long sand dyke (height: 5–6 m above NN).

Accretion along the tidal-flat facing coast of Borkum is minimal, and in this respect the island does not differ from the other East Frisian Islands.

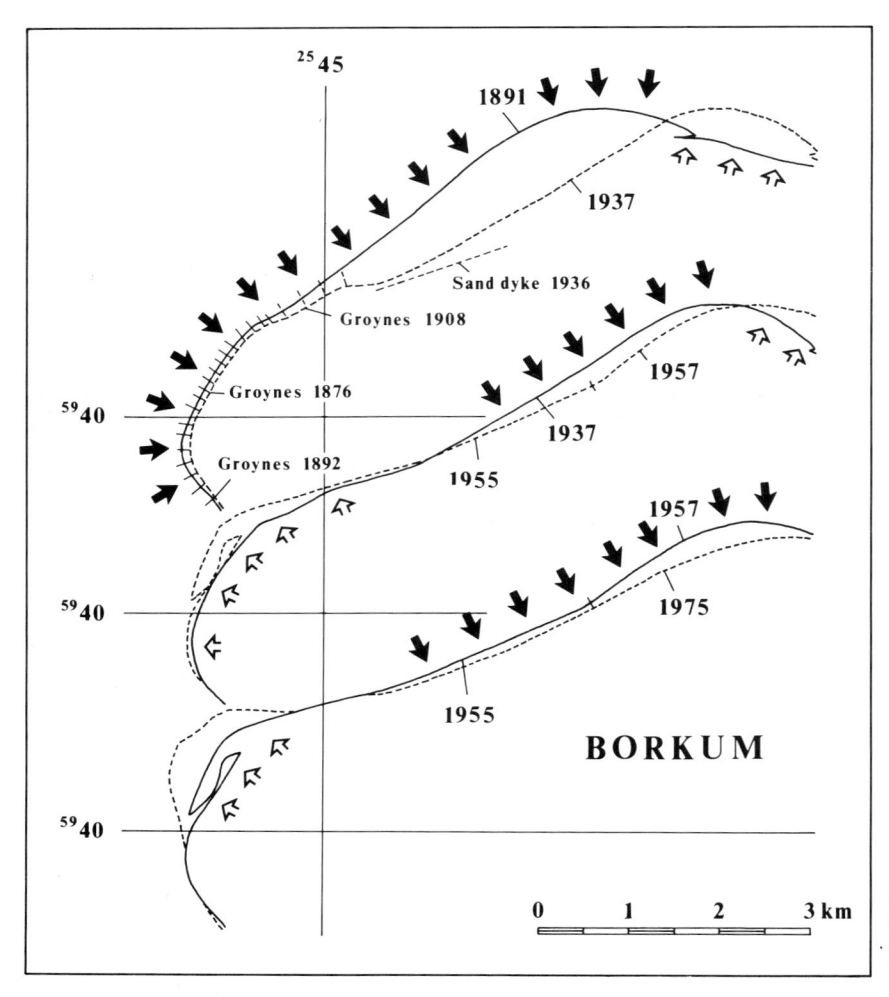

Figure 353: Coastal development of northwestern Borkum between 1891 and 1975 (after Homeier & Luck, 1977).

With the exception of a small *Salicornia* area in lee of the Reededamm, the salt marsh margin is defined by a low cliff and beach ridge (plate 23), both of which are indicators of prevailing erosion. A brushwood groyne system (Lahnungsfelder) has been built along the Reededamm in order to promote accretion, but it is no longer maintained.

Accretion in the east

From the earliest maps until the beginning of the 20th century the eastern dune core of Borkum has always been depicted as a complete arc (cf. map in Behrmann, 1919: 3). At present, however, the north coast north of the Tüskendör area is being eroded and in some parts the coast has retreated by more than 1 km since 1891 (fig. 353). Steep dune cliffs and fallen bunker remnants on the beach testify to the continuation of erosion here.

Nevertheless, eastern Borkum has a positive sand balance; although the accretion area has migrated eastwards to a position east of the middle of the Ostland dune arc. Here the beach is widening and recent dune ridges are forming in front of the old coastal dunes (fig. 354). Some of the easterly migrating sand supplied from the Hohes Riff shoals along the north coast is deposited here. At the eastern end of the dune complex recent dunes are intersected by a narrow washover channel but this is rarely flooded and is at present fully vegetated.

7.4 THE NETHERLANDS

7.4.1 ROTTUM

Rottum, or Rottumeroog, is the easternmost West

Figure 354: Recent dunes forming on eastern Borkum (1984).

Frisian Island (fig. 355). This small dune island, which was formerly used for agriculture, has been repeatedly displaced eastwards and progressively reduced in size.

A village was still in existence on the island in 1632 indicating a measure of stability at this time. The situation soon changed, however, because after the storm surge of 1717 Count Clancarty, the owner of the island, had to leave with all the inhabitants. It soon became obvious that the island could not be re-settled, and in 1738 it was sold to the States of Groningen. The old village was situated approximately in the recent position of the Rottumerplaat (fig. 355).

The displacement of Rottum has been recorded accurately since 1770 because the beacons on the island had to be repeatedly replaced and always re-surveyed. A small core of old, higher dunes persisted until the end of the 19th century in addition to the newly constructed artificial dune ridges. These dunes had migrated together with the island and reached a height of 11 - 13.5 m. The dunes still encircled an area of 25 ha in 1865, and were used for farming. Under the protection of the dunes in the southeast recent sandy clay was deposited, up to 30 cm thick, leading to some limited salt-marsh development (Isbary, 1936).

Erosion of the island accelerated after 1865. Until this time coastal retreat had been repeatedly interrupted by phases of consolidation, but after 1865 no such phases occurred (Isbary, 1936) and the migration rate increased from 25 m/year between 1770 and 1865 to 30 m/year after 1865. Simultaneously the seaward margin of the dunes began to retreat at a rate of about 1 m/year. A strandvogd (beach bailiff), who was employed to fix the dunes and, if possible, reclaim land, was stationed on Rottumeroog until 1965. After the retirement of the last strandvogd the island was effectively abandoned by the authorities (Toxopeus, 1981).

Rottumerplaat

Rottumerplaat is a sand bank, situated on the site of the former island of Bosch which was inhabited during the Middle Ages. Band island, which is also shown on old maps, later became part of the Frisian mainland (Bantelmann *et al.*, 1976: 224), but little is known of the fate of the other former islands of this area, Corenzand and Heffezand (fig. 356). It seems likely that they had been marsh islands which were finally completely eroded. On a map of 1584 they are no longer listed as islands but only as sand banks (Toxopeus, 1981: 12).

Dune formation started on Rottumerplaat in

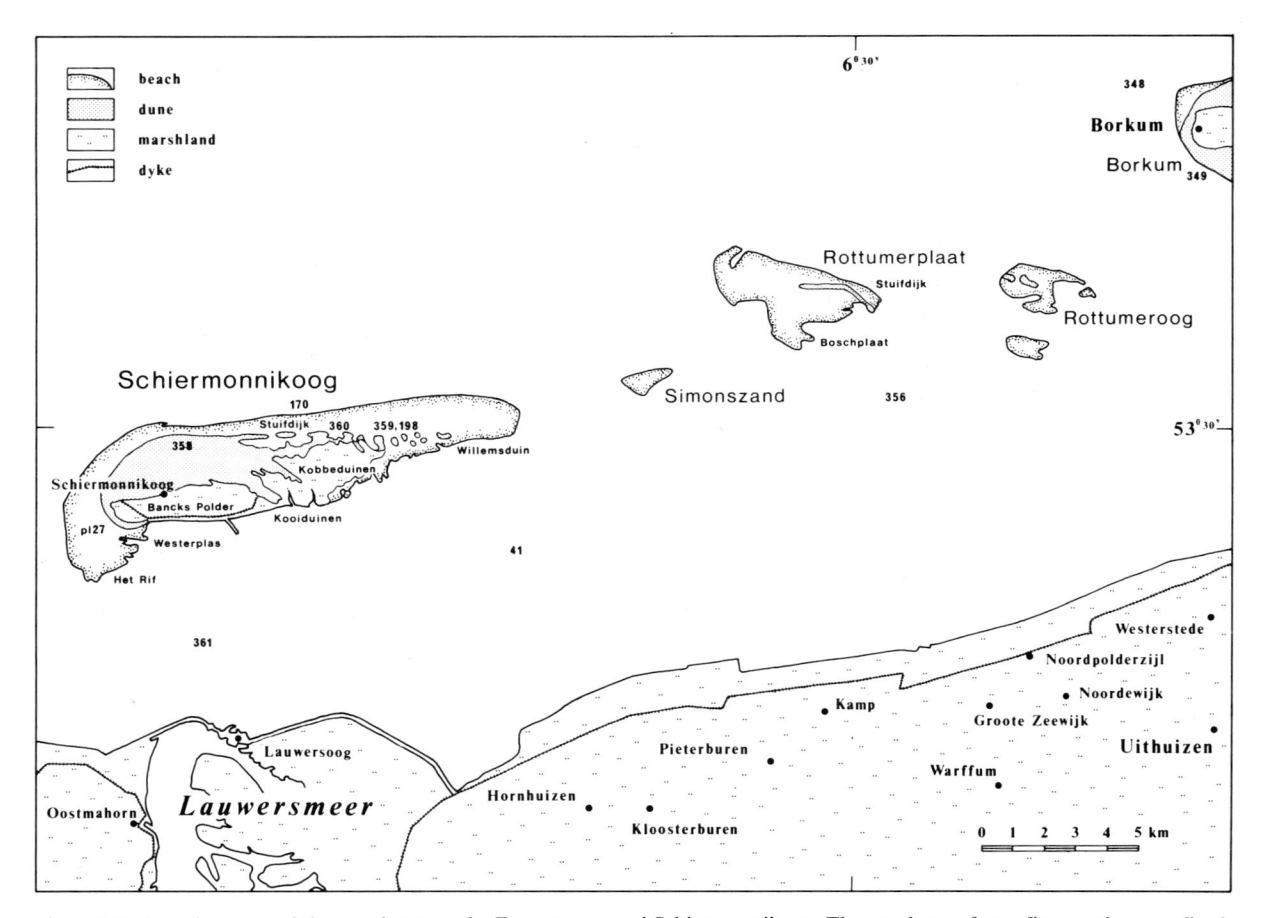

Figure 355: Location map of the area between the Ems estuary and Schiermonnikoog. The numbers refer to figures, those prefixed by 'pl' refer to plates.

1872. They were then eroded until 1888, but in 1916 dune formation re-started, and by 1933 an almost circular dune core had formed, 100 metres in diameter. The dunes were 3–4 m high, with their northwestern ends eroded. In the east, however, two low hummocks of *Agropyron junceum* had led to the development of new primary dunes (Isbary, 1936: 47).

An attempt was made to encourage accretion on Rottumerplaat by the construction of a stuifdijk (dune dyke) in 1951. Since the surface was still very low it took some time for the stuifdijk to develop. At present this 3 km long dyke has reached a height of 6 - 7 m above NAP (= mean sea level; Dutch ordnance datum), its maximum height being 8–9 m above NAP. In the northeast a natural dune core, 3–4 m high, has been accumulated, separated from the stuifdijk dunes by a washover area.

South of the stuifdijk a salt marsh area, about 70 ha in size, has formed, showing diverse vegetational development. From the southeast, however, the Lauwers Inlet is approaching the island, causing considerable erosion.

The Schild Inlet between Rottumerplaat and Rottum recently has started to decrease in size, giving rise to a positive sand budget on Rottum's northwest coast. It seems possible that the Schild will completely be filled, allowing Rottumerplaat and Rottum to form a single, major island (Reenders, personal communication).

Plate 32: Satellite image of the Niedersachsen coast between Langeoog and Schiermonnikoog; Landsat-5 TM image of 22.8.1984, Track 197, Frame 023 (DFVLR).

PLATE 32

PLATE 33

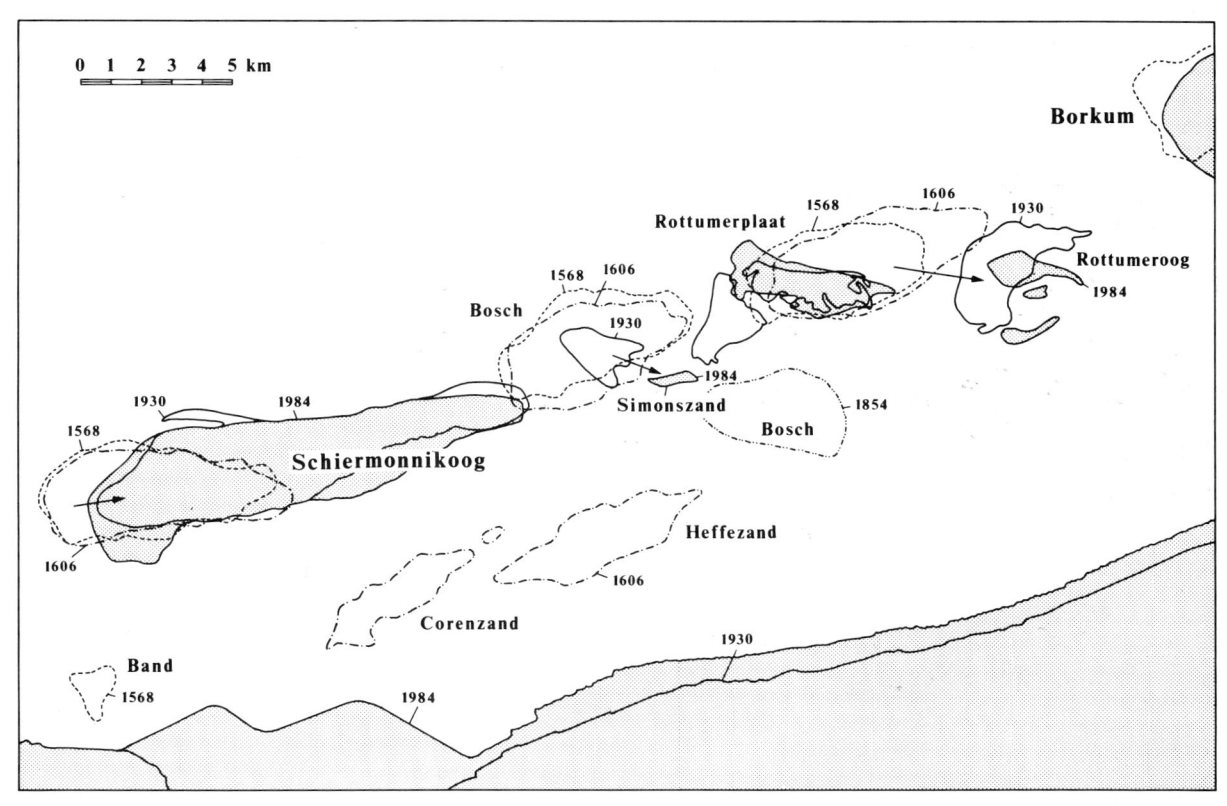

Figure 356: Map showing the displacement of the easternmost West Frisian Islands from Schiermonnikoog to Rottumeroog from 1568 to 1984. Sources: Isbary, 1936; Landsat 4 TM image of 1984.

Simonszand

The small Simonszand shoal between Rottumerplaat and Schiermonnikoog (fig. 356) resembles the mobile islands in the inner German Bight. It is rapidly migrating southwards and decreasing in size. Whilst the present Simonszand will soon be completely eroded, a new high shoal is forming on its seaward side recently, which will replace Simonszand as part of the barrier (Reenders, personal communication).

7.4.2 SCHIERMONNIKOOG

Schiermonnikoog is the easternmost inhabited West Frisian Island (fig. 357). Of all the North Sea barrier islands Schiermonnikoog has been the least altered by human interference, with the exception of the uninhabited outer sands. Coastal protection measures have not been necessary, and only minor damage has been inflicted by tourism, which has been expanding since 1960. The island can be broadly subdivided into the Westerstrand sand flat, the island core and the Oosterstrand.

The Westerstrand

The Westerstrand and its southern extension, het Rif, form a perfectly level sand flat, where a small field of primary dunes have developed in close proximity to the dunes of the island core (plate 27). A small salt marsh area has begun forming in the lee of these primary dunes south of the western end of the island core.

The island core

The dune core of Schiermonnikoog consists of a series of dune ridges, the most prominent being the 1.5 km long complex of former migrating dunes northeast of the village. These are not always easy to identify in the field, because part of

Plate 33: Satellite image of the Dutch North Sea coast between IJmuiden and Schiermonnikoog; Landsat-4 MS image of 14.3.1984, Track 198, Frame 023 (DFVLR).

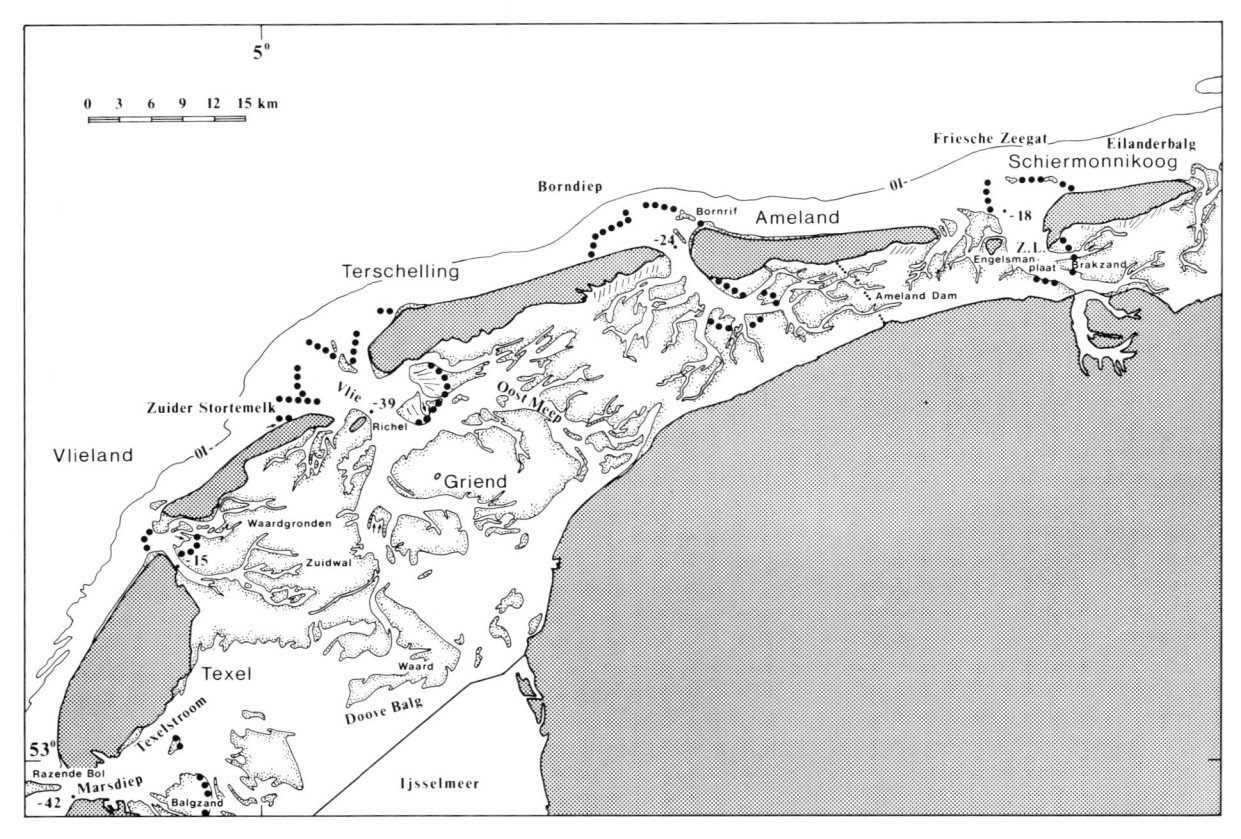

Figure 357: Tidal flats and inlets between Lauwersbalg and Marsdiep.

Figure 358: Pine forest in the dunes of Schiermonnikoog (1985).

the area has been afforested (fig. 358). Whilst both flanking dune ridges reach heights of 12.3 m in the north and 9.0 m in the south, the dune front has grown to a height of over 15 m. Their morphology has unfortunately been marred at this point by construction of the 'Wassermann' bunker during the German occupation; this is now used as an observation point.

Until the 19th century the inner dune margin of Schiermonnikoog was actively migrating. After 1809 the dune front migrated to a maximum of about 100 m southwards causing the destruction of seven houses north of the Noorderstreek path; these had to be demolished in order to avoid their burial by sand (Isbary, 1936: 21).

The WNW-ESE striking dune ridges to the east of the migrating dune complex formed later. On the map drawn in 1811 by Atthalin, the Kobbeduinen, Kooiduinen and other, lower dune ridges between are absent (cf. Abrahamse & Koning, 1983: 37). The 'Kobbe Duinen' were first represented on a map of the Ministerie van Oorlog (Ministry of War) in 1854 as a first dune ridge running eastnortheast from the Eendenkooi (Haan et al., 1983: 65). Recent dune migration is only found on two

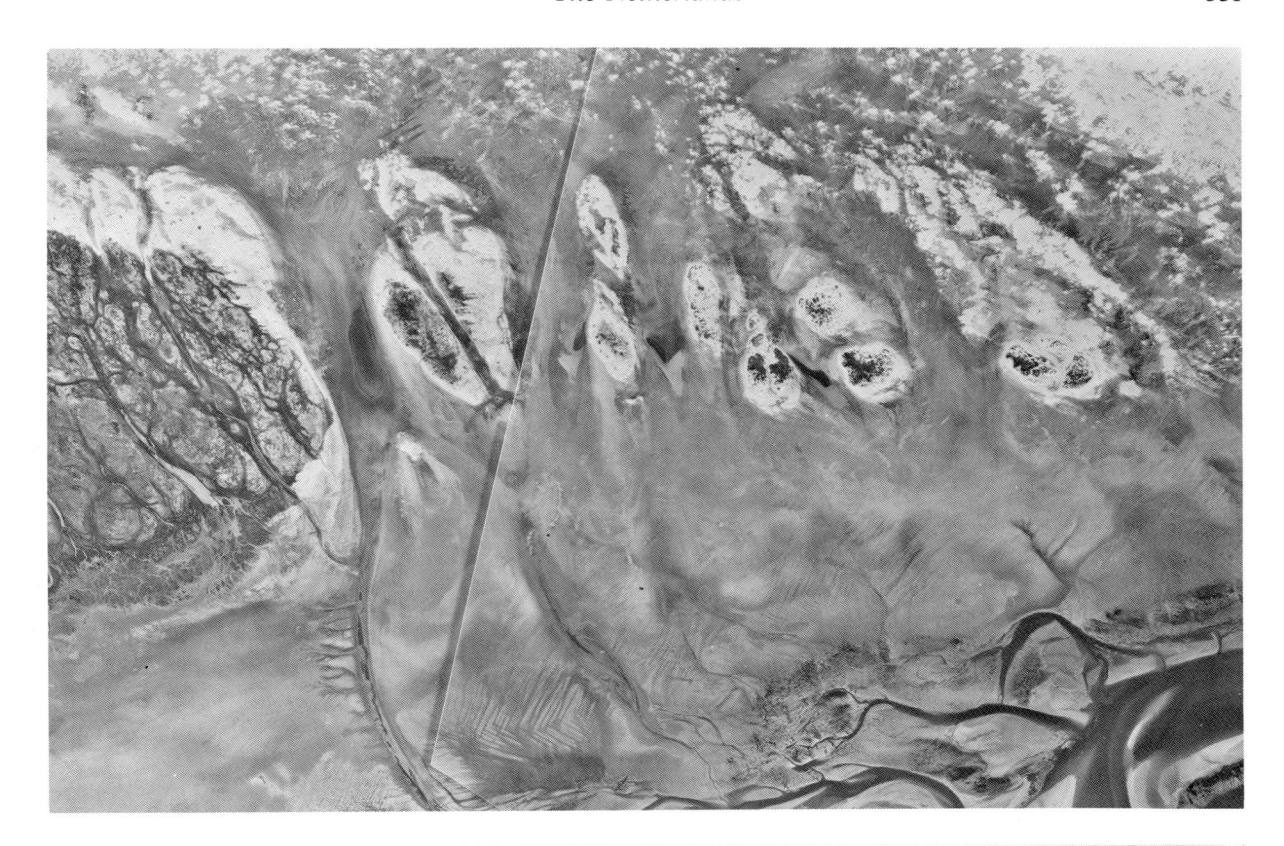

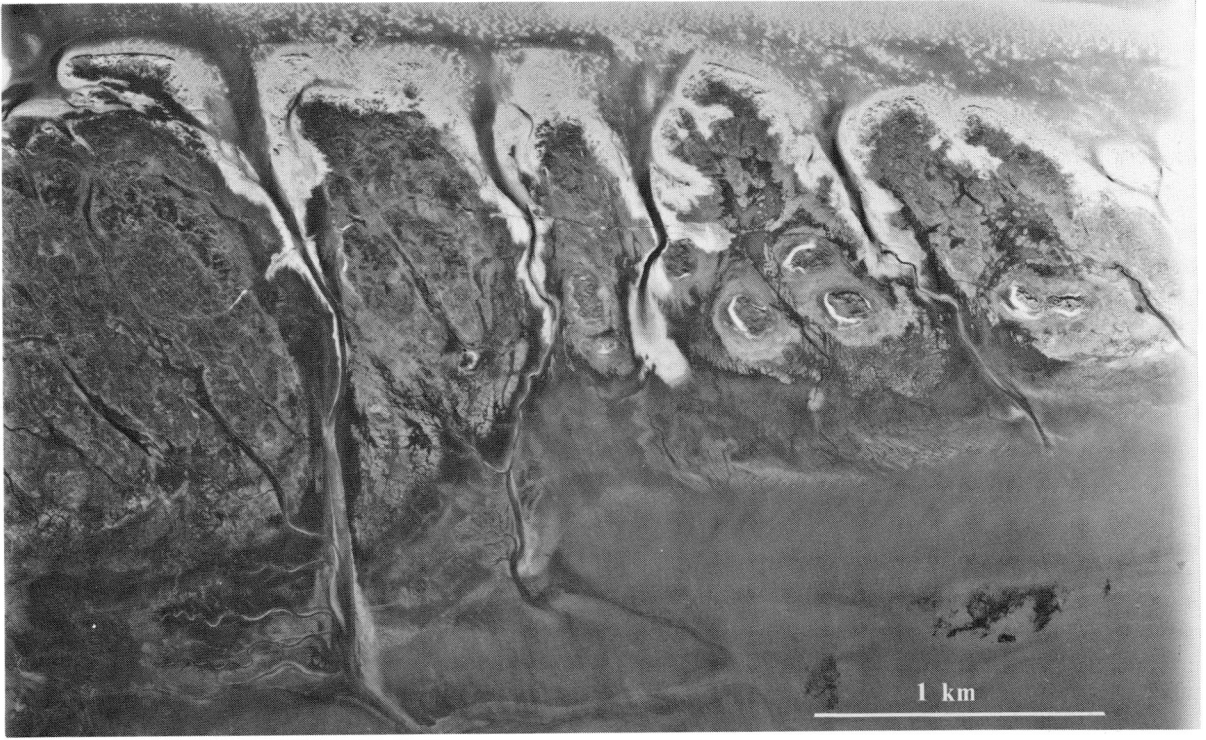

Figure 359: Schiermonnikoog, (a) the Oosterstrand before construction of the stuifdijk. Aerial photographs of the Fototheek Topografische Dienst, Emmen, of 1952, B2 Run V 91 and B2 Run VI 113; (b) the Oosterstrand of Schiermonnikoog today. Aerial photograph of the Fototheek Topografische Dienst, Emmen, of 11.3.1983, 02 H 08.

Figure 360: Present end of the continuous stuifdijk on the Oosterstrand of Schiermonnikoog; remnants of the bush fences used in the construction of the stuifdijk in 1956 can be seen in the foreground (1985).

places on Schiermonnikoog; in the Westerduinen area and northeast of the New Lighthouse.

Rather than being eroded, the dune core of Schiermonnikoog is actively being enlarged by new coastal dune ridges forming along the northern dune margin, promoted by brush fences and plantations of *Ammophila arenaria*. The backshore along the entire north coast is at least 200 m wide which ensures a sufficient sand supply for dune formation. At the same time a small salt marsh which had formed north of the New Lighthouse under the protection of the high beach has been connected to the island core by a low dune ridge.

A salt marsh, more than 1 km wide, had formed under the protection of the dune core of Schiermonnikoog on the Wadden Sea side of the island. This salt marsh was eroded during the early 19th century until it was finally dyked as Bancks Polder in 1860 (Mellema, 1981: 93). There has been no deposition since then; there are therefore no undyked salt marshes between 'het Rif' and the new ferry terminal. One small area of salt marsh has formed in the lee of the old ferry quay entirely restricted to its eastern side, demonstrating the predominantly easterly-directed overflow of the tidal flats under the influence of the prevailing westerly winds.

A narrow strip of salt marsh extends eastwards towards the Oosterstrand sand flat from east of the new ferry terminal, built in 1962. This salt marsh is currently being eroded, but further to the east some deposition is taking place under the protection of the recent dunes on the Oosterstrand.

The Oosterstrand

Prior to 1952 some small dune cores had formed on the highest parts of the Oosterstrand, the eastern sand flat of the island, separated from each other by broad washover areas. Salt marsh formation had started in the western part of the Oosterstrand (fig. 359 a). In order to encourage accretion, the Rijkswaterstaat began to construct a stuifdijk (dune dyke) to the easternmost dune complex, the 9 m high Willemsduin, in 1956 (Abrahamse & Luitwieler, 1985: 9). Its chosen course was situated over 200 m to the north of the natural dune cores.

The 7 km long sand dyke was short-lived, being quickly breached at Paal 10 (Pole 10) (fig. 360). Apart from the 3 km long western part of the stuifdijk, small fragments only have survived. The eastern parts of the stuifdijk have migrated southwards and some sections are now situated more than 50 m south of their original positions.

The eastern end of Schiermonnikoog has remained an active washover area with a few intercalated dune cores. In this area sand transport towards the tidal flats occurs during all higher tides (e.g. spring tides) so that the natural situation has prevailed. Broad, horseshoe-shaped dune cores have formed instead of linear coast-parallel dune ridges, in the lee of which salt marsh growth is active. Active dune formation and deposition is prevalent along almost the entire eastern flat of Schiermonnikoog today (fig. 359 b).

Migration of the island

Like Rottumeroog, Schiermonnikoog has been very unstable for a number of centuries, during which it has been displaced several kilometres eastwards (fig. 356). Abrahamse, Koning & Heuff (1983: 15), who compared the existing topography with the 1585 map of Aelbert Hayen, have calculated a displacement rate of 27.3 m per year. This, of course, is only true of the eastern end of the island; the western end has migrated at a much slower rate whilst the length of the island has increased.

The eastward migration of the island caused the relocation of the old settlement to its present position. Problems started in 1715 when the church in the old village had to be demolished because it became surrounded by aeolian sand. Heavy land losses in the west during the storm surges of 1717 and 1720 brought the village closer to the dune margin. Following the storm flood on September 7th, 1756 the Heerenhuis (Manor House) had to be abandoned; a new building was erected in the east in 1757 (the present 'Rijsbergen' youth hostel). The new village was called Oosterburen. The old village, Westerburen, was finally destroyed during the storm flood of December 26th, 1760 (Mellema, 1981).

The Lauwerszee originally drained through the Lauwersgat, east of Schiermonnikoog (fig. 361), but as Schiermonnikoog migrated eastwards the Lauwersgat was forced further and further away from its catchment area. Simultaneously the Friesche Zeegat in the west moved increasingly close. The watershed between Schiermonnikoog and Friesland was probably breached during the 14th century, and since then the Lauwerszee has drained northwestwards through the Friesche Zeegat (Haan, Ijbema & Reitsma, 1983: 10).

The damming of the Lauwerszee in 1969 caused the drainage pattern to alter again. It has reduced the tidal volume of the Friesche Zeegat by 150,000,000 m^3 to a mere 200,000,000 m^3. As a result the southern part of the watershed between Schiermonnikoog and the mainland has shifted eastwards by about 1–1.5 km, and the courses of the main tidal creeks have also started to change

Figure 361: Satellite image of Schiermonnikoog and the Lauwerszee. Landsat 4 TM image of 1.2.1983, Track 197, Frame 023, Q 1, Band 5, processed by ESA at Fucino.

(Abrahamse, Koning & Heuff, 1983: 15; Reenders, personal communication).

The western margin of the dunes on Schiermonnikoog has not moved significantly eastwards since the first accurate maps were published in 1811, and the same is true of the low water line. In fact, during the 20th century the Westerstrand has extended westwards, a trend which has continued since the damming of the Lauwerszee.

7.4.3 ENGELSMANPLAAT

There is a double tidal inlet between Ameland and Schiermonnikoog, consisting of the Pinkegat in the west and the Friesche Gat in the east. Engelsmanplaat has formed between both these inlets, constituting part of the flood deltas. Its development can be traced back to the 14th/15th centuries when the Isle of Schiermonnikoog migrated eastwards and the water divide between Schiermonnik-

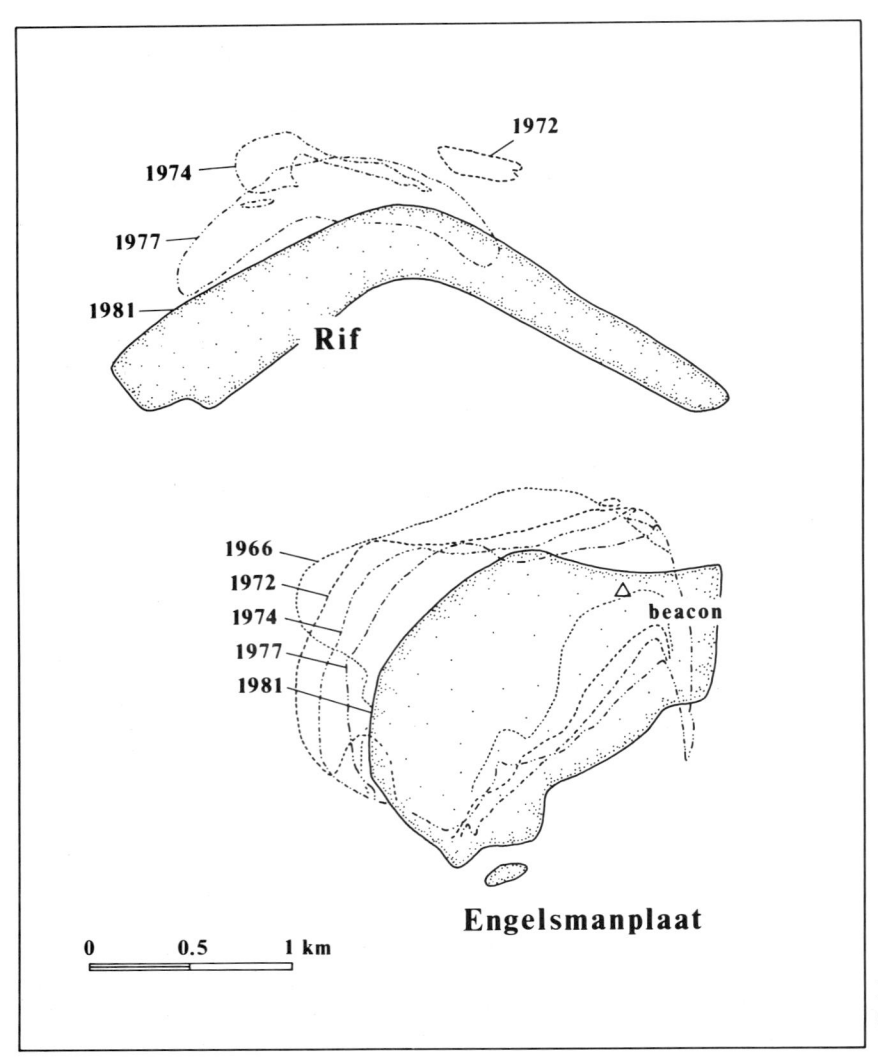

Figure 362: Recent migration of Engelsmanplaat. Sources: J.T. Postma, 1982; Mes *et al.*, 1980.

oog and Friesland was breached (Haan, Ijbema & Reitsma, 1983: 11).

Since the first publication of reliable maps the position of Engelsmanplaat has remained stable. Its eastern margin has always sloped steeply towards the Friesche Gat. The northern and western boundaries are subject to continuous changes. Haan, Ijbema & Reitsma (1983: 55) have studied the available maps since 1810 and conclude that there is some type of cyclic development every 90 - 100 years.

Accumulation on the sand bank has never been sufficient to instigate dune formation. Primary dunes about 2 m high have occasionally formed. Engelsmanplaat has recently declined in height whilst a new sand bank approaching from the north has grown considerably (fig. 362).

7.4.4 AMELAND

The dunes of Ameland (fig. 363) originally formed one continuous dune ridge providing the protection for the deposition of a continuous marine clay layer. Some of this clay has subsequently been eroded on the Wadden Sea side of the island. Ameland was mentioned as an island in a document as early as the 9th century (Bantelmann *et al.*, 1976: 223) and the oldest settlements on Ameland were presumably founded in the 10th/12th centuries. As the villages are all situated behind the existing dune cores it can be assumed that the cores were already in existence at that time. This also implies that the dissection of the continuous dune ridge had occurred by then (Isbary, 1936).

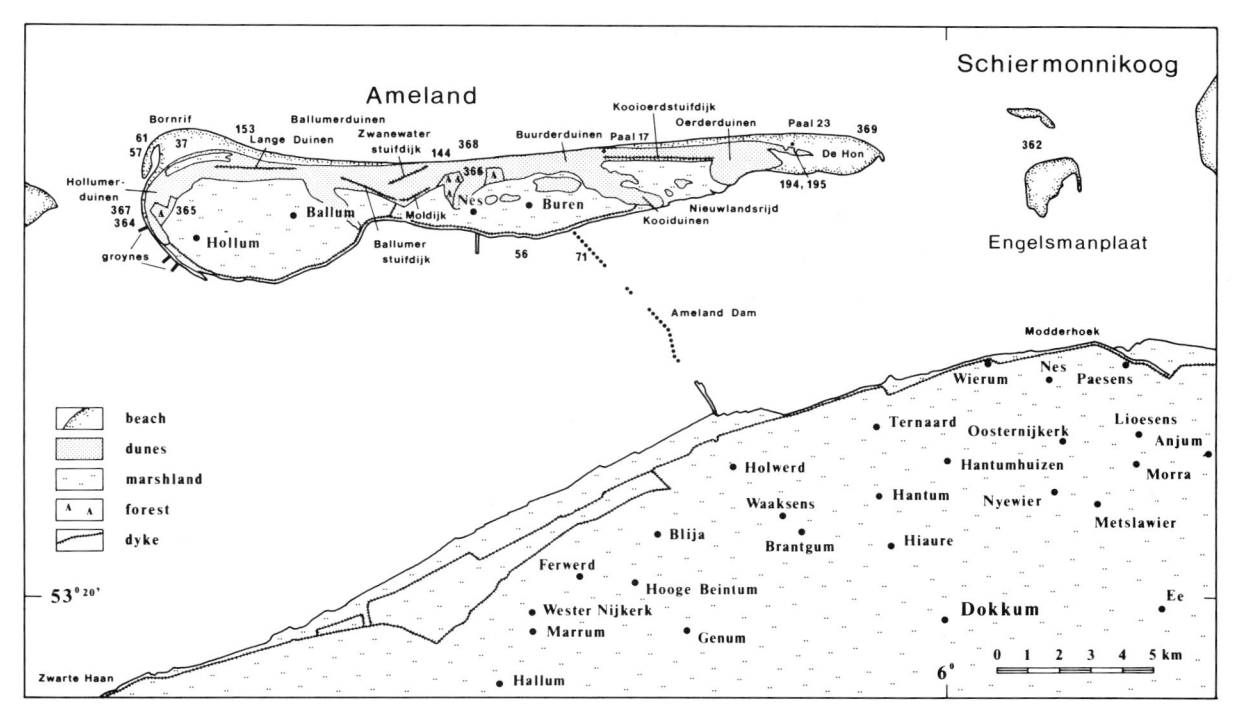

Figure 363: Location map of Engelsmanplaat and Ameland. The numbers refer to figures, those prefixed by 'pl' refer to plates.

Development of western Ameland

At the end of the 18th and the beginning of the l9th centuries the tidal-flat facing coast of Ameland in the Ballum Bight was heavily eroded by an approaching tidal creek called 'de Balg'. The washover area between the Ballum and Nes dune cores thus posed an increasing danger to the island. In 1807 a stuifdijk (dune dyke) was built across the gap in order to prevent a breach but the dyke was breached after only a year and had to be abandoned. In 1828 a second attempt was made 800 m further to the south; this stuifdijk was breached in 1842 and 1844, leading to the formation of a deep hollow. The dyke was eventually closed in 1846–1851. In the meantime the 'Balg' had eroded 2 km of the island's salt marshes but its approach was halted in 1843 with the construction of a stone bank protection.

Some migrating dunes were active in the Ballumerduinen as late as the 1930's. In the period from 1879 to 1927 they had migrated almost 250 m to the east (Isbary, 1936: 30), at a rate of about 5 m per year. Today the dunes are vegetated and inactive (fig. 364).

The Kooiduinen dune complex east of Ballum has largely disintegrated through blowouts. Isbary (1936: 31) reports that deflation in these dunes was encouraged by the grazing of horses, sheep and cows.

Farmland on Ameland was inundated during the storm surges of 12./13.3.1906 and 13./14.1.1916 (van Veen, 1956). After major land reform in 1916 (Bantelmann et al., 1976: 229) the western part of the salt marsh area was dyked. The eastern part around Buren was dyked about 10 years later (Bakker, 1973: 21).

The dunes in the east

The third large dune complex is situated at the eastern end of Ameland. The Oerderduinen are marked on the earliest maps of the island dating from 1567. A wide, open sand flat existed between the Oerd and the dunes of Buren (at Paal 17). This was regularly flooded during storm surges and for a period there was even a danger that the island itself might be breached (Kiewiet, 1985).

The Oerd village was sited in the protection of the Oerderduinen on the Wadden Sea side of the island. It had to be abandoned, however, when its arable land was ruined by shifting sand. The village was last referred to in documents dated 1558. In 1825 its remains were exhumed by a tidal creek

Figure 364: Deciduous forest in the dune area of Ameland (1981).

Figure 365: The western end of Ameland showing the coastal protection structures (centre) and sand accumulation area (to the north). Aerial photograph of the Fototheek Topografische Dienst, Emmen, of 12.3.1983, 01 E 21.

eroding the tidal-flat facing side of the Oerderduinen and subsequently the ruins themselves were eroded (Isbary, 1936: 41).

Before the 19th century there was no connection between the Oerder dune complex and the rest of the island. An early attempt to build a stuifdijk in the southern portion of the flat, the present Nieuwlandsrijd area, was a failure, but in 1882 a second attempt was made further to the north. The dyke was begun from both ends, from the Oerderduinen as well as from the Buurderduinen. The final gap was closed in 1893 in the area south of the present Paal 19 (Pole 19). Whilst the beach north of the dyke remained exposed to the sea, accretion intensified south of the dyke and an extensive salt marsh formed.

Recent dunes started to form north of the stuifdijk. As the tendency for dune formation continued in eastern Ameland, an attempt was made in 1960 to build an additional stuifdijk east of the Oerderduinen on the sand flat, connecting the recent dune cores. By 1962 the dyke was complete eastwards as far as Paal 23 and new dunes subsequently formed further to the east (Kiewiet, 1985: 12).

Coastal erosion

Isbary (1936: 41) has described the mechanism whereby the northwestern end of Ameland is supplied with sand. The Bornrif, the easternmost part of the Borndiep ebb delta, is the point where the sand approaches the shore, the sand being moved beachwards in the form of large, recurved shoals. When these begin to approach the beach, the separating channel becomes narrower and its erosive power is thus increased. After some time, however, the shoal connects with the beach and supplies large amounts of sand. The favourable sand budget in this area, however, cannot alleviate the

AMELAND

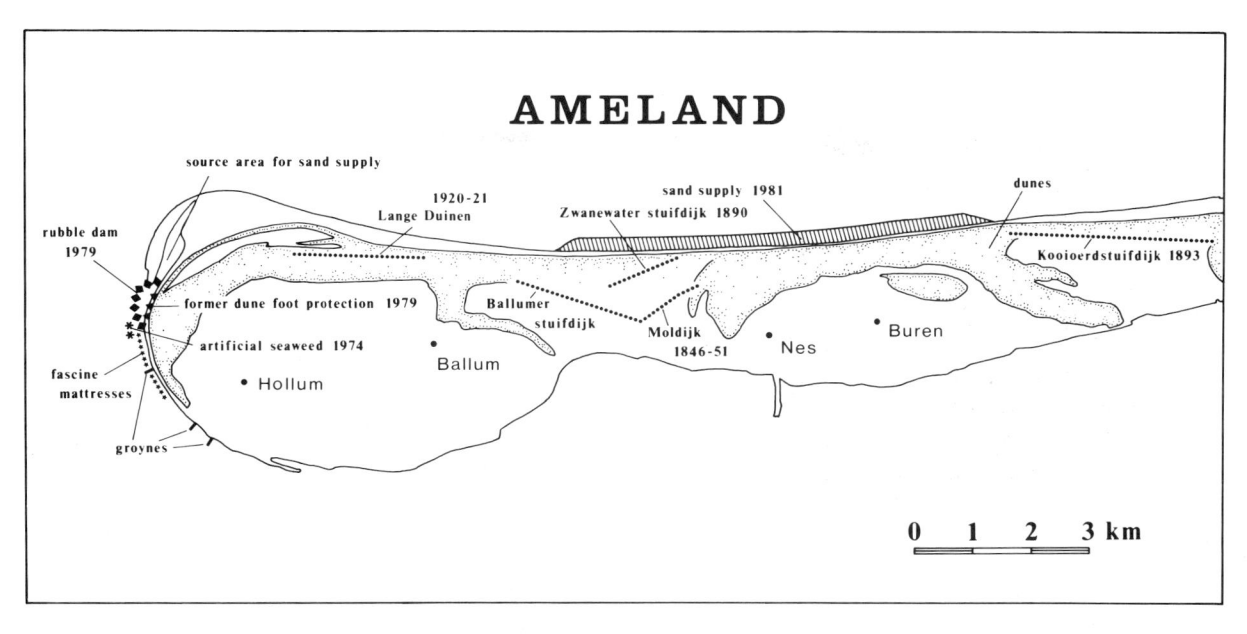

Figure 366: Map of the coastal protection structures on Ameland. Source: de Reus, 1983.

Figure 367: Rubble dam at the western end of Ameland, built for coastal protection; cf. fig. 364 (1981).

losses suffered a few hundred metres further to the south (fig. 365).

Like Schiermonnikoog and Rottumeroog, Ameland has been subject to constant changes of position. Whilst Terschelling has a stable western end and expands eastwards, the Borndiep inlet is forced close towards the coast of Ameland heavily eroding its western end (fig. 365). Since 1665 more than 1 km of beach and dunes has been lost at this point, about 300 m of this between 1935 and 1980 alone. In 1943 about 200 wells appeared on the beach west of Hollum. These marked the site of the former village of Sier which was covered by dunes in the 15th century. These remains together with the German bunkers dating from the Second World War have long since been eroded by the Borndiep (Bantelmann et al., 1976: 223).

Coastal erosion at the western end of Ameland became so severe in 1947 that protection measures became essential (fig. 366). At first three groynes

Figure 368: Beach replenishment on Ameland. The erosion cliff in the newly supplied sand is shaped by bulldozer into a dune-like slope (1981)

were constructed and the eastern flank of the Borndiep was stabilized by fascine mattresses. Additional fascine mattresses had to be repeatedly installed as erosion continued. A sand dyke had to be erected behind the dunes in 1959; later, part of this was also eroded. The fascine mattresses were supplemented by plastic sheets in 1974, and in 1979 the dune foot was paved over a length of 400 m. In the same year a rubble dam was built to a height of NAP−1 m at the northern end of the structures, behind which 300,000 m^3 of sand was supplied. This sand was removed from the Bornrif (Kooiker, 1981). None of this had any lasting effect, however, and by 1981 all the sand had been removed (fig. 367).

The north coast of Ameland has a negative sediment budget. It has been eroded between Paal 9 and Paal 16 at an average rate of 1 - 2.5 m/year since the 19th century. Between 1965 and 1975 alone about 300,000 m^3 of sand was lost from this area per year (Kooiker, 1981). Beach replenishment was necessary to counter dune erosion in 1981 (fig. 368).

Accretion in the east

Since 1731 the eastern end of Ameland has extended about 5.5 km eastwards. This growth has not been a continuous process, being interrupted by a number of phases of erosion, such as that occurring between 1870 and 1880 (Isbary, 1936: 42). Despite these minor oscillations, Ameland continued growing eastwards until 1940-1950. Since then, until 1980, about 2 km have been lost (fig. 369; J.T. Postma, 1982).

The 'Ameland Dam'

Until 1846 it was possible to walk on dry land from Ameland via the water divide to the mainland, and the route was dry enough to be used by horse-drawn coaches. Encouraged by these good ground conditions, plans were made to connect Ameland with the mainland by an embankment (van der Molen, 1978 a: 163). Construction started in 1871. The work progressed from both ends, from the Frisian coast as well as from the island, and the two halves met in August 1872.

The November storms of 1872 caused the first minor damage, which was quickly repaired. At this time the embankment was elevated to an altitude of NAP + 1.20 m. The storm surge of 14./15.10. 1881, however, created a number of breaches which were further widened by another flood in April 1882. Insufficient financial backing caused the embankment to be abandoned soon afterwards without further repair (Bakker, 1973: 79).

Remnants of the embankment can still be seen today in the tidal flats between Ameland and Holwerd (fig. 363). It is interesting to note how far to the west the watershed was situated in 1873. By 1935 it had already migrated to a position about 2 km east of the embankment (Isbary, 1936: 41), and it is found at present 3 - 5 km east of the 'Ameland Dam'.

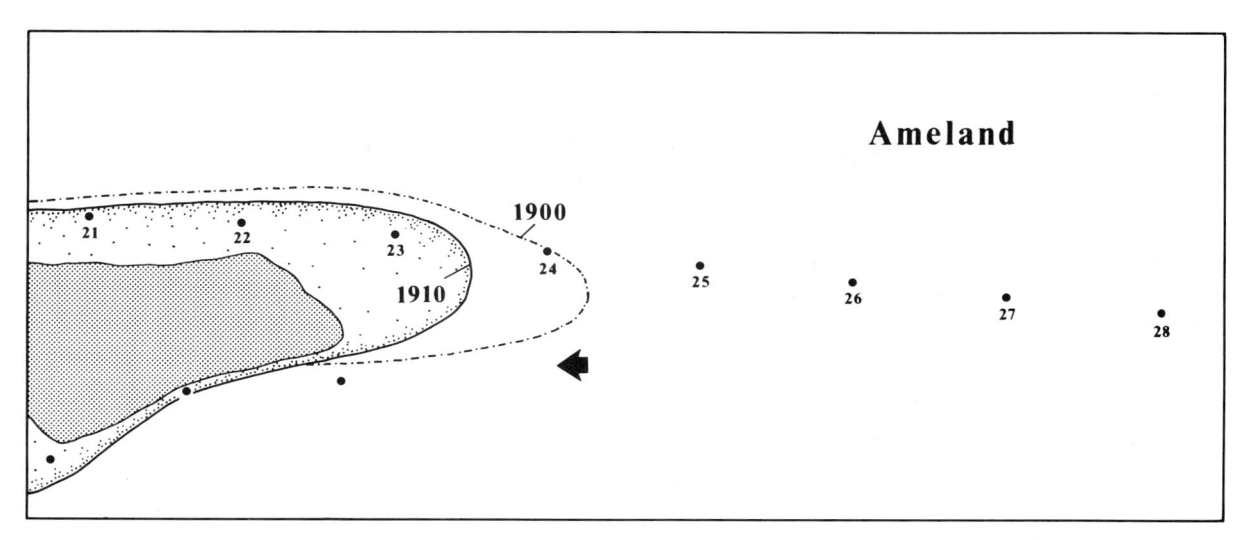

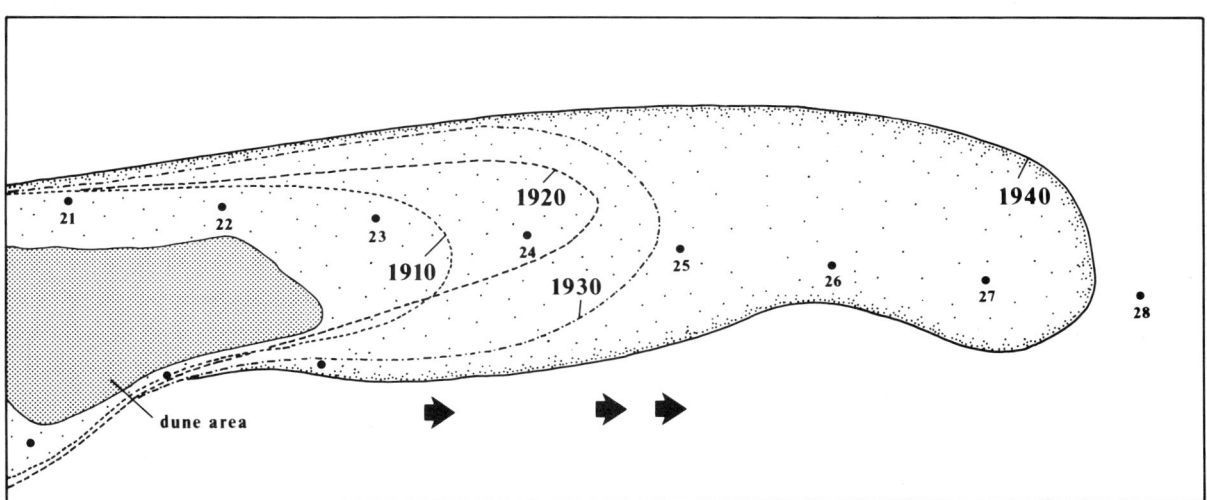

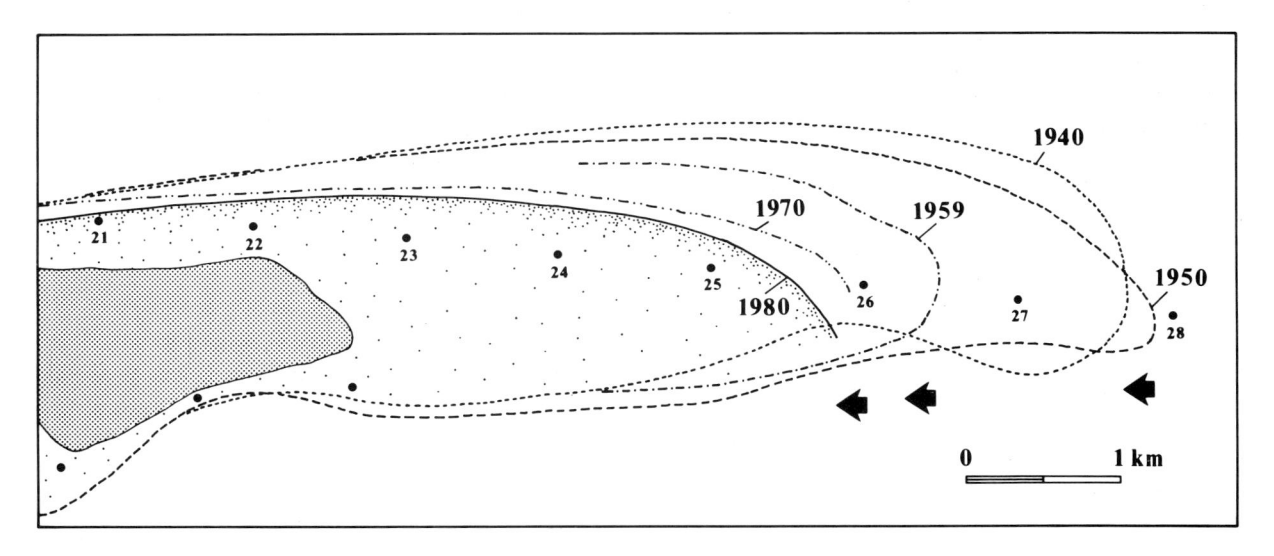

Figure 369: Development of the eastern end of Ameland between 1900 and 1980. Source: J.T. Postma (1982).

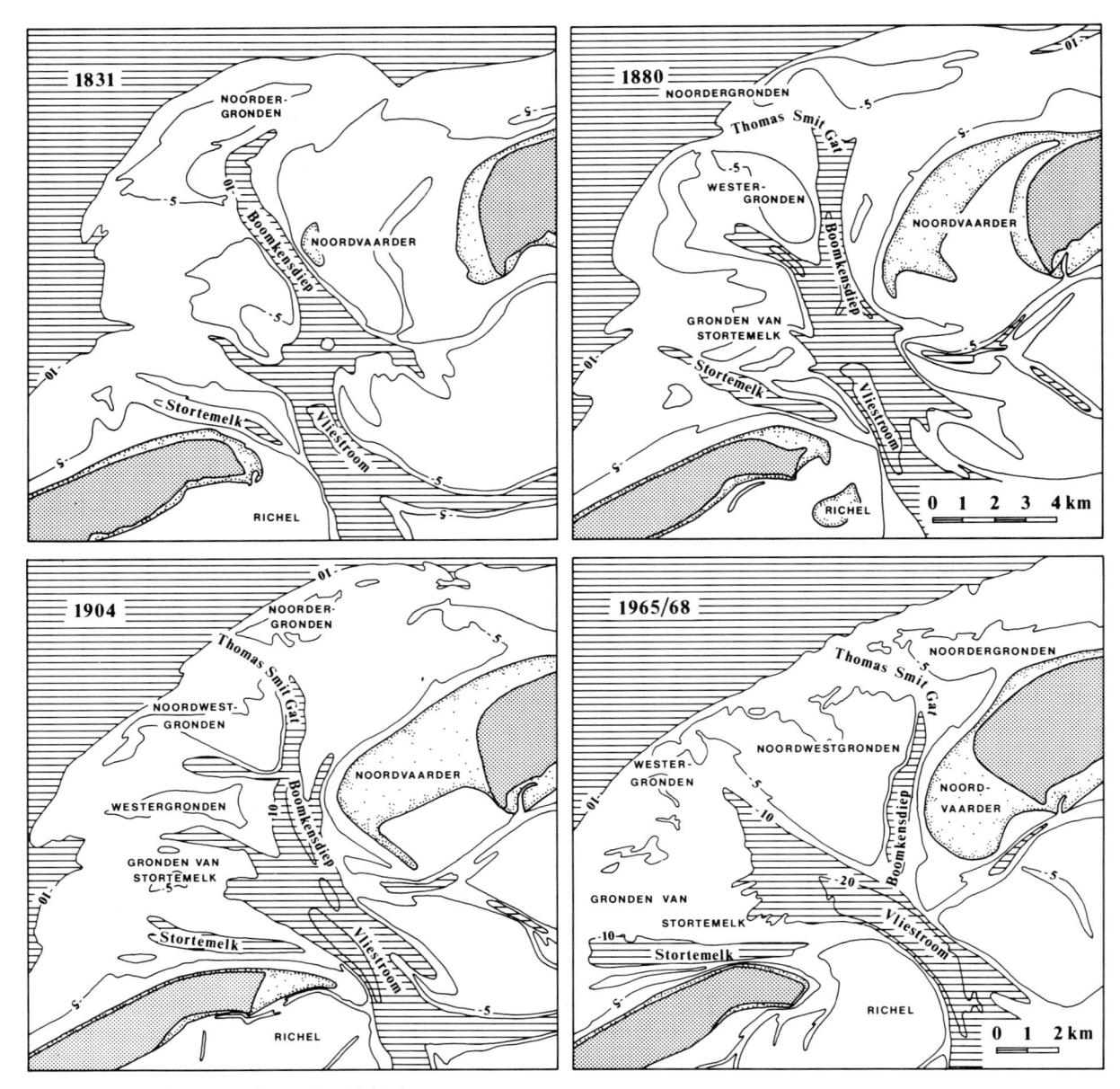

Figure 370: Development of the Vlie tidal inlet. Source: Joustra, 1971.

7.4.5 Terschelling

Terschelling is the largest of the barrier islands without a Pleistocene core. Hydrogeological investigations indicate that two freshwater lenses exist under the island. This suggests that the present single dune area has developed by the merging of two former discrete dune cores (Beukeboom, 1976).

The Noordvaarder

The three kilometre wide Noordvaarder sand flat forms the western end of the Isle of Terschelling. In 1780 the Noordvaarder was still separated from the western end of Terschelling, the so-called Robbezand, by the Westerboomsgat. This silted up within a few decades and subsequently the Noordvaarder became part of Terschelling (fig. 370).

The construction of stuifdijken (dune dykes) started on the Noordvaarder in 1920, commencing from the northwestern hook of the old dunes (fig.

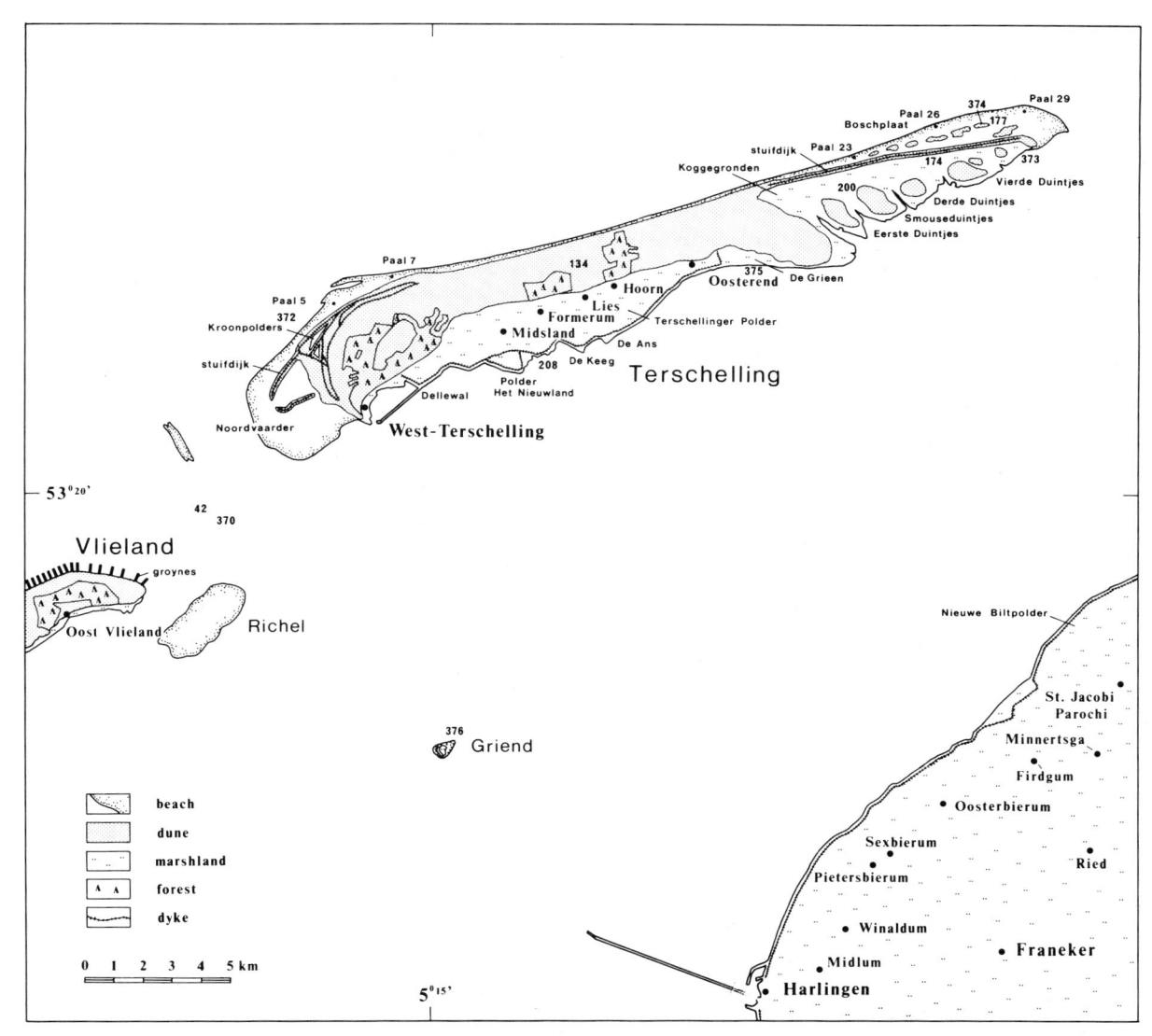

Figure 371: Location map of Terschelling and Griend. The numbers refer to figures, those prefixed by 'pl' refer to plates.

371). The stuifdijken prevented the dyked sand flats from inundation but simultaneously prevented further sand accumulation, leaving the Kroonpolders as low-lying wetlands (fig. 372).

In 1948 an attempt was also made to dyke a major part of the Noordvaarder sandflat in the west to reclaim land for use as an emergency air field. This stuifdijk, however, did not succeed, being breached in two places during the storm surge of February 1st, 1953. The breaches have never been repaired (Edelman, 1964: 56).

The Noordvaarder was originally part of the ebb delta of the Vlie tidal inlet. During its formation sand migrated to the westernmost end of Terschelling, but today this migration occurs further eastwards, in an area between Paal 5 and 7 (fig. 370). This has caused the Noordvaarder to be eroded from the west. A small lighthouse which was situated at the western end had to be re-sited eastwards three times within 40 years. By 1960 its total displacement was 1,300 m. This steel construction was finally demolished as erosion continued (van der Molen, 1978 b: 17).

The dunes

Whilst van Dieren (1934) described large areas of migrating dunes on Terschelling, the dunes are almost completely stabilized today. The first forest on Terschelling was an experimental plantation at

Figure 372: Wetland landscape in the Kroonpolders, Terschelling (1981).

Oosterend in 1885. When this succeeded, large scale *Pinus* afforestation started in 1911 after the dunes had become state property in 1910. Forest covers an area of almost 650 ha on the island today, and there have also been recent attempts to plant deciduous trees (Staatsbosbeheer, 1979).

The Boschplaat

During the 17th and early 18th centuries the present Boschplaat consisted of two completely unvegetated sand banks, the Groote- and Kleine Boschplaat, separated from the Isle of Terschelling by a small, 8–9 m deep tidal inlet, the Coggendiep (van der Burgt, 1936: 821; Isbary, 1936: 38). The Coggendiep increasingly filled with sand in the middle of the 18th century, connecting the two Boschplaats with the island. The resulting new Boschplaat was largely unvegetated until about 1900. Dune formation had started in the late l9th century with the development of the Smouseduintjes (Tweede Duintjes). Other dune islands soon followed (Staatsbosbeheer, 1979: 1; Joustra, 1971).

These dune cores (Eerste to Vierde Duintjes) have been eroded during the higher tides from the North Sea side as well as from the tidal flat side giving them their characteristic island-like shape. They are rather small equivalents of the large dune islands which form the cores of the West and East Frisian Islands.

As early as 1924 an area of 2,800 ha on the Boschplaat was declared a nature reserve, and in 1934 this area was expanded to 4,400 ha. Despite this an attempt was made in 1929 to connect the westernmost dune core on the Boschplaat, the Eerste Duintjes, with the main dune core of Terschelling using a stuifdijk. This project failed because it was sited too far to the south (Visser, 1947: 35). Remnants of the old stuifdijk can still be seen.

A more successful attempt was made two years later in a more northerly position. A stuifdijk almost 9 km long was built on the Boschplaat between 1931 and 1938 (Visser, 1947: 35), and, under its protection, a large salt marsh began to develop (figs. 373 a and b). There were even plans to use part of this area for agricultural purposes. Around 1950 there were proposals to dyke about 1,200 ha of dune and salt marsh territory to reclaim land for 40–45 farms, but fortunately these plans were not fulfilled (Staatsbosbeheer, 1979).

The construction of the stuifdijk has completely interrupted the natural development of the landscape. Being covered with dense vegetation like the former sand flats around them, the dune islands are unable to develop further. In addition, the marginal cliffs of the dune cores cannot be reactivated, because the washover processes have

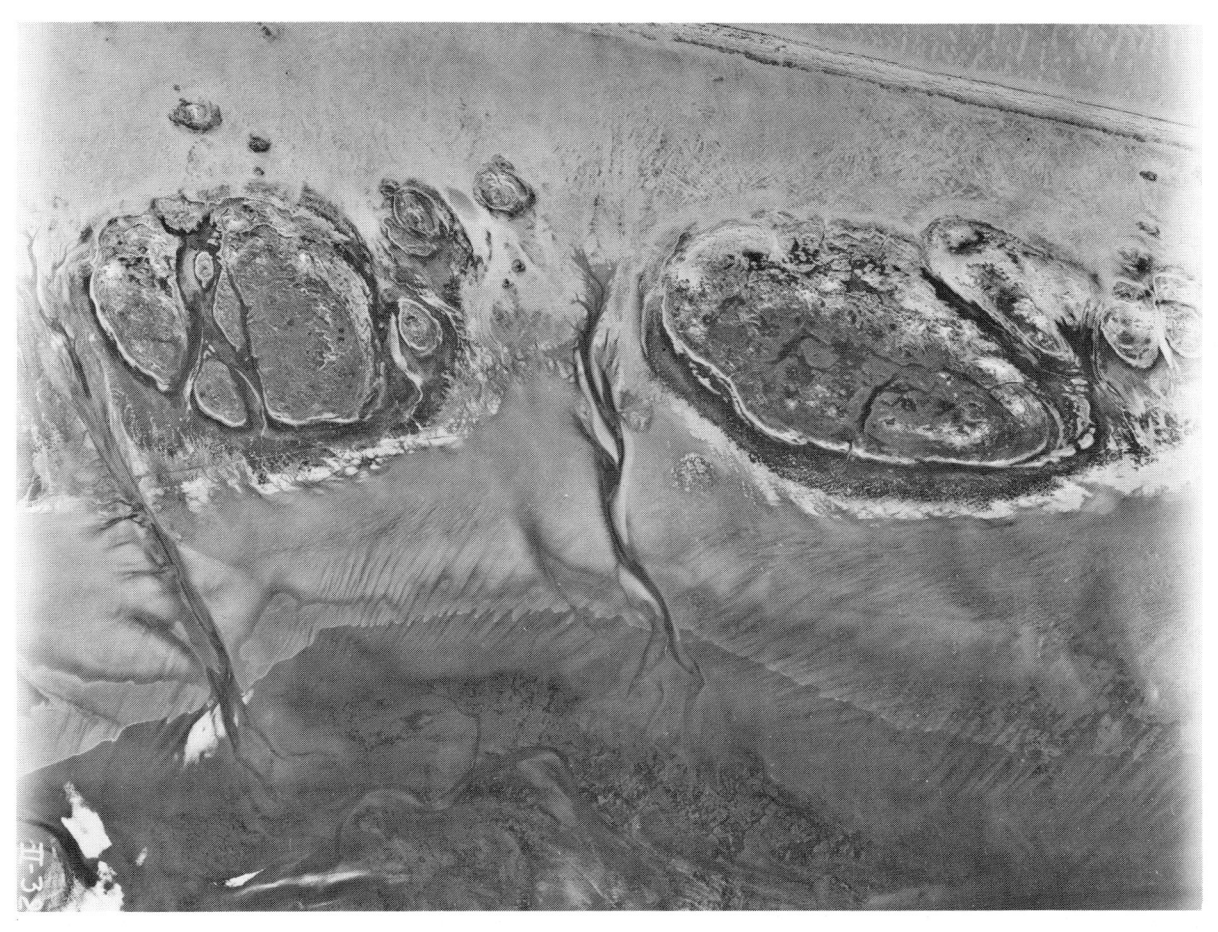

Figure 373: Aerial photographs showing the development of the Boschplaat, Terschelling, during the last 35 years. (a) the Boschplaat shortly after construction of the stuifdijk. Aerial photograph of the Fototheek Topografische Dienst, Emmen, of 1949, Bl Run II 32; (b) the Boschplaat today. Aerial photograph of the Fototheek Topografische Dienst, Emmen, of 12.3.1983, 01 E 17.

been halted (fig. 373 b).

Natural landscape development undisturbed by human interference is currently only occurring north of the stuifdijks between Paal 23 and 29 where an extensive new dune area has formed and developed far beyond the primary dune stage (fig. 374). This dune area is separated from the stuifdijk by a duneless area 200–300 m wide, which is open at both its western and eastern ends. Protected by the young seaward dunes salt marsh vegetation has spread in the duneless area, contributing to natural accretion. When visited by the author, this area was dissected by numerous tracks made by off-road vehicles used by the state authorities; these damage the natural vegetation and promote erosion.

Erosion on the leeward side

Like most of the other Wadden Sea islands Terschelling shows no signs of recent accretion on the tidal flat facing side with the single exception of the Boschplaat area. The extensive erosion of the Dellewal dune area south of West-Terschelling observed during earlier centuries was halted in the second half of the 17th century by the construction of a dam to fend off the approaching tidal creek (van der Burgt, 1936: 823). The old polders which appear to have been dyked earlier than 1506 (van der Molen, 1978 b: 15) are now protected by a strong dyke, which has been strengthened by a stone revetment in wind-exposed, west-facing places. Where there is no foreland the foot of the dyke has been reinforced by rubble and groynes.

In contrast, along the Terschellinger Polder only minor salt marsh areas remain, e.g. de Ans, de Oel,

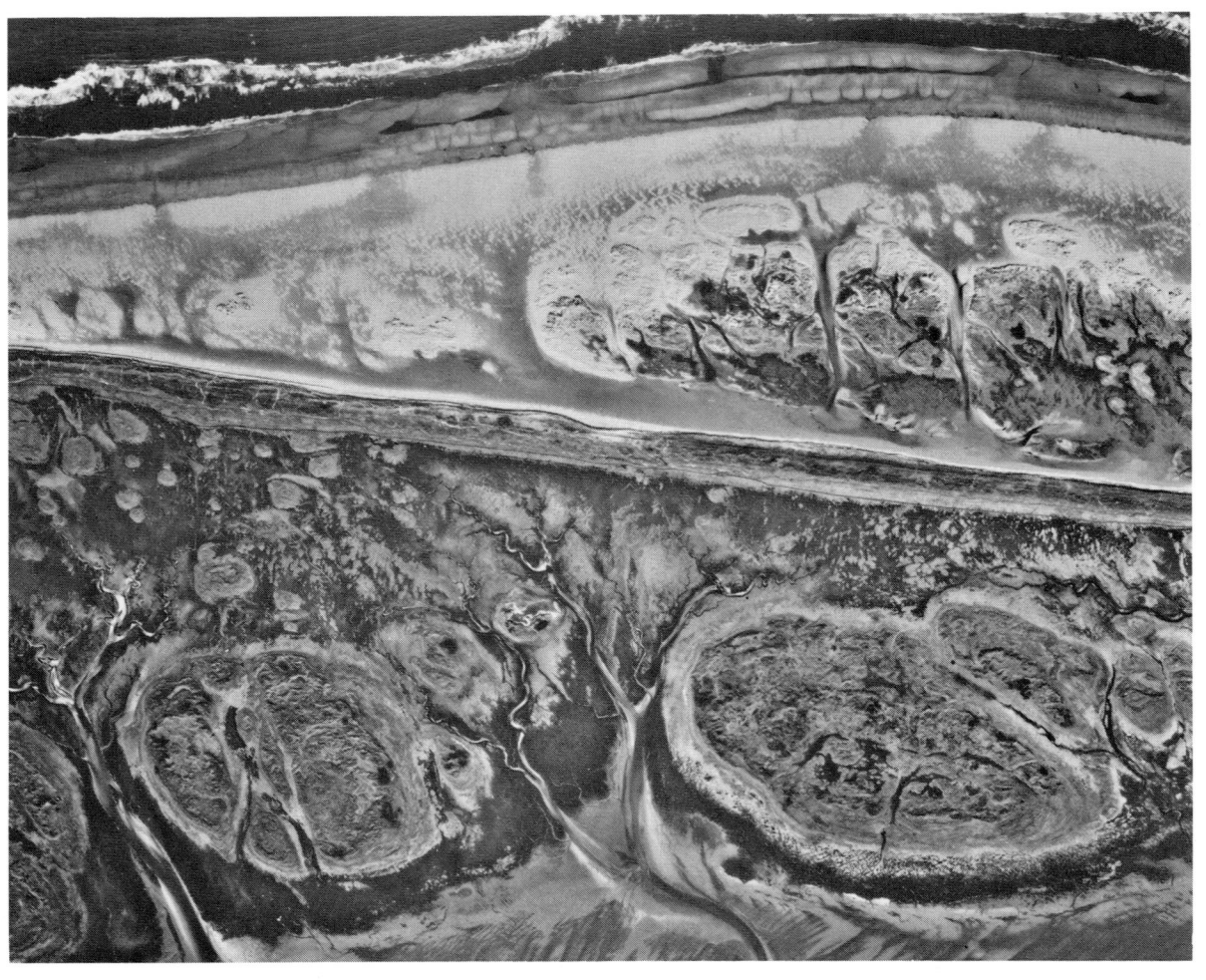

Figure 373 (cont.).

de Keach (van der Molen, 1978 b: 15), the larger salt marsh area of the Grieen survives east of Oosterend. The cliff edge of this area is progressively eroded during every high tide (fig. 375).

7.4.6 GRIEND

The Isle of Griend, situated in the Wadden Sea between Terschelling and the mainland, is a remnant of pre-existing larger land areas. Geological mapping of the Wadden Sea (van Staalduinen, 1977) indicates that it may have been connected with the Moerwaard areas southeast of Vlieland which appear as land on old maps e.g. of Sgrooten (1573) and Mercator (1592).

Historical development

The Lucia Storm Surge of 14.12.1287 inflicted heavy damage on Griend. The church was destroyed, and 'less than 20 houses' were left in the village. In 1399 the island still covered an area of about 165 ha, but coastal retreat was active. In 1611 4 houses remained on Griend, but by 1720 only one was left. The first accurate survey of the island in 1796 recorded an area of 24.7 ha, but by 1838 this had decreased to 23.4 ha. Between 1831

Plate 34: 'Groote Vlak' moist dune valley 2 km west of Den Hoorn, looking towards the northwest (1985).
Plate 35: Undyked salt marsh of the 'de Schorren' nature reserve, dissected by salt marsh creeks (1979).
Plates 36 and 37: View from the Eijerland Polder towards the 'Groote Zanddijk' of 1629/30 (1985).

PLATES 34-37

PLATES 38-40

Figure 374: Young dunes at the eastern end of the Boschplaat, north of the stuifdijk (1981).

Figure 375: Eroded salt marsh margin, Grieen, Terschelling (1981).

and 1865 three wooden beacons (Kaapen) were erected on Griend and the subsequent replacements of these beacons enables the migration of the island to be reconstructed. In 60 years it has migrated southeastwards at an average rate of 7 m per year (Abrahamse & Luitwieler, 1982 a: 87).

Following the completion of the Zuider Zee dam in 1932 the mean water level rose by several decimetres. This obviously had an adverse effect on Griend, which had remained relatively stable between 1850 to 1932. These problems were compounded by the extinction at the same time of *Zostera marina*. In the period 1920–1932 about 15,000 ha were occupied by *Zostera marina* in the Dutch Wadden Sea but after the outbreak of the so-called 'wasting disease' between 1950–1960 this area decreased to 400–450 ha (van den Hoek et al., 1979: 10). Every year prior to 1932 large quantities of *Zostera* had accumulated on the shore of Griend, enhancing the development of a

38	Plate 38: The low Pleistocene ridge at Den Hoorn, viewed from the south; maximum height + 6.8 m NAP (1985).
39	Plate 39: Tidal-flat facing dyke of the Prins Hendrik Polder, north of 't Hoorntje, looking towards the northeast (1985).
40	Plate 40: Coastal protection by revetment and sand replenishment at the northern end of Texel, seen looking towards the east from the car park west of the lighthouse (1985).

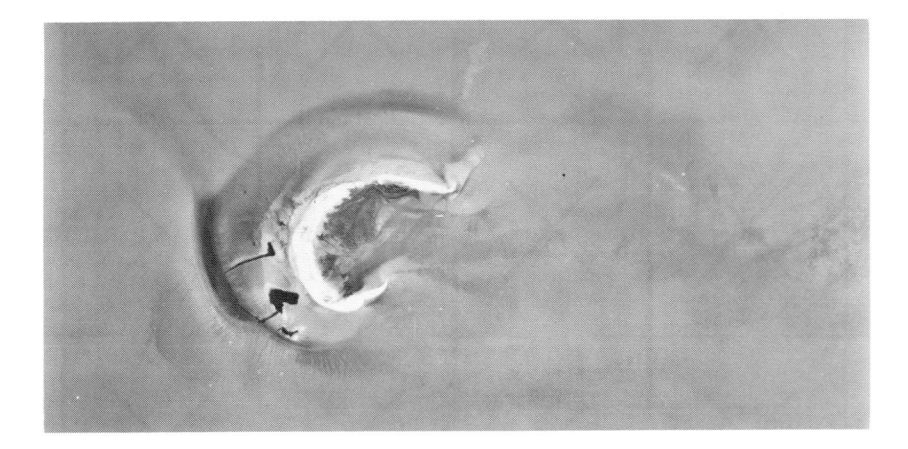

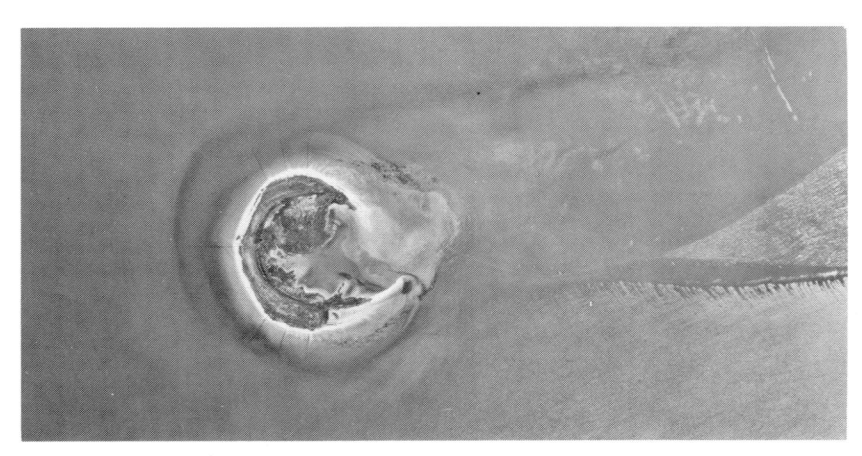

Figure 376: Recent development of the Isle of Griend. (a) Griend in 1959. Aerial photograph of the Fototheek Topografische Dienst, Emmen, of 1959, B5 Run G 93; (b) Griend in 1979. Aerial photograph of the Fototheek Topografische Dienst, Emmen, of 12.7.1979, 05 No. 502.

strong beach ridge (Brouwer et al., 1950: 125).

The subsequent changes in morphology and vegetation on the island have been scrupulously recorded by Brouwer et al. (1950). In 1937 the western beach ridge was breached in the centre, and in 1937/38 the sand was transported into the interior part of the island. In 1941 an attempt was made to close the gap with a brush fence, but this was breached in two places during the following winter.

Since then erosion has continued in the west at the same time as accretion has decreased in the east, almost ceasing completely during the last 10–15 years. The *Salicornia* zone which formerly characterized the transition zone between the salt marsh and the tidal flat has completely disappeared. The sand spits, which once formed the sickle-shaped western margin of the island as they do on Mellum, Scharhörn and Trischen, have increasingly encircled the island since 1959. By 1979 they almost completely surrounded the salt marsh (figs. 376 a and b).

Recent development

The recent development of Griend has been characterized by a change from a natural process to one dominated by artificial coastal protection measures. In the 1950's 20,000 m^3 of sand were pumped on to the western margin of Griend by the Rijkswaterstaat. In 1973 the island was further protected by a sand dyke in the west (height: NAP + 3 m, breadth: ca. 50 m) and seven pile groynes were built in order to reduce the current velocities in the crescentic channel west of the island.

By 1980 the width of the sand dyke had been reduced in one place to a mere 15 m thus risking a breach. Since the construction of this structure erosion has been a maximum of 5 m/year. In the autumn of 1981 the dyke was supported by sand replenishment and the foot of the dyke was reinforced using Gobi mats. In the meantime mud pumping has been considered as a means of promoting accretion on the leeside of the island (Abrahamse & Luitwieler, 1982 a: 89).

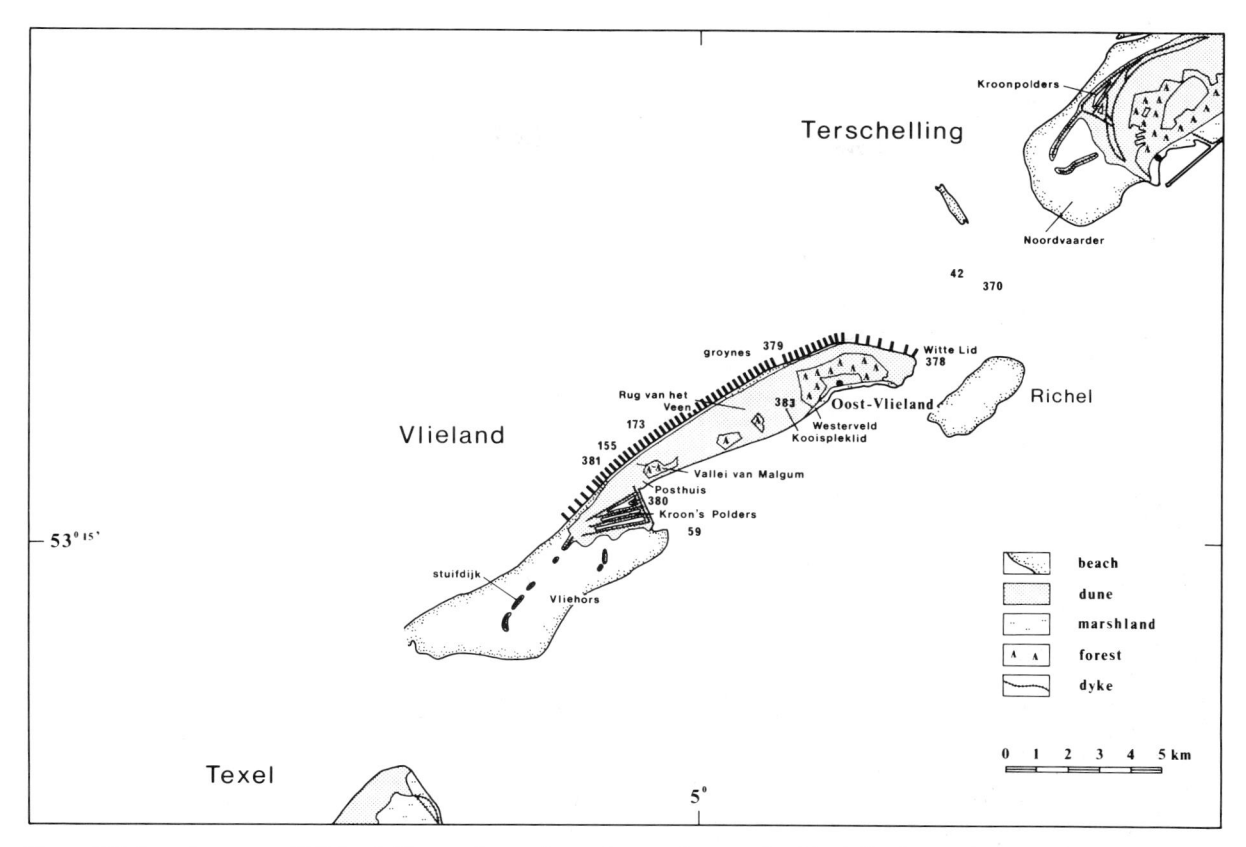

Figure 377: Location map of Vlieland. The numbers refer to figures, those prefixed by 'pl' refer to plates.

Without human intervention the Isle of Griend would almost certainly have vanished in recent years. Coastal protection measures allow the small island to be preserved as a bird sanctuary in the Wadden Sea. It has to be accepted, however, that the island now constitutes an artificial structure, not differing markedly from islands such as Minsener Oog which are created artificially for coastal engineering purposes.

7.4.7 VLIELAND

The Isle of Vlieland differs from all the other islands of the Wadden Sea in being almost entirely covered in dunes (fig. 377). The only salt marshes are the Westerveld area west of the settlement, today protected by a stone revetment, along with the recently formed salt marsh of the Posthuiswad.

Historial development

About 1230 the Isles of Vlieland and Eijerland, the latter now being part of Texel, formed one single island separated from Texel by a tidal inlet. According to legend, monks from the Ludinga cloister dug a canal from the Vlie towards the west in order to make a navigable connection with the Moerwaard, currently a submerged area in the Wadden Sea south of Vlieland. The canal (Monnikensloot) was allegedly connected to the North Sea between Vlieland and Eijerland. During the severe storm surge of 1296, in which the town of Griend was destroyed and in which the coastline approached the then 50-year old Harlingen, the gap between Vlieland and Eijerland was widened to form a tidal inlet (van der Burgt, 1936: 802).

Old documents demonstrate that Eijerland and Vlieland were already separate islands by 1314. Following the connection of Texel and Eijerland by a sand dyke in 1629/30, the importance of the Eijerlandse Gat between Eijerland and Vlieland decreased. Its catchment area was reduced in size and its ebb delta became smaller, which until this time had been the main seaward protection for the village of West Vlieland. Coastal erosion therefore

Figure 378: Recent dune erosion at Witte Lid, northeastern end of Vlieland (1981).

Figure 379: Young dunes on the beach of central Vlieland. The groynes which were built at the end of the 19th century have been completely covered by sand (1981).

occurred. The dunes off West Vlieland were breached for the first time in 1680 and the village, which had originally existed at the southern margin of the dunes adjoining the salt marshes, now approached the North Sea coast. By 1700 the number of houses had decreased from 388, in 1632, to a mere 150. Only about 100 houses were left by 1710.

During the storm surge of February 1714 the sea once again invaded the village and destroyed a great number of houses and the church. An attempt was subsequently made to close the gaps in the coastal dunes by constructing stuifdijks, but in vain. In 1727 West Vlieland was again inundated, and a new church and numerous houses were destroyed. The last houses in West Vlieland had to be

Figure 380: Aerial photograph showing the narrowest part of Vlieland at the Posthuis with the small recent salt marsh in the Posthuiswad and the Kroon's Polders. The uprooted groynes in the upper left of the picture are completely surrounded by the sea even at low tide. Aerial photograph of the Fototheek Topografische Dienst, Emmen, of 10.3.1983, 04 D 07.

abandoned in 1736 when the population was relocated to Oost Vlieland, the present village.

Recent coastal erosion

Like on Texel, but in contrast with the islands further to the east, the northeastern end of Vlieland is eroded. The dune belt has had to be reinforced and planted with *Ammophila arenaria* in order to stem this process. A recent dune cliff over 2 m high at Witte Lid testifies to this coastal retreat. The dune foot has been locally protected by a riprap of basalt blocks (fig. 378). The situation has not always been this acute. As a result of a prolonged period of positive sand budget it was possible in 1895 to construct stuifdijken (dune dykes) around the Kooremansvallei which then developed into wet meadow and was used as pasture. However, the situation soon changed, and the whole polder was lost between 1908 and 1917.

The entire northwestern beach of Vlieland was eroded during the middle of the last century. This led to the construction of groynes, starting in 1854, which stretch today from the Vliehors all along the coast to the northeastern end of the island (fig. 377). Whilst this negative sand budget has continued in most parts of the island until now, accretion has occurred along the central part of the coast during the last few decades (fig. 379). The 1:25,000 Topographical Map of Vlieland shows a broad beach between groynes 33 and 42. The old groynes are completely covered by sand in this area and the accretion area has broadened and extended to the northeast. A number of young dune ridges have formed in front of the coastal dunes; these are already over 50 m wide.

This area of positive sand budget occurs where the ebb delta of the Vlie impinges onto the island.

Figure 381: Old marine clay being eroded on the beach of Vlieland, north of the Posthuis (1981).

One of the main channels, the Zuider Stortemelk, which was for a long time situated directly off the coast, has started to shift northwards causing this accretion on the beach. Such displacements of the main channels of the Vlie have occurred repeatedly during the last few centuries (cf. Pronker & Eck, 1970).

The threat of a breaching of Vlieland

In the long term, coastal erosion has prevailed, exemplified by rootless groynes which were originally connected to the dune foot (fig. 380). The erosion of old marine clay can be seen north of the Posthuis on the present beach; this clay originally formed on the tidal flats behind the island (fig. 381). The coastal dunes have migrated across the whole breadth of the island.

The island is less than a kilometre wide at the Posthuis and the Vallei van Malgum, an old low-lying deflation area, occurs behind the coastal dunes. This area is therefore extremely susceptible to possible breaching. In fact it is the only place in all the Dutch Wadden Sea islands where breaching of an island appears likely. It remains to be seen if such a breach would have any serious consequences. There seems to be a general tendency towards the formation of larger barrier islands rather than towards creation of new tidal inlets in areas of smaller tidal range. As early as 1756 the Vliehors was cut off from Vlieland during a severe storm flood and a small tidal inlet, the Schildiep, developed between both islands. This breach did not last, however, the Schildiep being completely filled again during the 19th century.

The Kroon's Polders in southwestern Vlieland were dyked between 1905–1922 with the construction of a series of stuifdijks (Visser, 1947: 34). These help to prevent possible breaching of the narrowest part of the island (fig. 380). Because of their low elevation, the Kroon's Polders constitute a wetland landscape, which is used as a nature reserve (fig. 382).

The Vliehors

Before the destruction of West-Vlieland, the Vliehors formed part of the island core, bordered by a dune ridge against the sea. This was totally destroyed during the 18th and 19th centuries (van der Burgt, 1936: 820). At present the sand flat of the Vliehors, over 6 km long and 2.5 km wide, constitutes the eastern end of the ebb delta of the Eijerlandse Gat. In 1929 an attempt was made to initiate dune formation on the Vliehors by building a stuifdijk. This attempt failed, but remnants of the old stuifdijk can still be seen. The Vliehors is at present used as military training area and is therefore not accessible.

Some of the sand from the Vliehors migrates slowly eastwards around the southern end of Vlieland. The easterly margin of this sand sheet is clearly visible on aerial photographs (fig. 380).

Figure 382: View from the eastern dyke over Kroon's Polder, dyked in 1905 (1981).

Figure 383: Kooispleklid vegetated old dune ridge in central Vlieland (1981).

The dunes

The interior parts of Vlieland represent a dune landscape (fig. 383). The village of Oost Vlieland, which is surrounded by very high dunes, was threatened by aeolian sand drift as late as the 18th century. This was probably the result of overgrazing; a great number of goats and cattle used the dunes as pasture. Today the dunes are largely fixed by the planting of *Ammophila arenaria* and partly by afforestation. In 1892 the government started to afforest Vlieland, using amongst other trees *Alnus* and *Quercus*, to supply brushwood for the coastal protection works. In 1908 dune protection became a task for the Staatsbosbeheer (forest authority), who increased afforestation during the succeeding years. The dunes were finally fixed around 1910 (Staatsbosbeheer, 1980) and today aeolian processes are active in small areas only, such as the Rug van het Veen.

7.4.8 TEXEL

The Isle of Texel originally consisted of two separate islands Texel and Eijerland, separated from

352 *Regional descriptions*

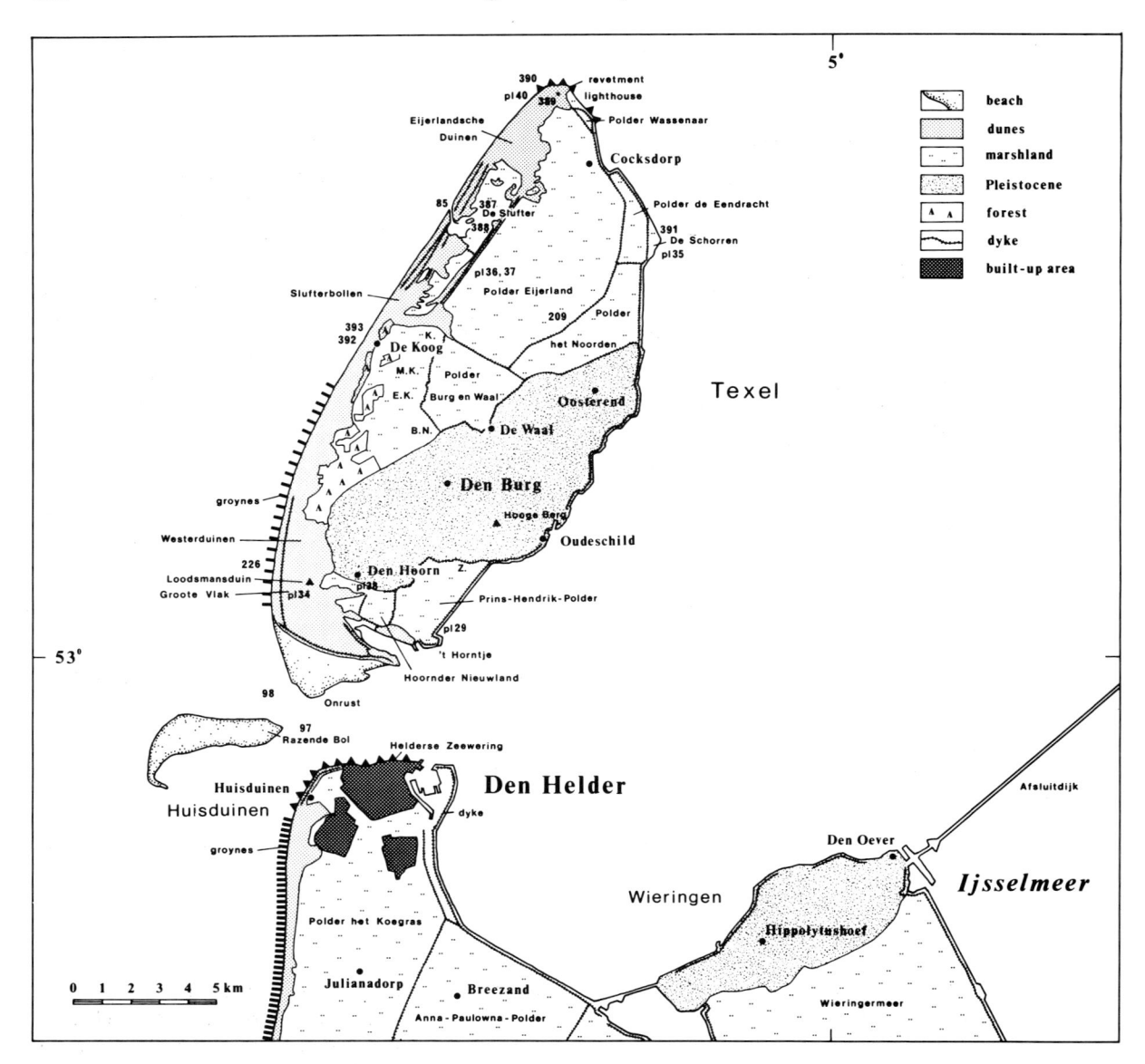

Figure 384: Location map of Texel. K = Koogerveld, M.K. = Middelste Koog, E.K. = Everste Koog, B.N. = Burger Nieuwland, Ż = Zuidelijke Zuidhaffel. The numbers refer to figures, those prefixed by 'pl' refer to plates.

each other by a minor tidal inlet. Texel has a Pleistocene core, the western end of which is covered by coastal dunes. This core forms a low ridge trending SW-NE, clearly visible in the otherwise flat polder landscape (plate 38). The highest point of the ridge, the Hooge Berg (High Mountain), reaches a height of 15.3 m above NAP. The remainder of the island consists of Holocene marshlands, bordered in the west by a belt of dunes (fig. 384).

In common with other islands with a Pleistocene core, Texel was settled in Prehistoric times. Settlement seems to have ended about 250 A.D., but the island was reoccupied in the 6th century after an interval of some 300 years (Bantelmann *et al.*, 1976: 223).

Texel was still part of the mainland in the early Middle Ages. However, storm floods between 1000 and 1300 A.D. breached the coastal barrier in North Holland in three places, forming the islands of Texel, Huisduinen and Callantsoog (fig. 385) (Hallewas, 1984: 301). Callantsoog was reconnected with North Holland by the dyking of the Zijpe in 1596 (Bremer, 1978: 552), and in 1610

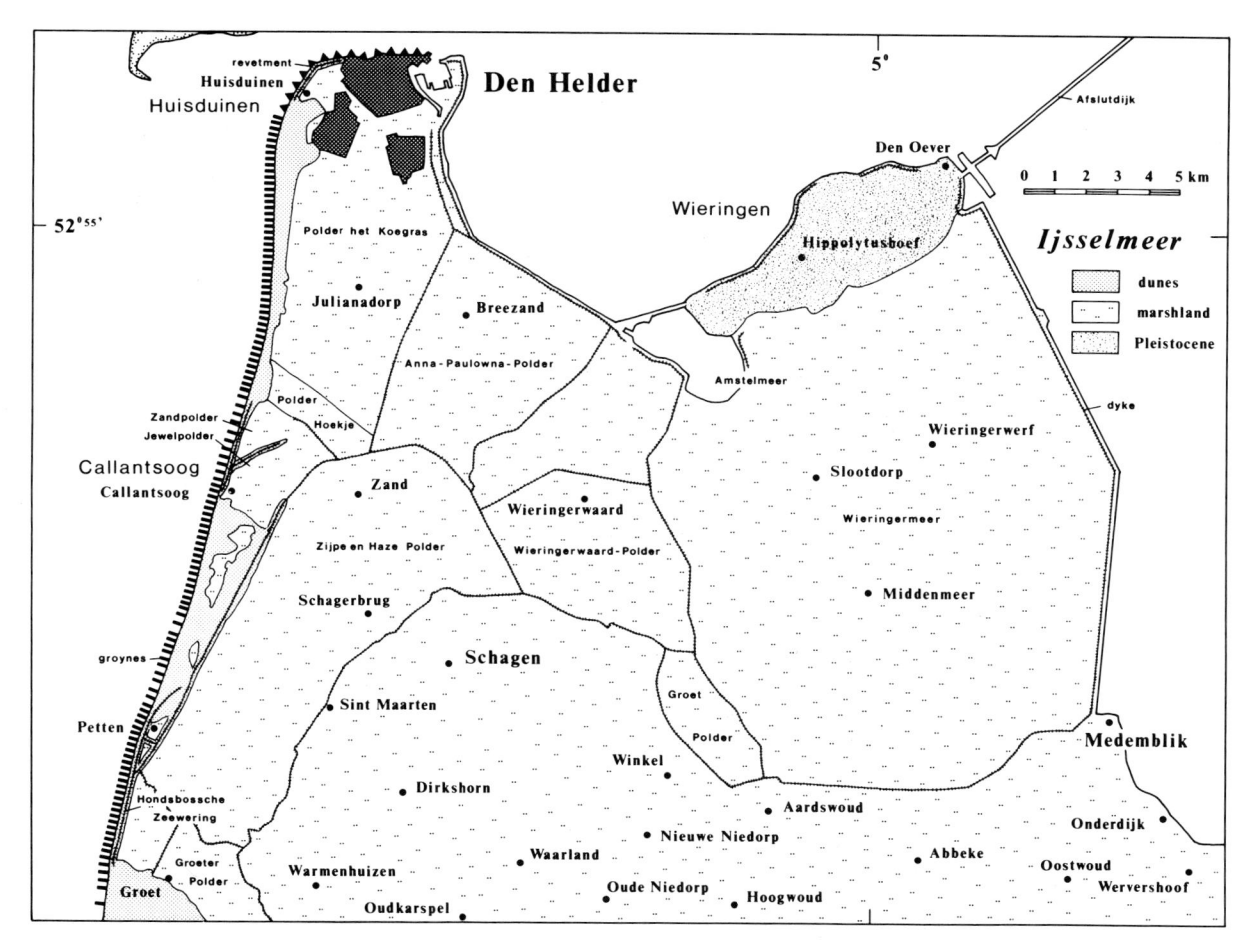

Figure 385: Location map of North Holland.

the Koegras salt marsh between Huisduinen and Callantsoog was dammed by a sand dyke, the Oldenbarneveldtsdijk (Westenberg, 1956: 226). The gap between Huisduinen and Texel, however, developed into the Marsdiep, the largest tidal inlet of the Wadden Sea.

History of coastal protection

The first low dykes on Texel were probably built as early as the 12th century. The Koogerveld, Middelste Koog and Everste Koog polders east of De Koog, and the Binnenspijk and Zuidelijke Zuidhaffel polders south of Den Burg, were all dyked during the 14th century. These early dykes were no higher than 2.70 m (Weijdt, 1980). The Waal en Burg polder was first dyked in 1488, only to be lost again in 1532 (Kleinkemm, 1910: 35).

This was only a minor setback, however. The Burger Nieuwland polder was dyked around 1500.

Renewed dyking of Waal en Burg took place in 1617, and in 1629/30 the small dune island of Eijerland was successfully linked with Texel by the construction of the 'Groote Zanddijk' (Great Sand Dyke; cf. plates 36 and 37). Rapid accretion occurred in lee of this dyke. In 1721 the 'Waterschap der 28 Gemeenschappelijke Polders' was founded, aiming by combined efforts to improve the dykes (Roeper, 1980: 60).

Until the 18th century much wood was used for dyking. A natural disaster struck all land reclamation works in 1731. The pile worm (*Teredo*) invaded the Netherlands, leading to the destruction of all wooden coastal protection structures within a few years (Dendermonde & Dibbits, 1956: 66). The dykes on Texel were similarly effected (Weijdt, 1980: 84).

A decade later, the construction of coastal protection works in 1743 at Den Helder, the so-called 'Helderse Zeeweering', halted the southward dis-

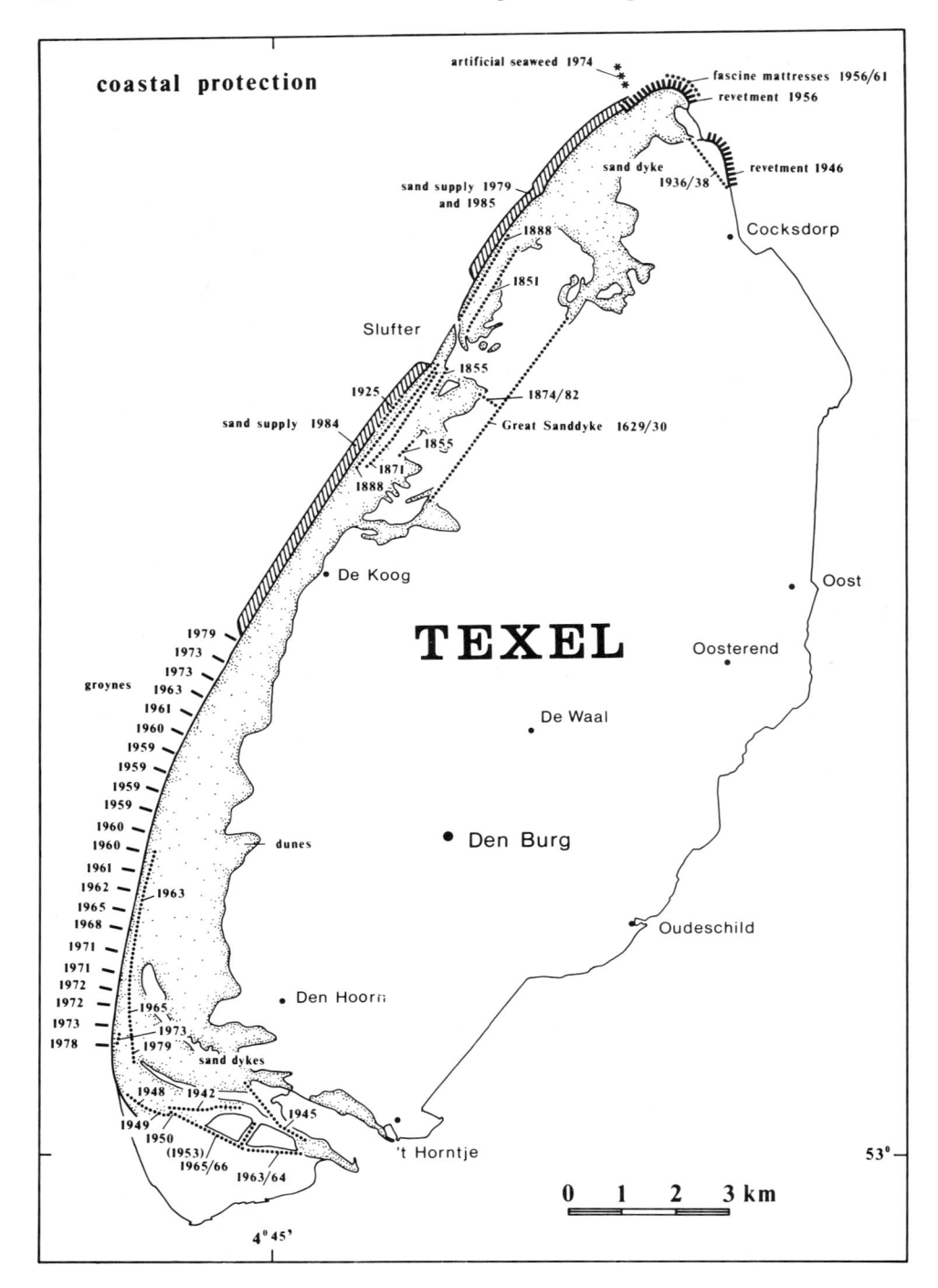

Figure 386: Coastal protection and dunes on Texel.

placement of the Marsdiep inlet. On Texel the Prins Hendrik Polder was dyked in 1768, but it was lost again in 1796 (Kleinkemm, 1910: 37). The severe storm flood of 1825 caused the breaching of the 'De Grie' polder, dyked in 1646.

The greatest reclamations on Texel took place during the 19th century:
1835: The Eijerland Polder was dyked (3,376 ha).
1846: Dyking of the 'De Eendracht' (247 ha)

Figure 387: The Slufter area, Isle of Texel. Aerial photograph of the Fototheek Topografische Dienst, Emmen, of 11.3.1983, 09 A 11.

and 'De Volharding' (16 ha) polders, and renewed dyking of the 'Prins Hendrik Polder' (460 ha; plate 39).

1876: Dyking of the 'Het Noorden' polder (765 ha).

Attempts were also made at this time to close the Slufter between Texel and Eijerland with a sand dyke (fig. 384). The damming of the Groote Slufter inlet succeeded, but it proved impossible to close the gap of the small Slufter as well. In 1874/82 the dune valleys adjoining to the south were protected by a sand dyke against further inundation (fig. 386), but the gap in the coastal dunes remained. Today the Slufter area is a nature reserve (figs. 387 and 388).

Scarcely any further salt marsh growth could be promoted during the 20th century. In 1910 the coastal protection work at 't Horntje had to be built to fend off the approaching Texelstroom. After the closure of the Zuider Zee in 1932 the tidal range at Oudeschild on Texel increased from 100 to 138 cm, effectively preventing any further accretion.

During the storm surge of 1953 the dyke of the 'De Eendracht' polder was breached. Using the experience of the 1953 flood, the dykes in the Netherlands have been successively strengthened and heightened, including those on Texel. The most recent reinforcements were completed in 1981 (Weijdt, 1980: 95).

The northern end of Texel

Like the northeastern end of Vlieland the northern end of Texel (fig. 389) has experienced frequent shoreline changes. On both islands accretion and erosion depend on the displacement of the main channels in the adjoining tidal inlets. The positive sediment budget on Texel at the beginning of the 19th century permitted the dyking of the 'De Volharding' polder in 1846, but the newly reclaimed land proved impossible to protect. The dyke was rebuilt further inland when it was breached in 1894 (Deuzeman, 1898: 34), but without success. Erosion continued, and the polder had to be finally abandoned in 1926 (Weijdt, 1980: 85). A map of

Figure 388: The Slufter area looking from the 'Groote Zanddijk' towards the northwest (1985).

1867 indicates a broad beach north of the lighthouse seawards of the dune belt, the so-called 'Engelsche Kerkhof' (English Churchyard). The coastline has retreated considerably since then (Roeper, 1980: 67).

The changes have been recorded by the Rijkswaterstaat since 1850 (fig. 390). From a fixed datum, a so-called 'Paal' (pole), the positions of the low water line, the high water line and the dune foot have been recorded. The dune foot data is the least accurate, because it is determined morphologically regardless of any absolute height. The measurements are accurate enough, however, to record all the major shoreline changes. Characteristically, the displacement of the dune foot always lags behind the changes of the high and low water lines.

Because of continuous coastal retreat a proposal to relocate the lighthouse was discussed, but turned down. Instead, a revetment was built in 1956, supplemented by groynes and reinforced by fascine mattresses. The situation changed once again after the 1962 storm surge when the sand budget became positive. In 1967 the beach near the lighthouse had reached a width of 700 m, and dunes began to form on it. However, erosion started again, soon followed by direct wave attack on the revetment. Strong leeside erosion led to the extension of the revetment for another 800 m towards the south. In 1971 3,000,000 m^3 sand was supplied to the beach south of the lighthouse (plate 40), dredged from the opposite bank of the Robbengat, in order to reduce the erosion at the northeastern end of Texel, especially on the Wassenaar Polder dyke (Harting, 1980).

The ebb current dominates in the Robbengat inlet. However, the flood current also affects the Eijerland coast. Two sandy spits, which have formed in a leeside position, are attached to the southern part of the Wassenaar Polder. The westerly spit can be seen on the aerial photograph of northern Texel (fig. 389). Under the protection of this spit a small foreland area has developed. The situation is comparable to the 'De Schorren' salt marsh area (plate 35). The development is similar to, but much slower than, that of the flood spits on northern Föhr. Aerial photographs indicate the slow rate of change (fig. 391 a and b); the tidal-flat facing beach ridge has hardly changed its position since the first survey.

The central dune area

The central area of Texel's North Sea coast is characterized by small, irregularly shaped areas of old dunes in addition to a maximum of three coast-parallel, more recent dune ridges (fig. 392). The old dunes have a height of over 20 m, while the younger dunes are on the whole lower. The most marked, outer, dune ridge has a maximum height of 17.5 m. The coast-parallel dunes are the result of seaward displacement of the coastline by about 300 m during the last 100 years. This displacement, however, has ceased today; the coastal dunes of Texel are experiencing erosion along the whole North Sea coast (fig. 393).

Whilst the dunes of Texel are now almost com-

Figure 389: The northern end of Texel. Aerial photograph of the Fototheek Topografische Dienst, Emmen, of 11.3.1983, 04 A 01.

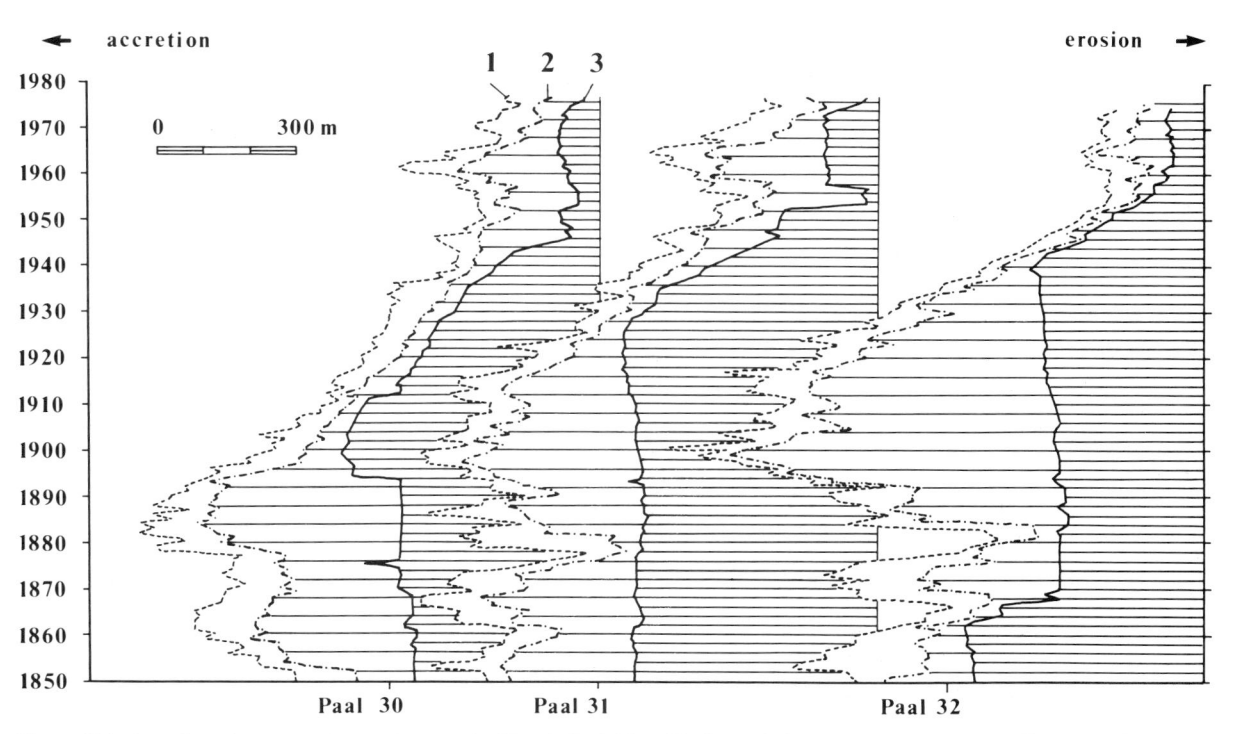

Figure 390: Coastline changes at the northern end of Texel, Paal 30 to 32, from 1850 to 1978; 1 = low water line, 2 = high water line, 3 = dune foot. Location see fig. 386. Source: Elorche (1982).

Figure 391: Recent development of the undyked 'de Schorren' salt marsh area on the northeastern side of Texel; (a) 1939. Aerial photograph of the Fototheek Topografische Dienst, Emmen, of 1939, B9 Run 6 No. 111; (b) 1983. Aerial photograph of the Fototheek Topografische Dienst, Emmen, of 11.3.1983, 09 C 12.

pletely vegetated, deflation played a major role until early in the 19th century. In 1894 a desire to fix the dunes led to planting of the first conifer forests on the island (mainly *Pinus sylvestris*). In the 20th century major dune areas have been made into nature reserves. Parts of these areas, exemplified by the 24.3 m high Loodsmansduin, the highest dune on Texel, were damaged by bunker building during the Second World War (Mantje, 1980).

Since 1959 attempts have been made to halt erosion along the southern part of the coast by building groynes. Measurements by the Rijkswaterstaat have demonstrated that coastal retreat in the protected area actually decreased, but at the expense of leeside erosion at both ends of the groyne field. More groynes have had to be constructed to stop this adverse effect. The number of groynes has now reached 21, the last one being built in 1979. The roots of the older groynes have now been washed free; their distance from the present dune foot is about 90 metres, a sign of the continuous erosion.

There is a tendency today to resort to a more flexible response to coastal erosion problems instead of building rigid coastal defence structures. Beach replenishment was attempted for the first time in 1979 between the Slufter and the revetment at the northern end of Texel, using 3,000,000 m^3 of sand. An extra 2,000,000 m^3 of sand was supplied to the sector between the Slufter and the northern end of the groynes in 1984.

The southern end of Texel

The coast-parallel dune ridges in the south have developed a fan of more than ten separate ridges. Seaward growth of the dune belt caused the groundwater to rise, and finally led to the formation of lakes in the primary dune valleys (plate 34). Between 1759 and 1950 the dune fan on Texel developed in association with the welding of two major shoals to the island (De Hors and De Onrust) (Klijn, 1981: 78). This process largely resembles the formation of the beach ridge and dune fans on Fanø and on the Eiderstedt peninsula.

Whilst the broad beach at the southernmost end of Texel promotes further dune formation, the westerly ends of the slightly older dune ridges are now being eroded. A sand dyke was constructed between 1963 and 1979 in order to prevent the sea from entering the dune valleys (fig. 386). Attempts were made at the same time to encourage new dune formation along the seaward coast, but without much success. This situation is unlikely to alter before the Razende Bol finally becomes attached to Texel, providing a new sand source and a broad western beach.

Figure 392: View from the outermost dune ridge at De Koog towards the second dune ridge. The built-up area does not extend beyond this inner dune ridge (1985).

Figure 393: The outermost dune ridge being eroded north of De Koog. Beach replenishment was undertaken in this area in 1984. An attempt is being made to broaden the coastal dune belt using brush fences (1985).

CHAPTER 8

SUMMARY

The formation of the Wadden Sea and of the island barrier is the result of predominantly landward directed sand transport. The development of the Isle of Trischen from a sandy shoal into a dune island serves as a model case to illustrate this process. Coast-normal sand transport is simultaneously overlain by coast-parallel sediment movement. In the West and East Frisian Islands this is directed from west to east, leading on most islands to erosion along the western ends and accretion at the eastern ends.

Sand transport from one island to the next is only imperfectly described by the often used term 'bar migration'. The migration rate of shoals, which was first measured by Homeier & Kramer (1957) in the Norderneyer Seegat ebb delta (405 m/year), can only be calculated reliably for the ends of the delta. Remodelling is so strong in the central area that even on aerial photographs taken at one month intervals, it is not possible in all cases to reconstruct shoal development.

Even in the marginal areas of the ebb delta the morphodynamics are more complicated than anticipated. While the groups of shoals migrate roughly parallel to the strike direction of the delta, single shoals can deviate markedly from this course, migrating much faster than the groups (up to 2 m each tide). The shoals resemble large ramps on which the sand migrates first upwards and then over the escarpment where it is deposited in giant foresets. The large delta-bedding structures which have been observed by Tietze (1983) on seismic profiles of the bars at the Hever mouth are formed in this way.

The deep bars at the mouth of tidal streams, like the Hever, are morphodynamically equivalent to the ebb deltas of the tidal inlets. Sand transport from one tidal flat area to the next is accomplished via these bars, which migrate at rates of about 20 m/year. Their surfaces are covered with megaripples.

The ebb delta shoals are also covered with megaripples, normally all ebb-orientated at low tide. Comparison of aerial photographs taken during four successive low tides has, however, revealed that only some of the megaripples migrate in the direction of the ebb current; the others are dominated by the flood current. They migrate at about 3–6 m each tide. The areas which are dominated by either the ebb or the flood current are not persistent and change from tide to tide.

The megaripples do not migrate along the shoal arc of the ebb delta, but deviate strongly from that direction by up to nearly 90° in the outermost part of the delta. This means that a considerable proportion of the sand transport occurs at right angles to the ebb delta. As the flow directions of the ebb and flood currents are not directly opposite, but deviate by a few degrees, this causes a long-term sand migration along the ebb delta. The admixture of additional material from the sea, as well as the loss of material, can significantly change the original composition of the sediments, as Veenstra's (1982, 1984) investigations have demonstrated.

The flood deltas only contribute to the redeposition of sediments in a very minor way. Comparison of aerial photographs, as well as field measurements, have demonstrated that the flood delta sand waves only migrate at a rate of about 10–20 cm each tide, which is less than one tenth of the megaripple migration rates measured in the ebb deltas. Provided that the ebb and flood deltas are of equal proportions, which in this case they are, this means that over 90 % of the sediment movement occurs in the ebb delta. The transport into and out of the tidal inlet plays a subordinate role.

Sediment transport across the tidal flats is accordingly low, but it cannot be completely ignored. Aerial photographs reveal the strong influence of wind-induced drift water on the tidal flats. In the course of these investigations it has not been possible to quantify this sediment movement. An indication of the rates to be expected can be gauged from the fact that the channel dredged across the Wangerooger Inselwatt in 1976 has not yet been infilled.

The island beaches provide the sediment supply for dune formation. The migration rates of wind ripples (5 cm/min), beach barchans (1.5 m/h) and large migrating dunes (4 m/year) have been measured, and the morphodynamic sequences from primary dunes to old, inactive and vegetated former migrating dunes, dissected by blowouts, demonstrated.

The island barrier adjusts itself to the respective sea level stages. A rising sea level results in landward migration of the barrier. Under natural conditions this is achieved by dune erosion and landward aeolian sand transport or by the breaching of the dune ridges and washover processes. Both these processes are largely prevented at present by coastal engineering countermeasures and the most exposed parts of the islands have been fixed and fortified against erosion. This has resulted in both undernourished beaches and lack of sediment on the tidal-flat facing sides of the islands. This tendency has had to be offset by extensive coastal protection in order to prevent an '*in-situ* drowning of the barrier', as envisaged by Leatherman (1982).

REFERENCES

Abele, R.W. 1977. Analysis of Short-Term Variations in Beach Morphology (and Concurrent Dynamic Processes) for Summer and Winter Periods, 1971-72, Plum Island, Massachusetts. *Miscellaneous Report* 77-5: 101 pp. Fort Belvoir: U.S. Army, Corps of Engineers, Coastal Engineering Research Center.

Abrahamse, J., J.D. Buwalda & L.M.J.U. van Straaten (eds.) 1965. *Het Waddenboek*: 234 pp. Zutphen: Thieme & Cie.

Abrahamse, J., W. Joentje & N. van Leeuwen-Seelt (eds.) 1976. *Wattenmeer. Ein Naturraum der Niederlande, Deutschlands und Dänemarks*: 371 pp. Neumünster: Wachholtz.

Abrahamse, J. & H. Koning (1983). *Schiermonnikoog*: 112 pp. Groningen: Uitgeverij Fotoplus.

Abrahamse, J. & F. Luitwieler 1982 a. Griend. *Waddenbulletin* 17 (2): 86-89.

Abrahamse, J. & F. Luitwieler 1982 b. Minsener Oldoog. *Waddenbulletin* 17 (3): 114-118.

Abrahamse, J. & F. Luitwieler 1984. Grosse Knechtsand. *Waddenbulletin* 19 (4): 175-178.

Abrahamse, J. & F. Luitwieler 1985. Schiermonnikoog groeit als kool. *Waddenbulletin* 20 (1): 8-10.

Aigner, Th. & H.-E. Reineck 1983. Seasonal variations of Wave-Base on the Shoreface of the Barrier Island Norderney, North Sea. *Senckenbergiana maritima* 15 (2/3): 87-92.

Allen, J.R.L. 1984 a. Sedimentary structures. Their character and physical basis. *Developments in Sedimentology* 30, Vol. 1: 593 pp. Amsterdam, Oxford, New York, Tokyo: Elsevier.

Allen, J.R.L. 1984 b. Sedimentary structures. Their character and physical basis. *Developments in Sedimentology* 30, Vol. 2: 663 pp. Amsterdam, Oxford, New York, Tokyo: Elsevier.

Allen, J.R.L. & P. Friend 1976. Changes in intertidal dunes during two spring-neap cycles, Lifeboat Station Bank, Wellsnext-the-Sea, Norfolk (England). *Sedimentology* 23: 329-346.

Andresen, C. 1984. Küstenschutz. Auf Sylt ist wahrhaftig alles anders. *Söl'ring Foriining, Jahresbericht* 1983: 7-8.

Andresen, F.H. 1980. Vorlandarbeiten und Deichbauplanungen. *Nordfriesisches Jahrbuch* N.F. 16: 61-67.

Andrews, J.T. 1982. Holocene Glacier Variations in the Eastern Canadian Arctic: A Review. *Striae* 18: 9-14.

Anwar, J. 1974. Der holozäne Meeressand im Seegebiet westlich von Sylt zwischen Kampen und Rantum (Ausgangsmaterial und Sedimentation). *Meyniana* 24: 43-55.

Arbeitsgruppe Küstenschutz des Küstenausschusses Nord- und Ostsee 1957. Gutachtliche Stellungnahme zur Anpassung der Warften auf den nordfriesischen Halligen an die heute möglichen Sturmfluthöhen. *Die Küste* 6 (1): 125-135.

Arends, F. 1833. *Physische Geschichte der Nordseeküste und deren Veränderungen durch Sturmfluthen seit der Cymbrischen Fluth bis jetzt.* 2 vols.: 384 pp. / 355 pp. Emden: Woortmann. Reprint 1974. Leer: Schuster.

Armon, J.W. 1979. Landward sediment transfers in a transgressive barrier island system, Canada. In S.P. Leatherman (ed.), *Barrier Islands – from the Gulf of St.Lawrence to the Gulf of Mexico*: 65-80. New York, San Francisco, London: Academic Press.

Aubrey, D.G. & P.E. Speer 1984. Updrift migration of tidal inlets. *Journal of Geology* 92: 531-545.

Ausschuß 'Küstenschutzwerke' der Deutschen Gesellschaft für Erd- und Grundbau e.V. sowie der Hafenbautechnischen Gesellschaft e.V. 1981. Empfehlungen für die Ausführung von Küstenschutzwerken – EAK 1981. *Die Küste* 36: 1-320.

Averdieck, F.-R. & P. Hummel 1974. Zum Alter und Aufbau des Heidsandes auf Sylt. *Meyniana* 24: 9-25.

Backhaus, H. 1943. Die ostfriesischen Inseln und ihre Entwicklung. Ein Beitrag zu den Problemen der Küstenbildung im südlichen Nordseegebiet. *Schriften der Wirtschaftswissenschaftlichen Gesellschaft zum Studium Niedersachsens e.V.*, Neue Folge 12: 143 pp.

Bäsemann, H. 1979. *Feinkiesanalytische und morphometrische Untersuchungen an Oberflächensedimenten der Deutschen Bucht*. Dissertation, Hamburg: 143 pp.

Bahr, M. 1963. *Die Entwicklung des Küstenvorfeldes zwischen Hever und Elbe seit dem Ende des 16. Jahrhunderts*. Unpublished Report, Küstenausschuß Nord- und Ostsee: 125 pp.

Bakker, H. 1973. *Ameland – Insel der Freien*: 96 pp. Haren: Knoop & Niemeijer.

Bakker, J.P. 1954. Relative sea-level changes in Northwest Friesland (Netherlands) since the prehistoric times. *Geologie en Mijnbouw* 16: 232-246.

Bantelmann, A. 1939. Das nordfriesische Wattenmeer, eine Kulturlandschaft der Vergangenheit. *Westküste* 2 (1): 39-115.

Bantelmann, A. 1960. Forschungsergebnisse der Marschenarchäologie zur Frage der Niveauveränderungen an der schleswig-holsteinischen Westküste. *Die Küste* 8: 45-65.

Bantelmann, A. 1966. Die Landschaftsentwicklung im nordfriesischen Küstengebiet, eine Funktionschronik durch fünf Jahrtausende. *Die Küste* 14 (2): 5-99.

Bantelmann, A. 1976. Die Frühgeschichte Eiderstedts auf Grund der Bodenfunde. In Eiderstedter Heimatbund (eds.) *Blick über Eiderstedt* 1: 9-16.

Bantelmann, A., S.J. van der Molen, A.H. Rasmussen & W. Reinhardt 1976. Geschichte des Wattenraumes. In J. Abrahamse, W. Joentje & N. van Leeuwen-Seelt (eds.), *Wattenmeer. Ein Naturraum der Niederlande, Deutschlands und Dänemarks*: 223-241.

Barckhausen, J. 1969. Entstehung und Entwicklung der Insel Langeoog. *Oldenburger Jahrbuch* 68: 239-281.

Barckhausen, J. 1970. *Geologische Karte von Niedersachsen 1:25,000, Erläuterungen zu Blatt Baltrum Nr. 2210 und Blatt Ostende-Langeoog Nr. 2211*: 44 pp. Hannover: Niedersächsisches Landesamt für Bodenforschung.

Barckhausen, J. & H. Streif 1978. *Geologische Karte von*

Niedersachsen 1:25,000, Erläuterungen zu Blatt Emden West Nr. 2608: 80 pp. Hannover: Niedersächsisches Landesamt für Bodenforschung.

Barckhausen, J., H. Streif & R. Vinken 1979. Räumliche Erfassung und Bilanzierung von holozänen Küstenablagerungen in Niedersachsen. Deutsche Forschungsgemeinschaft (ed.), *Sandbewegung im Küstenraum – Rückschau, Ergebnisse und Ausblick*: 49-63.

Bartholdy, J. & M. Pejrup 1980. Introduction to the Ho Bugt area. *Geografisk Tidsskrift* 80: 63.

Baschin, O. 1903. Dünenstudien. *Zeitschrift der Gesellschaft für Erdkunde zu Berlin* Bd. 38: 422-430.

de Bèaumont, E. 1845. Septième leçon: Levées de sable et de galet. In P. Bertrand (ed.), *Leçons de géologie practique*: 221-252. Paris.

Becker, F. 1976. *Daheim auf Inseln und Halligen*, 3rd edition: 128 pp. Hamburg: Christians.

Behre, K.-E. 1970. Die Entwicklungsgeschichte der natürlichen Vegetation im Gebiet der unteren Ems und ihre Abhängigkeit von den Bewegungen des Meeresspiegels. *Probleme der Küstenforschung im südlichen Nordseegebiet* 9: 13-47.

Behre, K.-E., B. Menke & H. Streif 1979. The Quaternary geological development of the German part of the North Sea. In E. Oele, R.T.E. Schüttenhelm & A.J. Wiggers (eds.), *The Quaternary History of the North Sea*, Acta Universitatis Upsaliensis, Symposia Universitatis Upsaliensis Annum Quingentesimum Celebrantis 2: 115-142.

Behrmann, W. 1919. Borkum – Strand- und Dünenstudien. *Meereskunde* 153: 40 pp.

Behrmann, W. 1921. Die ostfriesischen Inseln. *Annalen der Hydrographie und maritimen Meteorologie* 49 (3): 79-93.

Belderson, R.H., M.A. Johnson & N.H. Kenyon 1982. Bedforms. In A.H. Stride (ed.), *Offshore Tidal Sands*: 27-57.

Berg, J.H. van den 1981. Rhythmic seasonal layering in a mesotidal channel fill sequence, Oosterschelde Mouth, the Netherlands. In S.-D. Nio, R.T.E. Schüttenhelm & Tj.C.E. van Weering (eds.), *Holocene Marine Sedimentation in the North Sea Basin. Special Publication Number 5 of the International Association of Sedimentologists*: 147-159.

Beukema, J.J. 1976. Biomass and species richness of the macrobenthic animals living on the tidal flats of the Dutch Wadden Sea. *Netherlands Journal of Sea Research* 10: 236-261.

Beukeboom, T.J. 1976: *The Hydrology of the Frisian Islands*. Vrije Universiteit te Amsterdam (Thesis): 121 pp. Amsterdam: Rodopi.

Bird, E.C.F. 1985. *Coastline Changes. A Global Review*: 219 pp. Chichester, New York, Brisbane, Toronto, Singapore: Wiley.

Blaszyk, P. (ed.) 1975. *Naturschutzgebiete im Oldenburgerland: Mellum, Oldeoog, Wangerooge, Sager Meer, Dümmer*: 127 pp. Oldenburg: Holzberg.

Bobzin, E. 1926. Die Landschaften der Nordseeinsel Sylt mit besonderer Berücksichtigung ihrer Natur- und Kultur-Arbeitsformen. *Forschungen zur Deutschen Landes- und Volkskunde* 24 (3): 125-153.

Boer, P.L. de 1980. *Tussen Holland en Friesland. Het eiland Terschelling, voornamelijk in de 17e eeuw*: 303 pp. Thesis: Rijksuniversiteit Utrecht. Terschelling: Stichting Ons Schellingerland.

Boer, P.L. de 1981. Mechanical effects of micro-organisms on intertidal bedform migration. *Sedimentology* 28: 129-132.

Boothroyd, J.C. 1985. Mesotidal inlets and estuaries. In Davis, R.A. (ed.), *Coastal Sedimentary Environments*, 2nd. edition: 445-532.

Borchers 1929. Die Grodendeiche der Insel Wangeroog. In Landesverein Oldenburg für Heimatkunde und Heimatschutz (eds.), *Wangeroog wie es wurde, war und ist*: 173-178.

Borth-Hoffmann, B. 1980. *Flachseismische Untersuchung geologischer Strukturen in der östlichen Deutschen Bucht*: 93 pp. Universität Kiel: Unpublished Diplomarbeit.

Bowen, D.Q. 1984. Introduction. In D.Q. Bowen & A. Henry (eds.), *Quaternary Research Association Field Guide, April 1984, Wales: Gower, Preseli, Fforest Fawr*: 1-17.

Bowen, D.Q. & A. Henry (eds.) 1984. *Quaternary Research Association Field Guide, April 1984, Wales: Gower, Preseli, Fforest Fawr*: 102 pp.

Bradley, R.S. 1985. *Quaternary Paleoclimatology. Methods of Paleoclimatic Reconstruction*: 472 pp. Winchester (Mass.): Allen & Unwin.

Brandt, K. 1981. Siedlungsgeschichte und historische Entwicklung. In H. Streif, *Geologische Karte von Niedersachsen 1:25,000, Erläuterungen zu Blatt Nr. 2414 Wilhelmshaven*: 73-83. Hannover: Niedersächsisches Landesamt für Bodenforschung.

Brelie, G. von der 1959. Watt-Ablagerungen des Eem-Meeres im Raum von Norderney. *Zeitschrift der Deutschen Geologischen Gesellschaft* 111 (1): 1-7.

Bremer, J.T. 1978. Een zeventiende eeuws bedijkingsplan voor het Balgzandgebied. *Waddenbulletin* 13 (3): 552-554.

Brenninkmeyer, B. 1982. Minor beach features. In M.L. Schwartz (ed.), *Encyclopedia of Beaches and Coastal Environments*: 546-550.

Broome, R. & P.D. Komar 1979. Undular hydraulic jumps and the formation of backwash ripples on beaches. *Sedimentology* 26 (4): 543-559.

Brouwer, G.A., J.W. van Dieren, W. Feekes, G.W. Harmsen, J.G. ten Houten, W.J. Kabos, J.P. Mazure, A. Scheygrond, P. Tesch & A. van der Werff 1950. *Griend – het vogeleiland in de Waddenzee*: 310 pp. 's Gravenhage: Martinus Nijhoff.

Bruun, P. 1977. Practical solution to a beach erosion problem. *Coastal Engineering* 1: 3-16.

Bruun, P. 1978. Stability of Tidal Inlets – Theory and Engineering. *Developments in Geotechnical Engineering* 29: 510 pp. Amsterdam, Oxford, New York: Elsevier.

Bruun, P. 1983. Beach scraping – is it damaging to beach stability? *Coastal Engineering* 7: 167-173.

Bruun, P. & F. Gerritsen 1959. Natural by-passing of sand at coastal inlets. *Journal of the Waterways and Harbors Division, Proceedings of the American Society of Civil Engineers*: 75-107.

Bryan, R.B. & A.G. Price 1980. Recession of the Scarborough Bluffs, Ontario, Canada. *Zeitschrift für Geomorphologie* N.F., Suppl.-Bd. 34: 48-62.

Buchwald, K., G. Rincke & K.-H. Rudolph 1985. *Gutachterliche Stellungnahme zu den Umweltproblemen der Ostfriesischen Inseln - Schlußbericht*: 416 pp. Hannover, Zwingenberg, Nordhorn (unpublished).

Burgt, J.H. van der 1936. Veranderingen in den zeebodem van het Zeegat van het Vlie en in de kustlijn der waddeneilanden Vlieland en Terschelling. *Tijdschrift van het Koninklijk Nederlandsch Aardrijkskundig Genootschap*, Tweede Reeks LIII: 802-823.

Cepek, P. & H.-E. Reineck 1970. Form und Entstehung von Rieselmarken im Watt- und Strandbereich. *Senckenbergiana maritima* 2: 3-30.

Cornish, V. 1901. On sand waves in tidal currents. *Geographical Journal* 18 (2): 170-202.

Costello, J. & T. Hughes 1978. *Skagerrak. Deutschlands größte Seeschlacht*: 240 pp. Wien, München, Zürich, Innsbruck: Molden.

Cronin, L.E. (ed.) 1975. *Estuarine Research*, Vol.2: 587 pp. New York: Academic Press.

Cuvier, G. 1827. *Essay on the Theory of the Earth*: 550 pp. Edinburgh and London: Blackwood and Cadell.
Czock, H. & P. Wieland 1965. Naturnaher Küstenschutz am Beispiel der Hörnum-Düne auf der Insel Sylt nach der Sturmflut vom 16./17. Februar 1962. *Die Küste* 13: 61-72.
Dalrymple, R.W. 1984. Morphology and internal structure of sandwaves in the Bay of Fundy. *Sedimentology* 31: 365-382.
Dalrymple, R.W., R.J. Knight & G.V. Middleton 1975. Intertidal sand bars in Cobequid Bay (Bay of Fundy). In L.E. Cronin (ed.), *Estuarine Research*, Vol.2: 293-307.
Dankers, N., H. Kühl & W.J. Wolff (eds.) 1981. *Invertebrates of the Wadden Sea*: 221 pp. Rotterdam: Balkema.
Davies, J.L. 1964. A morphogenetic approach to world shorelines. *Zeitschrift für Geomorphologie, Sonderheft zum 70.Geburtstag Prof. H. Mortensen*: 127-142.
Davis, R.A. (ed.) 1985. *Coastal Sedimentary Environments*, 2nd edition: 716 pp. New York, Berlin, Heidelberg, Tokyo: Springer.
Davis, R.A. 1985. Beach and Nearshore Zone. In R.A. Davis (ed.), *Coastal Sedimentary Environments*: 379-444.
Dean, R.G. 1983. Principles of Beach Nourishment. In P.D. Komar (ed.), *CRC Handbook of Coastal Processes and Erosion*: 217-231.
Dechend, W. 1951. Das Eem im Raum Norderney–Hilgenriede. *Zeitschrift der Deutschen Geologischen Gesellschaft* 102 (1): 91-97.
Quartär der südlichen Nordseeküste. *Geologisches Jahrbuch* 68: 501-516.
Dechend, W. 1956. Der Ablauf der holozänen Nordsee-Transgression im oldenburgisch-ostfriesischen Raum, insbesondere im Gebiet von Jever i.O. *Geologisches Jahrbuch* 72: 295-314.
Degn, Chr. & U. Muuß 1979. *Topographischer Atlas Schleswig-Holstein und Hamburg*, 4th edition: 235 pp. Neumünster: Wachholtz.
Dekker, P. 1892. *Die Sturmfluthen am 3. und 4. Februar 1825 und ihre Verheerungen in Ostfriesland nebst Nachrichten über frühere und spätere Fluthen. Sonderabdruck aus der 'Ostfriesischen Zeitung'*: 112 pp. Emden: Hahn; Reprint 1983. Weener (Ems): Risius.
Dendermonde, M. & H.A.M.C. Dibbits 1956. *The Dutch and their dikes*: 175 pp. Amsterdam: De bezige Bij.
Dette, H.H. 1974. Wellenmessungen und Brandungsuntersuchungen vor Westerland/Sylt. *Leichtweiss-Institut für Wasserbau der Technischen Universität Braunschweig, Mitteilungen* 40: 285-330.
Deutsche Forschungsgemeinschaft (eds.) 1971. *Sandbewegung im Küstenraum. Neue Forschungsergebnisse zum Problem der Sandbewegung vor den Küsten der Ost- und Nordsee – Ein Zwischenbericht*: 75 pp. Wiesbaden: Steiner.
Deutsche Forschungsgemeinschaft (eds.) 1979. *Sandbewegung im Küstenraum – Rückschau, Ergebnisse und Ausblick*: 416 pp. Boppard: Harald Boldt.
Deutsche Forschungsgemeinschaft (eds.) 1984. *Archäologische und naturwssenschaftliche Untersuchungen an ländlichen und frühstädtischen Siedlungen im deutschen Küstengebiet vom 5. Jahrhundert v.Chr. bis zum 11. Jahrhundert n. Chr. – Band 1: Ländliche Siedlungen*: 461 pp. Acta humaniora. Weinheim: Verlag Chemie.
Deuzeman, P. (1898). *Aardrijkskunde van het eiland Texel*: 84 pp. Reprint 1978: Drukkerij bv v/h Langeveld en de Rooy.
Dewers, F. 1940. Probleme der geologischen Marschenforschung. *Probleme der Küstenforschung im südlichen Nordseegebiet* 1: 36-43.
Dewers, F. 1941. Das Alluvium. In K. Gripp, F. Dewers & F. Overbeck, *Das Känozoikum in Niedersachsen*: 268-454. Oldenburg: Stalling.
Dhonau, W. 1931. Über die Pflanzenwelt Spiekeroogs. In K. Ruhnau (ed.), *Die ostfriesische Insel Spiekeroog*: 27-33.
Dienemann, W. & W. Scharf 1931. Zur Frage der neuzeitlichen 'Küstensenkung' an der deutschen Nordseeküste. *Jahrbuch der Preußischen Geologischen Landesanstalt 1931*, 52: 317-390.
Dieren, J.W. van 1934. *Organogene Dünenbildung. Eine geomorphologische Analyse der Dünenlandschaft der West-Friesischen Insel Terschelling mit pflanzensoziologischen Methoden*: 304 pp. Den Haag: Martinus Nijhoff.
Dijkema, K.S. 1980. Large-scale geomorphologic pattern of the Wadden Sea area. In K.S. Dijkema, H.-E. Reineck & W.J. Wolff (eds.), *Geomorphology of the Wadden Sea*: 72-84.
Dijkema, K.S. 1983 a: The salt-marsh vegetation of the mainland coast, estuaries and Halligen. In K.S. Dijkema & W.J. Wolff (eds.), *Flora and vegetation of the Wadden Sea islands and coastal areas*: 185-220.
Dijkema, K.S. 1983 b: Use and management of mainland salt marshes and Halligen. In K.S. Dijkema & W.J. Wolff (eds.), *Flora and vegetation of the Wadden Sea islands and coastal areas*: 302-312.
Dijkema, K.S. & W.J. Wolff (eds.) 1983. *Flora and vegetation of the Wadden Sea islands and coastal areas*: 413 pp. Rotterdam: Balkema.
Dijkema, K.S., H.-E. Reineck & W.J. Wolff (eds.) 1980. *Geomorphology of the Wadden Sea*: 135 pp. Rotterdam: Balkema.
Dillo, H.-G. 1960. Sandwanderung in Tideflüssen. *Mitteilungen des Franzius-Instituts für Grund- und Wasserbau der Technischen Hochschule Hannover* 17: 135-253.
Dingler, J.R. & H.E. Clifton 1984. Tidal-cycle changes in oscillation ripples on the inner part of an estuarine sand flat. *Marine Geology* 60: 219-233.
Dionne, J.-C. 1968. Morphologie et sédimentologie glacielles, littoral sud du Saint Laurent. *Zeitschrift für Geomorphologie* Suppl.-Bd. 7: 56-84.
Dionne, J.-C. 1969. Tidal Flat Erosion by Ice at La Pocatière, St. Lawrence Estuary. *Journal of Sedimentary Petrology* 39 (3): 1174-1181.
Dittmer, E. 1938. Schichtenaufbau und Entwicklungsgeschichte des dithmarscher Alluviums. *Westküste* 1 (2): 105-150.
Dittmer, E. 1948. Die Küstensenkung an der schleswig-holsteinischen Westküste. *Forschungen und Fortschritte* 24 (17/18): 215-217.
Dittmer, E. 1952. Die nacheiszeitliche Entwicklung der schleswig-holsteinischen Westküste. *Meyniana* 1: 138-168.
Dittmer, E. 1960. Neue Beobachtungen und kritische Bemerkungen zur Frage der 'Küstensenkung'. *Die Küste* 8: 29-44.
Dolan, R. 1972. Barrier Dune System along the Outer Banks of North Carolina: A Reappraisal. *Science* 176: 286-288.
Duinker, J.C. (ed.) 1980 *Texel en de Zee. Een strijd van Eeuwen*: 101 pp. Texelse Museum Vereniging.
Duphorn, K. 1976. Gibt es Zusammenhänge zwischen extremen Nordsee-Sturmfluten und globalen Klimaänderungen? *Wasser und Boden* 28 (10): 273-275.
Easterbrook, D.J., P. Havlicek, K.-D. Jäger & F.W. Shotton (eds.) 1982. *IGCP-Project 73/1/24, Quaternary Glaciations in the Northern Hemisphere, Report No.7 on the Session in Kiel, F.R.G., September, 18-23, 1980*: 256 pp. Prag.
Edelman, I.T. 1964. De historische veranderingen in de natuurlijke gesteldheid van het Nederlandse waddengebied. In J. Abrahamse, J.D. Buwalda & L.M.J.U. van Straaten (eds.), *Het Waddenboek*: 37-65.

Ehlers, J. (ed.) 1983. *Glacial Deposits in North-west Europe*: 470 pp. Rotterdam: Balkema.
Ehlers, J. 1984. Platenwanderung an der ostfriesischen Küste? - Ergebnisse der Luftbildauswertung von zwei Befliegungen der Wichter Ee (zwischen Norderney und Baltrum) im Sommer 1982. *Mitteilungen aus dem Geologisch-Paläontologischen Institut der Universität Hamburg* 57: 123-129.
Ehlers, J. & H. Mensching 1982. *Erläuterungen zur Geomorphologischen Karte 1:25,000 der Bundesrepublik Deutschland, GMK 25 Blatt 10, 2213 Wangerooge*: 55 pp. Berlin.
Ehlers, J. & H. Mensching 1983. Besonderheiten geomorphologischer Kartierung im Wattenmeer, dargestellt am Beispiel des Blattes 10 der GMK 25 Wangerooge. *Zeitschrift für Geomorphologie* N.F.27 (4): 495-510.
Ehlers, J., K.-D. Meyer & H.-J. Stephan 1984. The pre-Weichselian Glaciations of North-West Europe. *Quaternary Science Reviews* 3 (1): 1-40.
Eiderstedter Heimatbund (eds.) 1976. *Blick über Eiderstedt* 1, 4th edition: 158 pp. Heide: Westholsteinische Verlagsanstalt Boyens & Co.
Ellenberg, L. & K. Hirakawa 1982. Die Packeisküste Japans. *Eiszeitalter und Gegenwart* 32: 1-12.
Elorche, M. 1982. *De kustontwikkeling van Texel*. Rijkswaterstaat, Adviesdienst Hoorn, Nota WWKZ-82.H011: 25 pp.
Elorche, M. 1983. *Morfologische ontwikkeling van het Zeegat van Ameland*. Rijkswaterstaat, Adviesdienst Hoorn, Nota WWKZ-83.H010: 13 pp.
Engel, H., W. Mauser & H.-J. Stibig 1984. Die 7-Kanal-Satelliten setzen Maßstäbe: Deutschland wird neu entdeckt. *Bild der Wissenschaft* 21 (10): 48-56.
Environmental Protection Agency 1983. *Projecting Future Sea Level Rise*: 121 pp. EPA 230-09-007, Washington, D.C.
Erchinger, H.F. 1971. Landgewinnung und Lahnungsbau im Wattgebiet. *Die Küste* 21: 102-108.
Evans, G. 1975. Intertidal Flat Deposits of the Wash, Western Margin of the North Sea. In R.N. Ginsburg (ed.), *Tidal Deposits - A Casebook of Recent Examples and Fossil Counterparts*: 13-20.
Falk, F.J. 1983. Die Grönlandfahrer der Nordseeinsel Rømø. *Studien und Materialien* 17: 279 pp. Bredstedt: Nordfriisk Instituut.
Falk, F.J. 1984. *Die Seefahrer von St.Johannis. Eine Föhrer Gemeinde zur Walfangzeit*: 112 pp. Bredstedt: Nordfriisk Instituut.
Feldmann, R.W. 1977. *Borkum in alten Ansichten*: 96 pp. Borkum: Selbstverlag.
Feldmann, R.W. 1980. *Grüsse aus Borkum – 100 Ansichten von anno dazumal*: 104 pp. Borkum: Burchana.
Felix-Henningsen, P. 1979. *Merkmale, Genese und Stratigraphie fossiler und reliktischer Bodenbildungen in saalezeitlichen Geschiebelehmen Schleswig-Holsteins und Süd-Dänemarks*: 219 pp. Kiel: Unpublished Thesis.
Figge, K. 1980. Das Elbe-Urstromtal im Bereich der Deutschen Bucht (Nordsee). *Eiszeitalter und Gegenwart* 30: 203-211.
Figge, K. 1983. Morainic deposits in the German Bight area of the North Sea. In J. Ehlers (ed.), *Glacial Deposits in North-west Europe*: 299-304.
Figge, K., R. Köster, H. Thiel & P. Wieland 1980. Schlickuntersuchungen im Wattenmeer der Deutschen Bucht - Zwischenbericht über ein Forschungsvorhaben des KFKI. *Die Küste* 35: 187-204.
Finley, R.F. 1976. Ebb-tidal delta morphology and sediment supply in relation to seasonal wave energy flux, North Inlet, South Carolina. *Journal of Sedimentary Petrology* 48 (1): 227-238.

Fischer, O. 1955. Nordfriesland. *Das Wasserwesen an der schleswig-holsteinischen Nordseeküste, Dritter Teil: Das Festland* 2: 416 pp. Berlin: Reimer.
Fischer, O. 1956. Eiderstedt. *Das Wasserwesen an der schleswig-holsteinischen Nordseeküste, Dritter Teil: Das Festland* 3: 328 pp. Berlin: Reimer.
Fischer, O. 1957. Dithmarschen. *Das Wasserwesen an der schleswig-holsteinischen Nordseeküste, Dritter Teil: Das Festland* 5: 328 pp. Berlin: Reimer.
Fischer, O. 1958. Stapelholm und Eiderniederung. *Das Wasserwesen an der schleswig-holsteinischen Nordseeküste, Dritter Teil: Das Festland* 4: 268 pp. Berlin: Reimer.
Fisher, J. 1968. Barrier island formation: Discussion. *Geological Society of America, Bulletin* 79: 1421-1426.
FitzGerald, D.M., S. Penland & D. Nummedal 1984. Control of barrier island shape by inlet sediment bypassing: East Frisian Islands, West Germany. *Marine Geology* 60: 355-376.
Flessner, G. 1981. Vorwort zu den Gutachten bezüglich der geplanten Küstenschutzmaßnahmen in der Nordstrander Bucht. In Der Minister für Ernährung, Landwirtschaft und Forsten des Landes Schleswig-Holstein (ed.), *Gutachten zur geplanten Vordeichung der Nordstrander Bucht: Schriftenreihe der Landesregierung Schleswig-Holstein* 12: 5-9.
Flinn, D. 1967. Ice front in the North Sea. *Nature* 215: 1151-1154.
Forchhammer, G. 1837. Über dauernde Niveauveränderungen und Spuren von Überfluthungen an der Westküste des Herzogthums Schleswig. *Neues Staatsbürgerliches Magazin* 6: 51 ff.
Frede, G. 1938. Die Arbeiten zur Verbesserung des Fahrwassers der Jade. *Jahrbuch der Hafenbautechnischen Gesellschaft* 16, 1937: 39-46. Berlin: Springer.
Frederiksen, P. 1977. Turistslitage i klitlandskab Skallingen 1976. *Geografisk Tidsskrift* 76: 68-77.
Fu, L.-L. & B. Holt 1982. Seasat Views Oceans and Sea Ice With Synthetic-Aperture Radar. *Jet Propulsion Laboratory Publication* 81-120: 200 pp.
Führböter, A. 1971. Über die Bedeutung des Lufteinschlages für die Energieumwandlung in Brandungszonen. *Die Küste* 21: 34-42.
Führböter, A. 1974. Einige Ergebnisse aus Naturuntersuchungen in Brandungszonen. *Leichtweiss-Institut für Wasserbau der Technischen Universität Braunschweig, Mitteilungen* 40: 331-371.
Führböter, A. 1983. Über mikrobiologische Einflüsse auf den Erosionsbeginn bei Sandwatten. *Wasser und Boden* 35 (3). 106-116.
Führböter, A. & J. Jensen 1985. Säkularänderungen der mittleren Tidewasserstände in der Deutschen Bucht. *Die Küste* 42: 78-100.
Führböter, A., R. Köster, J. Kramer, J. Schwitters & J. Sindern 1976. Beurteilung der Sandvorspülung 1972 und Empfehlungen für die künftige Stranderhaltung am Westrand der Insel Sylt. *Die Küste* 29: 23-95.
Fülscher, J. 1905. Über Schutzbauten zur Erhaltung der ost- und nordfriesischen Inseln. *Zeitschrift für Bauwesen* LV: 305-342, 528-562, 681-722.
Gabrielsen, E.K. & J. Iversen 1933. Die Flora von Skallingen. *Dansk Botanisk Tidsskrift* 42: 255-283.
Gast, R., R. Köster & K.-H. Runte 1984. Die Wattsedimente in der nördlichen und mittleren Meldorfer Bucht. *Die Küste* 40: 165-257.
Gaye, J. 1934. Entwicklung und Erhaltung der Ostfriesischen Inseln. *Zentralblatt der Bauverwaltung* 54 (22): 293-300.
Gaye, J. & F. Walther 1935. Die Wanderung der Sandriffe vor den ostfriesischen Inseln. *Die Bautechnik* XIII: 555-567.

Geerz, F. 1859. *Geschichte der geographischen Vermessungen und der Landkarten Nordalbingiens vom Ende des 15. Jahrhunderts bis zum Jahre 1859*: 277 pp. Berlin.

Gibbard, P. 1977. Pleistocene history of the Vale of St. Albans. *Philosophical Transactions of the Royal Society of London, B. Biological Sciences* 280 (975): 443-483.

Gierloff-Emden, H.G. 1954. Die morphologischen Wirkungen der Sturmflut vom 1. Februar 1953 in den Westniederlanden. *Hamburger Geographische Studien* 4: 23 pp.

Gierloff-Emden, H.G. 1961. *Luftbild und Küstengeographie am Beispiel der deutschen Nordseeküste*: 117 pp. Bad Godesberg: Institut für Landeskunde in der Bundesanstalt für Landeskunde und Raumforschung.

Gilbert, G.K. 1885. The Topographic Features of Lake Shores. *U.S. Geological Survey 5th Annual Report*: 87-88. Reprint in: Schwartz (1973).

Giller, H. 1981. *Morphodynamik der Außensände vor der Insel Amrum und der Halbinsel Eiderstedt – Ergebnisse quantitativ-morphometrischer Arbeitsmethoden und Kartenvergleiche*: 65 pp. Hamburg: Unpublished Diplomarbeit.

Ginsburg, R.N. (ed.) 1975. *Tidal Deposits – A Casebook of Recent Examples and Fossil Counterparts*: 428 pp. Berlin, Heidelberg, New York: Springer.

Göhren, H. 1965. Beitrag zur Morphologie der Jade- und Wesermündung. *Die Küste* 13: 140-146.

Göhren, H. 1968. Triftströmungen im Wattenmeer. *Mitteilungen des Franzius-Instituts für Grund- und Wasserbau der Technischen Universität Hannover* 30: 142-270.

Göhren, H. 1969 a. Die Strömungsverhältnisse im Elbmündungsgebiet. *Hamburger Küstenforschung* 6: 83 pp.

Göhren, H. 1969 b. Untersuchungen mit fluoreszierenden Leitstoffen im südlichen Außenelbegebiet. *Hamburger Küstenforschung* 10: 51 pp.

Göhren, H. 1970. Studien zur morphologischen Entwicklung des Elbmündungsgebietes. *Hamburger Küstenforschung* 14: 71 pp.

Göhren, H. 1971. Untersuchungen über die Sandbewegung im Elbmündungsgebiet. *Hamburger Küstenforschung* 19: 71 pp.

Göhren, H. 1974. Über Strömungsverhältnisse und Sandtransport in den Flachwassergebieten vor der südöstlichen Nordseeküste. *Hamburger Küstenforschung* 29: 50 pp.

Göhren, H. 1976. *Hafenprojekt Scharhörn – Bericht des Wissenschaftlichen Auschusses für gesamtökologische Fragen*: 51 pp. Hamburg: Strom- und Hafenbau.

Göhren, H. 1979. Gegenlaufige Restströmungen im Küstenmeer zwischen Amrum und Knechtsand und ihr Einfluß auf die Sandbewegung. In Deutsche Forschungsgemeinschaft (ed.), *Sandbewegung im Küstenraum – Rückschau, Ergebnisse und Ausblick*: 97-111.

Goethe, F. 1983 a. Shelduck (*Tadorna tadorna* (L.)). In C.J. Smit & W.J. Wolff (eds.), *Birds of the Wadden Sea*: 37-48.

Goethe, F. 1983 b. Black-headed Gull (*Larus ridibundus* (L.)). In C.J. Smit & W.J. Wolff (eds.), *Birds of the Wadden Sea*: 219-229.

Goethe, F. 1983 c. Herring Gull (*Larus argentatus* (P.)). In C.J. Smit & W.J. Wolff (eds.), *Birds of the Wadden Sea*: 238-250.

Goldsmith, V. 1985. Coastal Dunes. In R.A. Davis (ed.), *Coastal Sedimentary Environments*: 303-378.

Graaff, J. van de 1979. Kustmorfologie in Nederland. *De Ingenieur* 91 (18): 327-332.

Grafenstein, U. von 1984. Zur Aussagekraft von Oszillationsrippeln: Ereignisbezogene sedimentologische Untersuchungen in Gebieten mit unterschiedlichen Seegangsspektren in der Nord- und Ostsee. *Berichte – Reports, Geologisch-Paläontologisches Institut der Universität Kiel* 7: 127 pp.

Greene, H.G. 1970. Microrelief of an arctic beach. *Journal of Sedimentary Petrology* 40: 419-427.

Griede, J.W. 1978. *Het ontstaan van Frieslands noordhoek*: 186 pp. Amsterdam: Thesis. Amsterdam: Rodopi.

Gripp, K. 1944. Entstehung und künftige Entwicklung der Deutschen Bucht. *Aus dem Archiv der Deutschen Seewarte und des Marineobservatoriums* 63 (2): 45 pp.

Gripp, K. 1956. Das Watt; Begriff, Begrenzung und fossile Vorkommen. *Senckenbergiana lethaea* 37 (3/4): 149-181.

Gripp, K. 1963. Winter-Phänomene am Meeresstrand. *Zeitschrift für Geomorphologie* N.F. 7: 326-331.

Gripp, K. 1964. *Erdgeschichte von Schleswig-Holstein*: 411 pp. Neumünster: Wachholtz.

Gripp, K. 1968. Zur jüngsten Erdgeschichte von Hörnum/Sylt und Amrum mit einer Übersicht über die Entstehung der Dünen in Nordfriesland. *Die Küste* 16: 76-117.

Gripp, K., F. Dewers & F. Overbeck 1941. Das Känozoikum in Niedersachsen (Tertiär, Diluvium, Alluvium und Moore). *Schriften der wirtschaftswissenschaftlichen Gesellschaft zum Studium Niedersachsens e.V.*, Neue Folge 3: Geologie und Lagerstätten Niedersachsens, Dritter Band: 503 pp. Oldenburg: Stalling.

Gripp, K. & L. Martens (1963). *Wenn die Natur im Sande spielt..*: 54 pp. Hamburg: Verlag der Gesellschaft der Freunde des vaterländischen Schul- und Erziehungswesens e.V.

Gronwald, W. 1960. Welche Erkenntnisse zur Frage der vermuteten neuzeitlichen Nordseeküstensenkung hat die Wiederholung des Deutschen Nordseeküsten-Nivellements gebracht? *Die Küste* 8: 66-82.

Grube, F. 1982. Die Holstein-Warmzeit von Hamburg. In D.J. Easterbrook, P. Havlicek, K.-D. Jäger & F.W. Shotton (eds.), *IGCP-Project 73/1/24, Quaternary Glaciations in the Northern Hemisphere, Report No.7 on the Session in Kiel, F.R.G., September, 18-23, 1980*: 84-92.

Haage, R. 1899. *Die deutsche Nordseeküste in physikalisch-geographischer und morphologischer Hinsicht, nebst einer kartometrischen Bestimmung der deutschen Nordseewatten*: 83 pp. Leipzig: Thesis.

Haan, H. de, R. Ijbema & D.T. Reitsma 1983. *Engelsmanplaat - Geschiedenis van en gebeurtenissen rond een zandbank*. 2nd. edition: 168 pp. Moddergat: Stichting 't Fiskerhuske.

Haarnagel, W. 1961. Die Marschen im deutschen Küstengebiet der Nordsee und ihre Besiedlung. *Berichte zur Deutschen Landeskunde* 27: 203-219.

Hacht, U. von 1979. Neue Beobachtungen an Gesteinen aus Braderup, Sylt. *Natur und Museum* 109 (1): 10-17.

Häntzschel, W. 1936. Die Schichtungs-Formen rezenter Flachmeer-Ablagerungen im Jade-Gebiet. *Senckenbergiana* 18 (5/6): 316-356.

Häntzschel, W. 1938. Bau und Bildung von Groß-Rippeln im Wattenmeer. *Senckenbergiana* 20 (1/2): 1-42.

Hagedorn, H. 1965. Die morphologische Wirkung der Sturmflut vom 16./17. Februar 1962 auf die Küste bei Duhnen im Elb-Weser-Winkel. *Petermanns Geographische Mitteilungen* 109 (2): 103-112.

Hageman, B.P. 1969. Development of the Western Part of the Netherlands during the Holocene. *Geologie en Mijnbouw* 48 (4): 373-388.

Halbertsma, H. 1955. Enkele oudheidkundige aantekeningen over het ontstaan en de toeslijking van de Middelzee. *Tijdschrift van het Koninklijk Nederlandsch Aardrijkskundig Genootschap*, Tweede Reeks LXXII: 93-105.

Halem, F.W. von 1822. *Die Insel Norderney und ihr Seebad nach dem gegenwärtigen Standpuncte*: 240 pp. Hannover: Hahn. Reprint 1974 Leer: Schuster.

Hallewas, D.P. 1984. The interaction between man and his physical environment in the County of Holland between circa 1000 and 1300 AD: A dynamic relationship. *Geologie en Mijnbouw* 63 (3): 299-307.

Hamel, G. 1981. Erhaltung von Salzwasserbiotopen bei Baumaßnahmen im Küstengebiet. *Wasser und Boden* 33 (3): 121-124.

Hanisch, J. 1980. Neue Meeresspiegeldaten aus dem Raum Wangerooge. *Eiszeitalter und Gegenwart* 30: 221-228.

Hanisch, J. 1981. Sand transport in the tidal inlet between Wangerooge and Spiekeroog (W. Germany). In S.-D. Nio, R.T.E. Schüttenhelm & Tj.C.E. van Weering (eds.), *Holocene Marine Sedimentation in the North Sea Basin. Special Publication Number 5 of the International Association of Sedimentologists*: 175-185.

Hansen, C.P. 1859. *Die nordfriesische Insel Sylt, wie sie war und wie sie ist. Ein Handbuch für Badegäste und Reisende*: 196 pp. Leipzig: Weber. Reprint 1982 Schaan (FL): Sändig.

Hansen, C.P. 1865. *Das Schleswig'sche Wattenmeer und die friesischen Inseln*: 277 pp. Glogau: Flemming. Reprint 1972, Walluf bei Wiesbaden: Sändig.

Hansen, C.P. 1877. *Chronik der Friesischen Uthlande*, 2nd edition: 319 pp. Garding: Lühr & Diercks. Reprint 1984 Vaduz: Sändig.

Hansen, M. & N. (eds.) 1969. *Amrum – Geschichte und Gestalt einer Insel*, 2nd. edition: 191 pp. Itzehoe-Münsterdorf: Hansen & Hansen.

Hansen, N. 1814. Die Beschreibung von der Insel Gröde. In J. Lorenzen 1979, *Eine Beschreibung der Hallig Gröde aus dem Jahre 1814 von Nommen Hansen*: 1-22. Bredstedt: Nordfriisk Instituut.

Hansen, R. 1894 a. Beiträge zur Geschichte und Geographie Nordfrieslands im Mittelalter. *Zeitschrift der Gesellschaft für Schleswig-Holsteinisch-Lauenburgische Geschichte* 24.

Hansen, R. 1894 b. Die Hallig Habel. *Dr. A. Petermanns Mitteilungen aus Justus Perthes' Geographischer Anstalt* 40: 117.

Harck, O. 1974. Zur Datierung des Listlandes und der Hörnumer Halbinsel auf Sylt. *Meyniana* 24: 69-72.

Harting, C.P. 1980. De Zeereep in beweging. In J.C. Duinker (ed.), *Texel en de Zee. Een strijd van Eeuwen*: 41-57.

Hartung, W. 1964. Das Problem der sog. Küstensenkung - Schütte's wissenschaftliches Lebenswerk in seiner Bedeutung und der Sicht neuerer Forschung. *Oldenburger Jahrbuch* 63: 131-153.

Hartung, W. 1975. Mellum als eine werdende Nordseeinsel. In P. Blaszyk (ed.), *Naturschutzgebiete im Oldenburgerland*: 11-27.

Hartung, W. 1981. *Wasserbau und Naturschutz in Grenzsituationen - Vortrag bei der Festveranstaltung zum 75jährigen Jubiläum des Wasserwirtschaftsamtes Aurich der Niedersächsischen Wasserwirtschaftsverwaltung*: 12 pp. Aurich.

Hartung, W. 1983. Die Leybucht (Ostfriesland) – Probleme ihrer Erhaltung als Naturschutzgebiet. *Neues Archiv für Niedersachsen* 32 (4): 355-387.

Hayes, M.O. 1975. Morphology of sand accumulations in estuaries. In L.E. Cronin (ed.), *Estuarine Research*, Vol.2: 3-22.

Hayes, M.O. 1979. Barrier Island Morphology as a Function of Tidal and Wave Regime. In S.P. Leatherman (ed.), *Barrier Islands*: 1-27.

Heck, H.-L. 1937. Die nordfriesische neuzeitliche Küstensenkung als Folge diluvialer Tektonik. *Jahrbuch der Preußischen Geologischen Landesanstalt* 1936, 57 (1): 48-84.

Heck, H.-L. 1947. Der altinterglaziale Haffsee bei Bredstedt in Nordfriesland. *Geologie der Meere und Binnengewässer* 7.

Heide, G.D. van der 1955. Bewoningsfasen van het gebied van de latere Zuiderzee. *Tijdschrift van het Koninklijk Nederlandsch Aardrijkskundig Genootschap*, Tweede Reeks LXXII: 39-47.

Heimreich, M.A. 1819. *Nordfresische Chronik*, 2 vols.: 446 + 348 pp. Tondern: Falck. Reprint 1979. Wiesbaden: Sändig.

Hempel, L. 1980. Zur Genese von Dünengenerationen an Flachküsten. Beobachtungen auf den Nordseeinseln Wangerooge und Spiekeroog. *Zeitschrift für Geomorphologie* N.F. 24 (4): 428-447.

Hempel, L. 1983. Der Sandhaushalt als Hauptglied in der Geoökodynamik einer Ostfriesischen Insel. Abhängigkeiten von natürlichen und anthropogenen Kräften. *Geoökodynamik* 4: 87-104.

Hennig, R. 1897. *Untersuchungen über die Sturmfluten der Nordsee*: 42 pp. Friedrich-Wilhelms-Universität Berlin: Thesis. Berlin: Bernstein.

Hesp, P.A. 1981. The formation of shadow dunes. *Journal of Sedimentary Petrology* 51 (1): 101-112.

Heydemann, B. 1981 a. Ökologie und Schutz des Wattenmeeres. *Schriftenreihe des Bundesministers für Ernährung, Landwirtschaft und Forsten, Reihe A: Angewandte Wissenschaft* 255: 232 pp.

Heydemann, B. 1981 b. Biological Consequences of Diking on Saltmarshes and Mud and Sandy Flats. In S. Tougaard & C. Helweg Ovesen (eds.), *Environmental Problems of the Waddensea Region - Proceedings of the Scientific Symposium, Ribe, Denmark, 16-18 May 1979*: 31-71.

Higelke, B. 1978. Morphodynamik und Materialbilanz im Küstenvorfeld zwischen Hever und Elbe. Ergebnisse quantitativer Kartenanalysen für die Zeit von 1936 bis 1969. *Regensburger Geographische Schriften* 11: 167 pp.

Higelke, B. 1981. Bestandsaufnahme des Wattreliefs, Morphodynamik und Tendenzen morphologischer Veränderungen im Tidebecken der Norderhever und westlich der Insel Pellworm - Luftbildinterpretation. In Der Minister für Ernährung, Landwirtschaft und Forsten des Landes Schleswig-Holstein (ed.), *Gutachten zur geplanten Vordeichung der Nordstrander Bucht. Schriftenreihe der Landesregierung Schleswig-Holstein* 12: 155-174.

Higelke, B. 1982. Geographische Untersuchungen. In B. Higelke, D. Hoffmann & M. Müller-Wille (eds.), *Das Norderhever-Projekt. Beiträge zur Landschafts- und Siedlungsgeschichte der nordfriesischen Marschen und Watten*. Offa 39: 261-270.

Higelke, B., D. Hoffmann & M. Müller-Wille (eds.) 1982. *Das Norderhever-Projekt. Beiträge zur Landschafts- und Siedlungsgeschichte der nordfriesischen Marschen und Watten*. Offa 39: 245-270.

Hjulström, F. 1935. Studies of the morphological activity of rivers as illustrated by the River Fyris. *Bulletin of the Geological Institutions of the University of Uppsala* 25: 221-527.

Höfle, H.-Chr., J. Merkt & H. Müller 1985. Die Ausbreitung des Eem-Meeres in Nordwestdeutschland. *Eiszeitalter und Gegenwart* 35: 49-59.

Hoffmann, D. 1969. The marine Holocene of Sylt – Discussion of the age and facies. *Geologie en Mijnbouw* 48 (3): 343-347.

Hoffmann, D. 1974. Zum geologischen Aufbau der Hörnumer Halbinsel auf Sylt. *Meyniana* 24: 63-68.

Hoffmann, D. 1982. Ältere Landschaftsgeschichte. In B. Higelke, D. Hoffmann & M. Müller-Wille (eds.): *Das Norderhever-Projekt. Beiträge zur Landschafts- und Sied-*

lungsgeschichte der nordfriesischen Marschen und Watten. Offa 39: 245-254.
Hoffmann, D. 1984. The coastal zone of North Friesland. In INQUA Subcommission on Shorelines of Northwestern Europe (ed.), *Field Conference 1984, September 15–21, North Sea Coastal Zone between Jade Bay and Jammer Bight*: 59-71.
Hofmeister, B. & A. Steinecke (eds.) 1980. Beiträge zur Geomorphologie und Länderkunde – Prof. Dr. Hartmut Valentin zum Gedächtnis. *Berliner Geographische Studien* 7: 350 pp.
Holden, M.K. 1980. *Sand Wave Development and Textural Variation on a Flood Tidal Delta*: 280 pp. Unpublished MS Thesis, Boston College Graduate School.
Holdgate, M.W. & M.J. Woodman 1978. *The Breakdown and Restoration of Ecosystems*. Proceedings of the Conference on the Rehabilitation of Severely Damaged Land and Freshwater Ecosystems in Temperate Zones held in Reykjavik, Iceland, July 4-10, 1976: 496 pp. Plenum: New York.
Holstein, U. 1979. *Esbjerg – Et tyngdepunkt i det tyske vestkystforsvar 1940-1945*: 94 pp. Zac.
Homann, H. (ca. 1973). *Borkum*: 103 pp. Münster: Coppenrath.
Homeier, H. 1962. Historisches Kartenwerk 1:50,000 der niedersächsischen Küste. *Forschungsstelle Norderney, Jahresbericht* 1961, XIII: 11-29.
Homeier, H. 1963 a. Die Strandinsel Lütje Hörn an der Osterems. *Forschungsstelle Norderney, Jahresbericht* 1962, XIV: 41-46.
Homeier, H. 1963 b. *Beiheft zu: Niedersächsische Küste, Historische Karte 1:50,000 Nr.6*: 29 pp. Norderney: Forschungsstelle Norderney.
Homeier, H. 1964. *Beiheft zu: Niedersächsische Küste, Historische Karte 1:50,000 Nr.5*: 28 pp. Norderney: Forschungsstelle Norderney.
Homeier, H. 1965. Historisch-morphologische Untersuchungen der Forschungsstelle Norderney über langfristige Gestaltungsvorgänge im Bereich der niedersächsischen Küste. *Forschungsstelle Norderney, Jahresbericht* 1964, XVI: 7-40.
Homeier, H. 1968. Die Strandentwicklung der Insel Memmert. *Forschungsstelle Norderney, Jahresbericht* 1966, XVIII: 9-36.
Homeier, H. 1969. Das Wurster Watt – eine historisch-morphologische Untersuchung des Küsten- und Wattgebietes von der Weser- bis zur Elbmündung. *Forschungsstelle Norderney, Jahresbericht* 1967, XIX: 31-120.
Homeier, H. 1971. Untersuchung der Strandentwicklung Borkums unter besonderer Berücksichtigung der jüngsten Strandabbrüche im Bereich der Süddünen. *Forschungsstelle für Insel- und Küstenschutz, Jahresbericht* 1969, XXI: 7-22.
Homeier, H. 1973. Die morphologische Entwicklung im Bereich der Harle und ihre Auswirkungen auf das Westende von Wangerooge. *Forschungsstelle für Insel- und Küstenschutz, Jahresbericht* 1972, XXIV: 15-44.
Homeier, H. 1974. *Beiheft zu: Niedersächsische Küste, Historische Karte 1:50,000 Nr. 8*: 20 pp. Norderney: Forschungsstelle für Insel- und Küstenschutz.
Homeier, H. 1976. Die Auswirkungen schwerer Sturmtiden auf die Ostfriesischen Inselstrände und Randdünen. *Forschungsstelle für Insel- und Küstenschutz, Jahresbericht* 1975, XXVII: 107-122.
Homeier, H. 1979 a. Die Verlandung der Harlebucht bis 1600 auf der Grundlage neuer Befunde. *Forschungsstelle für Insel- und Küstenschutz, Jahresbericht* 1978, XXX: 106-115.
Homeier, H. 1979 b. *Beiheft zu: Niedersächsische Küste, Historische Karte 1:50,000 Nr. 1*: 34 pp. Norderney: Forschungsstelle für Insel- und Küstenschutz.
Homeier, H. 1980. Morphologische Entwicklung der Insel Baltrum und Veränderungen im Bereich der Wichter Ee. *Forschungsstelle für Insel- und Küstenschutz, Jahresbericht* 1979, 31: 11-36.
Homeier, H. & J. Kramer 1957. Verlagerung der Platen im Riffbogen vor Norderney und ihre Anlandung an den Strand. *Forschungsstelle Norderney, Jahresbericht* 1956, VIII: 37-60.
Homeier, H. & G. Luck 1971. Untersuchung morphologischer Gestaltungsvorgänge im Bereich der Accumer Ee als Grundlage für die Beurteilung der Strand- und Dünenentwicklung im Westen und Nordwesten Langeoogs. *Forschungsstelle für Insel- und Küstenschutz, Jahresbericht* 1970, XXII: 7-42.
Homeier, H. & G. Luck 1977. Untersuchungen zur Nordstrandentwicklung von Borkum als Grundlage für den Inselschutz. *Forschungsstelle für Insel- und Küstenschutz, Jahresbericht* 1976, XXVIII: 83-100.
Hoyt, J. 1967. Barrier island formation. *Geological Society of America, Bulletin* 78: 1125-1136.
Hubbard, D.K. 1975. Morphology and Hydrodynamics of the Merrimack River Ebb-Tidal Delta. In L.E. Cronin (ed.), *Estuarine Research*, Vol. 2: 253-266.
Hübbe, H. 1861. Von der Beschaffenheit und dem Verhalten des Sandes. *Zeitschrift für Bauwesen* XI: 19-42, 183-204.
Hülsemann, J. 1955. Großrippeln und Schrägschichtungs-Gefüge im Nordsee-Watt und in der Molasse. *Senckenbergiana lethaea* 36 (5/6): 359-388.
Hummel, P. & E. Cordes 1969. Holozäne Sedimentation und Faziesdifferenzierung beim Aufbau der Lundener Nehrung (Norderdithmarschen). *Meyniana* 19: 103-112.
Hylgaard, T. 1980. Recovery of plant communities on coastal sand dunes disturbed by human trampling. *Biological Conservation* 19: 15-25.
Hylgaard, T. & M.J. Liddle 1981. The effect of human trampling on a sand dune ecosystem dominated by *Empetrum nigrum*. *Journal of Applied Ecology* 18: 559-569.
INQUA Subcommission on Shorelines of Northwestern Europe (eds.) 1984. *Field Conference 1984, September 15–21, North Sea Coastal Zone between Jade Bay and Jammer Bight*: 105 pp. Hannover: Niedersächsisches Landesamt für Bodenforschung.
Isbary, G. 1936. Das Inselgebiet von Ameland bis Rottumeroog. Morphologische und hydrographische Beiträge zur Entwicklungsgeschichte der friesischen Inseln. *Aus dem Archiv der Deutschen Seewarte* 56 (3): 55 pp.
Iversen, J. 1936. *Biologische Pflanzentypen als Hilfsmittel in der Vegetationsforschung*. København: Dissertation.
Iversen, J. 1953. The Zonation of the Salt Marsh Vegetation of Skallingen in 1931-34 and in 1952. *Geografisk Tidsskrift* 52: 113-118.
Jacobsen, N.K. 1953. Mandø. Et klit-marskø i Vadehavet. *Geografisk Tidsskrift* 52: 134-146.
Jacobsen, N.K. 1969. Landskabsformerne. *Meddelelser fra Skalling-Laboratoriet* XXII: 3-22.
Jacobsen, N.K. 1976. Stormfloderne langs Vadehavskysten januar 1976. *Bygd* 7 (1): 3-11.
Jacobsen, N.K. 1977. Rekreation i det Danske Vadehav. *Geografisk Tidsskrift* 76: 52-58.
Jacobsen, N.K. 1978. The Balance between Agriculture, Forestry, Urbanisation and Conservation: Optimal Pattern of Land Use. In M.W. Holdgate & M.J. Woodman (eds.), *The*

Breakdown and Restoration of Ecosystems: 391-412.
Jacobsen, N.K. 1980. Form Elements of the Wadden Sea Area. In K.S. Dijkema, H.-E. Reineck & W.J. Wolff (eds.), *Geomorphology of the Wadden Sea*: 50-71.
Jäkel, D. 1980. Die Bildung von Barchanen in Faya-Largeau / Rep. du Tchad. *Zeitschrift für Geomorphologie* N.F. 24 (2): 141-159.
Jakobsen, B. 1953. Landskabsudviklingen i Skallingsmarsken. *Geografisk Tidsskrift* 52: 147-158.
Jakobsen, B. 1962. Morfologiske og hydrografiske undersøgelser af flod- og ebbeskar i tidevandsrender. *Geografisk Tidsskrift* 61: 119-141.
Jakobsen, B. 1964. Vadehavets Morfologi. En geografisk analyse af vadelandskabets formudvikling med særlig hensyntagen til Juvre-Dybs tidevandsområde. *Folia Geographica Danica* XI (1): 176 pp.
Jansen, J.H.F. 1976. Late Pleistocene and Holocene History of the Northern North Sea, Based on Acoustic Reflection Records. *Netherlands Journal of Sea Research* 10 (1): 1-43.
Jansen, J.H.F., J.W.C. Doppert, K. Hoogendoorn-Toering, J. de Jong & G. Spaink 1979 a. Late Pleistocene and Holocene Deposits in the Witch and Fladen Ground Area, Northern North Sea. *Netherlands Journal of Sea Research* 13 (1): 1-39.
Jansen, J.H.F., T.C.E. van Weering & D. Eisma 1979 b. Late Quaternary Sedimentation in the North Sea. In E. Oele, R.T.E. Schüttenhelm & A.J. Wiggers (eds.), *The Quaternary History of the North Sea, Acta Universitatis Upsaliensis, Symposia Universitatis Upsaliensis Annum Quingentesimum Celebrantis* 2: 175-187.
Janssen, Th. 1962. *Die Sturmflut am 16. Februar 1962 im Bereich der Wasser- und Schiffahrtsdirektion Aurich – eine Unterlagensammlung*: 42 pp. Aurich: Wasser- und Schiffahrtsdirektion (unpublished).
Janßen, Th. 1933. *Über die Kräfte, die die ostfriesischen Inseln, insbesondere den östlichen Sandstrand der Insel Spiekeroog, gestalten*. Thesis Hannover. Schweidnitz.
Jardine, W.G. 1979. The western (United Kingdom) shore of the North Sea in Late Pleistocene and Holocene times. In E. Oele, R.T.E. Schüttenhelm & A.J. Wiggers (eds.), *The Quaternary History of the North Sea, Acta Universitatis Upsaliensis, Symposia Universitatis Upsaliensis Annum Quingentesimum Celebrantis* 2: 159-174.
Jaritz, W. 1973. Zur Entstehung der Salzstrukturen Nordwestdeutschlands. *Geologisches Jahrbuch* A 10: 77 pp.
Jelgersma, S. 1961. Holocene sea level changes in the Netherlands. *Mededelingen van de Geologische Stichting* C VI (7): 100 pp., Maastricht.
Jelgersma, S. 1979. Sea level changes in the North Sea basin. In E. Oele, R.T.E. Schüttenhelm & A.J. Wiggers (eds.), *The Quaternary History of the North Sea, Acta Universitatis Upsaliensis, Symposia Universitatis Upsaliensis Annum Quingentesimum Celebrantis* 2: 115-142.
Jelgersma, S. 1983. The Bergen Inlet, Transgressive and Regressive Holocene Shoreline Deposits in the Northwestern Netherlands. *Geologie en Mijnbouw* 62 (3): 471-486.
Jelgersma, S. & P.J. Ente 1977. Genese van het Holoceen. In C.J. van Staalduinen (ed.), *Geologisch onderzoek van het Nederlandse Waddengebied*: 23-36.
Jelgersma, S., E. Oele & A.J. Wiggers 1979. Depositional History and coastal development in the Netherlands and the adjacent North Sea since the Eemian. In E. Oele, R.T.E. Schüttenhelm & A.J. Wiggers (eds.), *The Quaternary History of the North Sea, Acta Universitatis Upsaliensis, Symposia Universitatis Upsaliensis Annum Quingentesimum Celebrantis* 2: 115-142.

Jensen, J. 1969. Die Geschichte der Insel Amrum. In M. & N. Hansen (eds.), *Amrum – Geschichte und Gestalt einer Insel*, 2nd. edition: 55-56, 65-72, 81-88, 97-106.
Jensen, J. 1984. Änderungen der mittleren Tidewasserstände an der Nordseeküste. *Mitteilungen des Leichtweiß-Instituts für Wasserbau der Technischen Universität Braunschweig* 83: 435-550.
Jensen, K.-D. 1933. *Das Eiderstedter Alluvium, unter besonderer Berücksichtigung der schleswig-holsteinischen Westküste*: 126 pp. Friedrich-Wilhelms-Universität Berlin: Dissertation.
Jepsen, P.U. 1977. *Jordsand, Vogelinsel im Wattenmeer*: 59 pp. Esbjerg: Bygd.
Jespersen, M. & E. Rasmussen 1984. Geomorphological effects of the Rømø Dam: development of a tidal channel and collapse of a dike. *Geografisk Tidsskrift* 84: 17-24.
Jespersen, M. & E. Rasmussen 1985. En beretning om havets angreb på en ø i det danske vadehav. *Skrifter fra Højer Mølle- og Marskmuseum Kiers Gaard* 2: 24 pp.
Jessen, O. 1914. Morphologische Beobachtungen an den Dünen von Amrum, Sylt und Röm. *Landeskundliche Forschungen* 21: 231-365. München: Geographische Gesellschaft.
Jessen, O. 1922. *Die Verlegung der Flussmündungen und Gezeitentiefs an der festländischen Nordseeküste in jungalluvialer Zeit*: 181 pp. Stuttgart: Enke.
Jessen, W. 1933. Die postdiluviale Entwicklung Amrums und seine subfossilen und rezenten Muschelpflaster (unter Berücksichtigung der gleichen Vorgänge auf den Inseln Sylt und Föhr). *Jahrbuch der Preußischen Geologischen Landesanstalt* 53: 1-69.
Johnson, D.W. 1919. *Shore processes and shoreline development*: 584 pp. New York: Wiley.
Jong, J.D. de 1984. Age and vegetational history of the coastal dunes in the Frisian islands, The Netherlands. *Geologie en Mijnbouw* 63 (3): 269-275.
Jong, J.D. de & G.C. Maarleveld 1983. The glacial history of the Netherlands. In J. Ehlers (ed.), *Glacial Deposits in North-west Europe*: 353-356.
Joustra, D.Sj. 1971. *Geulbeweging in de buitendelta's van de Waddenzee*. Rijkswaterstaat, Studierapport W.W.K. 71-14: 27 pp.
Jürgens, H.-J. 1970. *Der 25. April 1945 – Bombenangriff auf Wangerooge*: 24 pp. Bremerhaven.
Jürgens, H.-J. 1974. Wangerooge muß der Jade wegen erhalten bleiben. *Wangerooger Inselbote* 1974 (9/10): 1-3.
Jürgens, H.-J. 1977. *Wangerooge – Zeugnisse aus alter Zeit*: 112 pp. Harsefeld-Marienwinkel: Rhode.
Kachholz, K.-D. 1982. *Statistische Bearbeitung von Probendaten aus Vorstrandbereichen sandiger Brandungsküsten mit verschiedener Intensität der Energieumwandlung*: 381 pp. Kiel: Unpublished Thesis.
Kachholz, K.-D. 1984. Vergleich einiger sandiger Brandungsküsten Schleswig-Holsteins. *Meyniana* 36: 93-119.
Kaiser, K. 1971. Beobachtungen über Fließmarken an leeseitigen Barchan-Hängen. *Kölner Geographische Arbeiten, Festschrift für K. Kayser* (Sonderband): 65-71.
Kamps, L.F. 1963. Mud distribution and land reclamation in the Eastern Wadden Shallows. *International Institute for Land Reclamation and Improvement, Publication* 9: 91 pp.
Kannenberg, E.G. 1951. Die Steilufer der schleswig-holsteinischen Ostseeküste. *Schriften des Geographischen Instituts der Universität Kiel* XIV (1).
Karff, F. 1968. *Nordstrand. Geschichte einer nordfriesischen Insel*: 332 pp. Flensburg: Wolff.
Keil, G.-W. 1985. Die schrittweise Anpassung der Elbe an die

Entwicklung des Seeschiffsverkehrs. *Jahrbuch der Hafenbautechnischen Gesellschaft* 40, 1983/84: 47-60. Heidelberg, New York, Tokyo: Springer.

Keilhack, K. 1925. *Geologische Karte von Preußen und benachbarten deutschen Ländern 1:25,000, Erläuterungen zu Lieferung 259, Blatt Borkum, Juist-West, Juist-Ost und Norderney*: 33 pp. Berlin: Preußische Geologische Landesanstalt.

Kemstedt, D. 1981. *Untersuchung der Morphodynamik im Rummelloch West und auf den angrenzenden Wattflächen*: 85 pp. Hamburg: Unpublished Diplomarbeit.

Kenyon, N.H., R.H. Belderson, A.H. Stride & M.A. Johnson 1981. Offshore tidal sand-banks as indicators of net sand transport and as potential deposits. In S.-D. Nio, R.T.E. Schüttenhelm & Tj.C.E. van Weering (eds.), *Holocene Marine Sedimentation in the North Sea Basin. Special Publication Number 5 of the International Association of Sedimentologists*: 257-268.

Kersten, K. 1969. Die Vorzeit der Insel Amrum. In M. & N. Hansen (eds.), *Amrum – Geschichte und Gestalt einer Insel*: 7-11.

Kielholt, H. (ca. 1430). Silter Antiquitäten. In M.A. Heimreich 1819, *Nordfresische Chronik*, Vol. 2: 343-348. Tondern: Falck. Reprint 1979. Wiesbaden: Sändig.

Kiewiet, R.T. 1985. Het Oerd en de Hon op Ameland. *Waddenbulletin* 20 (1): 11-13.

Kirchner, H. 1974. Die Sedimentverteilung des strandnahen Seebereiches vor Westerland/Sylt. *Meyniana* 24: 57-62.

Klein, G. deVries 1977. *Clastic Tidal Facies*: 149 pp. Champaign (Ill.): Continuing Education Publication Company.

Kleinkemm, H. 1910. *Die Insel Texel*: 102 pp. Giessen: Thesis.

Klijn, J.A. 1981. *Nederlandse kustduinen. Geomorfologie en bodems*: 188 pp. Wageningen: Pudoc.

Klok, B. & K.M. Schalkers 1980. *De veranderingen in de Waddenzee ten gevolge van de afsluiting van de Zuiderzee*. Rijkswaterstaat, Studiedienst Hoorn, Notitie 78.H238: 12 pp.

Klug, H. & B. Higelke 1979. Ergebnisse geomorphologischer Seekartenanalysen zur Erfassung der Reliefentwicklung und des Materialumsatzes im Küstenvorfeld zwischen Hever und Elbe 1936-1969. In Deutsche Forschungsgemeinschaft (eds.), *Sandbewegung im Küstenraum – Rückschau, Ergebnisse und Ausblick*: 125-145.

Knop, F. 1961. Untersuchungen über Gezeitenbewegung und morphologische Veränderungen im nordfriesischen Wattgebiet als Vorarbeiten für Dammbauten. *Mitteilungen aus dem Leichtweiß-Institut für Wasserbau und Grundbau der Technischen Hochschule Braunschweig* 1961 (1): 123 pp.

Knop, F. 1963. Küsten- und Wattveränderngen Nordfrieslands - Methoden und Ergebnisse ihrer Überwachung. *Die Küste* 11: 1-33.

Kobbe, Th. von & W. Cornelius 1841. *Wanderungen an der Nord- und Ostsee*: 116 + 128 pp. Leipzig: Wigand. Reprint 1973 Hildesheim, New York: Olms Presse.

Koch, M. & G. Luck 1973. Untersuchungen zur Erfassung der Strömungsverhältnisse auf den östlichen Weserwatten. *Forschungsstelle für Insel- und Küstenschutz, Jahresbericht* 1972, XXIV: 69-75.

Koch, M. & H.D. Niemeyer 1980. Strömungsmessungen im Bereich der Wattwasserscheiden von Norderney und Baltrum sowie im Seegat Wichter Ee. *Forschungsstelle für Insel- und Küstenschutz, Jahresbericht* 1979, XXXI: 37-55.

König, D. 1972. Deutung von Luftbildern des schleswig-holsteinischen Wattenmeeres, Beispiele und Probleme. *Die Küste* 22: 29-74.

Köster, R. 1971. Dreidimensionale Kartierung des Seegrundes vor den Nordfriesischen Inseln. In Deutsche Forschungsgemeinschaft (eds.) 1971, *Sandbewegung im Küstenraum. Neue Forschungsergebnisse zum Problem der Sandbewegung vor den Küsten der Ost- und Nordsee – Ein Zwischenbericht*: 25-33.

Köster, R. 1974. Geologie des Seegrundes vor den Nordfriesischen Inseln Sylt und Amrum. *Meyniana* 24: 27-41.

Köster, R. 1979. Dreidimensionale Kartierung des Seegrundes vor den Nordfriesischen Inseln. In Deutsche Forschungsgemeinschaft (eds.) 1979, *Sandbewegung im Küstenraum – Rückschau, Ergebnisse und Ausblick*: 146-168.

Köster, R. 1981. Geologisches Gutachten zu den geplanten Küstenschutzmaßnahmen im südlichen nordfriesischen Wattenmeer. In Der Minister für Ernährung, Landwirtschaft und Forsten des Landes Schleswig-Holstein (ed.) 1981, *Gutachten zur geplanten Vordeichung der Nordstrander Bucht*: 89-131.

Kolb, A. (ed.) 1962. Sturmflut 17. Februar 1962 – Morphologie der Deich- und Flurbeschädigungen zwischen Moorburg und Cranz. *Hamburger Geographische Studien* 16: 27 pp.

Kolumbe, E. 1943. Über die Entwicklung einiger Dünenkleinformen auf Sylt. *Zeitschrift für Geschiebeforschung und Flachlandsgeologie* 18 (2): 116-125, 232-235.

Komar, P.D. 1976. *Beach Processes and Sedimentation*: 429 pp. Englewood Cliffs: Prentice-Hall.

Komar, P.D. (ed.) 1983. *CRC Handbook of Coastal Processes and Erosion*: 305 pp. Baton Rouge (Fla.): CRC Press.

Kooiker, E.P. 1981. *Noordzeewering Ameland*: 14 pp. Rijkswaterstaat, Directie Friesland. Studiedagen Kring van Zeewerende Ingenieurs, 26 en 27 Oktober 1981, Ameland.

Kool, G., R. Peereboom, M.F. Lieshout & M. de Boer 1984. *Waterbeweging westelijke Waddenzee: Verloop natte en droge oppervlakten en komberginger*. Rijkswaterstaat, Nota WWKZ - 84.H009: 13 pp.

Küstenausschuß Nord- und Ostsee. (ehem.) Arbeitsgruppe Schutzwerke an sandigen Küsten 1981. Schutz sandiger Küsten an Nord- und Ostsee – Bestandsaufnahme und kritische Wertung. *Die Küste* 36: 321-364.

Krämer, L. (1984). *Sylt im Kampf mit Meer und Sand*: 32 pp. Söl'ring Foriining e.V.: Keitum.

Kramer, J. 1959. Die Strandaufspülung Norderney 1951–1952 und ein Plan zu ihrer Fortführung. *Die Küste* 7: 107-139.

Kramer, J. 1961. Natürliche Entwicklung des Großen Knechtsandes und seine Bedeutung für den Küstenschutz. *Forschungsstelle Norderney, Jahresbericht* 1960, XII: 81-86.

Kramer, J. 1983. *Sturmfluten – Küstenschutz zwischen Ems und Weser*: 172 pp. Norden: Soltau Verlag.

Krog, H. 1979. Late Pleistocene and Holocene shorelines in Western Denmark. In E. Oele, R.T.E. Schüttenhelm & A.J. Wiggers (eds.), *The Quaternary History of the North Sea, Acta Universitatis Upsaliensis, Symposia Universitatis Upsaliensis Annum Quingentesimum Celebrantis* 2: 75-83.

Krohn, H. 1949. *Die Bevölkerung der Insel Sylt*. Kiel: Dissertation. Published by Noordfriisk Instituut, Studien und Materialien Nr. 14: 195 pp. Bredstedt, 1984.

Krüger, W. 1911. Meer und Küste bei Wangeroog und die Kräfte, die auf ihre Gestaltung einwirken. *Zeitschrift für Bauwesen* LXI: 451-464, 584-610.

Krüger, W. 1929. Die heutige Insel Wangeroog, ein Ergebnis des Seebaues. In Landesverein Oldenburg für Heimatkunde und Heimatschutz (eds.), *Wangeroog wie es wurde, war und ist*: 179-224.

Krüger, W. 1937. Die Entwicklung der Harlebucht und ihr Einfluß auf die Außenjade. *Abhandlungen herausgegeben vom*

Naturwissenschaftlichen Verein zu Bremen XXX (1/2): 197-208.

Krüger, W. 1938. Die Entwicklung der Harlebucht und ihr Einfluß auf die Außenjade. *Jahrbuch der Hafenbautechnischen Gesellschaft* 16, 1937: 47-55. Berlin: Springer.

Krümmel, O. 1889. Über Erosion durch Gezeitenströme. *Dr. A. Petermanns Mitteilungen aus Justus Perthes' Geographischer Anstalt* 35 (VI): 129-138.

Kruse, A. 1981. *Die Dünen der Insel Sylt – Genese und ökologische Bedeutung*: 96 pp. Universität Hamburg, Hausarbeit im Fach Geographie (unpublished).

Kulinat, K. 1969. *Geographische Untersuchungen über den Fremdenverkehr der niedersächsischen Küste*: 140 pp. Göttingen: Wurm.

Lamb. H.H. 1982. *Climate, history and the modern world*: 387 pp. Methuen: London & New York.

Lamprecht, H.-O. 1955. Brandung und Uferveränderungen an der Westküste von Sylt. *Mitteilungen der Hannoverschen Versuchsanstalt für Grundbau und Wasserbau, Franzius-Institut der Technischen Hochschule Hannover* 8: 80-163.

Lamprecht, H.-O. 1957 a. Uferveränderungen und Küstenschutz auf Sylt. *Die Küste* 6 (2): 39-93.

Lamprecht, H.-O. 1957 b. Wirkungsweise von Küstenschutzbauwerken auf Sylt. *Die Wasserwirtschaft* 47 (5): 109-117.

Landesverein Oldenburg für Heimatkunde und Heimatschutz (eds.) 1929. *Wangeroog wie es wurde, war und ist*: 232 pp. Bremen: Leuwer.

Lang, A.W. 1970. Untersuchungen zur morphologischen Entwicklung des südlichen Elbe-Ästuars von 1650 bis 1960. *Hamburger Küstenforschung* 12: 195 pp.

Lang, A.W. 1975. Untersuchungen zur morphologischen Entwicklung des Dithmarscher Watts von der Mitte des 16. Jahrhunderts bis zur Gegenwart. *Hamburger Küstenforschung* 31: 154 pp.

Larsen, E.B. 1953. Studies on the Soil Fauna of Skallingen. *Oikos* 3 (II) 1951: 166-192.

Larsen, E.B. 1953. Successionsstudier i et havrendingsomrade, Skomagersletten, Skallingen. *Geografisk Tidsskrift* 52: 182-200.

Lassen, H., G. Linke & H.W. Braasch 1984. Säkularer Meeresspiegelanstieg und tektonische Senkungsvorgänge an der Nordseeküste. *Vermessungswesen und Raumordnung* 46 (2): 106-126.

Leatherman, S.P. (ed.) 1979. *Barrier Islands – from the Gulf of St. Lawrence to the Gulf of Mexico*: 325 pp. New York, San Francisco, London: Academic Press.

Leatherman, S.P. 1982. *Barrier Island Handbook*, 2nd edition: 109 pp.

Leatherman, S.P., T.E. Rice & V. Goldsmith 1982. Virginia Barrier Island Configuration: A Reappraisal. *Science* 215: 285-287.

Leberl, F.W. 1983. Photogrammetric Aspects of Remote Sensing with Imaging Radar. *Remote Sensing Reviews* 1: 71-158.

Leege, O. 1935: Werdendes Land in der Nordsee. *Schriften des Deutschen Naturkundevereins*, Neue Folge 2: 84 pp. Oehringen: Rau.

Liese, R. 1979. Veränderungen von Tidehochwasser, Tideniedrigwasser und Tidehub seit 1946. *Forschungsstelle für Insel- und Küstenschutz, Jahresbericht* 1978, XXX: 71-82.

Lillesand, T.M. & R.W. Kiefer 1979. *Remote Sensing and Image Interpretation*: 612 pp. New York, Chichester, Brisbane, Toronto, Singapore: John Wiley & Sons.

Linke, O. 1939. Die Biota des Jadebusenwatts. *Helgoländer Meeresuntersuchungen* 1: 201-348.

Linke, O. 1947. Pflanzen im Kampf gegen das Meer. *Kosmos* 1947 (1): 9-13.

Linke, G. 1969. Die Entstehung der Insel Scharhörn und ihre Bedeutung für die Überlegungen zur Sandbewegung in der Deutschen Bucht. *Hamburger Küstenforschung* 11: 45-84.

Linke, G. 1970. Über die geologischen Verhältnisse im Gebiet Neuwerk/Scharhörn. *Hamburger Küstenforschung* 17: 17-58.

Linke, G. 1979. Ergebnisse geologischer Untersuchungen im Küstenbereich südlich Cuxhaven – ein Beitrag zur Diskussion holozäner Fragen. In. *Probleme der Küstenforschung im südlichen Nordseegebiet* 13: 39-83.

Linke, G. 1981. Ergebnisse und Aspekte zur Klimaentwicklung im Holozän. *Geologische Rundschau* 70: 774-783.

Linke, G. 1982. Der Ablauf der holozänen Transgression der Nordsee aufgrund von Ergebnissen aus dem Gebiet Neuwerk/Scharhörn. *Probleme der Küstenforschung im südlichen Nordseegebiet* 14: 123-157.

Loor, G.P. de 1983. Introduction and Some General Aspects of Image Formation in Radar Remote Sensing. *Remote Sensing Reviews* 1: 3-18.

Lorenzen, J. 1979. *Eine Beschreibung der Hallig Gröde aus dem Jahre 1814 von Nommen Hansen*: 24 pp. Bredstedt: Nordfriisk Instituut.

Luck, G. 1968. Hydrologische Ursachen der Strand- und Dünenabbrüche im Westen Memmerts. *Forschungsstelle Norderney, Jahresbericht* 1966, XVIII: 37-41.

Luck, G. 1974. Untersuchungen der Forschungsstelle Norderney zu den Dünenabbrüchen auf der ostfriesischen Insel Langeoog. *Die Küste* 25: 46-52.

Luck, G. 1975. Der Einfluß der Schutzwerke der ostfriesischen Inseln auf die morphologischen Vorgänge im Bereich der Seegaten und ihrer Einzugsgebiete. *Mitteilungen aus dem Leichtweiß-Institut für Wasserbau der Technischen Universität Braunschweig* 47: 1-122.

Luck, G. 1976. Protection of the littoral and seabed against erosion – Fallstudie Norderney. *Forschungsstelle für Insel- und Küstenschutz, Jahresbericht* 1975, XXVII: 9-78.

Luck, G. 1979. Untersuchungen zur Sedimentbewegung mit Hilfe einer Unterwasserfernsehanlage. In Deutsche Forschungsgemeinschaft (eds.), *Sandbewegung im Küstenraum - Rückschau, Ergebnisse und Ausblick*: 186-206.

Luck, G. 1983. Von der Natur und den Menschen geformt. *Draußen* 25: 6-17.

Luck, G. & H.-J. Stephan 1983. Verlagerung morphologischer Großformen nördlich der Osterems und deren Einfluß auf das Westende von Juist. *Forschungsstelle für Insel- und Küstenschutz, Jahresbericht* 1982, XXXIV: 11-29.

Luck, G. & H.-H. Witte 1974. Erfassung morphologischer Vorgänge der ostfriesischen Riffbögen in Luftbildern. *Forschungsstelle für Insel- und Küstenschutz, Jahresbericht* 1973, XXV: 33-54.

Ludwig, G., H. Müller & H. Streif 1979. Neuere Daten zum holozänen Meeresspiegelanstieg im Bereich der Deutschen Bucht. *Geologisches Jahrbuch*, D 32: 3-22.

Ludwig, G., H. Müller & H. Streif 1981: New dates on Holocene sea-level changes in the German Bight. In S.-D. Nio, R.T.E. Schüttenhelm & Tj.C.E. van Weering (eds.), *Holocene Marine Sedimentation in the North Sea Basin. Special Publication Number 5 of the International Association of Sedimentologists*: 211-219.

Lüders, K. 1937. Die Zerstörung der Oberahneschen Felder im Jadebusen. *Abhandlungen herausgegeben vom Naturwissenschaftlichen Verein zu Bremen* XXX (1/2): 5-20.

Lüders, K. 1952. Die Wirkung der Buhne H in Wangerooge-

West auf das Seegat 'Harle'. *Die Küste* 1 (1): 21-26.
Lüders, K. 1977. 'Wangerooch hett'n hooge toren, ...'. *Forschungsstelle für Insel- und Küstenschutz, Jahresbericht 1976*, XXVIII: 11-38.
Lüders, K., A. Führböter & W. Rodloff 1972. Neuartige Dünen- und Strandsicherung im Nordwesten der Insel Langeoog. *Die Küste* 23: 63-111.
Lüpkes, H. & H.P. Siemens 1940. Planung und Stand der Maßnahmen zur Sicherung des Ellenbogens auf Sylt. *Westküste* 2 (2/3): 6-23.
Lundqvist, J. 1980. The deglaciation of Sweden after 10,000 B.P. *Boreas* 9: 229-238.
Lundqvist, J. 1981. Weichselian (Vistulian) Stratigraphy of Sweden. *Quaternary Studies in Poland* 3: 57-59.
Mager, F. 1927. Der Abbruch der Insel Sylt durch die Nordsee. Eine historisch-geographische Untersuchung. *Veröffentlichungen der Schleswig-Holsteinischen Universitätsgesellschaft* 8: 199 pp.
Mantje, M. 1980. Duinen en duinbeheer. In J.C. Duinker (ed.), *Texel en de Zee. Een strijd van Eeuwen*: 25-39.
Marschalleck, K.H. 1973. Die Salzgewinnung an der friesischen Nordseeküste. *Probleme der Küstenforschung im südlichen Nordseegebiet* 10: 127-150.
Martens, P. 1927. Morphologie der schleswig-holsteinischen Ostseeküste. *Veröffentlichungen der schleswig-holsteinischen Universitätsgesellschaft* 7: 72 pp.
Meesenburg, H. 1972. *Spartina*s kolonisation og udbredelse langs Ho Bugt. *Geografisk Tidsskrift* 71: 37-45.
Meesenburg, H. 1978. *Rømø – Natur, Mensch und Landschaft*: 63 pp. Esbjerg: Bygd.
Meesenburg, H., J. Termansen, S. Tougaard, & P. Uhd Jepsen 1977. *Fanø – Mensch und Landschaft*: 75 pp. Esbjerg: Bygd.
Meier, H. 1863. *Die Nordsee-Insel Borkum. Ein Handbuch für Reisende und Badegäste*: 167 pp. Leipzig: Weber. Reprint 1979 Leer: Schuster.
Mellema, L. 1981. *Schiermonnikoog – lytje pole*. 2nd. edition: 280 pp. Knoop Haren BV.
Menke, B. 1969. Vegetationskundliche und vegetationsgeschichtliche Untersuchungen an Strandwällen (mit Beiträgen zur Vegetationsgeschichte sowie zur Erd- und Siedlungsgeschichte West-Eiderstedts). *Mitteilungen der Floristisch-Soziologischen Arbeitsgemeinschaft* NF 14: 95-120.
Menke, B. 1976. Befunde und Überlegungen zum nacheiszeitlichen Meeresspiegelanstieg (Dithmarschen und Eiderstedt, Schleswig-Holstein). *Probleme der Küstenforschung im südlichen Nordseegebiet* 11: 145-161.
Menke, B. 1984. Holocene shorelines and barrier systems in Dithmarschen and Western Eiderstedt. In INQUA Subcommission on Shorelines of Northwestern Europe (eds.), *Field Conference 1984, September 15 – 21, North Sea Coastal Zone between Jade Bay and Jammer Bight*: 47-58.
Mentz, A. 1952. Halvøen Skallingen og dens Fredning. *Naturens Verden* XXIV: 413-423.
Merkt, J. 1984. Shoreline displacement and lake development in the Bederkesa area. In INQUA Subcommission on Shorelines of Northwestern Europe (ed.), *Field Conference 1984, September 15 - 21, North Sea Coastal Zone between Jade Bay and Jammer Bight*: 32-41. Hannover: Niedersächsisches Landesamt für Bodenforschung.
Mes, R., R. Schuckard & H. Smit 1980. *Flora en fauna van de Engelsmanplaat*: 95 pp. Leiden: Stichting Veth tot steun aan Waddenonderzoek.
Meyer, K.-D. 1970. Zur Geschiebeführung des Ostfriesisch-Oldenburgischen Geestrückens. *Abhandlungen des naturwissenschaftlichen Vereins zu Bremen* 37 (3/2): 227-246.

Meyer-Deepen, J. & M.P.D. Meijering 1979. *Spiekeroog - Naturkunde einer ostfriesischen Insel*: 223 pp. Spiekeroog: Kurverwaltung.
Meyer-Deepen, J. & M.P.D. Meijering 1983. *Spiekeroog - Geschichte einer ostfriesischen Insel*, 2nd edition: 188 pp. Spiekeroog: Kurverwaltung.
Meyn, L. 1876. Geognostische Beschreibung der Insel Sylt und ihrer Umgebung. *Abhandlungen zur geologischen Spezialkarte von Preußen* 1 (4): 605-759.
Michaelis, H. 1984. Studien zur ökologischen Fortentwicklung der Leybucht und zur Auswirkung des geplanten Bauvorhabens Leyhörn. *Forschungsstelle für Insel- und Küstenschutz, Jahresbericht 1983*, XXXV: 135-161.
Miller, M.C. & P.D. Komar 1980 a. Oscillation sand ripples generated by laboratory apparatus. *Journal of Sedimentary Petrology* 50 (1): 173-182.
Miller, M.C. & P.D. Komar 1980 b. A field investigation of the relationship between oscillation ripple spacing and the near-bottom water orbital motions. *Journal of Sedimentary Petrology* 50 (1): 183-191.
Minister für Ernährung, Landwirtschaft und Forsten des Landes Schleswig-Holstein (ed.) 1977. *Generalplan Deichverstärkung, Deichverkürzung und Küstenschutz in Schleswig-Holstein vom 20.12.1963. Fortschreibung*.
Minister für Ernährung, Landwirtschaft und Forsten des Landes Schleswig-Holstein (ed.) 1981. *Gutachten zur geplanten Vordeichung der Nordstrander Bucht*. Schriftenreihe der Landesregierung Schleswig-Holstein 12: 315 pp.
Möbius, R. 1931. Über die Gestaltung der Inseloberfläche durch den Menschen. In K. Ruhnau (ed.), *Die ostfriesische Insel Spiekeroog*: 8-9.
Molen, S.J. van der 1978 a. *O, welk een ontzettende waterplas - Vergeten epistels over de Waddenzee*: 230 pp. Baarn: van Kampen & Zn.
Molen, S.J. van der 1978 b. *Terschelling – van Noordsvaarder tot Bosplaat*: 94 pp. 's Gravenhage: Boekencentrum B.V.
Moore, J.N., W.J. Fritz & R.S. Futch 1984. Occurrence of megaripples in a ridge and runnel system, Sapelo Island, Georgia: Morphology and processes. *Journal of Sedimentary Petrology* 54 (2): 615-625.
Moritz, E. 1903. Die Nordseeinsel Röm. *Mitteilungen der Geographischen Gesellschaft in Hamburg* XIX: 2-210.
Mortensen, H. 1921. Die Morphologie der samländischen Steilküste auf Grund einer physiologisch-morphologischen Kartierung des Gebietes. *Veröffentlichungen des Geographischen Instituts der Albertus-Universität zu Königsberg* III: 70 pp.
Mroczek, P. 1980. Zu einer Karte der Veränderungen der Uferlinie der deutschen Nordseeküste in den letzten 100 Jahren. *Berliner Geographische Studien* 7: 39-57.
Müller, F. 1917 a. *Die Halligen*. Das Wasserwesen an der schleswig-holsteinischen Nordseeküste, Erster Teil, Bd. I: 377 pp. Berlin: Reimer.
Müller, F. 1917 b. *Die Halligen*. Das Wasserwesen an der schleswig-holsteinischen Nordseeküste, Erster Teil, Bd. II: 428 pp. Berlin: Reimer.
Müller, F. & O. Fischer 1936 a. Alt Nordstrand. *Das Wasserwesen an der schleswig-holsteinischen Nordseeküste, Zweiter Teil: Die Inseln* 2: 224 pp. Berlin: Reimer. Reprint 1982 Berlin: Reimer.
Müller, F. & O. Fischer 1936 b. Nordstrand. *Das Wasserwesen an der schleswig-holsteinischen Nordseeküste, Zweiter Teil: Die Inseln* 3: 316 pp. Berlin: Reimer. Reprint 1983 Berlin: Reimer.
Müller, F. & O. Fischer 1936 c. Pellworm. *Das Wasserwesen an der schleswig-holsteinischen Nordseeküste, Zweiter Teil:*

Die Inseln 4: 402 pp. Berlin: Reimer. Reprint 1982 Berlin: Reimer.
Müller, F. & O. Fischer 1937 a. Amrum. *Das Wasserwesen an der schleswig-holsteinischen Nordseeküste, Zweiter Teil: Die Inseln* 5: 237 pp. Berlin: Reimer. Reprint 1984 Berlin: Reimer.
Müller, F. & O. Fischer 1937 b. Föhr. *Das Wasserwesen an der schleswig-holsteinischen Nordseeküste, Zweiter Teil: Die Inseln* 6: 327 pp. Berlin: Reimer. Reprint 1983 Berlin: Reimer.
Müller, F. & O. Fischer 1938. Sylt. *Das Wasserwesen an der schleswig-holsteinischen Nordseeküste, Zweiter Teil: Die Inseln* 7: 304 pp. Berlin: Reimer. Reprint 1983 Berlin: Reimer.
Müller, F. & O. Fischer 1955. Kartenmappe. *Das Wasserwesen an der schleswig-holsteinischen Nordseeküste, Dritter Teil: Das Festland*: 41 maps. Berlin: Reimer.
Müller, W. 1962. Der Ablauf der holozänen Meerestransgression an der südlichen Nordseeküste und Folgerungen in bezug auf eine geochronologische Holozängliederung. *Eiszeitalter und Gegenwart* 13: 197-226.
Müller-Wille, M. 1982. Archäologische Untersuchungen. In B. Higelke, D. Hoffmann & M. Müller-Wille (eds.), *Das Norderhever-Projekt. Beiträge zur Landschafts- und Siedlungsgeschichte der nordfriesischen Marschen und Watten*. Offa 39: 254-260.
Mungenas, A. 1934. *Dynamisch-morphologische Untersuchung der Seegaten, Watten und des Vorlandes im Bereich der Ostfriesischen Außenküste*: 81 pp. Giessen.
National Research Council 1985. *Glaciers, Ice Sheets, and Sea Level: Effect of a CO_2-induced Climatic Change*. Report of a Workshop Held in Seattle, Washington September 13-15, 1984: 330 pp. Washington: United States Department of Energy.
Naumann, K.-E. & H. Laucht 1985. Zwanzig Jahre Planen und Bauen für den Hamburger Hafen–Aus der Arbeit des 'Strom- und Hafenbau' 1960 bis 1980. *Jahrbuch der Hafenbautechnischen Gesellschaft* 40, 1983/84: 109-171. Heidelberg, New York, Tokyo: Springer.
Nichols, R.L. 1953. Marine and lacustrine ice-pushed ridges. *Journal of Glaciology* 2 (13): 172-175.
Nicolaisen, H.-D. 1985. *Die Flakhelfer*: 304 pp. Ullstein.
Niedersächsisches Staatsbad Norderney (eds.) 1980. *Nordseeheilbad Norderney. Chronik einer Insel*. 2nd edition: (38 pp.). Norderney: Soltau.
Newton, R.S. & F. Werner 1969. Luftbildanalyse und Sedimentgefüge als Hilfsmittel für das Sandtransportproblem im Wattgebiet vor Cuxhaven. *Hamburger Küstenforschung* 8: 46 pp.
Niemeyer, G. 1972. Ostfriesische Inseln. *Sammlung Geographischer Führer* 8: 189 pp. Berlin, Stuttgart: Borntraeger.
Niemeyer, H.D. 1977. Seegangsmessungen auf Deichvorländern. *Forschungsstelle für Insel- und Küstenschutz, Jahresbericht* 1976, XXVIII: 113-139.
Niemeyer, H.D. 1979. Untersuchungen zum Seegangsklima im Bereich der Ostfriesischen Inseln und Küste. *Die Küste* 34: 53-70.
Nio, S.-D. & C. Siegenthaler 1979. A Holocene Mesotidal Channel Fill Sequence in the Oosterschelde Mouth, SW Netherlands. *International Meeting on Holocene Marine Sedimentation in the North Sea Basin, Texel, the Netherlands, September 17-23, 1979, Abstracts*: 26.
Nio, S.-D., R.T.E. Schüttenhelm & Tj.C.E. van Weering (eds.) 1981. Holocene Marine Sedimentation in the North Sea Basin. *Special Publication Number 5 of the International Association of Sedimentologists*: 515 pp. Oxford, London, Edinburgh, Boston, Melbourne: Blackwell.

Nommensen, B. 1982. *Die Sedimente des südlichen Nordfriesischen Wattenmeeres (Deutsche Bucht). Ergebnisse Geologisch-Sedimentologischer Untersuchungen an Pleistozänen und Holozänen Sedimenten und an Schwebstoffen der Gezeitenströme*: 268 pp. Kiel: Unpublished Thesis.
Nummedal, D. & S. Penland 1981. Sediment dispersal in Norderneyer Seegat, West Germany. In S.-D. Nio, R.T.E. Schüttenhelm & Tj.C.E. van Weering (eds.), *Holocene Marine Sedimentation in the North Sea Basin. Special Publication Number 5 of the International Association of Sedimentologists*: 187-210.
Oele, E., R.T.E. Schüttenhelm & A.J. Wiggers (eds.) 1979. The Quaternary History of the North Sea. *Acta Universitatis Upsaliensis, Symposia Universitatis Upsaliensis Annum Quingentesimum Celebrantis* 2: 248 pp.
Oesau, W. 1955. *Hamburgs Grönlandfahrt*: 316 pp. Hamburg: Augustin.
Oestreich, H. 1976. *Der Fremdenverkehr der Insel Sylt*: 603 pp. Studien und Materialien 9. Bredstedt: Nordfriisk Instituut.
Ordemann, W. 1912. Beiträge zur morphologischen Entwicklungsgeschichte der deutschen Nordseeküste mit besonderer Berücksichtigung der Dünen tragenden Inseln. *Mitteilungen der Geographischen Gesellschaft (für Thüringen) zu Jena* 30: 15-150.
Otvos, E.G. 1964. Observations on rhomboid beach marks. *Journal of Sedimentary Petrology* 34 (9): 683-687.
Otvos, E.G. 1965. Types of rhomboid beach surface patterns. *American Journal of Science* 263: 271-276.
Otvos, E.G. 1970. Development and migration of barrier islands, northern Gulf of Mexico. *Geological Society of America, Bulletin* 81: 241-246.
Otvos, E.G. 1979. Barrier island evolution and history of migration, north-central Gulf Coast. In S.P. Leatherman (ed.), *Barrier Islands–from the Gulf of St.Lawrence to the Gulf of Mexico*: 291-319.
Otvos, E.G. 1981. Barrier island formation through nearshore aggradation–stratigraphic and field evidence. *Marine Geology* 43: 195-243.
Otvos, E.G. 1985. Barrier Island Genesis–Questions of Alternatives for the Apalachicola Coast, Northeastern Gulf of Mexico. *Journal of Coastal Research* 1 (3): 267-278.
Özsoy, E. 1977. *Flow separation and related phenomena at tidal inlets*: 327 pp. Unpublished Ph.D. Thesis, The University of Florida.
Pätzold, U. 1982. Bilanz und Verformung aufgespülter Sandstrände. *Forschungsstelle für Insel- und Küstenschutz, Jahresbericht* 1980, XXXII: 41-53.
Pahl, M. & P. Carstensen 1973. *Hörnum–Heimat am Horn*: (146 pp.). Hamburg: Löding.
Panten, A.A. 1979. Die patronymische Namensgebung in Nordfriesland. In J. Lorenzen, *Eine Beschreibung der Hallig Gröde aus dem Jahre 1814 von Nommen Hansen*: II-V. Bredstedt: Nordfriisk Instituut.
Park, D.W. 1974. Flußmorphologie im Süßwasserbereich der Unterelbe (Elbe km 650-670). *Hamburger Geographische Studien*, Sonderheft: 173 pp.
Partensky, H.W. & R. Dieckmann 1981. Stabilitätsuntersuchungen für das südliche nordfriesische Wattenmeer. In Der Minister für Ernährung, Landwirtschaft und Forsten des Landes Schleswig-Holstein (ed.), *Gutachten zur geplanten Vordeichung der Nordstrander Bucht*: 35-62.
Partensky, H.W. & H. Schwarze 1981 a. Wissenschaftliches Gutachten zu den hydrologischen und morphologischen Auswirkungen der geplanten Baumaßnahmen in der Nordstrander Bucht. In Der Minister für Ernährung, Landwirt-

schaft und Forsten des Landes Schleswig-Holstein (ed.), *Gutachten zur geplanten Vordeichung der Nordstrander Bucht*: 11-34.
Partensky, H.W. & H. Schwarze 1981 b. Modellversuche für die Nordstrander Bucht – zusammenfassende Darstellung der Ergebnisse des Versuchsberichtes vom 24.10.1967. In Der Minister für Ernährung, Landwirtschaft und Forsten des Landes Schleswig-Holstein (ed.), *Gutachten zur geplanten Vordeichung der Nordstrander Bucht*: 63-88.
Penck, A. 1894. *Morphologie der Erdoberfläche*, 2. Teil: 696 pp. Stuttgart: Engelhorn.
Penck, A. 1933. Eustatische Bewegungen des Meeresspiegels während der Eiszeit. *Geographische Zeitschrift* 39: 329-339.
Petersen, M. 1978. Der Heverstrom – Schicksalsstrom Nordfrieslands. *Nordfriesisches Jahrbuch* N.F. 14: 13-44.
Petersen, M. & H. Rohde 1977. *Sturmflut – Die großen Fluten an den Küsten Schleswig-Holsteins und in der Elbe*: 148 pp. Neumünster: Wachholtz.
Pettijohn, F.J. 1975. *Sedimentary Rocks*. 3rd edition: 628 pp. New York, Evanston, San Francisco, London: Harper & Row.
Pettijohn, F.J. & P.E. Potter 1964. *Atlas and glossary of primary sedimentary structures*: 370 pp. Berlin, Göttingen, Heidelberg, New York: Springer.
Pötter, W. 1982. *Wir entdecken die Nordseeinsel Borkum*: 194 pp. Borkum: Selbstverlag.
Pontoppidan, C. 1785. *Hval- og Robbefangsten*. København.
Poppen, H. 1912. Die Sandbänke an der Küste der Deutschen Bucht der Nordsee. *Annalen der Hydrographie und Maritimen Meteorologie* XXXX (VI, VII, VIII): 273-302, 352-364, 406-420.
Postma, H. 1961. Transport and accumulation of suspended matter in the Dutch Wadden Sea. *Netherlands Journal of Sea Research* 1: 148-190.
Postma, H. (ed.) 1982. *Hydrography of the Wadden Sea: Movements and properties of water and particulate matter. Final report on 'Hydrography' of the Wadden Sea Working Group*: 75 pp. Rotterdam: Balkema.
Postma, J.T. 1982. *Ontwikkeling van het Pinkegat voor zover van belang voor de kustontwikkeling van Ameland*. Rijkswaterstaat, Directie Groningen, Meet- en Adviesdienst, Nota No. 82-30: 16 pp.
Prange, W. 1963. Das Holozän und seine Datierung in den Marschen des Arlau-Gebietes, Nordfriesland. *Meyniana* 13: 47-76.
Prange, W. 1968. Geologische Untersuchungen in den Marschen der alten Köge vor Bredstedt, Nordfriesland. *Neues Jahrbuch für Geologie und Paläontologie, Monatshefte* 1968 (10): 619-640.
Prange, W. 1982. Ribersalt. *Skalk* 2: 28-30.
Prasad, M. 1983. *Geologische Interpretation reflexionsseismischer Messungen aus der südlichen Deutschen Bucht*: 77 pp. Unpublished Diplomarbeit, Institut für Geophysik der Universität Kiel.
Pratje, O. 1951. Die Deutung der Steingründe in der Nordsee als Endmoränen. *Deutsche Hydrographische Zeitschrift* 4 (3): 106-114.
Priesmeier, K. 1970. Form und Genese der Dünen des Listlandes auf Sylt. *Schriften des Naturwissenschaftlichen Vereins für Schleswig-Holstein* 40: 11-51.
Pronker, T.J.F. & H.M. van Eck 1970. *Vlieland – van hors tot horn*: 134 pp. Haren: Knoop & Niemeijer.
Prügel, H. 1942. Die Sturmflutschäden an der schleswig-holsteinischen Westküste in ihrer meteorologischen und morphologischen Abhängigkeit. *Schriften des Geographischen Instituts der Universität Kiel* XI (3): 94 pp.

Quedens, G. 1960. Verliert Amrum seine Odde? - Küstenschutzprobleme einer nordfriesischen Geestinsel. *Nordfriesland* 1 (4): 36-40.
Quedens, G. 1961. Die Dünen von Amrum. Ein bemerkenswertes Kapitel modernen Naturschutzes. *Nordfriesland* 2 (6): 137-144.
Quedens, G. (1968). *Amrum – Aus alter Zeit*: (61 pp.). Münsterdorf: Hansen & Hansen.
Quedens, G. 1979. *Amrum*, 7th edition: 123 pp. Breklum: Breklumer Verlag.
Quedens, G. 1982. *Insel im Wattenmeer – Pellworm*: 127 pp. Breklum: Breklumer Verlag.
Quedens, G. 1984. *Nordstrand*, 3rd edition: 112 pp. Breklum: Breklumer Verlag.
Quedens, G. 1985. *Föhr*: 128 pp. Breklum: Breklumer Verlag.
Ragutzki, G. 1979. Zur Frage der biogenen Festigkeit von Wattsedimenten nach bodenphysikalischen Kriterien. *Forschungsstelle für Insel- und Küstenschutz, Jahresbericht 1978*, XXX: 117-140.
Ramsay, W. 1930. Changes of Sea-Level resulting from the Increase and Decrease of Glaciation. *Fennia* 52 (5): 62 pp.
Rankama, K. (ed.) 1967. *The Quaternary*, Volume 2: 477 pp. New York, London, Sydney: Interscience.
Reineck, H.-E. 1956. Wattenmeer im Winter. *Senckenbergiana lethaea* 37: 129-146.
Reineck, H.-E. 1961 a. Über Sandverlagerungen im Bereich des nassen Strandes. *Forschungsstelle Norderney, Jahresbericht 1960*, XII: 13-26.
Reineck, H.-E. 1961 b. Sedimentbewegungen an Kleinrippeln im Watt. *Senckenbergiana lethaea* 42 (1/2): 51-67.
Reineck, H.-E. 1962. Schichtungsarten in Wattböden. *Zeitschrift für Pflanzenernährung, Düngung, Bodenkunde* 99 (2/3): 154-159.
Reineck, H.-E. 1963. Sedimentgefüge im Bereich der südlichen Nordsee. *Abhandlungen der Senckenbergischen Naturforschenden Gesellschaft* 505: 138 pp.
Reineck, H.-E. 1975. German North Sea Tidal Flats. In R.N. Ginsburg (ed.), *Tidal Deposits – A Casebook of Recent Examples and Fossil Counterparts*: 5-12.
Reineck, H.-E. 1976. Drift Ice Action on Tidal Flats, North Sea. *Rev. Geogr. Montr.* XXX (1/2): 197-200.
Reineck, H.-E. (ed.) 1978 a. *Das Watt – Ablagerungs- und Lebensraum*. 2nd edition: 185 pp. Frankfurt: Kramer.
Reineck, H.-E. 1978 b. Die Watten der deutschen Nordseeküste. *Die Küste* 32: 66-83.
Reineck, H.-E. 1979 a. Sandversatzzonen im Vorstrand von Norderney. In Deutsche Forschungsgemeinschaft (ed.), *Sandbewegung im Küstenraum – Rückschau, Ergebnisse und Ausblick*: 293-298.
Reineck, H.-E. 1979 b. Rezente und fossile Algenmatten und Wurzelhorizonte. *Natur und Museum* 109 (9): 290-296.
Reineck, H.-E. & I.B. Singh 1975. *Depositional Sedimentary Environments*: 439 pp. Berlin, Heidelberg, New York: Springer.
Reineck, H.-E. & F. Wunderlich 1967. Zeitmessungen an Gezeitenschichten. *Natur und Museum* 97 (6): 193-197.
Reinhard, R. 1974 a. Ein Verfahren zur quantitativen Erfassung von Sandwanderungsvorgängen bei Messungen mit radioaktiven Tracern (I. Teil). *Die Küste* 26: 25-54.
Reinhard, R. 1974 b. Quantitative Messung der Sandwanderung an der Brandungsküste vor Westerland/Sylt (III. Teil). *Die Küste* 26: 77-82.
Reinhardt 1979. Küstenentwicklung und Deichbau während des Mittelalters zwischen Maade, Jade und Jadebusen. *Jahrbuch der Gesellschaft für Bildende Kunst und Vaterländische Altertümer Emden* 59: 17-61.

Reinke 1909. Die ostfriesischen Inseln. *Wissenschaftliche Meeresuntersuchungen* X.

Reus, J.H. de 1980. *Ontwikkeling Zeegat van Texel*. Rijkswaterstaat, Studiedienst Hoorn, Notitie WWKZ-80.H248: 21 pp.

Reus, J.H. de 1983. *Kustverdediging Ameland-west*. Rijkswaterstaat, Adviesdienst Hoorn, Nòta WWKZ-83.H012: 34 pp.

Reus, J.H. de 1985. *Ontwikkeling Noorderhaaks*. Rijkswaterstaat, Adviesdienst Hoorn, Notitie WWKZ-85.H209: 16 pp.

Reynolds, W.J. 1982. *Beach dynamics and the societal response to beach erosion at Marco Island, Florida*: 345 pp. Unpublished Ph.D. Thesis, Rutgers University, New Jersey.

Rice, T.E. & S.P. Leatherman 1983. Barrier island dynamics: the eastern shore of Virginia. *Southeastern Geology* 24 (3): 125-137.

Richardsen, A. 1902. *Die Marsch- und Halligwirtschaft Nordfrieslands und der gegenwärtige Stand der Seebauten im nordfriesischen Wattenmeer*: 85 pp. Halle a.S.: Thesis (Jena).

Riecken, G. 1982. *Die Halligen im Wandel*: 160 pp. Husum: Husum Druck- und Verlagsgesellschaft.

Rijkswaterstaat (ed.) 1981. *Zandwinning in de Waddenzee - Resultaten van een hydrografisch-sedimentologisch onderzoek*: 48 pp. Leeuwarden: Rijkswaterstaat.

Rodloff, W. 1970. Über Wattwasserläufe. *Mitteilungen des Franzius-Instituts für Grund- und Wasserbau der Technischen Universität Hannover* 34: 1-88.

Roeleveld, W. 1971. The Morphology of the Pleistocene Surface in the Marine-Clay District of Groningen (The Netherlands). *Berichten van de Rijksdienst voor het Oudheidskundig Bodemonderzoek* 20-21: 7-25.

Roeleveld, W. 1974. The Holocene Evolution of the Groningen Marine-Clay District. *Berichten van de Rijksdienst voor het Oudheidskundig Bodemonderzoek* 24 (Suppl.): 132 pp.

Roeloffs, B.C. 1985. *Von der Seefahrt zur Landwirtschaft. Ein Beitrag zur Geschichte der Insel Föhr*, 2nd edition: 383 pp. Neumünster: Wachholtz.

Roep, Th. B. 1984. Progradation, erosion and changing coastal gradient in the coastal barrier deposits of the western Netherlands. *Geologie en Mijnbouw* 63 (3): 249-258.

Roeper, M. 1980. Waterschappen en Polders. In J.C. Duinker (ed.), *Texel en de Zee. Een strijd van Eeuwen*: 59-71.

Rohde, H. 1976. Eiderstedt und die Sturmfluten. In Eiderstedter Heimatbund (eds.), *Blick über Eiderstedt* 1: 71-83.

Rohde, H. 1977. Sturmfluten und säkularer Wasserstandsanstieg an der deutschen Nordseeküste. *Die Küste* 30: 52-143.

Rolf, R. 1983. *Der Atlantikwall. Perlenschnur aus Stahlbeton*: 222 pp. Beesterzwaag: AMA.

Roterberg, P. 1983. *Die Nordseeinsel Spiekeroog*: 79 pp. Hamburg: Christians.

Samu, G. 1972. Morphologische und granulometrische Untersuchungen im Seegebiet vor Borkum. *Die Küste* 23: 150-188.

Samu, G. 1982. Zur Morphogenese des Seegebietes vor Borkum und des Südweststrandes der Insel. *Die Küste* 37: 37-57.

Schäfer, W. 1941. Mellum, eine Düneninsel an der deutschen Nordsee-Küste. *Abhandlungen der Senckenbergischen Naturforschenden Gesellschaft* 457: 34-54.

Schäfer, W. 1954. Mellum: Inselentwicklung und Biotopwandel. *Abhandlungen des naturwissenschaftlichen Vereins zu Bremen* 33 (3): 391-406.

Schalkers, K.M. 1984. *Kustachteruitgang Noord-Holland km 1-40 door de eeuwen*. Rijkswaterstaat, Adviesdienst Hoorn, Notitie WWKZ-84.H225: 9 pp.

Scherz, C.F. (ca. 1900). *Die Nordseeinsel Juist und ihr Seebad*, 2nd edition: 192 pp. Norden: Soltau. Reprint 1979 Leer: Schuster.

Scheurlen, U. 1974. *Über Handel und Seeraub im 14. und 15. Jahrhundert an der ostfriesischen Küste*: 187 pp. Unpublished thesis, Universität Hamburg.

Schirrmacher, G. 1983. *Hallig Hooge, die Königin der Halligen*, 5th edition: 176 pp. Breklum: Breklumer Verlag.

Schmedes, Th. 1929. Die Insel Wangeroog nach ihrem früheren und gegenwärtigen Zustande (1855). In Landesverein Oldenburg für Heimatkunde und Heimatschutz (eds.), *Wangeroog wie es wurde, war und ist*: 140-172.

Schmidt, R. 1937. Inselschutz vor der deutschen Nordseeküste. In G. Wüst (ed.), Werdendes Land am Meer – Landerhaltung und Landgewinnung an der Nordseeküste. *Das Meer in volkstümlichen Darstellungen* 5: 71-105.

Schmidt-Eppendorf, P. 1971. Der Sylter Chronist Hans Kielholt. *Sylter Beiträge* 5: 48 pp. Münsterdorf: Hansen & Hansen.

Schmidt-Thomé, P. 1982. Geologische Karte von Helgoland mit Erläuterungen. *Geologisches Jahrbuch* A 62: 17 pp.

Schubert, K. 1970. Ems und Jade. *Die Küste* 19: 29-67.

Schucht, F. 1903. Beitrag zur Geologie der Wesermarschen. *Zeitschrift für Naturwissenschaften* 76: 1-80.

Schucht, F. 1910. Die Frage der neuzeitlichen Senkung der deutschen Nordseeküste. *Monatsberichte der Deutschen Geologischen Gesellschaft* 1910 (2): 101-102.

Schütte, H. 1905. Ein neu entstandenes Düneneiland zwischen Aussenjade und Aussenweser. *Jahrbuch des Vereins für Naturkunde an der Unterweser* 1903/1904: 31-42.

Schütte, H. 1924. Mellum als Neuland. *Schriftenreihe des Heimat-, Natur- und Vogelschutzvereins Wilhelmshaven-Rüstringen* 1: 12-15.

Schütte, H. 1927. Krustenbewegungen an der deutschen Nordseeküste. *Aus der Heimat* XXXX (11): 325-356.

Schütte, H. 1929. Nordfrieslands geologischer Werdegang. *Nordfriesland. Heimatbuch für die Kreise Husum und Südtondern*: 39-59. Husum: Delff.

Schütte, H. 1931. Der Aufbau des Weser-Jade-Alluviums. *Schriften des Vereins für Naturkunde an der Unterweser*, N.F. V: 1-40.

Schütte, H. 1933. Der geologische Aufbau des Jever- und Harlingerlandes und die erste Marschbesiedlung. *Oldenburger Jahrbuch des Vereins für Landesgeschichte und Altertumskunde* 37: 1-39.

Schütte, H. 1939. Sinkendes Land an der Nordsee? – Zur Küstengeschichte Nordwestdeutschlands. *Schriften des Deutschen Naturkundevereins* N.F. 9: 144 pp. Öhringen: Rau.

Schulz, H. 1946. *Beschreibung der Hallig Hooge und der benachbarten Vogelinsel Norderoog*: 22 pp. Hamburg: Lettenbauer.

Schwartz, M.L. (ed.) 1972. *Spits and Bars*. Benchmark Papers in Geology: 452 pp. Stroudsburg, Pa.: Dowden, Ross & Hutchinson.

Schwartz, M.L. (ed.) 1973. *Barrier Islands*. Benchmark Papers in Geology: 451 pp. Stroudsburg, Pa.: Dowden, Hutchinson & Ross.

Schwartz, M.L. (ed.) 1982. *Encyclopedia of Beaches and Coastal Environments*: 940 pp. Stroudsburg, Pa.: Dowden, Hutchinson & Ross.

Schwedhelm, E. 1984. *Schwermetalle – Bioelemente in den Nordseewatten und der Jade und die Tonmineralverteilung in den Sedimenten der südöstlichen Nordsee*: 177 pp. Heidelberg: Dissertation.

Seliger, J. 1983. *Das Sturmflutgeschehen an der deutschen Nordseeküste. Eine Untersuchung über die Wechselwirkung zwischen Klimaelementen und der Sturmfluthäufigkeit seit*

Beginn regelmäßiger Pegelmessungen (1842): 154 pp. Kiel: Thesis.

Shepard, F.P. 1963. *Submarine Geology*, 2nd edition: 555 pp. New York, Evanston, London: Harper & Row.

Siefert, W. 1969. Seegangsbestimmung mit Radar und nach Luftbildern. *Hamburger Küstenforschung* 7: 79 pp.

Siefert, W. 1971. Untersuchung des Seegangs in flachem Wasser. *Die Küste* 21: 17-28.

Siefert, W. 1974. Über den Seegang in Flachwassergebieten. *Mitteilungen aus dem Leichtweiß-Institut für Wasserbau der Technischen Universität Braunschweig* 40: 1-243.

Siefert, W. & H. Lassen 1985. Gesamtdarstellung der Wasserstandsverhältnisse im Küstenvorfeld der Deutschen Bucht nach neuen Pegelauswertungen. *Die Küste* 42: 1-77.

Simon, W.G. 1959. Beobachtungen an Strombänken auf trockenfallenden Sandflächen im Gezeitenbereich der Elbe 1950/1956. *Abhandlungen und Verhandlungen des Naturwissenschaftlichen Vereins in Hamburg* N.F. III: 27-36.

Sindowski, K.-H. 1965. Das Eem im ostfriesischen Küstengebiet. *Zeitschrift der Deutschen Geologischen Gesellschaft* 115 (1): 163-166.

Sindowski, K.-H. 1969. *Geologische Karte von Niedersachsen 1:25,000, Erläuterungen zu Blatt Wangerooge Nr. 2213*: 49 pp. Hannover: Niedersächsisches Landesamt für Bodenforschung.

Sindowski, K.-H. 1970. *Geologische Karte von Niedersachsen 1:25,000, Erläuterungen zu Blatt Spiekeroog Nr. 2212*: 56 pp. Hannover: Niedersächsisches Landesamt für Bodenforschung.

Sindowski, K.-H. 1973. Das ostfriesische Küstengebiet. Inseln, Watten und Marschen. *Sammlung Geologischer Führer* 57: 162 pp. Berlin, Stuttgart: Borntraeger.

Sindowski, K.-H. & H. Streif 1974. Die Geschichte der Nordsee am Ende der letzten Eiszeit und im Holozän. In P. Woldstedt & K. Duphorn (eds.), *Norddeutschland und angrenzende Gebiete im Eiszeitalter*: 411-431.

Sjørring, S. 1983. The glacial history of Denmark. In J. Ehlers (ed.), *Glacial Deposits in North-west Europe*: 163-179.

Smit, G. 1971. *De agrarisch-maritieme struktuur van Terschelling omstreeks het midden van de negentiende eeuw*: 218 pp. Groningen: Sasland; Leeuwarden: Miedema Pers.

Smit, C.J. & W.J. Wolff (eds.) 1983. *Birds of the Wadden Sea*: 308 pp. Rotterdam: Balkema.

Sørensen, H.E. 1982. *Rømøs Historie*: 288 pp. Skærbek: Melbyhus.

Sokolow, N.A. 1894. *Die Dünen – Bildung, Entwickelung und innerer Bau*: 298 pp. Berlin: Springer.

Solger, F. 1908. Über Parabeldünen. *Monatsberichte der Deutschen Geologischen Gesellschaft, 1908* (3): 54-59.

Solger, F., P. Graebner, J. Thienemann, P. Speiser & W.O. Schulze 1910. *Dünenbuch – Werden und Wandern der Dünen, Pflanzen- und Tierleben auf den Dünen, Dünenbau*: 404 pp. Stuttgart: Enke.

Staalduinen, C.J. van (ed.) 1977. *Geologisch onderzoek van het Nederlandse Waddengebied*: 77 pp. Haarlem: Rijks Geologische Dienst.

Staatsbosbeheer (eds.) 1979. De Boschplaat – Staatsnatuurmonument en Europees natuurreservaat. *Landloperreeks* 12: 7 pp.

Staatsbosbeheer (eds.) 1980. De bossen van Vlieland. *Landloperreeks* 22: 7 pp.

Staderman, R. 1937. Landerhaltung und Landgewinnung an der deutschen Nordseeküste. In G. Wüst (ed.), Werdendes Land am Meer - Landerhaltung und Landgewinnung an der Nordseeküste. – *Das Meer in volkstümlichen Darstellungen* 5: 42-70.

Stephen, M.F. 1981. *Effects of seawall construction on beach and inlet morphology and dynamics at Caxambas Pass, Florida*: 196 pp. Unpublished Ph.D. Thesis, University of South Carolina.

Straaten, L.M.J.U. van 1950. Giant ripples in tidal channels. Wadden symposium. *Tijdschrift van het Koninklijk Nederlandsch Aardrijkskundig Genootschaap* LXVII: 336-341.

Straaten, L.M.J.U. van 1961. Directional effects of winds, waves and currents along the Dutch North Sea coast. *Geologie en Mijnbouw* 40: 333-346, 363-391.

Straaten, L.M.J.U. van 1965 a. De bodem der Waddenzee. In J. Abrahamse, J.D. Buwalda & L.M.J.U. van Straaten (eds.), *Het Waddenboek*: 75-151.

Straaten, L.M.J.U. van 1965 b. Coastal barrier deposits in South- and North-Holland, in particular in the areas around Scheveningen and Ijmuiden. *Mededelingen Geologische Stichting* N.S.17: 41-75.

Straaten, L.M.J.U. van 1979. Surface Geology of Texel, Marsdiep and Balgzand. *International Association of Sedimentologists, International Meeting on Holocene Marine Sedimentation in the North Sea Basin, Texel, The Netherlands, September 17-23, 1979, Guidebook for Excursions*: 1-8.

Straaten, L.M.J.U. van & Ph.H. Kuenen 1957. Accumulation of fine grained sediments in the Dutch Wadden Sea. *Geologie en Mijnbouw* 19: 329-354.

Streif, H. 1971. Stratigraphie und Faziesentwicklung im Küstengebiet von Woltzeten in Ostfriesland. *Beihefte zum Geologischen Jahrbuch* 119: 59 pp.

Streif, H. 1975. Versuch einer Bilanzierung der Sedimentation im Küstenholozän Ostfrieslands. *Geologisches Jahrbuch* A 28: 3-14.

Streif, H. 1978. Geologie des Küstenraumes. In H.-E. Reineck (ed.), *Das Watt – Ablagerungs- und Lebensraum*, 2nd edition: 24-30.

Streif, H. 1981. *Geologische Karte von Niedersachsen 1:25,000, Erläuterungen zu Blatt Nr. 2414 Wilhelmshaven*: 111 pp. Hannover: Niedersächsisches Landesamt für Bodenforschung.

Stride, A.H. (ed.) 1982. *Offshore Tidal Sands*: 222 pp. London, New York: Chapman and Hall.

Suhr, H. 1964. Generalplan Deichverstärkung, Deichverkürzung und Küstenschutz in Schleswig-Holstein vom 20. Dezember 1963. *Wasser und Boden* 16 (8): 249-254.

Taubert, A. 1982. Wohin wandern die Außensände? Formänderungen der nordfriesischen Außensände und deren küstengeographische Beurteilung. *Nordfriesland* 16 (61/62): 37-48.

Terwindt, J.H.J. 1981. Origin and sequences of sedimentary structures in inshore mesotidal deposits of the North Sea. In S.-D. Nio, R.T.E. Schüttenhelm & Tj.C.E. van Weering (eds.), *Holocene Marine Sedimentation in the North Sea Basin. Special Publication Number 5 of the International Association of Sedimentologists*: 4-26.

Thiel, H., M. Grossmann & H. Spychala 1984. Quantitative Erhebungen über die Makrofauna in einem Testfeld im Büsumer Watt und Abschätzung ihrer Auswirkungen auf den Sedimentverband. *Die Küste* 40: 259-314.

Thiesen, G. 1977. Studier af Turistbesøget pa Skallingen 1975-1976. *Geografisk Tidsskrift* 76: 78-83.

Thilo, R. & G. Kurzak 1952. Die Ursachen der Abbruchserscheinungen am West- und Nordweststrand der Insel Norderney. *Die Küste* 1 (1): 1-20.

Tietze, G. 1983. *Das Jungpleistozän und marine Holozän nach seismischen Messungen nordwestlich Eiderstedts, Schleswig-Holstein*: 118 pp. Dissertation, Universität Kiel.

Tiniakos, L. & K.-D. Kachholz 1984. Transportdifferenzierung von Korngrößenspektren klastischer Sedimente Schleswig-Holsteins. *Meyniana* 36: 51-92.
Tongers, J. 1975. *Unser Langeoog, wie es wurde*, 3rd edition: 151 pp. Rhauderfehn: Ostendorp.
Tooley, M.J. 1978. *Sea-level changes in North-West England during the Flandrian Stage*: 232 pp. Oxford: Oxford University Press.
Tougaard, S. & J. Frikke 1984. Ho Bugt. *Waddenbulletin* 19 (3): 132-134.
Tougaard, S. & H. Meesenburg 1974. Vadehavet fra luften. *Bygd* 5 (1): 32 pp.
Tougaard, S. & H. Meesenburg 1978. *Den jyske vestkyst. En geografisk fortælling i flyfoto og kort, 2nd edition: (90 pp.)*. Esbjerg: Bygd.
Tougaard, S. & C. Helweg Ovesen (eds.) 1979. *Environmental Problems of the Waddensea Region – Proceedings of the Scientific Symposium, Ribe, Denmark, 16-18 May 1979*: 149 pp. Esbjerg: Fiskeri- og Søfartsmuseet.
Toxopeus, W. 1981. *Ik ben van Rottum. – Erinneringen aan de strandvoogd en het eiland Rottum*: 119 pp. Groningen: Uitgeverij Kemper.
Traeger, E. 1892. *Die Halligen der Nordsee*: 117 pp. Stuttgart: Engelhorn.
Ulrich, J. 1973. Die Verbreitung submariner Riesen- und Großrippeln in der Deutschen Bucht. *Ergänzungsheft zur Deutschen Hydrographischen Zeitschrift* B 14: 31 pp.
Ulrich, J. 1979. Bodenrippeln als Indikatoren für Sandbewegung. In Deutsche Forschungsgemeinschaft (ed.), *Sandbewegung im Küstenraum – Rückschau, Ergebnisse und Ausblick*: 333-350.
Ulrich, J. & H. Pasenau 1973. Morphologische Untersuchungen zum Problem der tidebedingten Sandbewegung im Lister Tief. *Die Küste* 24: 95-112.
Ulrich, K. 1936. Die Morphologie des Roten Kliffs auf Sylt. *Aus dem Archiv der Deutschen Seewarte* 56 (1): 26 pp.
US Army Corps of Engineers (eds.) 1984. *Shore Protection Manual*, 4th edition, 2 Vols: 608 + 614 pp. Vicksburg: Coastal Engineering Research Center.
US Department of Energy (eds.) 1985. *Glaciers, Ice Sheets and Sea Level: Effect of a CO_2-induced Climatic Change. Report of a Workshop Held in Seattle, Washington, September 13-15, 1984*: 330 pp. Washington: Office of Energy Research.
Valentin, H. 1957. Die Grenze der letzten Vereisung im Nordseeraum. *Abhandlungen des Deutschen Geographentages in Hamburg* 30: 359-366.
Veen, J. van 1936. *Onderzoekingen in de Hoofden*: 252 pp. Den Haag: Landsdrukkerij.
Veen, J. van 1950. Eb- en Vloedschaar systemen in de Nederlandse Getijwateren. *Tijdschrift van het Koninklijk Nederlandsch Aardrijkskundig Genootschaap*, Tweede Reeks LXVII: 303-325.
Veen, J. van 1956. Overstromingen tijdens de negen stormvloeden sinds 1877. *Tijdschrift van het Koninklijk Nederlandsch Aardrijkskundig Genootschap*, Tweede Reeks LXXIII: 1-6.
Veenstra, H.J. 1976. Struktur und Dynamik des Gezeitenraumes. In J. Abrahamse, W. Joentje & N. van Leeuwen-Seelt (eds.), *Wattenmeer. Ein Naturraum der Niederlande, Deutschlands und Dänemarks*: 19-45.
Veenstra, H.J. 1982. Size, Shape and Origin of the Sands of the East Frisian Islands (North Sea, Germany). *Geologie en Mijnbouw* 61 (2): 141-146.
Veenstra, H.J. 1984. Size and shape-sorting of coastal sands in the eastern part of the German Bight (North Sea). *Geologie en Mijnbouw* 63 (1): 47-54.
Veenstra, H.J. & A.M. Winkelmolen 1976. Size, Shape and Density Sorting around two Barrier Islands along the North Coast of Holland. *Geologie en Mijnbouw* 55 (1/2): 87-104.
Venzke, J.-F. 1985. Witterung und Eisverhältnisse an der Deutschen Nordseeküste im Winter 1985. *Bericht der Naturhistorischen Gesellschaft Hannover* 128: 247-260.
Verwey, J. 1981 a. The blue mussel *Mytilus edulis*. In N. Dankers, H. Kühl & W.J. Wolff (eds.), *Invertebrates of the Wadden Sea*: 114-115.
Verwey, J. 1981 b. The cockle *Cerastoderma edule*. In N. Dankers, H. Kühl & W.J. Wolff (eds.), *Invertebrates of the Wadden Sea*: 115-116.
Visser, J.C. 1947. Stuifdijken op Vlieland en Terschelling. *Tijdschrift van het Koninklijk Nederlandsch Aardrijkskundig Genootschap*, Tweede Reeks LXIV: 31-39.
Visser, G. 1978. *Onderzoek ribbelparameters Westelijke Waddenzee en zeegaten*. Rijkswaterstaat, Studiedienst Hoorn, Nota 78H011: 11 pp.
Visser, M.J. 1980. Neap-spring cycles reflected in Holocene subtidal large-scale bedform deposits: A preliminary note. *Geology* 8 (11): 543-546.
Voigt, H. 1969. Die Insel Amrum: Landschaft und Entwicklung. In M. & N. Hansen, (eds.), *Amrum – Geschichte und Gestalt einer Insel*, 2nd edition: 11-20, 29-36, 49-54.
Voigt, H. 1977. *Der Sylter Weg ins Dritte Reich*: 160 pp. Münsterdorf: Hansen & Hansen.
Vortisch, W. & M. Lindström 1980. Surface structures formed by wind activity on a sandy beach. *Geological Magazine* 117 (5): 491-496.
Wadsworth, J.R. 1980. *Geomorphic characteristics of tidal drainage networks in the Duplin River System, Sapelo Island, Georgia*: 247 pp. Unpublished Ph.D. Thesis, University of Georgia, Athens.
Wahnschaffe, F. 1901. *Die Ursachen der Oberflächengestaltung des norddeutschen Flachlandes*: 258 pp. Stuttgart: Engelhorn.
Walther, F. 1934. Die Gezeiten und Meeresströmungen im Norderneyer Seegat. *Die Bautechnik* 12 (13): 141-153.
Wasmund, E. 1936. Chemikalisch-physikalische Daten der Alttertiärtone um Fehmarn und der 'Innere Küstenzerfall'. *Kieler Meeresforschungen* 1: 243-263.
Wasser- und Schiffahrtsdirektion Aurich 1968. *Nordseeinsel Borkum. Maßnahmen der Wasser- und Schiffahrtsverwaltung des Bundes für den Küsten- und Inselschutz zur Sicherung der Seewasserstraße Ems, Schutzhafen Borkum, Signalstelle Borkum*: 47 pp. Wasser- und Schiffahrtsamt Emden, Baubüro Borkum (Unpublished Report).
Wee, M.W. ter 1983. The Elsterian Glaciation in the Netherlands. In J. Ehlers (ed.), *Glacial Deposits in North-west Europe*: 413-415.
Wegmann, H.P. 1982. *Borkum – Geschichte der Insel*, 4th edition: 100 pp. Leer: Gerhard Rautenberg.
Weijdt, L.J. 1980. Oude en Nieuwe dijken. In J.C. Duinker (ed.), *Texel en de Zee. Een strijd van Eeuwen*: 73-97.
Wenzel, D. 1979. Strand- und Vorstrandentwicklung in Westerland nach der Sandvorspülung 1972. *Die Küste* 34: 140-149.
Werkgroep Rekreatie van de Landelijke Vereniging tot Behoud van de Waddenzee 1977. *Eilanden onder de voet*: 124 pp. Harlingen.
Werner, F. & R.S. Newton 1975. The pattern of large scale bed forms in the Langeland Belt (Baltic Sea). *Marine Geology* 19: 29-59.
West, R. 1967. The Quaternary of the British Isles. In K.

Rankama (ed.), *The Quaternary*, Volume 2: 1-87.
Westenberg, J. 1956. Oude kaarten en de geschiednis van het voormalige eiland Huisduinen. *Tijdschrift van het Koninklijk Nederlandsch Aardrijkskundig Genootschap*, Tweede Reeks LXXIII: 223-240.
Wieczorek, U. 1982. Methodische Untersuchungen zur Analyse der Wattmorphologie aus Luftbildern mit Hilfe eines Verfahrens der digitalen Bildstrukturanalyse. *Münchener Geographische Abhandlungen* 27: 149 pp.
Wieland, P. 1972. Untersuchung zur geomorphologischen Entwicklungstendenz des Außensandes 'Blauort'. *Die Küste* 23: 122-149.
Wiermann, R. 1962. Botanisch-moorkundliche Untersuchungen in Nordfriesland. *Meyniana* 12: 97-146.
Wildvang, D. 1911. *Eine prähistorische Katastrophe an der deutschen Nordseeküste und ihr Einfluß auf die spätere Gestaltung der Alluviallandschaft zwischen der Ley und dem Dollart*: 67 pp. Emden & Borkum: Haynel.
Wildvang, D. 1938. Die Geologie Ostfrieslands. *Abhandlungen der Preußischen Geologischen Landesanstalt* N.F. 181: 211 pp.
Williams, R.S. & W.D. Carter (eds.) 1976: ERTS-1 – A new window on our planet. *Geological Survey Professional Paper* 929: 362 pp. Washington: United States Government Printing Office.
Windberg, F. 1931. Die Dünen von Juist. *Annalen der Hydrographie und Maritimen Meteorologie* 59 (2): 53-63.
Wißmann, W. 1981. *Untersuchung der Morphodynamik im Wattstrom Süderaue*: 94 pp. Hamburg: Unpublished Diplomarbeit.
Wohlenberg, E. 1950. Entstehung und Untergang der Insel Trischen. *Mitteilungen der Geographischen Gesellschaft in Hamburg* XLIX: 158-187.
Wohlenberg, E. 1969. *Die Halligen Nordfrieslands*: 32 pp. Heide: Westholsteinische Verlagsanstalt Boyens & Co.
Woldstedt, P. 1958. *Das Eiszeitalter – Grundlinien einer Geologie des Quartärs*, Band 2: 438 pp. Stuttgart: Enke.
Woldstedt, P. & K. Duphorn 1974. *Norddeutschland und angrenzende Gebiete im Eiszeitalter*: 500 pp. Stuttgart: Köhler.
Wolff, E.J. (ed.) 1983. *Ecology of the Wadden Sea*, 3 Vols.: 1678 pp. Rotterdam: Balkema.
Wolff, W. 1910. Erwiderung auf die Ausführungen des Herrn Gagel. *Monatsberichte der Deutschen Geologischen Gesellschaft* 1910 (1): 63-64.
Wolff, W. 1922. *Erdgeschichte und Bodenaufbau Schleswig-Holsteins*: 166 pp. Hamburg: Friederichsen & Co.
Wolff, W. 1923. Über die Beziehungen zwischen Moor und Marsch im Lande Hadeln – Ein Beitrag zur Senkungsgechichte der Nordseeküste. *Jahrbuch der Preußischen Geologischen Landesanstalt* 1922, XLIII: 266-272.
Wolff, W. 1939. Present Day Geological Problems of the German North Sea Coast. *Research and Progress* V (5): 288-296.
Wrage, W. 1930. Das Wattenmeer zwischen Trischen und Friedrichskoog. *Aus dem Archiv der Deutschen Seewarte* 48 (5): 128 pp.
Wrage, W. 1958. Luftbild und Wattforschung. *Petermanns Geographische Mitteilungen* 102 (1): 6-12.
Wüst, G. (ed.) 1937. Werdendes Land am Meer – Landerhaltung und Landgewinnung an der Nordseeküste. *Das Meer in volkstümlichen Darstellungen* 5: 132 pp.
Wunderlich, F. 1973. Sekundäre Schichtdeformationen unter Eisauflast. *Senckenbergiana maritima* 5: 153-159.
Wunderlich, F. 1978. Marken. In H.-E. Reineck (ed.), *Das Watt - Ablagerungs- und Lebensraum*, 2nd edition: 95-105.
Wunderlich, F. 1979. Die Insel Mellum (Südliche Nordsee): Dynamische Prozesse und Sedimentgefüge. I. Südwatt, Übergangszone und Hochfläche. *Senckenbergiana maritima* 11 (1/2): 59-113.
Wunderlich, F. 1983. Sturmbedingter Sandversatz vor den Ostfriesischen Inseln und im Gebiet des Großen Knechtsandes, Deutsche Bucht, Nordsee. *Senckenbergiana maritima* 15 (4/6): 199-217.
Zagwijn, W.H. 1979. Early and Middle Pleistocene coastlines in the southern North Sea basin. In: E. Oele, R.T.E. Schüttenhelm & A.J. Wiggers (eds.), *The Quaternary History of the North Sea*: 31-42.
Zagwijn, W.H. 1984. The formation of the Younger Dunes on the west coast of The Netherlands (AD 1000-1600). *Geologie en Mijnbouw* 63 (3): 259-268.
Zagwijn, W.H. & C.J. van Staalduinen (eds.) 1975. *Toelichting bij Geologische Overzichtskaarten van Nederland*: 134 pp. Haarlem: Rijks Geologische Dienst.
Zandstra, J.G. 1977. Geologische opbouw van het Pleistoceen. In C.J. van Staalduinen (ed.), *Geologisch onderzoek van het Nederlandse Waddengebied*: 37-56. Haarlem: Rijks Geologische Dienst.
Zandstra, J.G. 1983. Fine gravel, heavy mineral and grain-size analyses of Pleistocene, mainly glacigenic deposits in the Netherlands. In J. Ehlers (ed.), *Glacial Deposits in Northwest Europe*: 361-377.
Zenius, M. 1983. *Mandø i hundrede år*: 125 pp. Esbjerg: Bygd.
Ziegler, P.A. 1975. Geologic Evolution of North Sea and it tectonic framework. *The American Association of Petrologists Bulletin* 59 (7): 1073-1097.
Ziegler, P.A. 1982. *Geological Atlas of Western and Central Europe*: 130 pp. The Hague: Shell Internationale Petroleum Maatschappij B.V.
Ziegler, P.A. & C.J. Louwerens 1979. Tectonics of the North Sea. In E. Oele, R.T.E. Schüttenhelm & A.J. Wiggers (eds.), *Quaternary History of the North Sea*: 7-22.
Zitscher, Fr.-F. 1960. Schutz des Westrandes der Insel Sylt durch Flachbuhnen. *Wasser und Boden* 12 (9): 300-302.
Zitscher, Fr.-F. 1985. Aktuelle Aufgaben und Probleme des Küstenschutzes in Schleswig-Holstein. *Jahrbuch der Hafenbautechnischen Gesellschaft* 40, 1983/84: 290-296.
Zitscher, Fr.-F., R. Scherenberg & U. Carow 1979. Die Sturmflut vom 3. und 21. Januar 1976 an den Küsten Schleswig-Holsteins. *Die Küste* 33: 71-100.

INDEX

The following abbreviations have been used in this index: f - figure, m – map, ph – photograph, pl – plate.

abrasion 190, 238f
Accumer Ee 27f, 35f, 77f, 80m, 83, 92, 138, 290m, 306, 307
   ebb delta 102
   bar migration 90
   inlet migration 85
   morphological development 306f
adhesion
   ripples 148, 148ph, 149ph
   warts 148
Adolf-Hitler-Koog (= Dieksander Koog) 215, 282
aeolian breccia 158
afforestation
   Ameland 336ph
   Amrum 249
   Texel 358
   Vlieland 351
Afsluitdijk 15, 352m, 353m
Agropyron junceum 328
air raids 216
airfield 218, 305
alder 240
algae 60, 220
Alkersum, Föhr 247m
Alkmaar 24
Alnus 351
Alte Harle 84, 301
Alte Herst Warft, Gröde 267
Alte Hever 110m, 259m
Alter Koog, Nordstrand 260m, 276
Alter Koog, Pellworm 260m, 275
Alter Westturm, Wangerooge 204m
Altona 212, 212m
Alt-Helmsand 282
Alt-Scharhörn 285
Ameland 2m, 14, 14m, 17, 26f, 32f, 52, 148, 170, 178, 190m, 212m, 214, 218, 330m, 333, 334, 335, 335m, 336, 337
   beach, backwash ripples 143ph
   beach replenishment 338ph
   coastal protection structures 337m
   comb-like rillmarks on the beach 148ph
   Dam 72, 72ph, 330m, 335m, 338
   morphological development of the eastern end 339m
   Mytilus colony south of 60ph
   washover channel 175ph
   western end, aerial photograph 336ph
   Young Dunes 24
Amelandse Gat 15
Ammophila arenaria 151, 160, 161f, 169, 243, 286, 299ph, 303, 310, 312 pl, 318, 332, 349, 351
Amrum 2m, 10m, 17, 18, 26f, 29, 40, 133, 134, 150, 152, 163, 186, 190, 212m, 214, 222m, 233, 238, 247, 248, 249, 250, 251, 252, 253, 259, 268
   deflation on the beach 158ph
   foam on the beach 220ph
   Landsat-5 image 231ph
   location map 247m
   micro mushroom forms on the beach 158ph
   morphodynamic subdivision 248m
   Young Dunes 25
Amrumbank 222m, 259m
Amrum Odde 247m, 248m, 251, 253
   aerial photographs 250ph
   coastal erosion 250ph, 252ph
Amsinck-Koog 268
Amsteldiep 15
Amstelmeer 353m
Amsterdam 2m, 8m, 13m, 32f, 190m, 213
Angles and Saxons 276
Anjum 335m
Anna-Paulowna-Polder 352m, 353m
Antarctic ice sheet 12
antidunes 141, 143ph
Appelland 266ph, 267
archaeology
   Amrum 249
   Borkum 321
   Sylt 243
   Texel 352
Archsum, Sylt 239m
Arenicola marina 60, 61ph, 244
Arensch 25, 45, 191, 283m
   tidal flats west of 71ph
Artemisia maritima 73 pl, 182, 194
artificial seaweed
   Ameland 337m
   Texel 354m
Aster tripolium 182
Atlantic period 11, 11f, 25
Atlantikwall 215, 216
Atriplex littoralis 73pl
Außensände 29, 152, 258
Aussichtsdüne, Baltrum 302m
avalanche tongues on dune slope 160, 160ph, 162, 164ph

Backenswarft, Hooge 260m
   water reservoir 271ph
backshore 135f, 136, 149, 150, 152, 332
   dunes 154
backwash
   marks 146
   ripples 141, 143ph
Balgzand 330m
Ballonplate 290m
Ballum, Ameland 229m, 335, 335m, 337m
   Bight 335
Ballumer stuifdijk 335m, 337m
Ballumerduinen 335, 335m
Baltic Sea 211, 214, 268
   cliffs 190
Baltrum 1, 2m, 26f, 27f, 29, 30, 33, 34, 34f, 35f, 77f, 80m, 83, 84, 84f, 85, 93, 96, 97, 99ph, 101ph, 101m, 114m, 118m, 129m, 130m, 135, 138, 146ph, 168, 168ph, 171ph, 172ph, 180, 185pl, 190m, 200, 207, 218, 290m, 301, 302m, 304, 305, 306, 306f, 307, 308, 309, 309ph, 310ph, 311ph
   foreshore shoal 97, 102
   morphological development 306f
   sand wave on the foreshore 140ph
   sawtooth bars 138
   tidal current and drift current 38f
   wind direction and velocity 35f
Baltrumer Balje 96, 96ph, 114m, 115
Baltrumer Inselwatt 72ph, 95, 96, 114m, 115
   barchan-like forms 49ph, 50ph
   sand wave migration 96m, 97m, 115ph
Bancks Polder, Schiermonnikoog 328m, 332
Band 329m
Bandixwarft

381

Hooge, destroyed 269, 269m
Langeneß 247m
Bant 321
bar migration 26, 27, 90, 109, 139, 280, 361
barchan migration on Terschelling 165
barchans, beach 45
barchan-like forms on the tidal flats 49ph, 50ph
barchan-like sand waves on the beach 152pl
Bargumer Koog 20
barrier island length 17
basal peat 10
Bassens 289m
Bay of Fundy 112
beach and foreshore
 Juist 137ph
 Sylt 135f, 207f
beach barchans 45, 152 pl, 154, 156, 156ph, 157ph, 158, 249, 362
 migration rates of 167
beach barriers, Eiderstedt 108, 183, 184ph, 185ph, 186, 277ph, 278ph, 279
beach development on Borkum 322, 324
beach dunes 152
beach erosion, effect of groynes 200
beach profile, general 135f
beach replenishment 204
 Ameland 338, 338ph
 De Koog, Texel 359ph
 Hörnum, Sylt 241
 Kampen, Sylt 241
 Norderney 205, 206ph, 208f, 313, 314
 Rantum, Sylt 241
 Sylt 205
 Texel 358
 Wangerooge 205, 295
 Wenningstedt, Sylt 241
 Westerland, Sylt 241
beach response to storm surges 150, 150f
beach ridge 24, 25, 41, 176f
 Fanø 226, 227
 Föhr 253, 257ph
 Griend 346
 Japsand and Norderoogsand 86m
 Jordsand 237
 Norderney 97, 99m
 Norderoog 272
 tidal-flat facing 226ph
beach ridges, dune-capped, Ellenbogen 189
beach ridges, fossil, Eiderstedt 276
beach terminology 135f
beach width 135
Bederkesa 8m
Beenshallig 261, 262
Benknolde dunes 223m, 225ph
Bensersiel 70, 289m, 302m
 sand flats north of 70ph, 71ph
benthic biomass 60
Berensch 283m

Bergen 2m
Inlet 24
berm 146
 runnel 135f
Bielshöft 282
Bielshövensand 259m
bifurcating rillmarks 147, 148
Bill, Juist 316, 317
Billdünen 317m
Billum 223m
bioturbation 46, 60
bird sanctuary
 Griend 347
 Jordsand 237
 Memmert 320
Blåvand 221, 222, 223m, 224m
Blåvands Fyr 221, 222, 223m
Blåvandshuk 1, 2m, 174, 221, 223m, 224m, 224ph, 276
black dunes 310
black-headed gull (= *Larus ridibundus*) 61
Blaue Balje 27f, 65m, 76, 132, 290, 290m, 291, 291m
Blauort 2m, 14, 26f, 29, 32f, 259m, 260m, 279m, 284
Blauort-Steert 78f
Blidsel, Sylt 188m, 244
 Bight 244
 lee-side erosion 209ph
 revetment 209ph
blowouts 150, 160, 162, 169, 172, 362
 Ameland 335
 Amrum 249
 Baltrum 310
 Fanø 152pl, 170ph, 227, 229
 in grey dune 168ph
 in former migrating dune 169f
 in old dunes 168
 Rømø 235
 secondary 173ph
 Spiekeroog 301
Böhl 260m
Boldixum, Föhr 247m, 255
Bollen, Marsdiep 103m, 104m, 105m, 106m, 107
bomb craters on Wangerooge 296ph, 297
Boomkensdiep 340m
Boragh, Amrum 247m, 251
Bordelumer Koog 21
Bordrup 221, 223m, 224m
Borgsum, Föhr 247m
Borkum 2m, 18, 19ph, 26f, 37, 102, 135, 154, 157ph, 158ph, 160ph, 161ph, 162, 164ph, 178, 190m, 200, 212m, 214, 215, 216, 218, 290m, 292, 317m, 320, 322, 323ph, 324, 328m, 329m
 aerial photograph of the north-western end 322ph
 boundary between positive and negative sand balance 324ph
 built-up areas 216m, 217m
 coastal development 326f
 coastal protection structures 323ph

exposed revetment base 325ph
recent dune formation in the east 327ph
Riffgrund 26, 290m
salt dome 6, 7
salt marsh erosion 280pl
tidal currents east of 41f
wind ripple migration 153ph
Borkum-Nord salt dome 7
Borndiep 330m, 336, 337, 338
Bornrif, Ameland 330m, 335m, 336, 338
borrow pits
 in tidal flats 73, 73ph
 south of Terschelling 186ph
Bosch 327, 329m
Boschplaat
 Rottumerplaat 328m
 Terschelling 29, 178, 341m, 342
 dune cores 180ph
 recent dunes 162ph, 345ph
 stuifdijk 160, 161f
Botterrug, Marsdiep 106m, 107
box cores from Harle ebb delta 128
Boy Ocksens Halg 272
Braderup, Sylt 244, 245, 239m
Brakzand 330m
branching rill marks 148
breakers 40f, 52
breakwaters, diffraction leewards of 52
breccia, aeolian 158
Bredstedt 7m, 9
Bredstedter Koog 21
Breewijd, Marsdiep 106m
Breezand 352m, 353m
Brekklumer Koog 21
Bremen 2m, 13m, 28f, 32f, 212m
Bremerhaven 214
Bronze Age 25, 195, 249
Brunsbüttel 212
brushwood groynes 186, 187, 198, 222
 Amrum 251, 253ph
 Baltrum 309
 Borkum 326
 Gröde 267
 Hamburger Hallig 268
 Hooge 270
 Langeneß 264
 Mandø 229
 Nordstrandischmoor 270
 Sylt 241, 245
 Wangerooge 291
Büsum 43f, 260m, 279m, 282, 284
Buise 30, 84, 311
Buisetief 311
bulkhead groynes 200
bulkhead, Spiekeroog 300
bulkheads 207
bunkers
 Ameland 337
 Blåvandshuk 222, 224ph
 Borkum 325, 326
 Fanø 227
 Kampen, Sylt 243
 Schiermonnikoog 330

Texel 358
Wangerooge 296
Bunter 7
Sandstone 5
Buntje-Ballum 229m, 239m
Buphever Koog, Pellworm 260m, 275
Buren, Ameland 335, 335m, 337m
Burger Nieuwland, Texel 352m, 353
buried channels 84
Buschsand, Trischen 29
Butjadingen 283m
Butterloch 263
Butwehl 264, 267
Buurderduinen, Ameland 335m, 336
By
   Dyke 229m, 231, 232, 233
   Koog 233

Cadmium 220
Cafe Cornelius, Norderney 302m, 311
Calais Transgression 11, 11f, 12, 25
Callantsoog 17, 352, 353, 353m
Calluna vulgaris 151, 170
Camperduin 28
Canterbury 303
Cappel 283m
Cappel-Neufeld 69ph, 283m
carbonate in mud flats 68
Cardium edule (= *Cerastoderma edule*) 59, 263
Carolinensiel 215, 289m
Carolinensieler Balje 75m, 76
cavernous beach sand 149ph, 150ph
Cecilienkoog 260m, 271
Cerastoderma edule 59, 263
Christians-Koog, Nordstrand 21, 275
clay balls 193, 193ph
coastal engineering 206, 207, 219, 362
coastal erosion 190m
   Ameland 336, 337
   Amrum Odde 252ph
   Hörnum Odde, Sylt 246ph
   Juist 317
   Mandø 229
   Rantum, Sylt 246ph
   Rottumeroog 327
   Texel 356
   Vlieland 349
coastal protection 188, 204, 215
   Ameland 337m
   Baltrum 306, 307, 308, 309
   Borkum 325
   costs of 207
   Föhr 255
   Griend 346
   Habel 268
   Halligen 264, 272, 273
   Hooge 270
   Jordsand 238
   Juist 316
   Langeoog 305
   Neuwerk 288
   Norderney 311, 313ph
   Norderoog 271ph, 272
   Nordstrand 276
   passive 204
   Spiekeroog 299
   Südfall 272
   Sylt 240, 241
   Texel 345pl, 354m, 355
   Wangerooge 202, 203, 204, 204m, 292, 296
coastline
   changes at Texel 357f
   development on Juist 318f
coast-normal sand transport 361
coast-parallel dune ridges 152, 160, 161, 168, 186, 227
   Amrum 249
   block diagram 163f
   Langeoog 163f
   Texel 356, 358, 359ph
coast-parallel
   sand transport 83, 97
   tidal current 81
cockle 59, 60
Cocksdorp, Texel 352m, 354m
Coggendiep, Terschelling 342
cohesiveness of the tidal flat surface 60
comb-like rill marks 148, 148ph
comet marks 72, 72ph, 73
commandeurs 212
commercial shipping 213
compaction of sediments 12
conical rill marks 148
Copenhagen 212
Copper 220
Corenzand 327, 329m
Corynephorus canescens 170, 312pl
crescent bars 80m, 81, 102
   megaripple migration on 125, 126, 127f
   megaripples on 52
   Wichter Ee ebb delta 97, 98, 100f, 101m, 118m, 128ph
Cretaceous 7
cryoturbation 251
currency reform 213f
current
   crescents 58, 58ph
   drift 34, 36, 44
   groynes 200
   lineations 138, 140, 142ph
   measurements 33, 36, 36f
   patterns 38
   residual 33, 34
   ripples 144
   stripes on tidal flats 71ph
   tidal 33, 34, 36, 38, 40
   tidal, in the Lister Tief 39f
   velocities 33, 37
Cuxhaven 2m, 12, 13m, 17, 25, 28f, 32f, 38, 40, 43f, 190m, 191, 214, 279m

Dagebüll 18, 190m, 239m, 247m, 256, 267
Dammkoog, Eiderstedt 276
De Ans salt marsh area, Terschelling 341m, 343
De Balg, Ameland 335
De Grie Polder, Texel 354
De Grieen salt marsh area, Terschelling 341m, 343
De Hon, Ameland 335m
De Hors megashoal, Texel 359
De Keach salt marsh area, Terschelling 343
De Keeg, Terschelling 341m
De Koog
   Terschelling 341m
   Texel 214, 352m, 354m
      dune erosion 359ph
De Oel salt marsh area, Terschelling 343
De Schorren salt marsh, Texel 352m, 356, 358ph
De Slufter, Texel 352m
De Waal, Texel 352m
debris apron 189, 191, 191ph
deciduous forest, Ameland 336ph
deep bars 108, 109f, 110m, 361
deflation 152, 159, 159ph, 162, 168
   landscape on Fanø 152pl
   on the beach 158ph, 161ph
Dellewal, Terschelling 341m, 343
delta bedding 361
   in deep bars 109
delta formed by rain-runoff 172, 173ph
Den Burg, Texel 352m, 353, 354m
Den Haag 24
Den Helder 1, 2m, 13, 13m, 14m, 15, 17, 32f, 89, 107, 174, 190m, 212m, 213, 214, 216, 276, 352m, 353, 353m
Den Hoorn, Texel 345pl, 352m, 354m
Den Oever 352m, 353m
dendritic
   drainage pattern on tidal flats 60, 62ph, 320f
   salt marsh drainage pattern 187
   tributaries to rill marks 147
Derde Duintjes, Terschelling 341m
desert
   barchans 162
   dunes 151
desiccation cracks in mud 278ph
diatom mats 61, 61ph, 62ph
diatoms 60
Dieksand 281, 282
Dieksander Koog 21, 215, 279m, 282
diffraction 52
Dithmarschen 5, 13m, 108, 280, 286
Doberan 214
Dogger 7
   Bank 6m, 9
Dokkum 335m
Dollart 2m, 20, 21, 22, 26f, 180, 182, 187, 190m
Domäne, Langeoog 302m, 303
Doove Balg 330m
Dorfgroden, Wangerooge 292
Dornum 302m
Dornumersiel 35f
Dorum 283m
Dorumer Neufeld 283m
Dove Harle 292

drainage
  of salt marshes 197
  problems, Fanø 227
dredging 25, 291
Dreebargen, Langeoog 302, 302m
drift current 34, 36, 38f, 44 , 262, 285
drift ice 38, 40, 42ph, 43f, 54, 56, 57, 58
  around Iceland 29
  sediment transport by 59
Dünemeyers
  Juist 316
  Langeoog 303
Duhnen 26f, 279m
dune cores 174, 198
  Borkum 320
  Boschplaat, Terschelling 342, 343ph
  erosion of 178
  Juist 316
  Langeoog 163f, 302
  Norderney 176f, 311, 314
  Rottumeroog 327
  Rottumerplaat 328
  Schiermonnikoog 186, 329, 331ph
  Spiekeroog, aerial photograph 298ph
  Terschelling 180ph, 340
dune development
  Fanø 225m
  Juist 318f
  Norderney and Baltrum 306f
  Schiermonnikoog 331ph, 332
dune dykes (= stuifdijken) 160, 178, 198
dune erosion 169, 192, 193
  by surface runoff 172
  on Ameland 337
  on Baltrum 306, 307
  on Langeoog 304
  on Norderney 315ph
  on Rottumeroog 327
  on Spiekeroog, Lütjeoog Dunes 299ph
  on Sylt 192ph, 243
  on Texel 356, 359ph
  on Vlieland 348, 348ph, 349
  on Wangerooge 294ph, 295ph
dune fan on southern Texel 359
dune formation 150, 362
  central Vlieland 348ph
  Boschplaat, Terschelling 342, 343, 345ph
  eastern Ameland 336
  eastern Baltrum 310
  eastern Borkum 326, 327ph
  eastern Schiermonnikoog 332
  Harlehörn, Wangerooge 297
  Juist 317
  Knechtsand 288
  Langeoog 302, 303
  Memmert 320
  Norderney Ostplate 314
  phases of 24
  Rottumerplaat 327, 328
  Schiermonnikoog 333
  Spiekeroog 299ph

  Vlieland 349
  Wangerooge 294ph, 296
dune generations on Juist 317
dune migration 206
  Amrum 249
  Fanø 229
  Sylt 166, 244
dune plantations on Terschelling 341
dune protection 168, 198, 219, 296
  Baltrum 310ph
  Juist 316
  Langeoog 303, 305
  Spiekeroog 301
  Sylt 240
  Vlieland 351
  Wangerooge 296
dune ridges
  coast-parallel 152, 162
  drowning of 23
  fan-like 224
dune slopes, avalanche tongues on 164ph
dune stripes 156
  on the beach 155ph
dune valleys, Texel 358
dune vegetation 151
Dune, Helgoland 21
dunes
  at Büsum 282
  beach 152
  desert 151
  grey 167
  negative shadow 160, 162ph
  old 152, 167ph
  on Ameland 334
  on Bielshöft 282
  on Blauort 284
  on Eiderstedt 278
  on Neuwerk 288
  on Rømø 235
  on Scharhörn 286
  on Texel 356
  on Trischen 31f, 281
  parabolic 150
  primary 151
  secondary 151
  shadow 160
dune-slope erosion by heavy rain 173ph
Dunkerque Transgression 11, 11f, 12, 24, 25
Dunkerque-I-Transgresion 195
Dunsum, Föhr 239m, 247m, 257ph
dwelling mounds 195, 196
  Dieksand 281
  Eiderstedt 276
  Gröde 267
  Habel 268
  Halligen 262
  Helmsand 282
  Hooge 268
  Jordsand 237
  Langeneß 261ph, 262ph, 263, 264
  Nordstrandischmoor 270
  Norderoog 272
  Oland 265

  protection of 273, 274
  Südfall 272
dyke
  breaches 20
  construction 197, 198
  connecting Büsum with the mainland 282
  effects of storm surges 18
  on Ameland 335
  on Amrum 251
  on Borkum 325
  on Dieksand 281
  on Fanø 229
  on Föhr 255
  on Mandø 231
  on Neuwerk 288
  on Norderney 314
  on Rømø 233
  on Schiermonnikoog 332
  on Spiekeroog 298, 299
  on Sylt 240
  on Terschelling 343
  on Texel 345pl, 353, 354, 355
  on Trischen 281
  on Wangerooge 204m, 292
dyking
  on Eiderstedt, historical development 277m
  projects 188
D-Steert 259m, 279m, 285

Easington 8m
East Anglia 6m, 9
ebb
  channel 33, 80m, 81f, 82, 83, 86m, 98
  current, erosion by 130m
  delta 37, 77f, 78, 79m, 81, 83, 89, 96, 102, 146, 308ph, 361
  shield 80m, 82
  spur 80m, 82
Ebbevej 229m, 232
ebb-orientated
  bedforms 95
  megaripples 99ph
Eem River 8m
Eemian
  drainage pattern 84, 85
  sea level 9
  transgression 8m
Eemshaven 317m
Eendenkooi, Schiermonnikoog 330
Eerste Duintjes, Terschelling 341m, 342
Eggstedt 7m
Egmond 24
Eider 20, 25, 27, 40, 78, 108, 259m, 260m
  Dam 2m
  Estuary 78, 78f, 185
Eiderstedt 2m, 6, 17, 22, 25, 26, 40, 108, 110m, 134, 180, 182, 183, 185, 186, 187, 218, 229, 233, 260m, 276, 277m, 278, 278ph, 279, 280, 359
  beach barrier at St. Peter-Ording 278ph

Eider-Sperrwerk 260m
Eidum church, Sylt 240
Eijerland, Texel 89, 178, 187, 198, 347, 351, 356
　old drainage pattern 187ph
　Polder 354
Eijerlandsche Duinen, Texel 352m
Eijerlandse Gat 15, 89, 92, 187, 347, 350
　SEASAT image 51ph
Eiland, Norderney 362m, 314
Eilanderbalg 131, 330m
Elbe 2m, 9, 13m, 21, 26f, 27, 32f, 33, 40, 108, 131, 190m, 220, 259m, 279m, 280, 283m, 285, 288
　ice-marginal valley 9
Elisabethgroden 289m
Elisabeth-Außengroden 75m, 194ph, 289 m
Elisabeth-Sophien-Koog, Nordstrand 183m, 260m, 263, 276
Ellenbogen Peninsula, Sylt 188m, 189, 239m, 243ph, 244
　revetment 201
Elmshorn 212
Elsterian 8, 9, 238
　buried channels 8, 9
　ice margin 6m
Emden 2m, 13m, 21, 25, 28f, 32f, 190m, 212m, 214, 216
emerging shoal, Trischen 29
Emmapolder 317m
Emmelsbüll 239m
Emmerlev 188, 239m
　Klint 18, 190, 229m, 239m
Empetrum nigrum 151, 169, 170, 195, 312pl
Ems 2m, 21, 27, 32f, 190m, 317m
Emsland 9
Engelsche Kerkhof, Texel 356
Engelsmanplaat 2m, 330m, 333, 334, 335m
　morphological changes 334m
Englandshallig, Nordstrand 183m
Englischer Sand 272
erosional rill marks 147
Esbjerg 2m, 13, 13m, 28f, 32f, 190, 190m, 194, 213, 214, 216, 223m, 225m
Esens 302m
Esing 260m, 277ph
Essex Estuary, Massachussetts 111
eustasy 5
Everschop 260m, 276, 277m
Everste Koog, Texel 352m, 353
expansion of tidal deltas 130

Fahretoft 247m, 267
Fallstief
　Eiderstedt 276
　Nordstrand 262
Fanø 2m, 13, 17, 26f, 29, 150, 154, 163, 169, 213, 216, 218, 222m, 224, 226, 227, 229, 230, 232m, 233, 235, 236, 259, 359
　aerial photograph of migrating dunes 170ph
　beach barchans 152pl
　deflation landscape 152pl
　development of the northern end 225m
　erosion of a beach ridge 226ph
　location map 223m
　migrating dune 152pl
　morphological development 228ph
　tidal-flat facing beach ridge 226ph
　Vesterhavsbad 223m, 227
fascine mattresses 291, 337m, 338, 354m
faults 192
Ferwerd 335m
Festuca rubra 169, 182
fetch 53
Finkhaushalligkoog 260m, 278
fishery 211, 214
Flensburg 28f, 32f, 190m
Flinthörn Dunes, Langeoog 178, 305
Flinthörn Peninsula, Langeoog 302m, 304ph, 305, 305ph
flood
　channel 33, 80m, 81, 81f, 82, 83, 86m
　cliffs 179
　deltas 79m, 81, 94, 131m, 361
　　extension of 131
　-orientated bedforms 95
　ramp 80m, 82, 96, 114, 115ph, 133
　spits 29, 80m, 94, 250, 314
flounder groyne, Sylt 241
fluorescent tracers 36
foam
　created by algae 220, 220ph
　stripes 37, 38, 39ph, 41f
Föhr 2m, 10m, 134, 190, 192, 212, 212m, 214, 218, 229, 238, 239m, 248, 251, 252, 253, 254, 255, 255m, 256, 259, 264, 356
　location map 247m
　Goting Kliff 192ph, 256ph, 256m, 257ph
　spit migration at the north coast 258m
Föhrer
　Ley 255m
　Schulter 222m, 255m
foredune construction, Sylt 240
foreshore 27, 135f, 136f, 137ph, 138, 141f, 205, 207f, 241
　bar 80m
　ridges 52, 144, 149, 150
　shoal 101m, 101ph
　　Baltrum 97, 102
Formerum, Terschelling 341m
France 216
Franeker 341m
Franse Bankje, Marsdiep 103m, 104m, 105m, 106m
Franzosensand 190m, 279m, 282
Frederik-VII-Koog (= Friedrichskoog) 282
freshwater
　lenses under Terschelling 340
　marsh 195
Friederikensiel 289m
Friedrichsgabekoog 279m
Friedrichskoog 279m, 282
Friedrichstadt 277m
Friedrich-Wilhelm-Lübke-Koog 239m
Friesche Zeegat 330m, 333, 334
Friesland 13m, 84, 333
fringy rill marks 148
frozen ground 54, 55, 174ph
frozen sand on the beach 55ph
Fuhle Schlot 263
Fulkum 302m

Gaestaenacka 259
Galgedyb 224, 227, 228ph, 230
Galgerev 222m, 232m
Galmsbüll 211
Galmsbüllkoog 239m
Gammel Mandø 229, 229m, 230, 231
Gammeltoft Bjerge 223m, 227
Garding, Eiderstedt 260m, 276, 277m
　beach barrier 17, 277ph
garland-like pattern 44, 45ph
Gat vom Wrack 75m
Gelbsand 259m
Georgshöhe, Norderney 302m
German Bight 27, 30, 33, 188
German-Danish War of 1864 240, 276
giant ripples 112
glacial landforms 9
Glückstadt 212
Golden Era 212, 214, 235, 269
Goting, Föhr 247m, 253, 256m
Goting Kliff 190, 192, 192ph, 247m, 253, 256m, 256ph, 257ph
Grådyb 222m, 226, 227
Grårup 221, 223 m
grain-size
　analyses 26f, 27f, 65, 66f, 67f, 68
　characteristics 55
Granzin 7m
greenhouse effect 13
Greenland 212
　ice sheet 12
Greetsiel 317m
grey dunes 167, 168, 168ph, 172
　Baltrum 310, 310ph
　Spiekeroog 312pl
Grieen, Terschelling, salt marsh erosion 345ph
Griend 2m, 15, 134, 236, 237, 330m, 341m, 344, 345, 346ph, 347
Grimstrup 223m
Gröde 2m, 134, 247m, 258, 261, 262, 263, 264, 265, 266ph, 267, 274
Grønning, Fanø 223m, 226
Groet 353m
Groeter Polder 353m
Grohde, Norderney 302m
Grohde Polder, Norderney 314
Gronden van het Stortemelk 340m
Groningen 2m, 13m, 24, 28f, 32f, 84, 190m, 212m, 327
Groote Vlak, Texel 352m
　Zanddijk, Texel 353

Zeewijk 328m
Große
  Mandränke (= storm surge of 1362) 20
  Schlopp, Langeoog 178, 302m, 303, 305
  Süderwarft, Hooge, destroyed 269, 269m
Großer
  Knechtsand 2m, 14, 26f, 29, 32f, 43f, 131, 259m, 288
  Kohldammer Koog 20
  Koog, Pellworm 260m, 275
  Norderkoog, Pellworm 275
  Vogelsand 259m
Großes Oberahnesches Feld, Jade Bay 194
ground
  failure 268
  ice 54, 59
groundwater table 169, 227, 310, 315
groyne
  fields on Norderney 208f
  H, Wangerooge 75, 75m, 200, 203, 289m, 292, 293, 293m, 301
  heads, erosion at 73ph
  J, Wangerooge 200, 201ph, 202ph, 293
groynes 198, 205, 207, 219
  Ameland 335m, 337, 337m
  Amrum 247m, 251
  Baltrum 302m, 307, 308, 309
  beach 200
  Borkum 216m, 217m, 321, 325
  current 200
  diffraction leewards of 52
  effect on accretion 199f
  flounder 241
  Föhr 255
  Griend 346
  Juist 188, 317
  Norderney 302m, 311, 313, 314
  North Holland coast 353m
  Oland 267
  Skallingen 222, 223m
  Spiekeroog 289m
  steel bulkhead 200
  stone 198
  Sylt 239m, 240
  Terschelling 343
  Texel 352m, 354m, 356, 358
  Trischen 281
  underwater 83, 200
  Vlieland 341m, 347m, 349
  Wangerooge 202, 204m, 289m, 292
Grüne Insel 260m
Grüppen (= ditches) 186
Guldager 223m
Gulf of St.Lawrence 176

Haarlem 2m, 13m, 32f, 190m
Habel 2m, 134, 247m, 263, 267, 268
Hadeln 13m
Haefre 259
Hafendeich, Norderney 302m
Haferacker sand pit, Eiderstedt 277ph

Hage 302m
Haiddünen, Juist 316, 317m
Hainshallig 261
Hakensand 259m
Halen 223m
Halimione portulacoides 182
Hallge, Langeneß 247m, 264
Halligen 18, 21, 29, 134, 180, 187, 195, 196, 197, 212m, 229, 230, 236, 237, 256, 258, 260m, 261, 262, 263, 265, 267, 268, 269, 270, 271, 272, 273, 274, 307
Hallohmsglopp washover area, Juist 316, 317m
Hallum 335m
halokinetic movements 6, 7
halophytes 68, 181
Hamborg Dyb 227
Hamburg 2m, 9, 10m, 13m, 20, 32f, 190m, 211, 212m, 285, 288
Hamburger Hallig 2m, 190m, 247m, 260m, 263, 268, 275
Hammer
  Bucht 317
  Lake 316, 317, 317m, 318ph
    dune ridge north of 319ph
    washover area 316
Hannover 214
Hanseatic League 211
Hanswarft, Hooge 260m, 273m
Hantum 335m
Hantumhuizen 335m
Harle 27f, 54f, 65m, 75m, 76, 79m, 82ph, 83ph, 84, 86, 92, 102, 108, 116, 128, 132, 200, 203, 290m, 292, 293m, 297, 297m, 298, 300, 301
  Bight 20, 21, 22, 84, 182, 289m, 301
  inlet migration 85
Harlehörn Peninsula, Wangerooge 75m, 289m, 297, 297m, 298
Harleriff salt dome 7
Harlesiel 54f, 65m, 289m
  salt marsh erosion 194ph
Harlesieler Watt 75m
Harlingen 2m, 14m, 15, 32f, 134, 190m, 341m, 347
Hattstedter Alter Koog 20
Hauen 317m
Hauener Hooge 182, 317m
Hauke-Haien-Koog 247m, 263
Hauptdamm
  Minsener Oog 290, 291, 291m,
  Wangerooge 204m
Hav Dyke, Mandø 229m, 231, 232, 233
Havneby 229m, 234m, 236
  Koog 229m, 233
Havsand, Rømø 29, 39f, 229m, 233, 234m, 236, 249
Havsande dunes 221, 223m
Havsands Lo 233, 234m
Havside Bjerge, Fanø 223m, 227
heavy metal pollution 220
Hedwigenkoog 260m, 279m
Heffezand 327, 329m
Heimliche Liebe, Borkum 317m, 325

Helderse Zeeweering 352m, 353
Helgoland 2m, 5, 7, 10m, 21, 43f, 211, 259, 259m, 303
Helmsand 258, 279m, 282
Helsdeur, Marsdiep 103m, 104m, 105m
Hemmingstedt oil refinery 38
Hensebek-Koog, Pellworm 275
Heppenser Fahrwasser, Jade 112, 283m
Hermann-Göring-Koog 215
Hermann-Lietz-School, Spiekeroog 289m, 300
Herrenhallig, Nordstrand 183, 183m
Herrenhus Dunes 302
herring 211
  gull (= *Larus argentatus*) 61
Herst 268
Herzeele 7m
Het Rif
  Engelsmanplaat 332, 334m
  Schiermonnikoog 328m, 329
Hever 1, 108, 109, 109f, 110m, 183m, 361
  Pleistocene core 26
Hevermündung 18
Heverstrom 259m
Hiaure 335m
Hilgenriedersiel 302m
Hilligenley, Langeneß 247m, 262ph, 264
Hindenburg Damm, Sylt 180, 215, 239m
Hippolytushoef 352m, 353m
Hjerpsted 229m, 236, 239m
Hjerting 223m
Ho 223m, 224m
  Bight 223m, 224m, 226
Hobo Dyb 221, 224m, 226
Højer 239m
Hønen, Fanø 223m, 227, 228ph
Hörnum, Sylt 18, 189m, 215, 216, 239m, 241, 245, 247m
  breached dunes 247ph
  dune erosion 192ph
  rill mark formation 147ph
  tetrapod groyne 203ph, 204ph
  dune stabilisation 240
Hörnum
  Odde 239m, 246ph, 247m, 248
  Peninsula 189, 189m, 240, 243, 245, 247, 250
Hörnumtief 189, 222m, 253
Hoher
  Rücken 75m
  Weg 259m, 283m, 288
Hohes Riff, Borkum 110, 290m, 317m, 321, 322, 322ph, 326
holiday apartments 218
Holland, North 13m
Hollum, Ameland 335m, 337, 337m
Hollumerduinen 335m
Holm 259
Holmer Fähre, Nordstrand 183, 263
Holocene 17, 302
  transgression 8, 9, 10m, 11f, 85, 88,

134
Holsteinian transgression 7m, 9
Holwerd 335m, 338
Hondsbossche Zeeweering 353m
Honkenswarft, Langeneß 247m
Hooge 2m, 76m, 86m, 133, 134, 187, 195, 258, 260m, 261, 265, 268, 269, 270, 274, 275
  former farmland south of 197ph
  morphological changes on tidal flat 76m
  old marsh surface south of 269m
  structural change 273m
  summer dyke 270ph
  tidal flats southwest of 76
  water reservoir (= Fething) 271ph
Hooge
  Beintum 335m
  Berg, Texel 352, 352m
Hoogeloch 76m, 77, 86, 86m, 262
Hoogwoud 353m
Hoorn, Terschelling 341m
Hoornder Nieuwland, Texel 352m
Hornhuizen 328m
Horns Rev 222m
Hornsbjerge dunes 221, 223m
Horsbüll 239m
Horumersiel 289m
Hospiz, Langeoog 302m, 303
Hotel Germania, Wangerooge 289m, 292
Hoyerschleuse 215
Hufschlag 302m
Huisduinen 17, 107, 352, 352m, 353, 353m
Hullbalje 131, 131m
Hullplate 75m
Hummelsbüttel 7m
Hundsand 282
Hunnenkoog, Pellworm 260m, 275
Hunnenswarft, Langeneß 247m
Hunningensände 39f
Hunze Valley 8m
Husum 2m, 18, 28f, 32f, 185, 190m, 212m
Hvidbjerg 26f
Hwaelae
  major 259
  minor 259, 268
Hypnum cupressiforme 169

ice
  blocks on the beach 58, 59, 59ph
  crystals, imprints of 55
  -dammed lake 9
    in the North Sea 6m
  floes 40, 42ph, 43f
  ground- 59
  jam 38
  sheets 8, 9, 10, 11, 12
IJmuiden 8m, 213
IJssel 2m, 32f, 190m
IJsselmeer 2m, 14m, 24, 330m, 352m, 353m
  Dam 15
Imsum 283m

inflation 213f
initial
  dunes on Norderney 176f
  ripples 157ph
inlet
  formation, potential for 88
  migration 83, 85
  trunk 81
Innenquage 110m
intertidal area 40, 82
Ipke Sandt 268
Ipkens Warft, Hooge 260m, 268
Iron Age 195
Isenhage 39f
Isern Hinnerk 78f, 285
island migration 83
Itzehoe 212, 212m
Itzendorf-Plate 290m

Jade 27, 54f, 65m, 108, 109, 215, 259m, 283m, 288, 290, 290m, 291, 291m
  Eemian deposits 8m
  Bay 2m, 5, 13, 17, 22, 25, 26f, 78, 180, 182, 187, 194, 292
    formation of 20
    maximum extension 21
  channel, old course 85
Jadebusen (= Jade Bay) 9
Jade-Weser area 5
Japsand 2m, 17, 32f, 76m, 231ph, 258, 259m, 260m, 269
  migration of 77, 86, 86m
Jennelt 317m
Jewelpolder 353m
Johannes-Heimreich-Koog, Pellworm 260m, 275
joints in till 192
Jordsand 39f, 134, 229m, 236, 237, 238, 239m, 259
  cross section 238f
  erosion in the west 237f
  morphological changes 236m
Jordsands Flak 39f, 222m, 236
Juist 2m, 26, 26f, 27f, 29, 37, 84, 85, 89, 90, 150, 178, 188, 214, 215, 218, 290m, 292, 302m, 311, 313, 316, 317, 317m, 321
  aerial photograph of the western end 319ph
  beach and foreshore 137ph
  development of dunes and coastline 317f
  dune ridge north of Hammer Lake 319ph
  Hammer Lake 318ph
Juist-Ost salt dome 7
Juist-West salt dome 7
Julianadorp 352m, 353m
Jungnamensand 222m
Jutland (=Jylland) 13
Juvre, Rømø 229m
Juvre
  Dyb 232m, 233
  Koog 198, 229m, 233
    dyke after breaching 198ph

partially eroded dyke 184pl
aerial photograph 197ph
Priel 198
Sand 29, 229m, 233, 234m, 249
Jylland 13m

Kaap Dunes, Langeoog 302, 302m
Kachelotplate 320
Kaiserin-Auguste-Viktoria-Koog 279m
Kaiser-Wilhelm-Koog 279m, 282
Kalfamerdünen, Juist 316, 317m
Kamp 328m
Kampen, Sylt 171, 189, 191, 218, 238, 239m, 243, 244
  salt marsh erosion 193ph
  secondary blowouts 173ph
  block diagram 245f
kaolin sands, Sylt 238, 245
Karolinenkoog 260m
Kating 260m
Katwijk 24
Keitum, Sylt 239m
Keizerbult, Marsdiep 106m
Keldsand 222m, 227
Kersig-Siedlung, Sylt 239m, 241
Ketelswarft, Langeneß 247m
Keuper 7
Kiel 32f, 190m, 212, 212m
Kirkeby 236
Klanxbüll 239m
Klappholttal, Sylt 239m, 240
Kleine
  Schlopp, Langeoog 302m, 303, 305
  Süderwarft, Hooge, destroyed 269
Kleiner
  Hafen, List, Sylt 188m
  Koog, Pellworm 260m, 275
Klixbüll 21
Kloosterburen 328m
Knechtsand 283m, 288
Kniephafen, Amrum 248m
Kniepsand, Amrum 29, 47ph, 247m, 248, 248m, 249, 250, 252, 254ph
  primary dunes 249ph
Knudedyb 222m, 227, 232m
Knudtswarft, Gröde 247m, 267
Kobbeduinen, Schiermonnikoog 328m, 330
Koegras salt marsh 353
Königshafen, Sylt 188m, 189, 239m
  salt marsh erosion 194ph
Koggegronden, Terschelling 341m
Koldby 229m, 239m
Kollmar 212
Kongeå 223m
  mouth, salt marsh erosion north of 195f
Kongsmark 229m, 236
Koogerveld 353
Kooiduinen
  Ameland 335, 335m
  Schiermonnikoog 328m, 330
Kooioerdstuifdijk, Ameland 335m, 337m
Kooispleklid, Vlieland 347m, 351
Kooremansvallei, Vlieland 29, 349

Koresand 32f, 222m, 230, 231ph, 258, 221
   morphological changes 232m
Kornkoog 247m
Kravnsø 223m
Kronprinzenkoog 282
Kroonpolders, Terschelling 341, 341m, 342ph, 347m
Kroon's Polders, Vlieland 347m, 350
   wetland area 351ph
   aerial photograph 349ph
Krügermauer seawall, Wangerooge 295
Kugelbake, Norderney 302m, 313
Kuham, Föhr 247m, 255
Kurhaus, Wangerooge 203, 289m, 292, 295

Lake Flevo 24
Lakolk, Rømø 214, 229m, 236
Lammelæger 39f, 222m
land reclamation 14, 15, 20, 21, 180, 187, 198, 215
   Amrum 251 252ph
   Eiderstedt 276
   Föhr 254, 255
   Hamburger Hallig 268
   Langeneß 265ph
   Mandø 232
   Nordstrand 276
   Pellworm 275
   Texel 354
   Trischen 281
Landsat-3-RBV image 164
Landsat-4
   MS image of the Dutch Wadden Sea 329pl
   TM image of Schiermonnikoog 333ph
Landsat-5 TM 75
   coast between Norderney and Ameland 328pl
   East Frisian Islands 313pl
   Rømø, Sylt and Amrum 231ph
   Schleswig-Holstein 232pl
landslides 189, 191
Landunter 262
Lange Duinen, Ameland 335m, 337m
Langen 283m
Langeneß 2m, 133, 133m, 134, 262, 263, 264, 265, 274, 275
   bank protection at Norderhörn 263ph
   during Landunter 280pl
   embankment to Oland 265ph
   Hallig surface with dwelling mound 261ph
   latest dwelling mound 196
   location map 247m
   salt marsh stratification 274ph
   traces of old salt peat digging 196ph
Langenhorner Alter Koog 20
Langeoog 2m, 26f, 27f, 35f, 77f, 80m, 83, 85, 97, 136, 161, 178, 214, 217, 218, 289m, 290m, 300, 302, 302m, 303, 304, 305, 313
   aerial photograph of western end 304ph
   beach and foreshore 136f
   block diagram of dune area 163f
   costs of coastal protection 207
   development of the built-up area 219m
   Flinthörn dunes 178
   geological history 302
   Große Slop washover area 179ph
   Hospiz 219m
   northwestern beach 303ph
   old dyke behind recent dunes 305ph
   salt dome 7
Langeooger Balje 305
Langer Jan 75m
Langli 221, 223m, 224m
Langwarden 283m
Låningsvej 229m, 232
Larus argentatus 61
   resting mould of 62ph
Larus ridibundus 61
Lauenburg 9
Lauwers 290m
   Inlet 84, 328
Lauwersgat 333
Lauwersoog 328m
Lauwerszee 2m, 9, 22, 84, 131, 182, 328m, 333
Lead 220
lee-side erosion 200, 201, 244, 243, 245, 251, 255
   Baltrum 308
   Norderney 311, 314, 358
   Sylt 209ph, 241, 242ph
   Texel 199f, 356, 358
   Wangerooge 292, 293, 295
Leeuwaarden 14m, 20
levées 94, 262
   in the Hoogeloch area 86m
   of the Lower Ems 195
   of salt-marsh creeks 182
   on Süderoog 261m
   on Pohnshallig 182ph
levelling by the surf 52, 53ph
Ley Bight 2m, 20, 180, 182, 187, 190m, 311, 317m
Liassic 7
Lies, Terschelling 165, 341m
Limonite Sandstone 190, 245
Limonium vulgare 182
linear sand ridges 16, 16f
lingoid
   ripples 144ph
   small ripples 144
Linnenplate 260m, 285
Lioesens 335m
List, Sylt 172, 215, 229m, 236, 239m, 241, 244
   damaged revetment 244ph
   revetment 209ph
   salt marsh erosion 194ph
Lister Tief 37, 222m, 233
   sand wave migration 112
   tidal currents 39f
Listland, Sylt 163, 198, 215, 239m, 243, 245
   block diagram 165f
   dune migration 166f
   dune stabilisation 240
   dune-slope erosion by rain 173ph
   migrating dunes 151ph
Litorina Cliff, Amrum 248
Little Ice Age 29, 239
littoral drift divide 17
load casts, formed by ice blocks 58
London 8m, 10m
longshore bars 80m, 126, 135f
   Juist 137ph
   megaripple migration on 111
   Trischen 31f
longshore ridges 97
Loodsmansduin, Texel 352m, 358
low bars (= deep bars) 78, 108
   block diagram of 78f
   migration of 111
Ludinga cloister 347
Lübeck 8m, 32f, 190m, 212m
Lühesand, Elbe 111
Lüneburg 211
Lütetsburg 302m
Lütje Hörn 32f, 290m, 317m, 320
   dune formation 320
   morphological changes 321m
Lütjeoog Dunes, Spiekeroog 289m, 299ph, 300
lugworm (= *Arenicola marina*) 60
Lunden 277m
Lundenbergharde 276
Lundener Nehrung 25

Maasbüll 247m
macrotidal 16, 16f, 17
main ebb channel 80m, 81, 101m, 108, 110
   Marsdiep 106
   migration of 108
   Wichter Ee 102
mainland shore, influence of waves on the 53
Malpeque Islands, Canada 176
Mandø 2m, 222m, 229, 230, 232, 232m, 233, 258, 259
   aerial photograph 230ph
   By 229m, 231
   Flak 222m
   location map 229m
Mandränke 22
Mannemorsumstal, Sylt 239m, 244
Manø (= Mandø) 229
Manslagt 317m
March 8m
Marco Island, Florida 207
Mariendeich, Norderney 302m, 314
Marne 279m
Marner Plate 259m
Marrum 335m
Marsdiep 15, 89, 102, 106, 107, 108, 110, 213, 330m, 353, 354
   ebb delta 106
   development of 103, 104m, 105m, 106m

marsh islands 134
Martensplate 75m
Maxqueller 282
Mayenswarft, Langeneß 247m
meandering rill marks 148
Medemblik 353m
Medemsand 259m
megaripple
  bedding
    of sand ridges 52
    on Spiekeroog beach 52ph
  crests, strike direction of 130m
  fields in the Hoogeloch area 86, 86m
    Wichter Ee ebb delta 97
  migration 111, 113, 114, 116, 117, 119, 122, 123, 124, 125, 129m, 361
    influence of algae 60
    Wichter Ee ebb delta 118m, 119m, 120ph, 121f, 122ph, 123f, 124f, 126f, 127f
  reorientation 96
megaripples
  at the bottom of tidal inlets 111
  ebb-orientated 96ph, 99ph
  formation of 111
  on crescent bars 52
  on Hohes Riff megashoal, Borkum 322
  on Wangerooge beach 143ph
  on Wichter Ee ebb tidal delta 53ph
  rhomboid 145ph, 146
  trapezoid 119
  with spurs 96ph
megashoals 102, 106, 110
  Hohes Riff, Borkum 322
Meldorf 18
  Bight 2m, 13, 190m, 279m
Melkhörn Dunes, Langeoog 302, 302m
Mellum 2m, 14, 26f, 29, 32f, 43f, 259m, 283m, 288, 289m, 346
  aerial photograph 289ph
  cavernous beach sand 149ph
  primary dunes on the beach 159ph
  salt marshes 290
meltwater channel 6m
Memmert 2m, 29, 290m, 316, 317m, 318, 319ph, 320
  dendritic drainage system E of 320f
  formation of 23
Mercury 220
mesotidal 14, 16f, 17
Mesozoic strata 7
Metslawier 335m
Mica Clay 190, 245
microtidal 14, 16f, 17
micro-relief 44
  of sand flats 71
  of tidal flats 70
Middelste Koog, Texel 352m, 353
Middelzee 182
  formation of the 20
Middenmeer 353m
Middle Ages 211, 249, 256, 259, 268, 285, 327, 352

Midlum, Föhr 247m, 283m, 341m
Midsland, Terschelling 341m
migrating
  barrier off Dithmarschen 284, 285
  dunes 151, 152, 162, 163, 164, 168, 193, 227, 362
    Ameland 335
    Baltrum 307
    block diagram of 165f
    Fanø 152pl, 170ph, 227, 229
    Hörnum Peninsula, Sylt 245
    Listland, Sylt 151ph, 243, 244
    migration rate of 163, 165, 166f, 167
    Norderney 166f
    Schiermonnikoog 329, 330
    Sylt 239
    Terschelling 341
migration rates of shoals 110
military
  airfield, Langeoog 305
  training ground 218, 221
    Knechtsand 288
    Rømø 236
    Vlieland 350
Minnertsga 341m
Minsen 289m
  Spartina on mud flats north of 73pl
Minsener Oog 54f, 65m, 76, 289m, 290, 290m, 291, 347
  coastal engineering structures 291m
  mud flats with *Mytilus* 68f
Misselwarden 283m
Mitteldorf, Baltrum 307
Mittelhever 110m, 259m
Mittelkoog, Pellworm 275
Mittelster Koog, Pellworm 260m
Mitteltritt, Hooge 260m
Modderhoek 335m
Mølby 229m
Moerwaard 344
Möwenberg Dyke, Sylt 239m, 241
Möwendüne, Norderney 176f, 302m, 316ph
moisture content of sand 41
Moldijk, Ameland 335m, 337m
Molengat 102, 103m, 104m, 105m, 106m
  development of 107f
Monnikensloot 347
Moordeichhallig, Nordstrand 183m
Morra 335m
Morsum, Sylt 239m
Morsum Kliff 18, 190, 192, 239m, 245
Morsumkoog, Nordstrand 260m, 276
mud flat sediments 68
mud flats 65, 68, 70, 95
  drainage pattern of 70
  southeast of Wangerooge 72pl
  west of Cappel-Neufeld 69ph
Munkmarsch, Sylt 215, 239m
mussel bed 95
Mya arenaria 59, 263
Mytilus edulis 59, 68, 68f
  colony on muddy tidal flats 60ph
  mud accumulation by 59

Næsbjerg Kirkeby 223m
Näshörn, Föhr 247m, 255
Napoleonic wars 213f, 214
nature conservancy 218, 219
  Norderney 314
  Spiekeroog 301
  Sylt 244
  Terschelling 342
  Texel 355
  Vlieland 350
nature reserve de Schorren, Texel 344pl
neap tide 41
Nebel, Amrum 247m, 251
  coastal erosion 253ph
Nebeler Strandweg, Amrum 247m, 249
negative shadow dunes 160, 162ph
Nes, Ameland 214, 335, 335m, 337m
Nesse 302m
Neßmer
  Nacken 95, 114m
  Plate 95, 114m
  Watt 84f
Neßmersiel 35f, 84f, 302m
  tidal flats with *Spartina* 181ph
Neudeich, Wangerooge 289m
Neuenkirchen 260m
Neuer
  Koog, Nordstrand 183m, 260m, 276
  Westturm, Wangerooge 296
Neufelder Koog 21, 279m
Neugarmssiel 289m
Neuharlingersiel 289m
Neuherst Warft, Gröde 267
Neuwarft
  Langeneß 247m
  Nordstrandischmoor 260m
Neuwerk 2m, 33, 43f, 259m, 283m, 285, 288
  aerial photograph 287ph
  tidal flats 36, 37f
Neu-Peterswarft, Langeneß 196, 264
Neu-Scharhörn 285
Nieblum, Föhr 247m, 256m, 264
Niebüll 211
Nielönn, Sylt 239m
Nieuwe
  Biltpolder 341m
  Niedorp 353m
  Schulpengat, Marsdiep 107
Nieuwlandsrijd, Ameland 335m, 336
Nipsau 213
nitrate 220
nitrogen demand 181
Nösse
  Dyke, Sylt 239m, 241
  Peninsula 190, 239m, 241, 245
Noordergat, Marsdiep 102, 103m
Noordergronden 340m
Noorderhaaks, Marsdiep 102, 103m, 104m, 105m, 106m, 107f, 108, 110
Noordewijk 328m
Noordpolderzijl 328m
Noordvaarder, Terschelling 29, 83, 340m, 341, 341m, 347m

Noordwestgronden 110, 340m
Noordzeekanaal 213, 314
Nordby, Fanø 213, 223m, 225m, 229
  beach ridge north of 226ph
Norddeich, Dithmarschen 260m
Norddeich, Lower Saxony 215, 302m, 317m
  borrow pit in tidal flats 73
  railway line to 213f
Norddorf, Amrum 247m, 251
  land reclamation 252ph
Norden 302m, 317m
Norder Neuer Koog, Pellworm 275
Norderaue 133, 222m, 253
  flood delta 134
    morphological changes 133m
Norderelbe 259m
Norderhafen, Nordstrand 260m, 263
Norderhever 259m, 262, 263
  drowned Pleistocene core 17
Norderheverkoog 260m, 278
Norderhörn, Langeneß 247m
  bank protection 263ph
Norderkoog, Pellworm 260m
Norderney 1, 2m, 26f, 27f, 29, 33, 34f, 35f, 41, 49, 83, 84, 93, 96, 97, 99m, 109, 111, 114m, 116, 118m, 119, 129m, 130m, 138, 150, 156, 168, 178, 190m, 193, 200, 214, 215, 216, 218, 290m, 301, 302m, 306, 311, 313, 317m
  beach
    bedform migration 99m
    current lineation 142ph
    replenishment 205, 206ph, 208f
    response to storm surges 150f
    ripple troughs 51ph
    swash marks 142ph
  block diagram
    of blowouts 169f
    of dune cores 176f
  cavernous beach sand 150ph
  development of tourism 213f
  dune development 306f
  dune stripes on the beach 155ph
  eastern end of 44ph
  first groynes 198
  groyne fields 208f
  inactive migrating dunes 166f
  morphological development 306f
  narrow beach at western end 314ph
  oblique aerial photograph 313ph
  old migrating dunes 165
  Ostplate, path system 316ph
  Othello Shoal Shield 99ph
  rhomboid ripples 145ph
  Riffgat 38
  salt dome 7
  sand waves south of 74ph
  sawtooth bars on the foreshore 138ph
  seawall 203ph
  washover area 174ph
  washover channel 153pl, 176ph
  Weiße Düne 27f
  western end, aerial photograph 312ph
Norderneyer
  Inselwatt 94
  Profil 200, 309
  Seegat 27f, 52, 90, 92, 200, 205, 290m, 292, 311, 314, 361
    aerial photograph 312ph
    wave pattern 51ph
Norderoog 76m, 134, 195, 238, 258, 260m, 262, 268, 272
  coastal erosion 196m
  coastal protection 271ph
Norderoogsand 2m, 17, 26f, 32f, 76m, 258, 259m, 260m, 269, 272
  migration of 77, 86, 86m
Norderpiep 108, 109, 259m, 284
Norderquage 110m
Norderwarft
  Habel 268
  Hooge 260m, 273
  Nordstrandischmoor 260m
Nordholz 283m
Nordmann Channel 8m
Nordmarsch 258, 264
Nordschleswig 215
Nordstrand 2m, 21, 134, 182, 256, 258, 259, 263, 268, 270, 272, 275, 276, 277m
  age of dwelling mounds 196
  Dam 180
  Pohnshallig Koog 182ph
  situation in 1857 183m
Nordstrander Bucht 188, 190m, 263, 268, 271, 276
Nordstrandischmoor 2m, 134, 260m, 262, 263, 270, 271, 272, 274, 275
Nordufermauer, Wangerooge 204m
North
  Frisia 5, 13m
  Holland 13m, 102, 352
  Sea Coast Triangulation 6
Norway 216
Norwegian trough 10
nutrients
  leaching of 170
  over-supply of 220
Ny Mandø 229, 229m, 231
Nyewier 335m

oak 240
Oberahnesche Felder, Jade Bay 5, 193
Ochtersum 302m
Ockenswarft, Hooge 260m
Ockholm 247m
Ockholm-Koog 263
Odense 32m, 190m, 212m
Oerd, Ameland 335, 336
Oerderduinen, Ameland 335, 335m, 336
Østerby, Rømø 229m
  erosion east of 235ph
Oevenum, Föhr 247m, 252
Ofkebüll 258
oil pollution 220
Oksbøl 223m
Oksby 222, 223m, 224m
Oland 2m, 134, 258, 259, 262, 265, 267, 273, 274
  Eemian deposits 8m
  embankment to Langeneß 265ph
  location map 247m
Old Dunes 24, 25
Oldenbarneveldtsdijk 353
Oldenburg 214, 292
Oldenswort 260m
  salt dome 7
Oldsum, Föhr 239m, 247m, 253, 254, 258m
Omø 268
Onderdijk 353m
Onrust, Texel 102, 103m, 104m, 106, 352m, 359
Oost
  Meep 330m
  Vlieland 349, 351
Oost, Texel 354m
Oosterbierum 341m
Oosterend
  Terschelling 341m, 342, 343
  Texel 352m, 354m
Oosternijkerk 335m
Oosterschelde
Oosterstrand, Schiermonnikoog 329, 331ph, 332
Oostmahorn 328m
Oostwoud 353m
Oost-Vlieland 341m, 347m
Ording 259, 260m, 278
  Koog 278
organic content of mud flat sediments 68
Orkney Islands 6m, 9
oscillation ripples 140, 141, 143ph
Ostbake, Spiekeroog 289m, 301
Ostbalje, Wichter Ee 94, 114m
Ostdorf, Baltrum 302m, 306
Osterems 290m
Ostergroen, Spiekeroog 298
Osterhever 260m
Osterhook
  Baltrum 302m, 310
  Langeoog 302m
Osterkoog, Nordstrand 260m, 276
Osterkurzmaße, Föhr 255
Osterloog, Juist 316
Ostersiels-Koog, Pellworm 275
Ostertill 131, 132, 133, 259m, 285, 286
  flood delta 131m
Osterwarft, Südfall 272
Oster-Ems 292, 317m, 320, 321
  ebb delta 317, 320
  flood delta 320
Ostfriesland (= East Frisia) 5
Ostinnengroden, Wangerooge 292, 296
Ostland
  Borkum 18, 317m, 325
    dyke breach 19ph
  dunes
    Borkum 326
    Juist 316
Ostlandföhr 254
Ostplate

Norderney 302m, 314, 315, 315ph
  dune formation 314
  washover area 315ph
 Spiekeroog 29, 289m, 300, 301
  dune formation 299ph
 Wangerooge 289m, 296
  aerial photograph 295ph
Othello Plate (= Othello Shoal Shield) 96
Othello Shoal Shield 96, 97, 98, 99m, 112, 112m, 113, 116
  aerial photograph 100ph
  megaripples 99ph
  rhomboid megaripples 145ph
Otzumer Balje 27f, 84, 131, 133, 290m, 300
  flood delta 132m
  inlet migration 85
Oude
 Niedorp 353m
 Rijn 24
Oudeschild, Texel 352m, 354m, 355
Oudkarspel 353m
Outer Sands 86, 152, 188, 230, 258, 329
 between Fanø and Rømø 232m
 migration of 76m
 north of Rømø and S of Amrum 231ph
 of the Danish coast, migration of 230, 231
overgrazing 151

Padingbüttel 283m
Paesens 335m
Palaeozoic 5
palisade revetment on Baltrum 308, 309
palynology 302
parabolic dunes 24, 150, 151
 Blåvandshuk 221
path system on Norderney Ostplate 316ph
path systems, initiation of 315
peat
 basal 10
 digging 197
Peilbake
 Baltrum 302m, 310
 Norderney 176f
Pellworm 2m, 20, 21, 134, 256, 260m, 262, 263, 268, 269, 275
 age of dwelling mounds 196
 borrow pit in tidal flats 73, 73ph
periglacial structures 251
Peter Meyers Sand, Fanø 221, 222m, 227, 230, 231, 258
 morphological changes 232m
Peterswarft, Langeneß 247m, 264
Petten 353m
Phaeocystis pouchetii 220, 220ph
phosphate 220
Piepe Warft, Oland 267
Pieterburen 328m
Pietersbierum 341m
Pilsum 317m

pine forest in dunes, Schiermonnikoog 330ph
Pinkegat 333
Pinus
 afforestation on Terschelling 342
 montana 240
 sylvestris 358
piracy 211
Pirolatal, Langeoog 304, 305
planar bed 111
plastic tubes, sand-filled, Norderney 313
Pleistocene 193
 core 134, 233
  Amrum 248, 251
  Föhr 252, 255
  Texel 352
 glaciations 8
 ridge at Den Hoorn, Texel 345pl
 sands 191
 sediments, cliffs in 190
Pliocene kaolin sands 190, 238, 245
Plio-Pleistocene core, Sylt 17, 238, 239, 243
Pohnshallig 182, 183, 276
 in 1857 183m
Pohnshalligkoog, Nordstrand 182, 260m, 263
 aerial photograph 182ph
Pohnsley 182, 183, 183m
Polder
 Burg en Waal, Texel 352m
 De Eendracht, Texel 352m, 354, 355
 De Volharding, Texel 355
 Eijerland, Texel 352m
 Het Koegras 352m, 353m
 Het Nieuwland, Terschelling 341m
 Het Noorden, Texel 352m, 355
 Hoekje 353m
 Koogerveld, Texel 352m
 Wassenaar, Texel 352m
pollution 220
Postbake, Norderney 302m, 314
Posthuis, Vlieland 347m, 350
Posthuiswad 347
potholes 193
precipitation 172
Prehistoric times 352
pre-Holocene relief 84
primary
 dune valleys on Texel 358
 dunes 151, 152, 159, 159ph, 160, 362
  eastern Baltrum 310
  Engelsmanplaat 334
  Het Rif, Schiermonnikoog 329
  Knechtsand 288
  Kniepsand, Amrum 249ph
  Lütje Hörn 320
  Memmert 318
  Rømø 233, 235ph
  Rottumerplaat 328
  Schiermonnikoog 312pl
  southwestern Borkum 325
  Spiekeroog 301
Prins Hendrik Polder, Texel 345pl, 352m, 354, 355

Prussia 214, 292
Psamma arenaria 151
Puan Klent, Sylt 239m
Puccinella maritima 181, 182
pushed ridges, formed by ice blocks 58

Quaternary 7, 8
 glaciations, maximum extent of 6
 sediments in the North Sea basin 3m
Quercus 351
Quermarkenfeuer, Amrum 247m, 249

rabbits 301, 302, 315
radiocarbon dating 12, 24, 291, 302
Råhede 229m
raindrop impact marks 173ph
rainfall runoff 191
Randzel 290m
Rantum, Sylt 135f, 136, 165, 239m, 240, 241, 243, 245, 246
Rantum-Becken, Sylt 239m
Rattendüne, Norderney 302m, 314
Razende Bol 102, 104m, 105m, 106, 106m, 108, 110, 330m, 352m, 359
RBV satellite image 164, 165
recurved
 sand ribbons 47ph, 48ph
 shoals of Bornrif, Ameland 336
 spits 97, 185, 189
  Juist 317
  Norderney 41, 44ph, 99m
  small 45, 47ph
  Trischen 31f
Reededamm, Borkum 317m, 325, 326
refraction 49
regression 10, 11
Reichsdeich, Wangerooge 178, 202, 204m, 292
Reichsmauer seawall, Wangerooge 202, 204, 204m, 292, 295
Rejsby Stjert 222m
residual current 33, 34
resting mould 62ph
revetment
 Baltrum 309
 Blidsel, Sylt 209ph, 244
 Borkum 317m, 324ph, 325, 325ph
 Ellenbogen Peninsula, Sylt 239m, 241, 244
 Huisduinen 353m
 Gröde 267
 Keitum, Sylt 239m, 241
 List, Sylt 209ph, 239m, 244
 Langeneß, west of Hilligenley 262ph
 Neuwerk 288
 Norderney 313
 of plastic tubes, Langeoog 305
 Spiekeroog 289m, 299, 300
 Texel 345pl, 352m, 354m, 356
 Trischen 281
 Wangerooge 184pl, 203, 204, 204m, 289m, 293, 295, 296, 297m
 Westerland, Sylt 241

revetments 200, 201
  underwater 83
Rhine 21
rhomboid
  megaripples 145ph, 146
  rillmarks 146, 146ph
  ripples 144, 145ph, 146
Ribe 2m, 32f, 190m, 212, 212m, 213
  Dyke 229m, 232
Richel 330m, 340m, 341m, 347m
  Dyke, Spiekeroog 299
ridges
  of the foreshore 138, 149, 150
  scour-remnant 160
Ried 341m
Rif, Engelsmanplaat, morphological change 334m
rill marks 147
  bifurcating 147, 148
  branching 148
  comb-like 148, 148ph
  conical 148
  erosional 147
  formation of 147, 147ph
  fringy 148
  meandering 148
  rhomboid 146, 146ph
  tooth-shaped 148
  with accumulation tongues 148
Rindby, Fanø 223m, 227, 229
Rindby Strand 223m
ring dyke on Gröde 267
rip currents 138, 138ph
ripple
  crests 45
    flattened 44
  field, small 47ph, 48ph, 49ph
  height 113
  junctions 154
  length 113, 116
  marks, small 44
  migration
    laboratory experiments 111
    resulting 130f
  pattern, changes of 116
  tongues 144
  trough 44, 45, 51ph, 52
ripples
  adhesion 148
  backwash 141
  lingoid 144ph
    small 144
  oscillation 140
  rhomboid 144
  small 45, 140
  transport 140
  undulatory small 140
  wave 140
  wind 152, 154
riprap
  Vlieland 349
  Wangerooge 204, 204m, 205ph
Rissen 7m
Ristinge 8m
Risum 247m
  Gap, Amrum 247m, 251

Rixwarft, Langeneß 247m, 264, 280pl
Robbengat, Texel 83, 356
Robbenplate, Norderneyer Seegat 205
Robbezand, Terschelling 340
Robinsbalje 259m
Rocheley 263
Rochelsand 110m, 279
Rochelsteert 259m
Rodenæs 188, 239m
Rodenkirchen 195
Røgle 7m
Rømø 2m, 17, 26f, 29, 39f, 112, 134, 152, 163, 180, 186, 187, 198, 212m, 214, 215, 216, 218, 222m, 230, 232m, 233, 235, 236, 238, 239m, 259
  aerial photograph of Juvre Koog 197ph
  Dam 180, 229m, 236
  erosion at the east coast 235ph
  Kirkeby 229m
  Landsat-5 image 231ph
  location map 229m
  morphodynamic subdivision 234m
  partially eroded dyke of Juvre Koog 184pl
  primary dunes 235ph
  sand tongues on the beach 146ph
  the breached Juvre Koog Dyke 198ph
  Robbe radar station 218ph
Roman settlement 24
Rote Balje 300, 84, 301
Rotes Kliff, Sylt 188, 189, 190, 191, 192, 233pl, 238, 239, 239m, 240, 242, 242ph, 243
  erosion by rainfall runoff 191ph
  debris apron 191ph
Rottumeroog 2m, 32f, 102, 290m, 326, 327, 328, 328m, 333, 337
  morphological changes 329m
Rottumerplaat 2m, 32f, 290m, 327, 328m, 329
  morphological changes 329m
rubble dam on western Ameland 337m, 337ph, 338
Rütergat 259m
Rug van het Veen, Vlieland 347m
Rummelloch 262, 269m
  block diagram 81f
Rungholt 20
  Bay 21
runnel 97, 101m, 102, 135f, 136f, 140, 147, 200
runoff
  rills on sand flats 70ph, 71ph
  channels on tidal flats 73

Saalian 8, 9
  ice margin 6m
  till 191, 238
sag features 174, 174ph
Sahara 162
Sahlenburg 279m
Salicornia europaea 68, 181, 181ph, 185, 281, 282, 326, 346

Saline, Wangerooge 167ph, 289m, 295
salt 211
  crust 158
  in sea water 54
salt marsh
  accumulation 182
  development
    Baltrum 309, 310
    Boschplaat 342
    Memmert 318, 320
    Rottumeroog 327
    Schiermonnikoog 332, 333
    Spiekeroog 301
    Texel 355, 356
  drainage 197
    pattern 182, 186ph
    system 187, 274ph
  erosion 179, 193, 193ph, 194, 194ph
    Ameland 335
    Borkum 280pl, 326
    Danish mainland coast 195f
    Gröde 267
    Halligen 262
    Schiermonnikoog 332
    Terschelling 343, 345ph
    Texel 358ph
  exposed to wave action 53
  formation 29, 190m
    Ameland 336
    Eiderstedt 184
    Langeoog 302
    Rottumerplaat 328
    Schiermonnikoog 186, 329
    Spiekeroog 300ph
    Trischen 281
  morphology, Halligen 262
  stratification, Langeneß 274ph
salt marshes 134
  Amrum 251
  Blauort 284
  'de Schorren, Texel 344pl
  Fanø 226, 227
  Föhr 253, 257ph
  Griend 346
  Gröde 267
  Jordsand 237
  Kampen, Sylt 245f
  Langeoog 136f
  Mandø 229
  Mellum 290
  Neuwerk 288
  Norderney 41
  Nordstrandischmoor 270
  Skallingen 221
  Spiekeroog 298, 300
  Sylt, Nielönn 244, 245
  Trischen 31f
  Vlieland 347
  Wangerooge 297
salt peat 211
  digging 197
  on Bant 321
  south of Langeneß 264
  traces of 196ph
saltation 157
salt-encrusted dune surface 159ph

salt-water lenses, unfrozen 55
Salzsand, Sylt 39f
sand dykes 178
  Borkum 325
  Griend 346
  Juist 316, 317
  Texel 354m
sand fence 161f
  Vlieland 161ph
sand fencing on Scharhörn 286
sand flats 65, 68, 69ph, 70, 72pl
  drainage pattern 70
  micro-relief 70ph, 71ph
sand gaper (= *Mya arenaria*) 59, 60
sand keels, erosional 162ph
sand migration 361
  model 28f
sand pumping 325
sand replenishment
  Griend 346
  Texel 345pl, 356
sand ribbons 44, 45, 47ph, 49ph
sand ridges, megaripple bedding of 52
sand shadows 161
sand stripes 96
sand supply from the sea 27
sand tails 160
sand tongues 44, 45, 47ph, 48ph, 146, 146ph
sand transport
  across tidal flats 361
  coast-parallel 83, 97, 136
sand waves 73, 102
  at the bottom of tidal inlets 111
  flood-orientated 115ph
  in the Baltrum ridge system 140ph
  in the Hoogeloch area 86m
  migration 96, 111, 112, 114, 115, 116, 129m
    Lister Tief inlet 112
    Wichter Ee flood delta 97m, 116f
  on the beach of Juist 137ph
  on the foreshore 138
  reorientation of 96
  south of Norderney 74, 74ph
  tide-induced oscillation of 112
sand wave shoals, migration rate 129m
Sandflod Hede, Fanø 223m, 227, 229
sand-flow tongues 164ph
satellite image 38, 40, 42ph, 75, 230
Satteldüne, Amrum 247m, 249
sawtooth bars 136
  Baltrum 97, 102, 138, 140ph
  Langeoog 136f
  migration rate of 129m, 138
  Norderney 138ph
  Terschelling 137ph
  Wangerooge 138, 139f
Schagen 353m
Schagerbrug 353m
Scharhörn 2m, 7, 7m, 14, 29, 32f, 43f, 131, 237, 259m, 283m, 285, 286, 288, 346
  foreshore area 141f
  aerial photograph 284ph
  beachward migration of ridges 138

  morphological development 286m
  sand waves on tidal flats 73
Scharhörn-Eversand-Mellum salt dome 7
Scharhörnriff 259m
Scheveningen 28
Schiermonnikoog 2m, 17, 26f, 45, 83, 84, 170, 178, 180, 186, 214, 328m, 329, 330, 330m, 332, 333, 335m, 337
  eroded end of stuifdijk 332ph
  fragmented dune surface 159ph
  Landsat-4 TM image 333ph
  morphological changes 329m
  pine forest in the dunes 330ph
  primary dunes 312pl
  shifting of water divide 333
  tidal flats southeast of 50ph
  washover areas 177ph
  Young Dunes 24
Schild 290m, 328
Schildiep, Vlieland 350
Schillbalje 131, 131m, 301
Schillig 289m
Schlüttsiel 247m, 274
Schmaltief 259m
Schulpengat 103m, 104m, 105m, 106, 107
Schweinshammer washover area, Juist 316, 317m
Schweinsrücken, west of Langeneß 133m
scour-remnant ridges 160
sea-level
  change 258, 362
  rise 5, 10, 11, 12, 13, 23, 134, 195, 206
sea water, freezing point of 54
SEASAT 50, 136
  image of Eijerlandse Gat and Vlie 51ph
  image of Terschelling 137
sea-side resorts 214
seawalls 200, 201, 207, 219
  Amrum 247m, 250, 251, 252
  Baltrum 185pl, 302m, 308, 309
  Borkum 321, 324, 324ph
  Föhr 247m, 255, 256
  Juist 188, 317
  Norderney 203ph, 302m, 311, 313
  Sylt 185pl, 200, 239m, 240, 241
  Wangerooge 202, 203, 204m, 292, 293, 296
secondary
  blowouts on dune crests 173ph
  dunes 151
  inlet on Juist 319ph
sediment
  compaction 12
  migration, Neuwerk tidal flats 37f
  transport, cross-channel 27
seepage 192
seismic
  investigations 9, 108
  profiles 361

semi-high bank protection, Wangerooge 293, 295
Sexbierum 341m
shadow dunes 160, 161ph
  negative 160, 162ph
shallow seismic investigations 108
sheet flow 147
sheet-pile groyne, Nordstrandischmoor 271
shelduck 61
shell fragments in
  mud flat sediments 68
  thin section 72pl
shield deltas 82
shipbuilding 211
shoal 90, 129m
  arc 81, 90, 97
  complex 129m
  emerging 23, 31
  migration 90, 129m
  subunits 90, 97, 98, 129m
    migration rate 98, 129m
shoreface 135, 135f, 136, 150
shoreline configuration 16
Sier, Ameland 337
Sievern 283m
Sievertswarft, Hooge, destroyed 269
Simonszand 290m, 328m, 329, 329m
Sint Maarten 353m
Skærumhede 8m
Skallingen 26f, 29, 32f, 169, 178, 180, 190m, 216, 221, 222m, 223, 224, 224m
  Benknolde 225ph
  erosion and accretion 224m
  location map 223m
Skast 223m, 229m, 239m
slab failure 193
Slootdorp 353m
slope failure 193
Slufter, Texel 87ph, 89, 355, 355ph, 356ph, 358
Slufterbollen, Texel 352m
small ripple field
  flattening of 46ph
  reshaping of 49ph
small ripple orientation 89
  on megaripples 117f
  on sand waves 116f
small ripples 45, 47ph, 48ph, 140, 141, 144ph
  on sand flats 72pl
  ebb- and flood-orientated 87ph
  ebb-orientated 89
  flood-orientated 89, 90
  formation of 111
  lingoid 89
Slufter Inlet, Texel 87ph
Smouseduintjes, Terschelling 341m, 342
snow 38
Sønder
  Farup 229m
  Sejerslev 229m, 239m
Sønderho, Fanø 213, 223m, 224, 227, 229

morphological development 228ph
Sønderstrand, Rømø 229m, 233
Sönke-Nissen-Koog 247m, 263, 276
Søren Jessens Sand 222m, 223m, 227
soil fertility, decline of, Hooge 270
Spartina anglica 73pl, 180, 181, 181ph, 185, 310
Speeton, England 8m
Speicherkoog
  Nord 279m
  Süd 279m
Spieka 283m
Spieka-Neufeld 283m
Spiekeroog 1, 2m, 26f, 27f, 29, 54f, 65m, 75m, 79m, 84, 85, 92, 131, 141, 150, 152, 154, 178, 180, 187, 200, 214, 218, 289m, 290m, 292, 293m, 296, 298, 299, 300, 301, 302, 305, 314
  aerial photograph of dune core 298ph
  beach, aeolian sand transport 152ph
  beach barchan migration 156ph
  eastern end 82ph, 83ph
  flooded Ostplate 153pl
  grey dune 312pl
  Lütjeoog Dunes, recent erosion 299ph
  megaripple bedding on beach 52ph
  negative shadow dunes 162ph
  old dune area 301 ph
  salt dome 7
  salt marsh formation 300ph
  white dunes 312pl
spits 17, 23, 25, 134, 258m
  flood, Trischen 29
  recurved 41, 44ph, 185, 189
Spitsbergen 212
spring tide 41
Sprogø 268
Steenodde, Amrum 190, 247m, 251
Steingrund 259m
Steinplate 84f
Stensigmose (=St.) 8m
steric effect 12
Störtebeker Dyke, Ley Bight 182, 317m
stone groynes 198
storage area, reduced by dyking 197
Store Darum 223m
storm overwash 176
storm surge
  frequency 17, 18, 314
  levels, historic 18
  periods 18
storm surges of
  1010 20
  1020 20
  1041 20
  1075 20
  1094 20
  1102 20
  1114 20
  1164, Julianen Flood 20
  1287, Lucia Flood 20, 344
  1296, Vlieland 347

  1338 20
  1362 20, 21, 84, 197, 211, 275, 278
  1380 20
  1387 20
  1391 20
  1393 20
  1395 20
  1436, Eiderstedt 276
  1509 21
  1510 21
  1511 21
  1532 21
  1570, All Saints Flood 21, 282, 299
  1625, Ice Flood 267
  1634, 2nd Mandränke 21, 197, 240, 267, 268, 270, 272, 275, 276, 278
  1714, Vlieland 348
  1717, Christmas Flood 21, 264, 282, 288, 303, 311, 316, 327
  1720, Sylvester Flood 264, 269
  1727, Vlieland 348
  1743, Juist 316
  1751 21, 275
  1756 275, 333
  1760, Schiermonnikoog 333
  1821 264, 272
  1824 21, 272
  1825, Hallig Flood 21, 264, 268, 272, 276, 278, 282, 299, 307, 317, 354
  1834, Föhr 255
  1839, Amrum 251
  1845, Amrum 251
  1850, Norderoog 272
  1854, Wangerooge 291
  1855 21, 215, 291, 299
  1873, Baltrum 307
  1881 264, 268, 271, 338
  1884 264, 308
  1895 237, 251
  1904, Oland 267
  1906 21, 240, 299, 335
  1909, Föhr 256
  1911 21, 240, 251
  1914 251, 267
  1916 233, 240, 335
  1917, Wangerooge 292
  1920, Sylt 241
  1928/29, Spiekeroog 301
  1936 21, 241, 281, 300, 321
  1942 29
  1949 21
  1953, Holland Flood 20, 21, 273, 341, 355
  1954, Borkum 321
  1962 19ph, 20, 21, 203, 213f, 238, 241, 245, 268, 274, 292, 293, 300, 309, 314, 325,
  1967, Sylt 241
  1973, Wangerooge 295, 297
  1976 238, 245, 253
  1981 229, 241, 242, 244, 268, 313
  1984, Sylt 242
storm surges, influence on dune formation 152, 165
storm-surge shelters, Halligen 274

Stortemelk 340m
Strand 21, 259, 263, 275
Strandbalje, Wangerooge 292, 293, 293m
striae, formed by drift ice 56ph, 57, 57ph
Strucklahnungshörn, Nordstrand 260m, 263
Struckum 260m
stuifdijken 160, 198
  Ameland 335, 336
  Rottumerplaat 328, 328m
  Schiermonnikoog 186, 328m, 332, 332ph
  Terschelling 160, 161f, 340, 341, 341m, 342, 344ph
  Vlieland 347m, 348, 349, 350
  Wangerooge 204m, 295ph, 296
Sturmeck, Spiekeroog 289m
Stutton 8m
St. Jacobi Parochi 341m
St. Peter 185, 259, 260m, 277, 278
St. Peter-Böhl 186
St. Peter-Ording 26, 26f, 183, 233, 250, 276, 279
  aerial photograph 185ph
  beach barrier 184ph, 278ph
Suaeda maritima 73pl, 181
Subatlantic 11
Subboreal 11, 25
submarine bar aggradation 30
subsidence 5, 6, 9, 23
subtidal area 40, 82
sub-shoals 109
Süddamm, Minsener Oog 291, 291m
Süddeich dyke on Spiekeroog 300
Süderaue 222m, 226
Süderdeich 260m
Süderdünen
  Langeoog 302
  Spiekeroog 300
Süderende, Föhr 247m
Südergroen, Spiekeroog 298
Süderhever 110m, 259m, 276
Süderkoog, Pellworm 260m, 275
Süderoog 2m, 134, 260m, 261f, 262, 272, 274
Süderoogsand 2m, 13, 17, 26, 26f, 32f, 110m, 258, 259m, 260m
Süderpiep 259m, 284
Südersand 75m
Süderstrand, Borkum 317m
Süderwarf, Gröde 263
Süderwarft
  Habel 268
  Südfall 272
Südfall 2m, 21, 134, 260m, 262, 272, 274, 277m
Südstrandpolder, Norderney 216, 302m, 314
Südwesthörn 18, 239m
Südwestmauer seawall, Wangerooge 204m, 295
summer dykes
  Hooge 270, 270ph
  Süderoog 272

summer houses 218
   Fanø 227
   Mandø 231
   Rømø 236
summer polders
   Gröde 267
   Hooge 269
surf 52, 53, 81
   currents 52
   erosion by 130m
   levelling by 52, 126
Suurhusen 21
swash 135f, 141, 144
   bars 81
   marks 138, 142ph
Sylt 2m, 10m, 17, 18, 20, 23, 26, 26f, 32f, 39f, 111, 112, 134, 136, 150, 152, 163, 165, 171, 188, 190, 190m, 192, 193, 198, 200, 207, 212m, 213, 214, 215, 218, 222m, 229m, 233, 235, 236, 238, 239, 240, 241, 242, 243, 245, 247, 247m, 248, 259
   beach and foreshore changes 207f
   beach replenishment 205
   block diagram of migrating dunes 165f
   coastal erosion 188m
   foreshore, block diagram of 135f
   Landsat-5 image 231ph
   lee-side erosion at Blidsel 209ph
   location map 239m
   migrating dunes 151ph
   morphological changes 189m
   revetments at
      Blidsel 209ph
      Ellenbogen peninsula 201
      List 209ph
      Rotes Kliff 233pl
   salt marsh erosion 194ph
   secondary blowouts on dune crests 173ph
   tetrapod groyne at Hörnum 203ph, 204ph
   tetrapod walls 201

Tadenswarft, Langeneß 247m
Tadorna tadorna 61, 288
Tanderup 223m
tanker accidents 220
Tarbeck 7m
Tarp 223m, 224m
Tating, Eiderstedt 259, 260m, 276, 277m
Tegeler Plate 259m
Telegraphenbalje 75m
Teredo 353
Ternaard 335m
Terschelling 2m, 14m, 15, 26f, 29, 83, 102, 134, 170, 178, 180, 213, 218, 330m, 337, 340, 341, 341m, 344, 347m
   barchan migration 165
   borrow pit 73
   dune formation 343
   Polder 186ph, 187, 341m, 343
   SEASAT image 137ph

stuifdijk development 161f
   Young Dunes 24
Tertiary 3m, 5, 7, 23
tertiary dunes 151
Tertius 2m, 26f, 29, 259m, 279m, 284
Tetenbüll 260m
tetrapod groyne at Hörnum, Sylt 189, 203ph, 204ph, 241, 245, 247
tetrapods 201, 219
   Sylt 239m
   Westerland, Sylt 241
Teutonic Knights 211
Texel 2m, 10m, 13, 14m, 17, 26f, 28, 32f, 51ph, 52, 83, 89, 92, 102, 106, 107f, 134, 170, 178, 187, 190m, 198, 200, 212m, 214, 216, 218, 330m, 347, 347m, 351, 352, 352m, 353, 354m
   aerial photograph of northern end 357ph
   coastal protection at northern end 345pl
   coastline changes 357f
   dyke of Prins Hendrik Polder 345pl
   effect of groynes 199f
   Groote Vlak moist dune valley 344pl
   Groote Zanddijk 344pl
   low Pleistocene ridge at Den Hoorn 345pl
   old salt marsh drainage pattern 187ph
   SEASAT image 51ph
   Slufter area 356ph
   Slufter area, aerial photograph 355ph
Texelstroom 330m, 355
Thames, River 8
thaw holes 55, 55ph, 56
The Wash 6m
Theeknobs 222m, 245
thin section of
   mud flat sediments 68, 70, 72pl
   sand flat sediments 72pl
Tholendorf 260m
   beach ridge 17
Thomas Smit Gat 340m
thrust moraine created by shoaling ice 57ph
thrust moraines 9
tidal currents 33, 34, 36, 38, 40, 56
   and drift current, Baltrum 38f
   between Borkum and Juist 41f
   coast-parallel 81
   direction and strength of 33f
   in the Lister Tief 39f
   velocities, Wichter Ee 34f
tidal
   delta 53, 78
   deltas, expansion of 130
   flats
      current stripes on 71ph
      erosion around Gröde 263
      increasing dissection of 252
      micro-relief of 70
      NE of Föhr, dissection of 255m

      run-off channels on 73
      sculptured by wind 71
      striated by drift ice 56
   inlets 53
      formation, potential for 89
      form elements of 80m, 81
      morphology 77f
      primary 84, 88
      secondary 88, 89
   jet 78
   range 13m, 14m, 16
   wave 262
tidal-flats facing beach ridges 176f, 194
till cliffs 192
Timmermannsgat, Baltrum 302m, 305, 306f, 307
Tinnum, Sylt 239m
Tipkenhoog 241
Tjæreborg 223m
Tønder 212
Tönning 260m, 277m
Tötel 283
Tofte Dyke, Mandø 231
Toftum, Föhr 239m, 247m
tongue hummocks 160
tongue-like avalanches on dune slopes 160ph
tooth-shaped rillmarks 148
tourism 214, 215, 217, 218, 219, 223, 240, 296
   Fanø 227
   Halligen 275
   Hooge 273m
   Juist 317
   Mandø 231, 232
   Pellworm 275
   Rømø 236
   Schiermonnikoog 329
   Spiekeroog 301
   Sylt 240
   Wangerooge 296
tourists, footsteps of 62, 63ph
tracers, fluorescent 36
trade 211
trampling 169, 170, 171, 218, 223
   moulds 61, 63ph
transgression
   Holocene 23, 88
   rate, horizontal 23
transparency of water 37
transport ripples 140
transverse
   bar migration 120m, 122, 124, 125f
   bars on the Othello Shoal Shield 100ph
trapezoid megaripples 119
Trendermarschkoog, Nordstrand 21, 260m, 276
Treuburg Warft, Langeneß 247m, 264
Trischen 2m, 14, 18, 26f, 29, 30f, 31f, 32f, 43f, 109, 134, 237, 258, 259m, 279m, 280, 281, 285, 346, 361
   aerial photograph 280
Trischendamm 279m
Trischenkoog 29, 281

Triticum junceum 151, 159
Tümlauer
　Bucht 2m, 185, 187, 190m, 260m, 276, 280
　Koog 215, 260m, 278
Tüskendör, Borkum 178, 317m, 326
Tüskendörsee 325
tunnel valleys 6m, 9
Turminsel, Knechtsand 288
Tuul 211
Tvismark 229m

Ual anj cliff, Amrum 18, 190, 247m, 251
Uelvesbüll 260m
Uelvesbüller Koog 260m, 278
Uithuizen 328m
United States, Gulf Coast 31
underwater
　groynes 200
　topography 40
　　revealed by SEASAT 136
undulatory small ripples 140
Unterer Wittsand 69ph
Urk 15
urstromtal 6m
Utarp 302m
Utermarker Koog, Pellworm 260m
Utersum, Föhr 247m, 253
Uthörn, Sylt 188m, 239m
Utholm 260m, 276, 277m, 278

Vallei van Malgum, Vlieland 347m, 350
Varde 223m
　Å 223m, 226
　Bakkeø 221, 223m
Vecht urstromtal 24
Vejers 223m
Veronica officinalis 170
Vester
　Nebel 223m
　Vedsted 229m
Viborg 6m, 9
Vierde Duintjes, Terschelling 341m
Viking Age 25, 276
Vindgab Bjerge, Fanø 223m, 227
　aerial photograph 170ph
Virginia, USA 188
Visquard 317m
Vitalienbrüder 211
Vlaardingen Culture 195
Vlie 15, 83, 102, 106, 108, 110, 213, 330m, 341, 347, 349, 350
　morphological development 340m
　SEASAT image 51ph
Vliehors 347m, 349, 350
Vlieland 2m, 14m, 26f, 29, 32f, 83, 92, 102, 170, 190m, 200, 212m, 213, 218, 330m, 341m, 344, 347m, 350, 355
　adhesion ripples on the beach 149ph
　dendritic drainage pattern 60
　formation of recent dunes 348ph
　narrowest part at Posthuis 349ph

　old clay being eroded at beach 350ph
　SEASAT image 51ph
　shadow dunes behind sand fence 161ph
Vliestroom 15, 340m
vortices 37
　seawards of the Wichter Ee 40f
Vortrapptief 40, 222m, 250
Voslapper Groden 283m

Waaksens 335m
Waal en Burg Polder, Texel 353
Waard 330m
Waardgronden 330m
Waarland 353m
Wacken 7m
Wadden Sea, definition 2
Wallsbüll 260m
Wangerooge 1, 2m, 21, 26f, 27f, 29, 43f, 54f, 55, 57, 65, 65m, 75m, 79m, 84, 85, 90, 92, 109, 134, 135, 150, 152, 160, 174, 178, 190 m, 200, 207, 211, 214, 215, 216, 217, 218, 259m, 289m, 290, 290 m, 291, 292, 293, 293m, 295, 296, 297, 298, 299, 300
　Alter Westturm 204m
　beach 142ph, 143ph
　　replenishment 204, 205
　coastal protection 202, 203, 204, 204m
　dune erosion 294ph
　dune formation 294ph
　dykes 204m
　eastern end, aerial photograph 295ph
　exposure in centre of the settlement 281pl
　first groynes 198
　grain-size analyses 66f, 67f
　groynes 204m
　morphological changes 134m
　mud flats southeast of 68ph, 72pl
　old marine clay 281pl
　recent morphological changes SW of 75f
　revetment at western end 184pl
　revetments 204m
　riprap 205ph
　salt dome 7
　sand accumulation on old dunes 167ph
　sand flats south of 72pl
　sawtooth bars 138
　shoreface, bar migration 139f
　small delta at the dune foot 174ph
　tidal flat morphology 75
　tidal flats, creeks on 77
　western end 82ph, 83ph
Wangerooger Inselwatt 289m, 361
Wardammkoog 260m, 282
Warffum 328m
Warften 196
Warmenhuizen 353m
washover areas 174, 187

Ameland 175ph
Baltrum 305
Borkum 320, 325, 326
Jordsand 238f
Juist 316
Langeoog 179ph, 303, 305
Norderney 41, 153pl, 174ph, 176ph, 176f, 314, 315ph
Rottumerplaat 328
Schiermonnikoog 177ph, 186, 331ph, 332
Skallingen 221
Spiekeroog 301
Terschelling 180ph, 343ph
Trischen 31f
　damming of 178
washover processes 29, 135, 206, 362
　on barrier beaches, Eiderstedt 185
Wash, The 9
Wassenaar Polder, Texel 356
Waterschap der 28 Gemeenschappelijke Polders 353
watersheds
　between drainage basins 83
　shifting of 74, 338
wave
　backwash 193
　effect, depth of 49
　erosion 189
　pattern 51ph, 136, 138ph
　ripples 140
waves on tidal flats 50
Weddewarden 283m
Wedel 7m
Weichselian 8, 9
　deglaciation 11
　ice margin 6m
Weiße
　Düne, Norderney 27f, 302m, 313, 314
　Dünen, Spiekeroog 289m
　Klippe, Helgoland 21
Weißes Kliff, Sylt 239m, 245
Wells-next-the-Sea 111
Welt 260m
Wenningstedt, Sylt 171, 232pl, 239m, 242
　aerial photograph 242ph
　beach, rhomboid rill marks 146ph
　cliff erosion 233pl
　Rotes Kliff 191ph
　secondary blowouts 173ph
Werversheof 353m
Weser 2m, 13m, 21, 26f, 27, 32f, 33, 40, 190m, 195, 220, 259m, 283m, 288
　estuary 108, 109, 131
　radar image 69ph
Wesselburen 260m
Wesselburener
　Koog 260m
　Loch, block diagram of 78f
West Vlieland 347
Westdorf
　Baltrum 302m, 306
　Juist 316

Westen, Wangerooge 289m
Wester
  Neuer Koog, Pellworm 275
  Nijkerk 335m
Westeraccumersiel 302m
Westerboomsgat, Terschelling 340
Westerbur 302m
Westerduinen
  Schiermonnikoog 332
  Texel 352m
Westerems 290m
Westergroen, Spiekeroog 298
Westergronden 340m
Westerheide, Sylt 239m, 244
Westerhever 260m, 276, 277m
Westerheversand 260m
Westerkoog, Pellworm 260m, 275
Westerland, Sylt 38, 188, 189, 191, 214, 215, 218, 233pl, 238, 239m, 240, 242
  seawall 20, 185pl, 200, 241
Westerplas, Schiermonnikoog 328m
Westerstede 328m
Westerstrand, Schiermonnikoog 329, 333
Westerveld, Vlieland 347, 347m
Wester-Ems 102, 110
  ebb delta 321
  flood delta 325
Westgat, Marsdiep 103m, 104m, 105m, 106, 106m, 107
Westgroden dyke, Wangerooge 292
Westinnengroden, Wangerooge 292
Westturm, Wangerooge 291, 292, 297m
Westufermauer, Wangerooge 204m
West-Terschelling 213, 341m, 343
whaling 212, 212m, 213, 235, 325
white dunes 160, 162, 193
  Baltrum 310
  Juist, north of Hammer Lake 319ph
  Spiekeroog 312pl
Wichter Ee 1, 27f, 33, 35f, 36, 41, 83, 84, 85, 86, 93, 94, 99m, 108, 112, 113, 129m, 138, 290m, 306, 307, 308
  aerial photographs 88ph, 89ph, 90ph, 91ph, 92ph, 93ph, 94ph, 95ph
  block diagram of 84f, 85f

breaker zone 40f
crescent bar formation 100f
development of catchment area 307f
ebb delta 96, 117, 118m, 121ph
  aerial photograph 308ph
  crescent bars 98, 101m, 128ph
  escarpments within 130m
  megaripple fields 53ph, 97, 112m
  megaripple migration 117f, 118m, 119m, 120m, 121f, 122ph, 124f, 126f, 127
  shoal migration 98m
flood delta 95, 96. 96m, 113ph, 117
  location map 114m
  sand wave migration 116f
main ebb channel 101m, 102
maximum current velocities 93
megaripple migration 123f
morphological development 306f
Othello Shoal Shield 100ph
salt dome 7
sand wave migration 97m
tidal current velocities 34f
transverse bar migration 125
vortices seawards of 40f
Wiedingharder Alter Koog 239m
Wierbalg 15
Wieringen 10m, 352m, 353m
Wieringermeer 352m, 353m
Wieringerwaard 353m
  Polder 353m
Wieringerwerf 353m
Wierum 335m
Wilhelmshaven 214, 216, 290, 292
Willemsduin, Schiermonnikoog 186, 328m, 332
Winaldum 341m
wind
  direction and velocity, Baltrum 35f
  funnels 240
  ribbon fields on tidal flats 71
  ripple migration 153ph, 154f, 155f, 157, 167
  ripples 152, 154, 162, 362
    on beach barchans 156
  shadows 160
  streaks on tidal flats 45, 50ph
wind-parallel stripes 45

Winkel 353m
Witsum, Föhr 247m
Wittdün, Amrum 214, 247m, 248m, 251, 252
  aerial photograph 254ph
Wittdün Peninsula 250, 251, 252, 253
Witte Lid, Vlieland 347m
  dune erosion 348ph
Witzwort 260m
Wobbenbüll 260m
Wöhrden 260m, 279m
World War
  I 213f, 214, 215, 251, 276, 290, 296
  II 215, 216, 217, 222, 227, 241, 244, 249, 271, 272, 291, 296, 304, 305, 311, 314, 337, 358
Wremen 283m
Wretton 8m
Wriakhörn, Amrum 247m, 249
Wrixum, Föhr 247m
Wüstes Moor (= Nordstrandischmoor) 270
Wyk, Föhr 214, 247m, 255, 256

yacht harbours 218
Young Dunes 24, 25
  Sylt 239

Zand 353m
Zandpolder 353m
Zechstein salt 7
Zijpe 352
  en Haze Polder 353m
Zinc 220
Zostera marina 345
Zoutkamperlaag Inlet 84
Zuidelijke Zuidhaffel, Texel 352m, 353
Zuider Stortemelk 330m, 350
Zuider Zee 9, 17, 22, 24, 108, 182, 213, 355
  dam 345
  damming of the 14, 15
  formation of the 20
Zuiderhaaks, Marsdiep 106m
Zuidwal 330m
Zwanewater stuifdijk, Ameland 335m, 337m
Zwarte Haan 335m